Handbook of organic food safety and quality

Related titles:

Improving the safety of fresh fruit and vegetables
(ISBN 978-1-85573-956-7)
Consumers are increasingly demanding 'fresh' foods with fewer preservatives. Manufacturers have responded by developing new 'minimal' processing techniques. These developments have, in turn, focused attention on the safety of raw materials such as fruit and vegetables and the way they are produced. With its distinguished editor and international team of contributors, this authoritative collection summarises best practice in ensuring the safety of fresh fruit and vegetables both during cultivation on the farm and in subsequent processing operations.

Food safety control in the poultry industry
(ISBN 978-1-85573-954-3)
Consumers' expectations about the safety of products such as poultry meat and eggs have never been higher. The need to improve food safety has led to renewed attention on controlling contamination at all stages of the supply chain 'from farm to fork'. This collection reviews the latest research and best practice in ensuring the safety of poultry meat and eggs, both on the farm and in subsequent processing operations.

Improving the safety of fresh meat
(ISBN 978-1-85573-955-0)
It is widely recognised that food safety depends on effective intervention at all stages in the food chain, including the production of raw materials. This book provides an authoritative summary of the wealth of research on reducing microbial and other hazards in raw and fresh red meat.

Details of these books and a complete list of Woodhead's titles can be obtained by:

- visiting our website at www.woodheadpublishing.com
- contacting Customer Services (e-mail: sales@woodhead-publishing.com;
  fax: +44 (0)1223 893694; tel.: +44 (0)1223 891358 ext. 30; address:
  Woodhead Publishing Ltd, Abington Hall, Abington, Cambridge CB21 6AH, England)

Handbook of organic food safety and quality

Edited by
Julia Cooper, Urs Niggli and Carlo Leifert

Nafferton Ecological Farming Group

CRC Press
Boca Raton Boston New York Washington, DC

WOODHEAD PUBLISHING LIMITED
Cambridge, England

Published by Woodhead Publishing Limited, Abington Hall, Abington,
Cambridge CB21 6AH, England
www.woodheadpublishing.com

Published in North America by CRC Press LLC, 6000 Broken Sound Parkway, NW,
Suite 300, Boca Raton, FL 33487, USA

First published 2007, Woodhead Publishing Limited and CRC Press LLC

British Library Cataloguing in Publication Data
A catalogue record for this book is available from the British Library.

Library of Congress Cataloging in Publication Data
A catalog record for this book is available from the Library of Congress.

Woodhead Publishing ISBN 978-1-84569-010-6 (book)
Woodhead Publishing ISBN 978-1-84569-341-1 (e-book)
CRC Press ISBN 978-0-8493-9154-5
CRC Press order number: WP9154

The publishers' policy is to use permanent paper from mills that operate a sustainable forestry policy, and which has been manufactured from pulp which is processed using acid-free and elementary chlorine-free practices. Furthermore, the publishers ensure that the text paper and cover board used have met acceptable environmental accreditation standards.

Typeset in India by Replika Press Pvt Ltd.
Printed by TJ International Limited, Padstow, Cornwall, England

Contents

Contributor contact details

(* = main point of contact)

Editors
J. Cooper
Nafferton Ecological Farming Group
Nafferton Farm
Newcastle University
Stocksfield
NE43 7XD
UK

Email: Julia.Cooper@nefg.net

U. Niggli
Research Institute of Organic Agriculture (FiBL)
CH-5070 Frick
Switzerland

Email: urs.niggli@fibl.org

and

C. Leifert
Nafferton Ecological Farming Group
Naffcrton Farm
Newcastle University
Stocksfield
NE43 7XD
UK

Email: C.Leifert@newcastle.ac.uk

Chapter 1
C. Leifert
Nafferton Ecological Farming Group
Nafferton Farm
Newcastle University
Stocksfield
NE43 7XD
UK

Email: C.Leifert@newcastle.ac.uk

Chapter 2
U. Niggli
Research Institute of Organic Agriculture (FiBL)
CH-5070 Frick
Switzerland

email: urs.niggli@fibl.org

Chapter 3

C. J. Seal* and K. Brandt
School of Agriculture, Food &
Rural Development
Newcastle University
Newcastle upon Tyne
NE1 7RU
UK

Email: chris.seal@ncl.ac.uk
kirsten.brandt@nel.ac.uk

Chapter 4

B. van Elzakker* and J. Neuendorff
P.O. BOX 63
NL-6720 AB Bennekom
The Netherlands

Email: b.vanelzakker@agroeco.nl

Chapter 5

J. Bloksma*
M. Northolt
M. Huber
G-J. van der Burgt and
L. van de Vijver
Hoofdstraat 24
3972 LA Driebergen
The Netherlands

Email: info@louisbolk.nl

Chapter 6

E. Oughton* and C. Ritson
Centre for Rural Economy,
Newcastle University,
Newcastle Upon Tyne
NE1 7RU
UK

Email: E.A.Oughton@newcastle.ac.uk
Christopher.Ritson@newcastle.ac.uk

Chapter 7

R. F. Weller*, C.L. Marley and
J. M. Moorby
IGER
Plas Gogerddan
Aberystwyth
SY23 3EB
UK

Email: richard.weller@bbsrc.ac.uk
christina.marley@bbsrc.ac.uk
jon.moorby@bbsrc.ac.uk

Chapter 8

H. Hirt and E. Zeltner*
Animal Husbandry and Animal
Breeding Division
Research Institute of Organic
Agriculture (FiBL)
Ackerstrasse, Postfach
CH-5070 Frick
Switzerland

Email: esther.zeltner@fibl.org

and

C. Leifert
Nafferton Ecological Farming
Group
Nafferton Farm
Newcastle University
Stocksfield
NE43 7XD
UK

Email: C.Leifert@newcastle.ac.uk

Chapter 9
A. Sundrum
University of Kassel
Dept. of Animal Nutrition and Animal Health
Nordbahnhofstr. 1a
Witzenhausen, D-37213
Germany

Email: sundrum@wiz.uni-kassel.de
sundrum@hrz.uni-kassel.de

Chapter 10
F. Diez-Gonzalez
1334 Eckles Avenue
St. Paul
MN 55108
USA

Email: fdiez@umn.edu

Chapter 11
P. Klocke* and M. Walkenhorst
Animal Health Division
Research Institute of Organic Agriculture (FiBL)
Ackerstrasse, Postfach
CH-5070 Frick
Switzerland

Email: peter.klocke@fibl.org

and

G. Butler
Nafferton Ecological Farming Group
Nafferton Farm
Newcastle University
Stocksfield
NE43 7XD
UK

Email: gillian.butler@ncl.ac.uk

Chapter 12
V. Maurer*, P. Hördegen and H. Hertzberg
Research Institute of Organic Agriculture (FiBL)
Ackerstrasse, Postfach
CH-5070 Frick
Switzerland

Email: veronika.maurer@fibl.ch

Chapter 13
B. Biavati* and C. Santini
Dipartimento di Scienze and Tecnologie Agroambientali
Bologna University
Via Fanin-42
40127 Bologna
Italy

Email: biavati@agrsci.unibo.it
ceciliasantini@virgilio.it

and

C. Leifert
Nafferton Ecological Farming Group
Nafferton Farm
Newcastle University
Stocksfield
NE43 7XD
UK

Email: C.Leifert@newcastle.ac.uk

Chapter 14

C. Benbrook
Chief Scientist
The Organic Center
90063 Troy Road
Enterprise
Oregon 97828
USA

Email: cbenbrook@organic-center.org

Chapter 15

E. A. S. Rosa*, R. N. Bennett
and A. Aires
Fitotecnia e Engenharia Rural
N3.49 (Ed. Ciências Agrárias)
UTAD – Apartado 1013 – 5001-
801 Vila Real
Portugal

Email: erosa@utad.pt
alfredoa@utad.pt

Chapter 16

F. P. Weibel* and T. Alföldi
Anbautechnik 'Mehrjährige
Kulturen'
Fachgruppenleiter
Forschungsinstitut für biologischen
Landbau (FiBL)
Ackerstrasse
CH-5070 Frick
Switzerland

Email: franco.weibel@fibl.ch

Chapter 17

U. Köpke*
Institute of Organic Agriculture (IOL)
University of Bonn
Katzenburgweg 3
53115 Bonn
Germany

Email: iol@uni-bonn.de
ukiol@uni-bonn.de

and

B. Thiel
Institute for Nutrition and Food
Science, Food Microbiology and
Hygiene
University of Bonn
Meckenheimer Allee 168
D53115 Bonn
Germany

and

S. Elmholt
Faculty of Agricultural Sciences
Department of Agroecology
University of Aarhus
PO Box 50
DK 8830 Tiele
Denmark

Email: Susanne.Elmholt@agrsci.dk

Chapter 18

R. Ghorbani
Department of Agronomy
Faculty of Agriculture
Ferdowsi University of Mashhad
Iran

Email: reza-ghorbani@um.ac.ir

and

S. Wilcockson*
School of Agriculture, Food and Rural Development
Newcastle University
Newcastle upon Tyne
NE1 7RU
UK

Email: s.j.wilcockson@ncl.ac.uk

Chapter 19
U. Köpke
Institute for Organic Agriculture (IOL)
University of Bonn
Katzenburgweg 3
53115 Bonn
Germany

Email: ukiol@uni-bonn.de
iol@uni-bonn.de

and

J Krämer
Institute for Nutrition and Food Science, Food Microbiology and Hygiene
University of Bonn
Meckenheimer Allee 168
D53115 Bonn
Germany

and

C. Leifert
Nafferton Ecological Farming Group
Nafferton Farm
Newcastle University
Stocksfield
NE43 7XD
UK

Email: C.Leifert@newcastle.ac.uk

Chapter 20
G. S. Johannessen
Section for Feed and Food Microbiology
National Veterinary Institute
PO Box 8156 Dep
0033-Oslo
Norway

Email: gro.johannessen@vetinst.no

Chapter 21
M. Bourlakis*
Brunel University
Business School
Elliott Jaques Building
Uxbridge
Middlesex
UB8 3PH
UK

Email: Michael.Bourlakis@brunel.ac.uk

and

C. Vizard
School of Population and Health Sciences
William Leech Building
Newcastle University
Newcastle upon Tyne
NE2 4HH
UK

Email: c.g.vizard@ncl.ac.uk

Chapter 22
R.C. Van Acker*
Department of Plant Agriculture
University of Guelph
Guelph
Ontario
Canada
N1G 2W1

Email: vanacker@uoguelph.ca

and

N. McLean
Department of Plant and Animal Sciences
Nova Scotia Agricultural College
Truro
NS
Canada
B2N 5E3

Email: nmclean@nsac.ca

and

R. C. Martin
Organic Agriculture Centre of Canada
Nova Scotia Agricultural College
Truro
NS
Canada
B2N 5E3

Email: rmartin@nsac.ca

Chapter 23

K. Brandt*
School of Agriculture, Food and Rural Development
Agriculture Building
Newcastle University
Newcastle upon Tyne
NE1 7RU
UK

Email: Kirsten.brandt@newcastle.ac.uk

and

U. Kjærnes
SIFO
P.O. BOX 4682
Nydalen
N-0405 Oslo
Norway

Email: unni.kjarnes@sifo.no

and

G. S. Wyss
Research Institute of Organic Agriculture (FiBL)
CH-5070 Frick
Switzerland

Email: gabriela.wyss@fibl.org

and

L. Lück
QualityLowInputFood
Nafferton Ecological Farming Group
Nafferton Farm
Newcastle University
Stocksfield
NE43 7XD
UK

Email: lorna.lueck@nefg.net

and

A. Hartvig Larsen
Aarstiderne A/s
Barritskowej 34
DK 7150
Barrit
Denmark

Email: ahl@openspace.dk

The **Nafferton Ecological Farming Group (NEFG)** is the main research centre for applied agricultural systems research at Newcastle University. It has around 20 staff and research focuses on the following main activities:

- **applied R & D** addressing agronomic and marketing issues/ challenges in 'low input', integrated and organic food production systems
- **strategic R & D** into the effects of agronomic practices on sustainability and environmental impact of agriculture and **food quality and safety**
- maintaining field research facilities (e.g. the long-term Nafferton factorial systems comparison field experiments) for **fundamental research** groups in the areas of soil, plant, animal, ecology, molecular and environmental science
- providing **technology transfer services** to the 'low-input' and organic food and farming industries

Within the UK the NEFG currently has projects on winter and spring cereals, pulses, field vegetables (brassicas, alliums, lettuce and carrots) and protected crops (tomato and cucumber). In collaboration with companies and academic institutions in Greece, Italy and Turkey the NEFG group also has ongoing R & D projects on olive, citrus, top fruit and winter greenhouse and vegetables production systems in the Mediterranean area. Projects have focused mainly on improving crop protection, fertilisation, rotational designs and variety, and on rootstock selection and irrigation systems used in organic and low-input production systems.

Livestock projects have focused on UK dairy, beef, sheep and pig production systems and have targeted improvements in feeding regimes and preventative health management systems (e.g. for the control of mastitis and endo-parasites and gastrointestinal diseases).

The **NEFG also provides consultancy** services to food producers, processors, multiple retailers and seed companies in the UK and mainland Europe and houses the **North East Organic Programme**, which is designed to provide agronomic, standards and marketing advice to organic farmers, processors and consumers in the North East of England.

The NEFG is based at Nafferton Farm, 15 miles west of Newcastle, in the Tyne Valley, off the A69. For further information on the Group and its activities, or to arrange a visit, please contact:

Ms Lois Bell
NEFG, Nafferton Farm, Stocksfield, Northumberland, NE43 7XD, UK
Tel. +44-1661-830-222 or 444
Fax. +44-1664-830-222
e-mail: l.e.bell@ncl.ac.uk
or
visit our website: www.ncl.ac.uk/tcoa

1

Introduction

Carlo Leifert, Newcastle University, UK

The production and consumption of organic food has increased rapidly over the last 20 years. In Europe the land area under organic management increased from less than 0.1 to more than 6 million ha between 1985 and 2004. It is estimated that more than 30 million ha of land are farmed organically across more than 600,000 organic farms, creating an organic food and drinks market of more than £16.5 billion (= €24 billion = US$ 32 billion) (Soil Association, 2006).

The increase in demand and consumption of organic (also known as ecological or biological) foods has mainly been due to an increasing number of consumers associating significant environmental, biodiversity, ethical (e.g. animal welfare, local, fair trade) and food quality and safety benefits with organic foods and/or food production systems. Since organic production systems are frequently associated with lower yields and higher costs, consumers currently have to pay significant price premiums for organic food. Future increases in demand will therefore rely on maintaining and/or improving consumer confidence in the benefits of organic foods (e.g. Soil Association, 2006).

The environmental and biodiversity benefits of organic farming practices are now widely accepted and have become the main rationale for government support of organic farming (e.g. DEFRA, 2006 a, b; Swiss Federal Office for Agriculture, 2006). Benefits include (a) reduced levels of pollution of surface and ground water from leaching and/or run-off of mineral nitrogen and phosphorus fertilisers and chemosynthetic crop protection products (Drinkwater *et al.*, 1998; Porter *et al.*, 1999; Stolze *et al.*, 2000; PAN 2002), (b) reduced energy use (and associated CO_2 carbon emissions) (Helsel, 1992; Fluck, 1992; Dubois *et al.*, 1999; Cormack, 2000; Pretty *et al.*, 2002) and (c) increased density and diversity of small mammal, bird, invertebrate and non-crop plant populations (Bengtsson *et al.*, 2005; Critchley *et al.*, 2004, 2006; Pyšek and Lepš, 1991; Pyšek *et al.*, 2005; Berry *et al.*, 2004; Stoate *et al.*, 2001). Also, there is evidence that soil biological activity, organic matter content structural stability/erosion resistance and inherent fertility (level of yield obtained per unit fertiliser input) has decreased in soils used for 'high input' crop production (Matson *et al.*, 1997; Reganold *et al.*, 1987, 2001; Fließbach and Mäder, 2000; Mäder *et al.*, 2002).

However, while the environmental and biodiversity benefits are well documented (see publications quoted in the paragraph above), there is still considerable controversy about nutritional and sensory qualities and safety aspects of organic foods from both livestock and crop production systems. For example, consumer studies show that an increasing number of consumers perceive organic foods as having higher sensory and nutritional quality and being safer with respect to microbial pathogens and agrochemical residues in foods (see Chapter 6 below).

With respect to nutritionally desirable compounds (essential micronutrients, vitamins, antioxidants and unsaturated fatty acids such as conjugated linolenic acid or omega 3 fatty acids) several recent scientific reviews conclude that there is a trend towards higher levels of nutritionally desirable micronutrients and certain antioxidants (e.g. vitamin C) in organic compared to conventional foods. However, they also concluded that the (a) variability and often conflicting results between studies and (b) the lack of scientifically sound human dietary intervention and cohort studies which clearly linked organic food consumption to positive health impacts, make it currently impossible to claim that organic foods have superior nutritional quality compared to conventional foods (Woese *et al.*, 1997; Brandt and Molgaard, 2001).

In considering pesticide residues, it was clearly shown that crops from organic production systems contain no or significantly lower levels of pesticide residues than crops from conventional systems (Baker *et al.*, 2002). However, while some scientists are concerned about the potential health impacts from such residues (Porter *et al.*, 1999; Benbrook, 2002), pesticide legislators maintain that current pesticide risk assessments and pesticide registration procedures are adequate and that residues below the current legal limits can not have a negative health impact in humans (e.g. PSD, 2006).

With respect to other undesirable compounds (e.g. mycotoxins, antibiotics) and microbiological food safety risks, some scientists have claimed that organic food production systems are associated with higher food safety risk (Avery, 1998; Trewavas, 2001). However, apart from identifying the more frequent use of manure-based fertility inputs in organic systems, they did not produce substantiating scientific evidence to underpin their claims. In contrast, a range of studies focused on quantifying risks at specific critical control points, indicated that the agronomy practices used in organic farming systems result in reductions in risks from mycotoxins and enteric pathogen contamination (Obst *et al.*, 1998; Diez-Gonzalez *et al.*, 1998; Wyss, 2005).

However, quantitative surveys comparing the incidences of foodborne diseases and/or mycotoxin related illnesses between consumers of organic and conventional food are currently not available. This makes it difficult to compare the relative risks associated with different organic and conventional production systems.

Given the controversy and uncertainties regarding the relative quality and safety of organic foods, and the lack of structured textbooks describing the current state of the art with respect to quality and safety assurance in organic food and farming systems, the *Handbook of organic food quality and safety* aims to:

- Critically evaluate the currently available evidence for differences in food quality and safety between foods from organic and conventional production systems;
- Describe (where possible) the production system components (e.g. crop or livestock health management practices, soil fertilisation methods, crop rotation designs, livestock feeding and husbandry regimes, crop varieties/livestock breeds used) responsible for differences in food quality and safety between production systems;
- Identify strategies to improve nutritional, processing and sensory quality and safety characteristics in organic farming systems;
- Review the current concepts and practices of quality assurance in organic production systems.

This handbook is subdivided into four main parts which are each subdivided into several chapters.

Part I provides an introduction to the concepts of quality and safety as recognised by different stakeholder groups involved in organic food supply chains, including organic/biodynamic farming pioneers (Chapter 2), nutrition and food scientists (Chapter 3), quality assurance specialists (Chapter 4), organic/biodynamic farming researchers (Chapter 5) and social scientists in qualitative and quantitative consumer research studies (Chapter 6).

Part II provides detailed information on the main quality and safety issues related to the production of organic livestock foods. This includes three chapters (Chapters 7 to 9) which review the effect of livestock husbandry on nutritional and sensory quality of livestock foods: including milk and dairy products (Chapter 7), poultry (Chapter 8) and pork (Chapter 9). It also includes four chapters (Chapters 10 to 13) which review the strategies used to minimise microbiological risks and antibiotic and veterinary medicine use in livestock production systems including safety of ruminants (Chapter 10), mastitis treatment in organic dairy production systems (Chapter 11), internal parasites (Chapter 12) and pigs and poultry (Chapter 13).

Part III provides detailed information on the main quality and safety issues relating to the production of organic crop foods. This includes two chapters (Chapters 15 and 16) which focus on the effects of agronomic methods used in crop production on the nutritional and sensory quality of crop foods, for phytochemicals (Chapter 15) and shelf-life of fruit (Chapter 16).

It also includes two chapters (Chapters 14 and 18) focusing on the effects of agronomic practices on pesticide residues and their potential health impacts (Chapter 14) and the strategies to reduce the use of copper-based fungicides, which are among the limited number of pesticides that can be used (under derogation) in organic farming systems: (Chapter 18).

Finally, Part III also includes two reviews (Chapters 17 and 19) of the relative risks and novel strategies available to reduce mycotoxin and enteric pathogen contamination in organic crop production systems, for mycotoxin and fungal alkaloid contamination in organic and conventional production systems (Chapter 17) and for the microbiological safety of fruit and vegetables (Chapter 19).

Part IV focuses on improving and assuring food quality and safety throughout the food supply chain. This includes Chapter 20 which focuses on novel post-harvest strategies to reduce enteric pathogen contamination in crop foods. It also includes Chapter 21 on trading behaviour in organic food supply chains, which is an organisational aspect related to food quality and safety. Finally there are two chapters (Chapters 22 and 23), which review quality assurance strategies and innovations specific to organic food supply chains, including protocols to prevent GM-contamination of organic crops (Chapter 22) and organic HACCP systems (Chapter 23).

Most of the authors of individual chapters are senior scientists and/or industry based quality assurance specialists who have participated in the EU-funded Integrated Project (IP) QualityLowInputFood. The project aims are: 'improving the quality and safety and reduction of cost in the organic and low input food supply chains'. Several chapters in handbook report preliminary results from the IP and/or associated/linked projects supported by national governments and/or industry (see IP website: www.qlif.org for further details).

Other chapters have been contributed by leading scientists and research groups in North America (see for example Chapters 10, 14 and 22 which focus on prevention of enteric pathogens, pesticide residues and prevention of GM-contamination, respectively).

References

Avery D T (1998). 'The hidden dangers of organic food'. *American Outlook*, (Fall) 19–22.

Baker B P, Benbrook C M, Groth E and Benbrook K L (2002). 'Pesticide residues in conventional, IPM-grown and organic foods: Insights from three U.S. data sets'. *Food Additives and Contaminants*, **19**(5), 427–446.

Benbrook C M (2002). 'Organochlorine residues pose surprisingly high dietary risks'. *Journal of Epidemiology and Community Health*, **56**, 822–823.

Bengtsson J, Ahnström J and Weibull A (2005). 'The effects of organic agriculture on biodiversity and abundance: a meta-analysis'. *Journal of Applied Ecology*, **42**, 261–269.

Berry P, Ogilvy S and Gardner S (2004). 'Integrated farming and biodiversity'. *English Nature Report 634*, English Nature, Agriculture Unit, Peterborough.

Brandt K and Molgaard J P (2001). 'Organic Agriculture: does it enhance or reduce the nutritional value of food plants'. *Journal of Science in Food and Agriculture*, **81**, 924–931.

Cormack W F (2000). Energy use in organic farming systems (OF0182) http://orgprints.org/8168

Critchley C N R, Allen D S, Fowbert J A, Mole A C and Gundrey A L (2004). 'Habitat establishment on arable land: assessment of an agri-environment scheme in England, UK'. *Biological Conservation*, **119**, 429–442.

Critchley C N R, Fowbert J A and Sherwood A J (2006). 'The effects of annual cultivation on plant community composition of uncropped arable field boundary strips'. *Agriculture, Ecosystems and Environment*, **113**, 196–205.

DEFRA (2006a). Countryside stewardship scheme (CSS) http://www.defra.gov.uk/erdp/schemes/css/default.htm

DEFRA (2006b). Organic farming and the Environment. http://www.defra.gov.uk/farm/organic/policy/actionplan/annex3.htm

Diez-Gonzalez F, Callaway T R, Kizoulis M G and Russell J B (1998). 'Grain feeding and the dissemination of acid-resistent *Escherichia coli* from cattle'. *Science*, **281**, 1666–1668.

Drinkwater L E, Wagoner P and Sarrantonio M (1998). 'Legume-based cropping systems have reduced carbon and nitrogen losses'. *Nature*, **396**, 262–264.

Dubois D, Gunst L, Fried P M, Stauffer W, Spiess E, Alföldi T, Fliessbach A, Frei R and Niggli U (1999). 'DOC-trial: yields and energy use efficiency'. *Agrarforschung*, **6**, 293–296.

Fließbach A and Mäder P (2000). 'Microbial biomass and size-density fractions differ between soils of organic and conventional agricultural systems'. *Soil Biology and Biochemistry*, **32**, 757–768.

Fluck R C (1992). 'Energy analysis of agricultural systems'. In Fluck R C, *Energy in World Agriculture,* Vol 6 Elsevier Science Publishing, Amsterdam, 45–52.

Helsel Z R (1992). 'Energy and alternatives for fertiliser and pesticide use'. In Fluck RC, *Energy in World Agriculture*, Vol 6 Elsevier Science Publishing, Amsterdam, 177–210.

Mäder P, Fliessbach A, Dubois D, Gunst L, Fried P and Niggli U (2002). 'Soil fertility and biodiversity in organic farming'. *Science*, **296**,1694–1697.

Matson P A, Parton W J, Power A G and Swift M J (1997). 'Agricultural intensification and ecosystem properties'. *Science*, **277** (5325), 504–509.

Obst A, Lepschy J, von Gleissenthall J and Beck R (1998). 'Die Desoxynivalenol-Belastung von Weizen durch *Fusarium graminearum* in Abhängigkeit von Pflanzenbaufaktoren und Witterung', in *Mitteilungen aus der Biologischen Budesanstalt für Land – und Forstwirtschaft*, Parey Buchverlag, Berlin Vol. 51, p. 85.

PAN (2002). Pesticide impacts in the environment. http://www.pan-europe.info/About%20pesticides/Impacts%20environment.shtm

Porter W P, Jaeger J W and Carlson I H (1999). 'Endocrine, immune, and behavioral effects of aldicarp (carbamate), atrazine (triazine) and nitrate (fertilizer) mixtures at groundwater concentrations'. *Toxicology and Industrial Health*, **15**,133–150.

Pretty J N, Ball A S, Xiaoyun Li and Ravindranath, N H (2002). 'The role of sustainable agriculture and renewable-resource management in reducing greenhouse-gas emissions and increasing sinks in China and India'. *Philosophical Transactions of the Royal Society A*, **360**, 1741–1761.

PSD (2006). 'Pesticide safety directorate'. http://www.pesticides.gov.uk

Pyšek P and Lepš J (1991). 'Response of a weed community to nitrogen fertilization: a multivariate analysis'. *Journal of Vegetation Science*, **2**, 237–244.

Pyšek P, Jarošik V, Kropáč Z, Chytry M, Wild J and Tichý L (2005). 'Effects of abiotic factors on species richness and cover in Central European weed communities'. *Agriculture, Ecosystems & Environment*, **109**, 1–8.

Reganold J P, Elliott L F and Unger Y L (1987). 'Long-term effects of organic and conventional farming on soil erosion'. *Nature*, **330**, 370–372.

Reganold J, Glover J, Andrews P and Hinman H R (2001). 'Sustainability of three apple production systems'. *Nature*, **410**, 926–930.

Soil Association (2006) *Organic Market Report 2006*. Soil Association, Bristol, UK.

Stoate C, Boatman N D, Borralho R J, Rio Carvalho C, de Snoo G R and Eden P (2001). 'Ecological impacts of arable intensification in Europe'. *Journal of Environmental Management*, **63**, 337–365.

Stolze M, Piorr A, Häring and A Dabbert (2000). *The Environmental Impacts of Organic Farming in Europe*. University of Hohenheim, Stuttgart.

Swiss Federal Office for Agriculture (2006). 'Verordnung über die Direktzahlungen an die Landwirtshaft'. 2. Kapitel: Extensive Produktion von Getreide und Raps. http://www.admin.ch/ch/d/as/1999/229.pdf

Trewavas A (2001). 'Urban myths of organic farming'. *Nature*, **410**, 409–410.

Woese K, Lange D, Boess C and Bögl KW (1997). 'A comparison of organically and conventionally grown foods – results of a review of the relevant literature'. *Journal of the Science in Food and Agriculture*, **74**, 281–293.

Wyss G (2005). 'Assessing the risk from mycotoxins for the organic food chain: results from Organic HACCP-project and other research'. In, Hovi M, Walkenhorst M and Padel S, *Systems development: Quality and Safety.* University of Wales, Aberystwyth, UK, 133–136.

Part I

Organic food safety and quality: introduction and overview

2

History and concepts of food quality and safety in organic food production and processing

Urs Niggli, FiBL, Switzerland

2.1 Introduction

Organic agriculture is a multifaceted phenomenon in the field of agriculture and food production. On the one hand, it is a low external input production technique originating from both traditional and alternative farming practices developed in the late 19th and early 20th century and from European and USA contexts of intensive agriculture. On the other hand, it reflects societal debates on the sustainability of agriculture, on food quality and nutritional habits and on ethical issues like animal welfare. A growing number of scientists and policy makers qualify organic agriculture as an efficient and holistic approach to reach the multiple goals of agriculture including food security, sustainable use of natural resources and the dignity of creatures (Jaber, 2000).

Organic farming is a food production method defined at great length in many international (e.g. *Codex Alimentarius*), supranational (e.g. EU Regulation on Organic Farming) and national (e.g. the US National Organic Program (NOP), the Japanese Agricultural Standard for Organic Products (JAS) or the Swiss Regulation on Organic Farming) standards.

In the developed world, crop production was intensified in the 19th and first half of the 20th century by the use of commercial fertilizers. Soluble phosphorus and nitrogen triggered a first increase in yield levels. The next step in the intensification of agriculture was the widespread use of insecticides, fungicides and herbicides, a practice that also made many conventional farmers feel uncomfortable. The pursuit of yield increases also took hold in livestock husbandry, leading to changes in feeding regimes, industrialized methods for keeping animals and increasing use (and misuse) of veterinary medicines (e.g. antibiotics, anthelmintics) and growth hormones. The arrival and continuous expansion of organic farming has to be seen against this background of continuous intensification of food production and the associated negative impact on environment and biodiversity (Stolze *et al.*, 2000; Stoate *et al.*, 2001; Pyček *et al.*, 2005).

Although it is perceived by the public as a rather uniform and regulated farming method, organic farming has had a range of origins and a multifaceted development until standardization started in Europe in the late 1980s. The most important of these historical food and farming concepts are described in this chapter. Although in some cases only of historical interest, these concepts reveal the background of modern organic farming and food processing and help to elucidate some of its characteristics.

Lately, the progress in organic farming has been dominated by standard setting, their harmonization and the introduction of equal certificates. These activities were driven by (a) fears among organic farmers that organic standards and principles may be compromised by competing strategies like integrated pest management (IPM) or integrated production (IP), (b) consumers who wanted protection from deceit and (c) emerging markets (in particular supermarket chains) in search of certified quality standards. In food markets worldwide, organic foods represented the first food standards, which defined, audited and certified a specific food production process (tracking) rather than specific product properties (e.g. size or colour of vegetables) or composition of the end product (tracing). Such a process-oriented approach in quality management was necessary as organic and conventional foods were difficult to distinguish.

2.2 History of different food concepts of organic farming

One of the earliest sources of inspiration for organic farming was the concept of naturalness of foods. It derived from different ecosocial movements of the early 20th century like the 'naturalist', the 'vegetarian' and the 'reform' philosophies. Of particular influence was the German *Lebensreform* movement, which became important during the time of the Weimar Republic (1919 to 1933). Deteriorations in the living conditions of people during the transition from an agrarian to an industrialised society were correlated with the 'unnaturalness' of the living conditions of the cities (Vogt, 2000). Back to nature was seen as an escape and alternative. Medical doctors and nutritionists like Werner Kollath, Max Bircher-Benner or Stefan Steinmetz propagated whole food (raw vegetables and fruits, whole meal bread or muesli). In this context, the pioneers of 'natural' husbandry and gardening, the Germans Julius Hensel, Heinrich Bauernfeind, Ewald Könemann or the Swiss Mina Hofstetter, experimented – among other farming and gardening techniques – with different rock powders as natural fertilizers to cure the negative effects of mineral sources of nutrients (Vogt, 2000). It can be concluded that 'natural' husbandry was the first concept of organic farming in Europe, which developed quickly from lifestyle movements in the 1920s to an alternative farming method based on the emerging soil and agricultural sciences and on practical farming and gardening experience in the 1930s.

Such idealistic 'back to nature' movements also developed in other parts of Europe. Almost contemporaneously, a group of British writers including Harold John Massingham, Adrian Bell and Rolf Gardiner, promoted their vision of a revitalised countryside (Moore-Colyer, 2001). Central to this vision was an agriculture based on organic principles and this movement became one of the origins of Soil Association which was founded in 1946.

The concept of the vitality of food was raised for the first time by Rudolf Steiner in his seven lectures in 1924 (Steiner, 1929). The emphasis of his lectures was less ecological or agronomical, but focused on describing his views on the deterioration of modern food quality. As part of a wider 'holistic' philosophy called anthroposophy which covered education, art, social theory and science, Steiner developed a spiritually based plant, animal and human nutrition theory, where the real quality of food was not linked to compounds and their metabolisms, but to the spiritual forces which are supposed to 'bound' to them. Many agricultural practices he introduced (e.g. biodynamic preparations, the consideration of lunar or cosmic rhythms when cultivating, sowing or harvesting) aimed to influence these spiritual forces, which were in Steiner's thinking vital for all organisms (Endres and Schad, 1997). Subsequently, anthroposophic scientists introduced the term 'vital quality' (Balzer-Graf and Balzer, 1991; Bloksma *et al.*, 2001).

The efficacy of the specific biodynamic agronomic measures introduced by Steiner has been studied extensively over the last 75 years, but focused mainly on investigations into the way that lunar cycles and biodynamic preparations affect yield, the composition and the nutritional quality of crops. The relative efficacy of these measures is often considered to be less than that of other agricultural measures like variety choice, the intensity of organic fertilization, soil tillage and/or other permitted plant protection measures. To conclude, the improvements achieved by these specific biodynamic techniques are small, often not reproducible and therefore, from a scientific point of view, obsolete. However disenchanting the lack of activity of these specific measures might be, the overall management approach taken by biodynamic farming as a whole is a surprisingly effective and efficient one. In addition, long-term biodynamic soil management has been shown to achieve greater improvements in soil biological activity, structural stability and inherent fertility than more 'mainstream' organic management practices in the long-term field trial DOK where bio*D*ynamic, *O*rganic and conventional (in German *K*onventionell) plots have been compared since 1977 (Mäder *et al.*, 2002).

Since Steiner's aim was to improve 'immaterial' qualities of foods, anthroposophic scientists have developed analytical methods, which aim to visualize this kind of 'inner' quality. This is done by preparing watery solutions of the plant, meat or milk (= juices) which are then brought into reaction with metallic salts like copper chloride (copper chloride crystallization method) or silver nitrate (two different capillary picture methods). The quality of the pictures is either interpreted by visual evaluation or by computerized image texture analysis (Meier-Ploeger *et al.*, 2003). Both interpretations are reproducible and the results are often correlated with standard food analytical quality parameters (e.g. for a case study comparing organic and conventional apples, see Weibel *et al.*, 2000). The main concept of analysing the pictures created by such methods is that crops grown under optimal biodynamic conditions should have a higher degree of 'order' and should be better organized and structured. However, there are currently no sound scientific data that validate and calibrate such methods against standard food composition and metabolic profiling analyses and no studies that demonstrate that consumption of food showing a greater level of 'order' when assessed by 'picture forming methods' results in improved animal or human health.

Another important concept introduced by Steiner was that of 'holism' or 'integrity' of food and farming (Steiner, 1929). Steiner saw a farm as an organism with an inner structure and functionality and not purely as a business with different lines of production. He stressed greatly the common bonds between physiological processes in soils, plants and livestock. This was one reason why organs of cattle (e.g. cow horns or bovine peritoneum) played an important role in the production of biodynamic preparations which aimed to improve soil fertility and plant quality. He believed that, like an organism, a farm has to be managed as a whole unit in its full complexity and integrity. Steiner was influenced by the theory of 'emergent properties' which was developed in the 19th century and which is still used today to characterize very complex systems and phenomena, in nature, physics or engineering (Fromm, 2004). An emergent property can appear when a number of simple subsystems operate as a collective and show more complex and often unexpected behaviours which cannot be explained by adding up the behaviour of the single subsystems. As a consequence, biodynamic farmers are very sceptical about isolated partial interventions (e.g. phytomedical treatments) and rely very much upon preventive and long-term strategies of farm management.

The concept of self regulating and healthy systems was introduced by the English pioneer Sir Albert Howard who stated in the 1930s: '[E]vidence for the view that a fertile soil means healthy crops, healthy animals, and healthy human beings is rapidly accumulating. At least half of the millions spent every year in trying to protect all three from disease in every form would be unnecessary the moment our soils are restored and our population is fed on the fresh produce of fertile land' (Howard, 1942). Lady Eve Balfour, the founder of the Soil Association in Great Britain later described the same concept: 'The health of soil, plant, animal and man is one and indivisible' (Balfour, 1943).

To some extent this concept of a self-regulating nature dovetailed with the idealisation of nature by the philosopher Jean-Jacques Rousseau. In Albert Howard's words:

> The crops and livestock look after themselves. Nature has never found it necessary to design the equivalent of the spraying machine and the poison spray for the control of insect and fungus pests. There is nothing in the nature of vaccines and serums for the protection of the livestock. It is true that all kinds of diseases are to be found here and there among the plants and animals of the forest, but these never assume large proportions. The principle followed is that the plants and animals can very well protect themselves even when such things as parasites are to be found in their midst. Nature's rule in these matters is to live and let live (Howard, 1943).

The idea that husbandry means to maintain a health chain from soil to plant to livestock to human being greatly stimulated the development of organic farming towards a modern, science-based agroecological concept. The scientific work of the Austrian–Hungarian Raoul Heinrich Francé (1874–1943) on soil biology, which has inspired all organic pioneers, was a key factor in the development of this concept. At the beginning of the 20th century, Francé was the first soil ecologist who described the soil as a complex network of organisms (he called it an 'edaphon'). His and – after his death – his wife Annie Francé-Harrar's work on humus in arable soils (Francé, 1922) was the very foundation for what later became the core concept of organic crop production. This concept was based on dynamic processes of composition and decomposition of organic matter, took into account the very complex food chains of millions of organisms and animals in the soil, and made use of the various ecological and agronomical functions of humus and soil microorganisms. In contrast, at very much the same time, conventional crop nutrition was still using 'simplistic' input–output models of macro- and micro-nutrient elements, based on the work of Justus von Liebig (1803–1873) and Eilhard Alfred Mitscherlich (1874–1956), which did not consider soil physical characteristics and biological processes (Von Liebig, 1840, 1855; Mitscherlich, 1952, 1954; Van der Ploeg *et al.*, 1999).

In Switzerland, Germany and Austria, organic farming was taken up by a significant number of farmers earlier than elsewhere in the world. These farms were economically viable and early collaborative forms of marketing and commercialisation emerged. It was the great teachers and motivators Hans Müller (1891–1988) and Maria Müller (1899–1969) who made this first boost possible, supported by scientific work on soil fertility by the microbiologist Hans Peter Rusch (1906–1977). Their approach to organic farming was very much influenced by Mina Hofstetter, Albert Howard and Raoul Francé. Müllers' and Rusch's concepts cannot be seen as genuinely new. However, what was surprisingly novel was the fact that they reduced complex theories to a manageable practice and were therefore extremely important for the spread of organic farming. Their approach was participative (farmers were researchers and advisors), they developed analytical tools which helped farmers to check the progress and the 'organic' quality of their farms (the so-called Rusch test) and they saw how important were the training and education of farmers.

In one way or another, all these historical concepts of organic farming were responses to undesirable or one-sided developments in mainstream intensive agriculture. These concepts are still influencing the framework of organic farming and its standard setting. In 2005, the worldwide umbrella organization of organic farmers, the International Federation of Organic Agriculture Movements (IFOAM), substantiated and updated these historical concepts in the form of four principles: the principle of health, the ecological principle, the principle of fairness and the principle of care (IFOAM, 2005), thus adding two important ethical concepts (fairness and care) to historical concepts.

The perception that organic agriculture and foods are characterised as more 'natural' is widespread among consumers and the concept of naturalness is used by organic farmers to distinguish their technique from those of their conventional colleagues (Verhoog *et al.*, 2002). As an educational strategy, Dutch farmers and advisors use three approaches in order to introduce and understand the concept of naturalness during the conversion from conventional to organic farming. The first step is to ban chemosynthetic pesticides and mineral fertiliser inputs in production and processing and to replace them, if possible, by more natural substances or techniques. In a second step, 'preventative strategies' are introduced to achieve agro-ecological stability (e.g. beetle banks are established to increase natural enemy populations). In a third step, the integrity of plants, animals, humans and ecosystems is explained and treated as an intrinsic value and thus influences the attitude of farmers (Baars, 2002). This practical experience from working with farmers in the Netherlands shows how important the conceptual background is in order to improve the sustainability of organic farming systems continuously.

2.3 Where are modern organic food and farming concepts heading?

There is no question that organic farming is mainly about the sustainable use of natural resources, including livestock, while at the same time reaching acceptable levels of productivity. Three indicators that appear most frequently in a definition of sustainable agriculture are (i) environmentally sound, (ii) economically viable and (iii) socially acceptable. The debate among scientists and policy makers about which concept of agriculture and food production best matches sustainability has already filled many books and conference proceedings and has remained controversial (Trewavas, 2001, 2004; Stoate *et al.*, 2001, Pretty *et al.*, 2003). However, there appears to be a wider consensus for the view that organic agriculture should primarily be analysed in the context of sustainable land use and food production, as it ranks high in all three criteria mentioned above.

By modelling soil erosion, Pimentel *et al.* (1995) showed that 30% of the world's arable land was lost from 1955 to 1995. As losses continue by 10 million hectares per year, the approaches taken by mainstream conventional agriculture so far to reverse or at least stop further soil degradation – e.g. integrated production (IP) and minimum tillage techniques – have obviously failed. Soil degradation processes are irreversible and trigger a chain of reactions of further decline such as reduced water-holding capacity in soils resulting in water shortage, loss of ecological habitats and biodiversity and finally migration of farmers from the land. Organic agriculture, on the other hand, was demonstrated to build-up soil fertility and to improve the physical and biological properties of soils (Mäder *et al.*, 2002) and was described as an avenue towards achievement of sustainability and long-term prosperity (Pretty *et al.*, 2003). The future advancement of the organic concept will depend on maximising its potential as a viable approach to sustainable and efficient use of natural resources. The key elements that should be further explored and exploited by scientific efforts are described below.

2.3.1 Organic crop husbandry – a model for sustainable crop systems

In organic crop production, prevention and recycling should be the predominant strategies used and further developed. Prevention is the preferential measure for the control of weeds, pests and disease pathogens. Recycling of organic matter maintains soil fertility and provides a balanced supply of nutrients to crops. How close agricultural production systems are to achieving crop health via preventive strategies and how closed their nutrient cycles are, are crucial criteria when evaluating ecological sustainability. Therefore, these two techniques should be the focus and the starting point of any crop research focused on sustainability (Lampkin, 1999).

Fertility management

Traditionally, organic farms are mixed ley/arable systems with relatively short nutrient cycles, based on fertility-building legume crops and application of fresh manure and immediate incorporation. This is the best way to recycle maximum amounts of nutrients back to the field production environment. The mixed farm is still the model for the efficient recycling of nutrients, but economic pressures often result in specialization of organic farms too. Therefore, highly specialized organic farms emulate closed cycles through cooperation with other farms. Recent research work on the use of farmyard manures has primarily targeted the reduction of nitrogen losses through leaching and gaseous emissions (Philipps and Stopes, 1995). This research has further reduced losses and improved the nutrient use efficiency of the whole manure and slurry chain from livestock to plant uptake (Fortune *et al.*, 2000; Von Fragstein, 1995). The traditional knowledge of organic farmers and gardeners of composting techniques combined with government support for the recycling of communal green waste and organic household waste as well as technological innovations and new composting machinery, has resulted in organic matter recycling based methods of fertilization which not only improve the productivity of organic horticultural production in particular, but are also increasingly adopted by conventional farms.

Nitrogen is the only nutrient element which can be supplied in sufficient quantities on each farm, either via manure and organic matter recycling approaches or by introducing grain legumes, legume-rich leys for fodder production and cover crops into arable and horticultural rotations. So-called 'catch crops' (Thorup-Kristensen *et al.*, 2003) are grown to catch available nitrogen in the soil in order to prevent nitrogen leaching losses and improve the nutrition of the succeeding main crops. Most frequently sown cover crops in central and northern Europe are crucifers (fodder radish, white mustard), monocots (ryegrasses, winter rye, oats) and legumes (hairy vetch, red, white, sweet and crimson clover, faba bean, field pea and alfalfa).

Advances in cover crop techniques have made even stockless organic systems a productive and economically viable option (Schmidt *et al.*, 1999; Welsh *et al.*, 2002). Despite the availability of efficient biological nitrogen fixation and conservation-based strategies, chemo-synthetic, mineral nitrogen fertilisers have remained the main nitrogen input in conventional farming. The production of mineral nitrogen fertilisers requires extremely high fossil fuel inputs and is therefore associated with substantial CO_2 emissions. Their continued use has to be questioned. For example, the production of 1 kg N requires 11 litre fuel (Finck, 1979), which means a 100 ha stockless arable farm (which applies 170 kg N/ha, a level commonly used for cereals) will use approximately 17,000 litres of fuel each year. This and the high level of nitrogen losses and associated environmental problems, make the sustainability of nitrogen fertilizer-based food production systems increasingly questionable. The nitrogen self-reliance of organic systems is a major advantage in times of fossil energy shortage (Cormack, 2000).

On the other hand, self-sustaining nitrogen supply in crops is a major innovation of organic farming and a step forward to making agriculture independent from fossil energy supply. In the near future, rising oil prices and decreasing oil reserves are expected to increase the economic competitiveness of the organic 'low input' approach to nitrogen supply and it is likely also to become the predominant approach in conventional food production systems.

Crop rotation

The key to successful organic crop production is a carefully designed crop rotation. It is recommended that organic rotations are not based on large, homogeneous fields, which are typically used by conventional farming for convenience of maximum economic efficiency. Instead smaller fields, careful monitoring of crops to allow optimum timing of field operations and increased labour inputs are used, owing to the need to fine tune field activities in response to weather and to specific environmental conditions at the field level.

It has been shown that an ideal sequence of crops can significantly reduce or alleviate many crop protection and fertility management problems in farming (Robson *et al.*, 2002). In conventional farming the rotations are mainly designed following relatively short-term economic considerations which often focus on maximising yields and reduction of, in particular, labour costs. However, convenience (the desire for a regular working day) and a reduction of the need for field by field, season by season, decision making are also thought to have been motivations for the adoption of conventional farming practice. This leads in most cases to a simple rotation (or even monoculture) focused on few cash crops. In contrast, organic crop rotations aim to build up and maintain soil fertility, producing nitrogen through legumes, 'organising' loss-free nitrogen transfer to demanding crops and, most important, minimising weed, disease and pest problems (see Robson *et al.*, 2002 for a detailed review).

For example, growing crops promotes the establishment of weeds with similar life cycles in the same field. Therefore, by alternating crops with different life cycles and periods of growth in the field, any one weed will find it more difficult to establish itself year after year, because of the differential effects (e.g. below and above ground growth pattern, canopy structure, spacing, allelopathic effects) expressed by the succession of crops in the rotation (Robson *et al.*, 2002). Also, in arable crops, ecosystem complexity (e.g. a minimum density and greater diversity of weed ground cover to maintain high levels of predatory beetles within the crop) can best be achieved by complex rotations resulting in habitats where the balance between beneficial and noxious insects can be better controlled (Gurr *et al.*, 2004). Finally, diverse crop rotations can efficiently prevent many problems associated with soilborne pathogens attacking plant roots (Cook and Baker, 1983). The combined use of rotations and organic matter-based fertility management practices may also reduce the incidence of foliar diseases by minimising the persistence and dispersal of diseases (Zadoks and Schein, 1979) and by inducing crop resistance (Van Loon *et al.*, 1998).

Crop rotations are supplemented by proven agronomic practices including the use of (a) tillage and mechanical weed control measures, (b) row spacing and planting dates, (c) cover-, intra- and intercrops and (d) pest and disease tolerant or resistant varieties. Such rotation-based agronomic systems are often summarized by the term 'habitat management', which also includes the establishment of botanically diverse field margins and hedgerows or wildflower strips, in order to increase both levels of natural enemy populations and biodiversity in agricultural ecosystems (El-Hage Scialabba and Hattam, 2002; Hole *et al.*, 2005; DEFRA, 2006). An overall logical framework describing the additive and/or synergistic effects of combining organic fertility management, rotation design and crop protection practices is published by Leifert *et al.* (2007).

Important food quality and safety issues associated with organic livestock production are addressed further in Part 2 of the book.

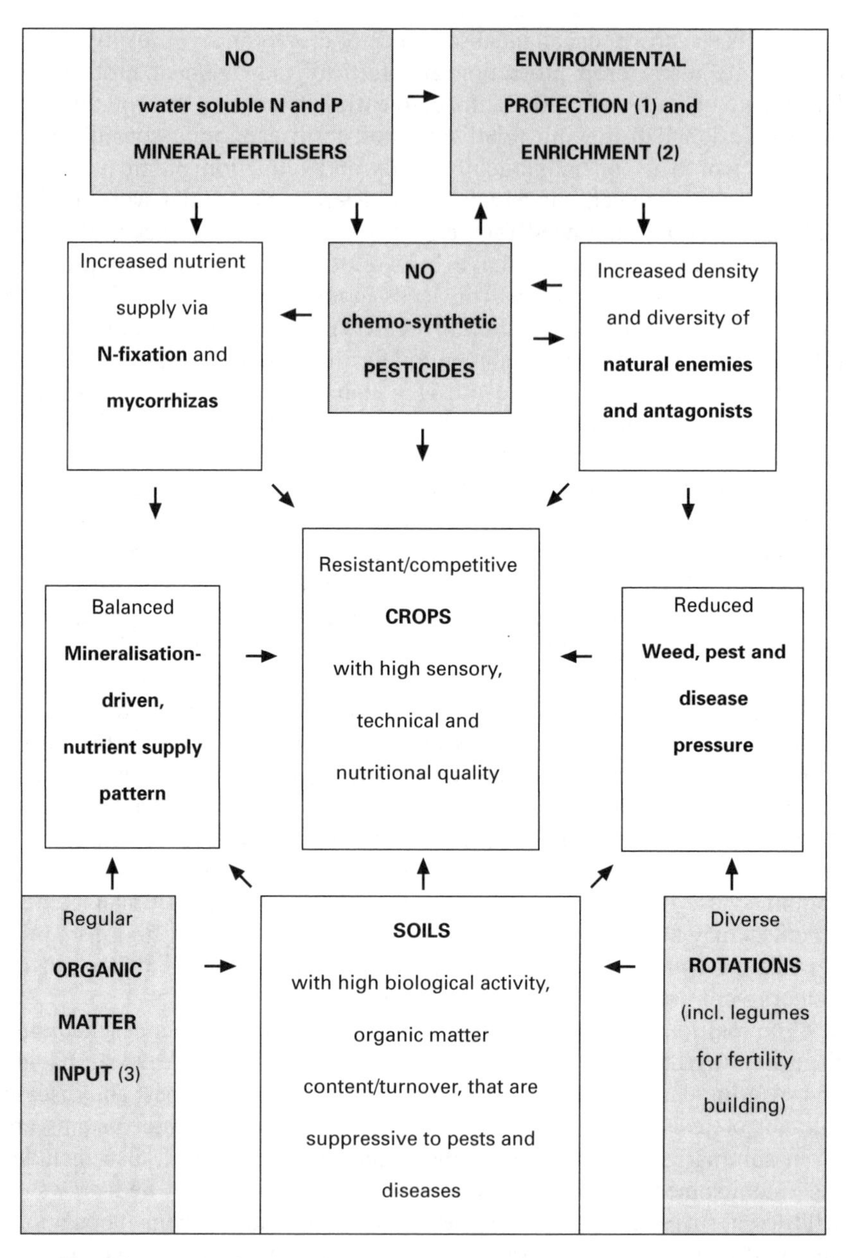

Fig. 2.1 Logical framework for organic production systems (redrawn with permission from Leifert *et al.*, 2007). Shaded area represents agronomic practices prescribed or recommended under organic and other low input farming standards. Unshaded area represent positive impacts on soils, crops and the environment associated with agronomic practices in organic and other low input food production systems. Bold arrows (all directions) represent positive effects or impacts. (1) Measures to minimise pollution (e.g. N-leaching and P-run off) and soil erosion events associated with agricultural activities (e.g. non-cropped field margins, fertility catch crops). (2) Measures taken to increase biodiversity on farms (introduction of non crop vegetation e.g. woodlands, hedges, non-cropped field margins, beetle banks). (3) Animal and green manures, manure and waste based composts, etc.

2.3.2 Organic livestock husbandry – does a satisfactory concept exist?

Livestock husbandry is not explicitly addressed in the historical concepts of organic farming. Livestock provided manure and was therefore an important link in the crop and soil fertility chain. Livestock played a major role on its own only in biodynamic farming, especially cattle whose organs are described as important catalysts of processes and transmitters of cosmic energies (Steiner, 1929). Nonetheless, even the pioneers of biodynamic farming did not consider livestock husbandry from an ethical or animal welfare point of view.

However, a strong link between the organic concept of 'naturalness' (which was previously used mainly for crops or for processed foods) and the concept of animal welfare can be constructed. Naturalness in the case of livestock means that livestock are kept in a way that they can express their natural behaviour patterns including natural reproduction and growth or production rates which do not lead to health problems, stress and behaviour disorders (Alrøe *et al.,* 2001). The Danish authors additionally named 'harmony' and 'care' as important principles of livestock husbandry on organic farms. 'Harmony' reflects all interactions among the different organisms and branches of a farm (e.g. livestock and crop husbandry, between the species kept on a farm or between the animals of a herd), but also between a farm and its ecological, economic or social environment. 'Care' also paraphrases the special responsibility that farmers, transporters and processors have towards domestic animals.

Unlike animal welfare, health concepts for organic herds are less developed and are considered more challenging to implement. Because diseases cause livestock to suffer and become weak, interventions with conventional allopathic drugs are not generally banned in most standards, including those of the EU. Rather, they are restricted to when and how frequently they may be used, and prescribe that the withholding periods between an animal being treated and livestock products being marketed, is at least double that prescribed for conventional farming systems. Good livestock health means not simply the absence of diseases and parasites, but also a high level of vigour and vitality, thus enhancing the animal's ability to resist infections, parasite attack and metabolic disorders, and to recover from injury.

However, the real situation in organic herds in most regions of the world does not yet reflect this ideal. Keatinge *et al.,* (2000) summarised the health status on organic farms as follows: 'Taken as a whole, the information indicates that within current standards the incidence of disease on organic farms is generally at acceptable levels, and at least in some specific circumstances, better than that on conventional farms. The data also indicates a significant use of allopathic medicine.' This leads to a divergence between what is claimed to be an essential feature of organic farming (no use of chemo-synthetic inputs) on one hand and the reality of organic livestock husbandry on the other. Organic dairy farmers in the USA have adopted a throwaway approach, either culling sick cows or moving them to conventional farms.

Most European producers prefer a more sustainable health management concept, but currently use preventative animal health management strategies still requiring significant further development. For example, diseases such as mastitis are multi-faceted problems and require well designed interdisciplinary research programmes, bringing together veterinarians, agronomists and economists. There is also a need for a trans-disciplinary approach, as farmers and practical veterinarians are the main actors in a successful health strategy (Vaarst *et al.,* 2004). To develop a concept of preventive health management of livestock is a high priority challenge for organic farming.

Important food quality and safety issues associated with organic livestock production are addressed further in Part III of the book.

2.3.3 Organic principles in the context of a changing lifestyle

Convenience is becoming an increasing consumer desire with respect to food. For example, in Great Britain, the value of household spending on eating out, surpassed spending on food sector products eaten at home in 2004 (National Statistics, 2006). Moreover, eating at home increasingly means preparing pre-prepared convenience food with consumers demanding the same types of food throughout the year. The British development is mirrored in many other countries of the developed world and has to do with a rapidly changing lifestyle linked to the status of women in society, family form and size, and other socioeconomic conditions. In view of these societal changes, organic concepts, focused mainly on primary 'on farm' production of the raw product, have to be questioned. If food qualities like 'naturalness', 'authenticity' or 'integrity' are lost during processing, transport and storage of fresh foods, these concepts may lose their relevance for the consumer.

There is no doubt that consumers of organic foods have specific expectations with respect to quality parameters of processed food (Zanoli and Naspetti, 2002). These may relate to the degree of processing, concern about specific additives, nutritional composition, integrity or whole food concepts, the degree of convenience, the level of energy use and transportation distances, but also to food safety. Consequently, Gallmann (2000) proposed use of the criteria 'freshness', 'minimal processing' and 'careful treatment' in order to characterize organic food quality. A guaranteed freshness includes aspects of maximum transport time, of single ingredients and split material flows, of food miles and of authenticity. Minimal processing means reducing or replacing additives and processing aids by using ingredients with functional properties and/or optimizing processes. And careful treatment includes diminishing mechanical stress and heat load during processing. With increasing amounts of organic foods being produced in the developing world, the principle of 'fair trade' will also have to be included in organic principles/standards.

As an increasing percentage of organic foods reach the consumer in a processed form, additional criteria become important, like packaging (renewable materials, recycling), industrial ecology, life cycle assessments of products and transport costs. Organic principles, which initially focused mainly on primary production, need to be expanded to include processing, packaging and transporting as well as catering-related principles and standards. Such principles and a code of practice were developed in the EU project Quality of Low Input Foods in the 6th Framework of the Commission (Schmid *et al.*, 2004; Beck, 2006).

Important food quality and safety issues associated with organic processing and quality assurance along the food chain are addressed further in Part 3 of the book.

2.4 Conclusions

The historical texts describing concepts and principles of organic farming may appear 'old fashioned' with their often flowery language and missionary approach. However, they turn out to represent an amazing 'foresight' given that they were mainly based on 'on-farm' observations and resulting 'inspirations' and developed before the current scientific methodologies and knowledge of soil, plant, animal and food science became available. Given the recent realization that energy use and CO_2 emissions are becoming the main challenges with respect to global sustainability, they appear surprisingly modern. In fact, after re-editing into modern language they provide the most exciting and integrative concepts for productive and sustainable and high quality agriculture.

Major challenges ahead include the (a) conceptual development of organic livestock (including fish farming), processing and/or even retail/supply chain standards, (b) the initiation of both crop and animal breeding programmes focused on the needs of organic systems and (c) coexistence with production systems which permit the use of genetically modified organisms.

2.5 References

Alrøe, H F, Vaarst, M and Kristensen, E S (2001) 'Does organic farming face distinctive livestock welfare issues? A conceptual analysis', *Journal of Agricultural and Environmental Ethics*, **14**, 275–299.

Baars, T (2002) *Reconciling Scientific Approaches for Organic Farming Research. Part I: Reflection on Research Methods in Organic Grassland and Animal Production at Louis Bolk Institute, The Netherlands*. CIP-Data Koninklike Bibliothek, Den Haag, The Netherland, 3–224.

Balfour, E B (1943) *The Living Soil*, Faber and Faber, London, UK.

Balzer-Graf, U and Balzer, F (1991) 'Steigbild und Kupferchloridkristallisation – Spiegel der Vitalaktivität von Lebensmitteln', in Meier-Ploeger, A and Vogtmann, H *Lebensmittelqualität – ganzheitliche Methoden und Konzepte*, 2nd edition. Verlag C.F. Müller, Karlsruhe, 163–210.

Beck, A (2006) *Code of Practice for Organic Food Processing*, (with contributions from Ursula Kretzschmar, Angelika Ploeger and Otto Schmid). Published by the Research Institute of Organic Agriculture (FiBL), Frick, Switzerland. Available as a web-publication on http://orgprints.org/7031.

Bloksma, J, Northolt, M and Huber, M (2001) *Parameters for Apple Quality and an Outline for a New Quality Concept*. Louis Bolk Institute, Driebergen, The Netherlands. Publication GVV 01. FQH Publication FQH 01, 56 pp.

Cook, R J and Baker, K F (1983) *The Nature and Practice of Biological Control of Plant Pathogens*, St Paul, MN, USA, American Phytopathological Society Press, 539 pp.

Cormack, W F (2000) *Energy use in Organic Farming Systems* (OF0182). Final Project Report to Ministry of Agriculture, Fisheries and Food, London, UK. http://orgprints.org/8169/

DEFRA (2006) Organic farming and the Environment http://www.defra.gov.uk/farm/organic/policy/actionplan/annex3.htm

El-Hage Scialabba, N and Hattam, C (2002) *Organic Agriculture, Environment and Food Security*, Environment and Natural Resources Series 4. Food and Agriculture Organisation of the United Nations, Rome, 252 pp.

Endres, K-P, Schad, W (1997) *Biologie des Mondes*, S Hirzel Verlag, Stuttgart - Leipzig, 308 pp.

Finck, A (1979) *Dünger und Düngung. Grundlagen, Anleitung zur Düngung der Kulturpflanzen*, Verlag Chemie, Weinheim, New York, 295 pp.

Fortune, S, Conway, J S, Philipps, L, Robinson, J S, Stockdale, E A and Watson, C (2000) 'N, P and K for some UK organic farming systems – implications for sustainability', in Rees, RM, Ball, B, Watson, C and Campbell, C, *Sustainable Management of Soil Organic Matter and Sustainability*, CABI, Wallingford, 286–293.

Francé, R H (1922) Das Leben im Boden. Das Edaphon. (Nachdruck) Edition Siebeneicher, Deukalion Fachverlag für Landwirtschaft und Ökologie. Uwe Hils Verlag, Holm, reprinted in 1995.

Fromm, J (2004) *The Emergence of Complexity*, Kassel University Press, Germany, 200 pp.

Gallmann, P U (2000) 'All natural and convenience products: a contradiction? The impact of food technology', in Stucki, B and Meier, U *Proceedings of the 1st International Seminar on Organic Food Processing*, held in Basel, 28 to 31 August 2000, published as FiBL (Research Institute of Organic Agriculture) report, Frick, Switzerland.

Gurr, G M, Wratten, S D and Altieri, M A (2004) *Ecological Engineering for Pest Management. Advances in Habitat Manipulation for Arthropods*, CABI, Wallingford, 232 pp.

Hole, D G, Perkins, A J, Wilson, J D, Alexander, I H, Grice, P V and Evans, A D (2005) Does organic farming benefit biodiversity? *Biological Conservation*, **122**, 113–130.

Howard, A (1942) 'Medical testament', in Grant, D, *Feeding the Family in War-time, Based on the New Knowledge of Nutrition*. Available as a web-publication on http://journeytoforever.org/farm_library/medtest/medtest_howard.html

Howard, A (1943) *An Agricultural Testament*, Oxford University Press, New York and London, Reprinted 2000.

IFOAM (2005) *The principles of organic farming*. Available as a web-publication on http://www.ifoam.org/organic_facts/principles/pdfs/Principles_Organic_Agriculture.pdf

Jaber, D (2000) 'Human dignity and the dignity of creatures', *Journal of Agricultural and Environmental Ethics*, **13** (1/March), 2000, 29–42.

Keatinge, R, Gray, D, Thamsborg, S M, Martini, A and Plate, P (2000) 'EU Regulation 1804/1999 – the implications of limiting allopathic treatment', in Hovi, M and Garcia Trujillo, R, *Diversity of Livestock Systems and Definition of Animal Welfare*. Proceedings of the Second NAHWOA Workshop, Cordoba, January 8–11, 2000. University of Reading, UK, 92–98.

Lampkin, N (1999) *Organic Farming*, Farming Press Miller Freeman, Tonbridge.

Leifert, C, Rembialkowska, E, Nielson J H, Cooper, J M, Butler, G and Lueck L (2007) Effects of organic and 'low input' production methods on food quality and safety. In Niggli, U, Leifert, C, Alföldi, T L, Lueck and Willer, H, *Improving Sustainability in Organic and Low Input Food Production Systems*. Published by the Research Institute of organic Agriculture (FiBL), Frick, Switzerland. Available as a web-publication on http://www.fibl.org/shop/pdf/hp-1455-organic-food-production.pdf

Mäder, P, Fliessbach, A, Dubois, D, Gunst, L, Fried P and Niggli, U (2002) 'Soil fertility and biodiversity in organic farming', *Science*, **296**, 1694–1697.

Meier-Ploeger, A, Kahl, J, Busscher, N, Mergardt, G, Strube, J, Mende, G, Negendank, C, Stolz, P, Böhm, B, Köhl-Gies, B, Staller, B, Merschel, M, Werries, A, Rahmann, G, Weirauch, K, Treutter, D, Degert, A and Kromidas, S (2003) *Ganzheitliche Untersuchungsmethoden zur Erfassung und Prüfung der Qualität ökologischer Lebensmittel: Stand der Entwicklung und Validierung*. Report published by Geschäftsstelle Bundesprogramm Ökologischer Landbau, Bundesanstalt für Landwirtschaft und Ernährung (BLE), Bonn, http://orgprints.org/4815/01/4815-02OE170-ble-uni-kassel-2003-validierung-qualitaet.pdf, 319 pp.

Mitscherlich, E A (1952) *Das Gesetz von abnehmenden Bodenertrag und was darunter zu verstehen ist*, Akademie-Verlag, Berlin.

Mitscherlich, E A (1954) *Bodenkunde für Landwirte, Forstwirte and Gärtner in pflanzenphysiologischer Ausrichtung und Ausrichtung*, Parey, Berlin/Hamburg.

Moore-Colyer, R J (2001) 'Back to basics: H J Massingham, Rolf Gardiner and "A Kinship of Husbandry"', *Rural History, Economy, Society and Culture*, **12**(1), 85–108.

National Statistics (2006) *United Kingdom Input–Output Analyses*, 2006 edition. Available as a web-only publication on www.statistics.gov.uk/inputoutput.

Philipps, L and Stopes, C (1995) 'The impact of rotational practice on nitrate leaching losses in organic farming systems in the United Kingdom', *Biological Agriculture and Horticulture*, **11**, 123–134.

Pimentel, D C, Harvey, P, Resosudarmo, K, Sinclair, D, Kurz, M, McNair, S, Crist, L, Shpritz, L, Fitton, L, Saffouri, R and Blair, R (1995) 'Environmental and economic costs of soil erosion and conservation benefits', *Science*, **267**, 1117–1123.

Pretty, J, Morison, J I L and Hine, R E (2003) 'Reducing food poverty by increasing agricultural sustainability in developing countries', *Agriculture, Ecosystems and Environment*, **95**, 217–234.

Pyšek P, Jarošik V, Kropáč Z, Chytrý M, Wild J and Tichý L (2005) 'Effects of abiotic factors on species richness and cover in Central European weed communities', *Agriculture, Ecosystems and Environment*, **109**, 1–8.

Robson, M C, Fowler, S M, Lampkin, N H, Leifert, C, Leitch, M, Robinson, D, Watson, C A and Litterick, A M (2002) 'The agronomic and economic potential of break crops for ley/arable rotations in temperate organic agriculture', *Journal of Agronomy*, **77**, 370–427.

Schmid, O, Beck, A and Kretzschmar, U (eds) (2004) *Underlying Principles in Organic and 'Low-Input Food' Processing – Literature Survey*, FiBL Report, Frick, Switzerland. Available as a web-publication on http://orgprints.org/00003234.

Schmidt, H, Philipps, L, Welsh, J P and von Fragstein, P (1999) 'Legume breaks in stockless organic farming rotations: Nitrogen accumulation and influence on the following crops', *Biological Agriculture and Horticulture*, **17**, 159–170.

Steiner, R (1929) *Geisteswissenschaftliche Grundlagen zum Gedeihen der Landwirtschaft*. Landwirtschaftlicher Kurs. Koberwitz bei Breslau 1924. Rudolf Steiner Verlag. 7th edition, 1984, 261 pp.

Stoate, C, Boatman, N D, Borralho, R J, Rio Carvalho, C, de Snoo, G R and Eden, P (2001) 'Ecological impacts of arable intensification in Europe', *Journal of Environmental Management*, **63**, 337–365.

Stolze, M, Piorr, A, Häring, A and Dabbert, A (2000) *The Environmental Impacts of*

Organic Farming in Europe, Economics and Policy, **6**, University of Hohenheim, Stuttgart.

Thorup-Kristensen, K, Magid, J and Jensen, L S (2003) 'Catch crops and green manures as biological tools in nitrogen management in temperate zones', *Advances in Agronomy*, **79**, 227–302.

Trewavas, A (2001) 'Urban myths of organic farming', *Nature*, **410**, 409–410.

Trewavas, A (2004) 'A critical assessment of organic farming-and-food assertions with particular respect to the UK and the potential environmental benefits of no-till agriculture', *Crop Protection*, **23**, 757–781.

Vaarst, M, Wemelsfelder, F, Seabrook, M, Boivin, X and Idel, A (2004) 'The role of humans in the management of organic herds', in Vaarst, M, Roderick, S, Lund, V and Lockeretz, W. *Animal Health and Welfare in Organic Agriculture*, CABI, Wallingford, UK, 419 pp.

Van Loon, L C, Bakker, P A H M and Pieterse, C M J (1998) 'Systemic resistance induced by rhizosphere bacteria', *Annual Review of Phytopathology*, **36**, 453–483.

Van der Ploeg, R R, Böhm, W and Kirkham, M B (1999) 'On the origin of the theory of mineral nutrition of plants and the law of minimum', *Soil Science of America Journal*, **63**, 1055–1062.

Verhoog, H, Matze, M, Lammerts van Bueren, E T and Baars, T (2002) 'The role of the concept of the natural (naturalness) in organic farming', in Lammerts van Bueren, E (2002). *Organic Plant Breeding and Propagation: Concepts and Strategies*. PhD, Wageningen University, 18–37.

Vogt, G (2000) *Entstehung und Entwicklung des ökologischen Landbaus*. Ökologische Konzepte 99. Stiftung Ökologie und Landbau, Bad Dürkheim, Germany, 399 pp.

Von Fragstein, P (1995) 'Manuring, manuring strategies, catch crops and N-fixation', *Biological Agriculture and Horticulture*, **11**, 275–287.

Von Liebig, J (1840) *Die organische Chemie in ihrer Anwendung auf Agricultur und Physiologie* (Organic Chemistry in its Applications to Agriculture and Physiology) Friedrich Vieweg und Sohn, Braunschweig, Germany.

Von Liebig, J (1855) *Die Grundsätze der Agricultur-Chemie mit Rücksicht auf die in England angestellten Untersuchungen von Herrn J.B. Lawes (1st & 2nd Ed.)* (The Relationship of Chemistry to Agriculture and the Agricultural Experiments of Mr J B Lawes), Friedrich Vieweg und Sohn, Braunschweig, Germany.

Weibel, F P, Bickel, R, Leuthold, S and Alföldi, T (2000) 'Are organically grown apple tastier and healthier? A comparative field study using conventional and alternative methods to measure fruit quality', in *Acta Horticulturae*, **517**, 417–427.

Welsh, J P, Philipps, L and Cormack, W F (2002) 'The long-term agronomic performance of organic stockless rotations', in Powell *et al.* (eds). *UK Organic Research 2002*: Proceedings of the COR Conference, 26–28 March 2002, Aberystwyth, 47–50.

Zadoks, J C and Schein, R D (1979) *Epidemiology and Plant Disease Management*, Oxford University Press, New York.

Zanoli, R and Naspetti, S (2002) 'Consumer motivations in the purchase of organic food', *British Food Journal*, **104**, 8–14.

3

Nutritional quality of foods

Chris J. Seal and Kirsten Brandt, Newcastle University, UK

3.1 Introduction

Diet plays an important role in most of the chronic diseases that are the largest causes of morbidity and mortality in the developed world. In a 'reductionist' approach, scientists have often made the role of individual nutrients in the maintenance of health the focus of their research. This approach, and in particular the discovery of essential nutrients and their roles in disease prevention, has been instrumental in the elimination of deficiency diseases in large parts of the world. However, nutrients are not consumed in isolation, but as components of whole foods and in an infinite number of combinations. In addition, foods contain a myriad of chemicals (or non-nutrients) which either serve no role in human metabolism or for which the role has not yet been elucidated. This introduces a significant level of complexity, which may be difficult to unravel.

Many nutrients are highly correlated with each other in their effect and many display interactive and synergistic effects. The influence of other lifestyle factors, such as social class, physical activity and smoking, adds a further layer of interaction. Nutritional science has been effective in recognising the impact of some of these factors only where extremes in intake can be identified, for example in micronutrient deficiencies, protein-energy malnutrition or toxicities. It has been less effective where subtle changes in dietary intake are involved. Despite this, governments worldwide have given credence to the fundamental importance of food and nutrition in determining health and have made these central to their public health strategies. Since the start of the 20th century for developed countries, the emphasis has moved from the provision of safe food (i.e. free from foodborne pathogens and contaminants) to the delivery of diets providing adequate nutrients for normal growth and health. The new millennium has seen a further progression to the concept of nutrition in disease prevention, given added impetus with the developing field of gene-nutrient interactions and recognition of the potential importance of individualised nutrition (Arab, 2004; German *et al.*, 2005). Recently, the concept of a 'polymeal', designed using foods identified from the literature as having cardio-protective effects, has been proposed as a mechanism for disease prevention (Franco *et al.*, 2004).

For the majority of the population in developed countries, the emphasis of nutritional advice is on balanced nutrition. Undernutrition can affect every system in the body and is mainly restricted to illnesses (whether poor nutrition leads to disease or disease adversely affects nutritional status) compared with unbalanced nutrition brought about by overconsumption. In the context of balanced nutrition, national surveys of food and nutrient consumption have shown that, in the majority of developed countries, total energy intake has declined since the mid-1970s (For UK data, see Prentice and Jebb, 1995). Over the same period, the relative contributions of carbohydrate and fat to total energy intake have fallen and risen, respectively. The consumption of fruits and vegetables by omnivores has remained generally low. Thus current recommendations have focussed on reducing the proportion of daily energy from fat, increasing the proportion of energy from carbohydrate and increasing fruit and vegetable consumption. In making recommendations on food and nutrient consumption, governments across international boundaries have generally adopted a similar approach with broadly similar objectives.

The traditional way for evaluating the nutritional adequacy of diets is based on comparisons of nutrient intakes with a recommended daily allowance (or in the UK a reference nutrient intake) (Kant, 1996). Among populations not affected by poverty, protein intake generally exceeds current recommendations for both men and women. For example, the latest data from the UK National Diet and Nutrition Survey shows that protein intake was 161% of the reference nutrient intake (RNI) for men and 140% of the RNI for women (Henderson *et al.*, 2003). For nutrient supply, therefore, changes in protein content or quality as a consequence of organic practices are unlikely to have a significant impact at the population level. Similarly, for carbohydrates in the diet, changes in the carbohydrate composition of foodstuffs *per se* are unlikely to result in a pattern of carbohydrate consumption which could be considered more healthful; increasing the proportion of energy derived from carbohydrates will be achieved through overall increased consumption of carbohydrate-rich foods and reduction in fat intake.

Current targets for a reduction in the total fat content of the diet are accompanied by recommendations to reduce the proportion of saturated fats in the diet in favour of unsaturated forms. In this context, organic practices, especially those for animal husbandry and feeding are more likely to result in products with lower fat content (see for example Hansson *et al.*, 2000, and Chapter 9, Section 9.5 of this publication). Manipulating the fatty acid profile of dairy products to reduce saturated fatty acid concentrations and increase the concentration of monounsaturated and polyunsaturated fatty acids may also be a possibility, but requires further study in the context of organic farming practices, since most of the methods which have been shown to be effective in this regard are more associated with intensive supplementation and use of 'non-organic' materials (Givens, 2005). Recently, interest has also focussed on conjugated linoleic acid (CLA) in milk fat because of its anti-cancer effects. Extended grazing on fresh forage, common in organic dairy farm management, is one strategy which has been shown to result in higher CLA concentrations in milk (Givens, 2005).

It is no coincidence that the increasing popularity of organic foods runs in parallel with increasing public concern about nutrition and health, food safety and environmental issues. There have been a large number of studies investigating the factors which influence consumer purchasing and consumption of organic foods. These studies show that perception of health benefits appears to be a more important influence on purchasing than perception of environmental benefits (Shepherd *et al.*, 2005; see also Ch 6). The drive for increased consumption of organic produce represents a shift in the balance of risk assessment in the context of the twin paradigms that high input products (both from the perspective of whole foods and processed foods) represent relatively *high* risk, compared with low-input or organic products which represent *low* risk. This has recently been reviewed (BMVEL (Bundesministerium für Verbraucherschutz Ernährung und Landwirtschaft; German Ministry of Consumer Protection Nutrition and Agriculture, 2003; Magkos *et al.*, 2003). Both reviews concluded that there is currently no hard evidence to support or refute the claim that organic food is healthier or safer than conventional food or vice versa. The majority of risk associated with agricultural food products lies in the area of food safety, contamination of foodstuffs with pesticides, microorganisms and other toxic agents being high on the agenda (Williams and Hammitt, 2001). Many of the benefits associated with the consumption of low-input or organic products involve reducing these risks rather than a perception that the food is more nutritious (Williams and Hammitt, 2001). The failure to relate to the nutritional value of these food sources is most likely due to the lack of information available on changes in nutrient composition of organic agricultural products and, in particular, the paucity of reliable studies which have investigated the nutritional value of low-input or organic foods in humans or animals.

3.2 Methods for determining changes in nutritional quality

Studies which have attempted to demonstrate beneficial effects of consumption of organic foods can broadly be divided into two types: (i) animal and human feeding studies with organic products, which include measures of changing health status and (ii) studies which have measured changes in the nutrient content of foods produced under defined organic conditions. The former have been fraught with difficulties in experimental design, choice of markers of health status and choice of animal model, and the data available from such studies is sparse, especially in humans. The latter make assumptions that changes in nutrient or metabolite concentrations can translate into a perceived health benefit, but for most nutrients this assumption cannot be justified by the data currently available in this area. Even in cases where this assumption is justified, another limitation of this is that estimates of nutrient concentrations *per se* do not give an indication of the bioavailability of the nutrient in animals or humans, nor how the nutrient may be metabolised once absorbed. The number of studies which have made direct comparisons of the bioavailability of individual nutrients from the same foods produced under organic or conventional production systems is extremely small.

3.2.1 Studies in humans

Many of the studies which have evaluated the effect of organically grown food on human health have relied on surrogate markers of health status such as semen quality in men working in the organic food industry. A number of articles have been published in which extensive measurements have been made on sperm number and quality in organic farmers or members of organic associations in comparison with working men from other industries (Table 3.1). In some cases small improvements have been noted for men with an 'organic' background, but the data are not at all clear and are severely confounded by uncontrolled differences in demographic variables. In only one study was the diet accurately quantified in terms of its organic content (Juhler *et al.*, 1999), but the results of this study were inconclusive and many of the changes observed may relate to differences in rural versus urban environments and in particular to environmental/work exposure to pesticides or industrial chemicals. In this context an organic diet may be of secondary importance to the measured outcome, especially since the pesticide contents of all the diets, organic and conventional, was very low.

A study in adolescents (Wendt, 1943) and another in adults (Reiter *et al.*, 1938) investigated the effects of consuming vegetables grown using different fertiliser systems, but neither reported any significant difference between the effects of the treatments on human health. In contrast, a report by Schuphan (1972), which is a transcript of a paper presented by the 'Expert Panel', 'Effects of intensive fertiliser use of the human environment', has been cited several times as demonstrating the benefit of organically grown food on infant weight and blood measures. In this report, Schuphan defines the biological value of a food as embracing 'the nutritional value of a food plant, its wholesomeness including flavour, and its value in the maintenance of human health'. The limited nutritional information provided refers to the superiority of vegetables fertilised with a mixture of 'farmyard manure +

Table 3.1 Select packaging factors for several packaging categories (a complete list may be found in Appendix IV of the chemistry guidance document)

Subjects	Dietary comparison	Response for organic compared with conventional	Source
Members of Danish Association of Organic Farmers	Comparison was between farmers who were members of the Association and working Danish men with occupational exposure. No measure of diet	Higher sperm concentration in association members	(Abell *et al.*, 1994)
Members of organic food associations, Denmark consuming 25% of diet as organic produce	Members of organic associations consuming 25% of diet as organic produce versus healthy men working for airline company with no exposure to agents known to affect semen quality	Sperm concentration 43% higher in members of organic association	(Jensen *et al.*, 1996)
Traditional and organic farmers	Traditional versus organic farmers. No measure of diet	No significant difference in conventional measures of semen quality found between organic and traditional farmers	(Larsen *et al.*, 1999)
Traditional and organic farmers	Traditional versus organic farmers, grouped according to proportion of organic grown fruit and vegetables in the diet (1%, 1–49% and 50–100%)	Significantly higher percentage of normal sperm in highest organic consumers. Dietary intake of 40 different pesticides could not explain the differences in semen quality	(Juhler *et al.*, 1999)
Children from anthroposophic schools (Steiner schools), children from neighbouring schools	Dietary detail not reported, but consumption of fermented vegetables and organic food assessed	Significant reduction in atopic disease in anthroposophic children	(Alm *et al.*, 1999)
Healthy non-smokers, 6 males, 10 females	Duplicate whole diet menus from organic or conventional sources, 22 day duration	Significant increase in urinary flavonoid excretion for organic diet. No difference in antioxidant status between organic or conventional diets.	(Grinder-Pedersen *et al.*, 2003)
Children from farm families, anthroposophic families (Steiner families), the PARSIFAL* study	Dietary detail not reported. Anthroposophic lifestyle likely to include intake of biodynamic food	Significant reduction in allergic symptoms and sensitisation between Steiner children and reference children	(Alfven *et al.*, 2006)

*PARSIFAL = prevention of allergy risk factors for sensitization in children related to farming and anthroposofic lifestyle.

NPK' compared with farmyard manure alone, as a result of the higher vitamin and mineral contents of the products. The superior biological value of the food to the infants refers to the 'average daily gain in weight, the contents of vitamin C and of carotene in the blood, to the teething, to the red blood picture and to the serum iron' (Schuphan, 1972). The report in fact summarises a number of studies completed some 30 years previously and published in German in the late 1930s/early 1940s (cited in Woese *et al.*, 1997) and includes the studies in adolescents and adults discussed above. The studies as a whole, however, provide very limited information on experimental design, used crops grown within a limited range of experimental conditions and involved a heterogeneous group of subjects. Information on dietary intakes is not reported and thus extrapolation to modern day crop production and dietary habits is perhaps not advisable.

There have been two more recent studies with children in which health improvements have been reported for those adopting a predominantly organic diet. In both cases the studies involved children from anthroposophic schools (based on Rudolf Steiner's philosophy) with comparisons against children from neighbouring public schools (Alm *et al.*, 1999; Alfven *et al.*, 2006). The anthroposophic lifestyle is characterised by a predominantly organic diet with substantial amounts of fermented vegetables (providing live *lactobacilli*) following a longer period of breastfeeding, and restricted use of antibiotics and vaccinations. Thus consumption of organic foods is only part of a wider lifestyle characteristic. In both studies incidence of atopy was significantly lower in the anthroposophic groups, indicative of an improvement in immune function in these children. The study by Alm *et al.* (1999) reported on children in Sweden, whereas the study by Alfven *et al.* (2006) was a much larger study involving children from four European countries, suggesting that these positive health benefits are seen in diverse communities and physical environments but following similar lifestyles.

The number of studies which have compared measures of nutritional status following acute or short-term consumption of organic versus conventional products is small and therefore it is difficult to make definitive conclusions about possible nutritional benefits. Table 3.1 summarises some of these studies. One study, which compared the response to consuming 100 g of either organic or conventional tomato puree for three weeks in healthy, non-smoking young women, was unable to show differences in plasma β-carotene or vitamin C concentrations between the groups (Caris-Veyrat *et al.*, 2004). In this experiment, the β-carotene content of the tomato puree fed to volunteers appeared to be lower in the organic supplement (1.71 versus 3.56 mg/100 g) although this was not statistically significant. However, the vitamin C content was almost twice as high in the organic supplement (39.9 versus 22.5 mg/100 g). The lack of response to the two different sources is perhaps not surprising since the differences in intakes of these micronutrients were small relative to the overall intake during the experimental period.

Different plant production methods may result in differences in the concentration of secondary plant metabolites, especially polyphenolic compounds which may have potential antioxidant or pseudo-oestrogenic properties (see for example de Whalley *et al.*, 1990). Probably the best controlled intervention study to investigate the effects of organic versus conventional foods on flavonoid excretion and antioxidant status was completed by Grinder-Pedersen *et al.* (2003). The study was a double-blind randomised, cross-over design with two intervention periods with a strict control of dietary intake in which 16 volunteers consumed each of the test diets for 22 days. The study is unique in providing all of the foods as either organic or conventional products (compared with the Caris-Veyrat study describe above for example, when only the source of one dietary component was changed). Whilst this limits the usefulness in examining the effects of a single product, it is perhaps more useful in providing data which can be compared with populations or individuals who adopt a 'total organic' lifestyle. There were some differences in flavonoid content between the two dietary regimes which may have been due to varietal differences between some of the fruits and vegetables used to formulate the two diets. However, the varieties used reflected those commonly available in Denmark at the time of the study and thus represent a realistic composition of the diet from a consumer's perspective. The organic diet contained significantly higher amounts of quercetin than the conventional diet (4.2 mg/10 MJ compared with 2.6 mg/10 MJ) and about twice the amount of kaempferol, although this was not significantly different from the amount in the conventional diet. For both flavonols, urinary output was significantly higher in subjects whilst consuming the organic diet, but the proportion of intake excreted in urine was not different. For the majority of markers of antioxidant status there was no difference between the two diets. Plasma glutathione reductase activity was increased in both diets and although 2-aminoadipic semialdehyde (a marker of oxidative damage to protein) was slightly higher for the organic diet compared with the baseline, there was no difference in this marker between the treatment groups. These differences were ascribed to changes in overall fruit and vegetable intake during both treatment periods compared with the baseline.

3.2.2 Studies in animals

The majority of animal studies which have compared the consumption of organically and conventionally grown feed on animal health were carried out some time ago and reports are scant in detail. A summary of some of the most commonly cited studies is shown in Table 3.2. Many of these experiments have used measures of fertility as the principal outcome measure, as an indirect marker of nutritional status. The earlier studies, reviewed by Hodges and Scofield (1983), suggest that, in some cases, intensive use of mineral fertilisers caused increased infertility in cattle. In the longest running experiment of its type, the 'Haughley Experiment' collected data over a 34-year period comparing cows reared in an 'organic section' with those grown in a 'mixed

Table 3.2 Animal studies comparing the effects of feeding diets based on organic feeds compared with diets based on conventional feeds. Developed from Worthington (1999)

Species	Feed comparison	Response for organic compared with conventional	Source
Pigeon (adult)	Manure-fertilised millet supplement versus chemically fertilised millet supplement	Less weight loss (22.4% versus 37.4%) over the study period for birds fed manure-fertilised millet. Longer survival (50 versus 33 days) in birds with polyneuritis fed the manure-fertilised millet	(McCarrrison, 1926)
Chicken	Biodynamic versus conventionally fertilised grain	Birds fed biodynamic grain began laying earlier (166 versus 181 days). Spoilage of eggs after six months lower for eggs from birds fed biodynamic grain (27% versus 60%)	(Pfeiffer, 1938, cited by Linder, 1973)
Chicken	Biodynamic versus commercially fertilised feed	Better weight gain after coccidial illness and fewer incidents of illness in chickens fed biodynamic feed. Significantly higher total egg and yolk weights in chickens fed biodynamic feed	(Plochberger, 1989)
Rat	Manure-fertilised millet supplement versus chemically fertilised millet supplement	Rats fed manure-fertilised wheat gained on average 10–17% more weight	(McCarrrison, 1926)
Rat	Biodynamic feed versus conventional feed	Fewer still births and perinatal deaths and better weight maintenance in lactating female rats fed biodynamic feed	(Velimirov *et al.*, 1992)
Rat	Manure-fertilized seed versus chemically fertilised seed	Greater weight gain in vitamin B-deficient rats fed manure-fertilised seed	(Rowlands and Wilkinson, 1930)
Rat	Traditional versus Improved (with fertilisers and pesticides) grown sorghum	Improved sorghum contained higher phytate and feeding resulted in lower incorporation of zinc and iron in bones compared with traditional sorghum	(Ali and Harland, 1991)
Rat	Organic versus conventionally grown wheat, fed at well-nourished or protein-energy-malnourished levels	*In vitro* lymphocyte proliferation response and acute phase proteins not different between organic and conventional wheat when fed at well nourished or protein-energy-malnourished levels.	(Finamore *et al.*, 2004)

Table 3.2 Continued

Species	Feed comparison	Response for organic compared with conventional	Source
		In vitro lymphocyte proliferation in media containing rat serum compared with foetal calf serum was lower in rats fed conventional wheat compared with organic wheat in protein-energy-malnourished rats. Conventional wheat represented higher risk for lymphocyte function than organic wheat in vulnerable conditions for the rat	
Mice	Biodynamic versus conventionally fertilised grain	Mortality rate at 9 weeks of age for 3 generations of mice lower in mice fed biodynamic grain (9% versus 17%)	(Pfeiffer, 1938, cited by Linder, 1973)
Mice	Manure-fertilised wheat *versus* conventional feed	Less testes degeneration and similar reproductive performance in mice fed manure-fertilised wheat.	(Scott *et al.*, 1960)
Rabbit	Biodynamic versus conventionally fertilised feed	Larger number of eggs and higher fertilisation rate in rabbits fed biodynamic diet	(Aehnelt and Hahn, 1978)
Rabbit	Biodynamic versus conventionally grown feed	Fertility rate of rabbits remained constant over three generations in rabbits fed biodynamic feed but declined in rabbits fed conventional feed	(Staiger, 1986, cited by Vogtmann, 1988)
Rabbits	Organic feed versus conventional feed and pellets	Mortality of newborn rabbits lower in animals fed organic feed compared with conventional feed or conventional pellets	(Gottschewski, 1975, cited by Vogtmann, 1988)
Cattle	Organic versus conventional fodder	Reduced sperm motility in bulls transferred from organic to conventional fodder; sperm motility restored when returned to organic fodder	(Aehnelt and Hahn, 1978)
Pigs	Organic versus conventional feeding	No effect of organic feed compared with conventional feed on measures of immune or stress status	(Millet *et al.*, 2005)

section' (Balfour, 1975). Overall, the results suggested that organically grown feed may have some benefit for animal health and performance, with reports of higher milk yield and greater productivity in cows fed organically. However, the study was not well controlled, had large numbers of variations and treatments and as a result statistical comparisons in the data were not possible. In addition, the study is now quite dated and many of the farming practices would not be comparable with current modern agriculture. It is questionable, therefore whether the data are applicable to other situations.

A better controlled and designed recent study of performance of dairy cows on Swiss farms with organic and integrated production was unable to show differences in fertility between the two production systems (Roesch *et al.*, 2005). In contrast, rabbits fed biodynamic feed produced more embryos, gave birth to larger litters and were less susceptible to infections than rabbits fed conventional feed, and chickens show higher fertility and less morbidity when fed organically (cited in Bourn and Prescott, 2002). In a well-controlled study on the effects of organic and conventional feeds on fertility in rats, Velimirov *et al.* (1992) showed that significantly fewer offspring were born dead in the organic group compared with the conventional group in the first generation. The data were somewhat inconsistent, however, between litters, with no difference in death rates in litters between groups in the second generation, although once again they were reduced in the organic group in the third generation. The overall conclusion from this study was that 'some aspects of rat fertility may be improved from organic feed' although no mechanisms were proposed.

In addition to the studies mentioned previously which demonstrated improved immune function in anthroposophic school children (Alm *et al.*, 1999; Alfven *et al.*, 2006) there are some studies in laboratory animals in which immune function was shown to be improved following consumption of diets based on organic ingredients. In one study, Finamore *et al.* (2004) used an *in vitro* assay and compared the proliferative response in stimulated and unstimulated lymphocytes from well-nourished or protein-energy-malnourished (PEM) rats fed organically and conventionally grown wheat. The proliferative responses of lymphocytes cultured in media containing foetal calf serum (FCS, a standard component of the incubation medium) were not affected by the level of feeding or the source of wheat. When the FCS was replaced with rat serum in order to mimic the *in vivo* response better, there was no difference in proliferative response for well-nourished rats for the different wheat sources. However, the proliferative response was inhibited in PEM rats fed conventional wheat compared with rats fed organic wheat. There was no difference in concentration of acute phase proteins between rats fed conventional or organic wheat. The authors concluded that the conventional wheat sample tested in this model system represented 'a higher risk' for lymphocyte function than the organic wheat sample, at least when the rats were in a vulnerable PEM state. It would be interesting to see whether the assay used in this study could be modified for use with lymphocytes isolated from human blood to develop a model which was more applicable to studies with humans.

3.2.3 Nutrient composition of organic and conventional foods

Perhaps the largest review of historic data comparing the nutrient composition of organically and conventionally grown foods was produced by Woese *et al.* (1997) in which they compared data published over a 70-year period from 1924 for more than 150 foods. The authors themselves point out the shortcomings of their results. In particular, a large proportion of the studies provided little information about the growing conditions (some of which were not rigorously controlled), often small numbers of samples were analysed and many were derived from markets where the authenticity of origin could not be confirmed. Because the studies were collated from such a wide time period it is also probable that changes in farming practice, especially in conventional farming and the use of fertilisers, would have a substantial effect on the results. Williams (2002) also points out that many of the publications in this review were in the form of dissertations and were published as reports in the German literature and thus the extent of rigorous peer review is debatable. Nevertheless, the authors concluded that despite the heterogeneity of the sample material, some differences in 'quality' between products from conventional and organic farming or foods produced from different fertilisation systems were identified. The effects of organic production systems on the nutrient composition of plant and animal foods are discussed elsewhere in this publication (see Chapters 7, 8, 9, 15 and 16). In general however, the changes reported in nutrient composition have been relatively small and inconsistent and more rigorously controlled studies are required in order to demonstrate significant benefits.

3.2.4 Organoleptic properties of organic foods

In its broadest definition the nutritional value of a foodstuff should also include consideration of the sensory properties of the product, since this is a principal driver in food choice and taste is one of the most important measures by which the consumer judges product quality (Torjusen *et al.*, 2001; see also Ch 6). Improved taste and flavour are often cited as being motivators for consumers in the purchase of organic foods, however, quantifying these factors has also been an area in which data have been inconsistent. Comparison of organic crop products with their non-organic counterparts requires all other aspects of the products to be equal. In particular, variation among different varieties, effects of degree of ripeness, freshness or length of storage of the same fruit or vegetable, have major impacts on organoleptic properties which may exceed differences caused by the method of production. In addition, the precise description of, and the different types of, sensory analysis techniques used in evaluating products is varied and makes comparison difficult (Bourn and Prescott, 2002). Thus much of the data are conflicting and inconclusive. For example, for potatoes, Hajšlová *et al.* (2005) were unable to demonstrate clear differences in the sensory characteristics of crops grown under organic or conventional conditions, in contrast to Thybo *et al.* (2002) who reported small differences in 'mealiness' and 'graininess' of potatoes grown using different methods of fertiliser application. The importance of how the potatoes were prepared before sensory evaluation was shown by Wszelaki *et al.* (2005). When skins were left on potato wedges before cooking, panellists detected differences between organic potatoes and conventional potatoes, although not between different organic treatments. When the skins were removed, however, the panellists could not distinguish between potatoes from the different treatments, suggesting that the principal factors affecting the sensory characteristics were located in the skin.

Some differences have been observed in the flavour of tomatoes grown under a variety of growing conditions such as changing the total ion concentration in nutrient solutions or by using ammonium as the dominant N-source in fertilisers, especially those conditions which result in less watery fruits (Heeb *et al.*, 2005). However, Johansson *et al.* (1999) found that there were no differences in taste characteristics of organic tomatoes, but tomatoes grown ecologically scored lower for firmness and juiciness regardless of variety compared with conventionally grown tomatoes. The same group previously suggested that different types of mulch had a significant effect on tomato flavour, with nutrient-rich substrates producing less sweet, less firm but more juicy fruit (Haglund *et al.*, 1997).

In some cases sensory attributes may be considered poorer for organic compared with conventional products. For example, while most studies of carrots found no difference or that the organic carrots were sweeter, in one study organically grown carrots were found to have a more bitter taste than conventionally grown carrots (Haglund *et al.*, 1999) and in one out of two years' crops, conventionally grown carrots had a sweeter taste and were crunchier than ecologically grown carrots, which were harder and had a pronounced aftertaste.

The organoleptic properties of fruits and vegetables grown under different agronomic conditions should be relatively easy to assess since they are either eaten fresh or require minimal processing (cooking) before consumption. However, foods requiring greater processing, or foods consisting of many different ingredients, are much more difficult to evaluate. In this instance the farming method may not only affect the organoleptic properties of the final product, but may also interact with the manufacturing process. For example, loaf volume and elasticity of wholemeal bread prepared from ecologically grown wheat compared with conventionally grown wheat was significantly affected by kneading intensity for conventional wheat and some, but not all, ecologically grown wheat (Haglund *et al.*, 1998). This difference was attributed to the lower protein content of the ecologically grown wheat. Interestingly, taste attributes of smell and acidulous taste for the different breads were not affected by the source of wheat but in all cases the loaves made from conventional flour had a greater volume and scored lower on dryness than loaves made with the organic flour. These were key determinants of consumer acceptability in principle component analysis of the sensory attributes of the breads (Haglund *et al.*, 1998).

In all of these studies the subjects have been blinded to the source of the products and thus the differences detected were only those related to the physicochemical composition of the food. However, this tends to underestimate greatly the actual preference for organic foods, since studies where consumers were asked to assess food with or without labels describing, for example, tomatoes as organic or conventional, have identified a substantial and systematic hedonic component (Schutz and Lorenz, 1976; Von Alvensleben and Meier, 1990; Johansson *et al.*, 1999). The hedonic preference for organic food, which can be measured as the perceived difference between identical foods presented with different labels, tends to be at least as large as the objective difference, the perceived difference between different products presented without labels or other identifiers. One of the effects of a hedonic component is that the taste of the preferred product tends to become the preferred taste (Johansson *et al.*, 1999). This means that consumers with a strong belief in the desirability of organic foods (e.g. based on environmental or health concerns) will tend to prefer any characteristic 'organic' taste that they have learned to associate with the preferred product. In marketing terms, organic food is a very strong brand, which can have power to change consumer preferences. No studies have been published on the taste preferences of dedicated organic consumers compared with consumers who are completely unfamiliar with organic food, but it is not unlikely that such studies would show systematic 'adaptation' of preferences to the preferred food, which would have the fortunate consequence that both groups of consumers are happy with the taste of the food they eat.

3.3 Conclusions

There is a widespread belief among advocates of the organic lifestyle that ecological, low input production systems result in foods of higher nutritional quality. Whilst it has been demonstrated in a large number of analytical investigations that the nutrient content of organic foods can be modified by growth conditions, there is a paucity of evidence showing that these changes can result in measurable health benefit, most importantly in humans. Current data are mostly confounded by poor experimental design; in particular, there is a lack of studies using single varieties of foods grown in similar localities under tightly controlled conditions and of assessment methods sensitive enough to detect the impact on human health of any diet which is still nutritionally adequate. In many cases small differences in nutrient composition are most likely to be due to varietal, seasonal and harvesting/storage differences which have not been recognised in experimental design and data analysis. If consumer perceptions of the nutritional benefits of organic foods are to be supported, well-designed and strictly controlled intervention studies are required in order to demonstrate specific health benefits of consuming organic foods.

3.4 References

Abell A, Ernst E and Bonde JP (1994) 'High sperm density among members of organic farmers association'. *Lancet*, **343**, 1498–1498.

Aehnelt E and Hahn J (1978) 'Animal fertility: a possibility for biological quality assay of fodder feeds'. *Biodynamics*, **25**, 36–47.

Alfven T, Braun-Fahrlander C, Brunekreef B, Mutius E, Riedler J, Scheynius A, Hage M, Wickman M, Benz MR, Budde J, Michels KB, Schram D, Ublagger E, Waser M and Pershagen G (2006) 'Allergic diseases and atopic sensitization in children related to farming and anthroposophic lifestyle – the PARSIFAL study'. *Allergy* **61**, 414–421.

Ali HI and Harland BF (1991) 'Effects of fiber and phytate in sorghum flour on iron and zinc in weanling rats – a pilot-study'. *Cereal Chemistry*, **68**, 234–238.

Alm JS, Swartz J, Lilja G, Scheynius A and Pershagen G (1999) 'Atopy in children of families with an anthroposophic lifestyle'. *Lancet*, **353**, 1485–1488.

Arab L (2004) 'Individualized nutritional recommendations: do we have the measurements needed to assess risk and make dietary recommendations?' *Proceedings of the Nutrition Society*, **63**, 167–172.

Balfour E (1975) *The Living Soil and the Haughley Experiment*, Faber and Faber, London.

BMVEL (Bundesministerium für Verbraucherschutz Ernährung und Landwirtschaft – German Ministry of Consumer Protection Nutrition and Agriculture) (2003) *Bewertung von Lebensmitteln verschiedener Produktionsverfahren* (Evaluation of foods from different production systems). Report. http://www.bmelv-forschung.de/download/tdm200306_bericht_030515.pdf. Date accessed: January, 2006.

Bourn D and Prescott J (2002) 'A comparison of the nutritional value, sensory qualities, and food safety of organically and conventionally produced foods'. *Critical Reviews in Food Science and Nutrition*, **42**, 1–34.

Caris-Veyrat C, Amiot MJ, Tyssandier V, Grasselly D, Buret M, Mikolajczak M, Guilland JC, Bouteloup-Demange C and Borel P (2004) 'Influence of organic versus conventional agricultural practice on the antioxidant microconstituent content of tomatoes and derived purees: Consequences on antioxidant plasma status in humans'. *Journal of Agricultural and Food Chemistry*, **52**, 6503–6509.

Finamore A, Britti MS, Roselli M, Bellovino D, Gaetani S and Mengheri E (2004) 'Novel approach for food safety evaluation. Results of a pilot experiment to evaluate organic and conventional foods'. *Journal of Agricultural and Food Chemistry*, **52**, 7425–7431.

Franco OH, Bonneux L, de Laet C, Peeters A, Steyerberg EW and Mackenbach JP (2004) 'The Polymeal: a more natural, safer, and probably tastier (than the Polypill) strategy to reduce cardiovascular disease by more than 75%'. *British Medical Journal*, **329**, 1447–1450.

German JB, Watkins SM and Fay L-B (2005) 'Metabolomics in practice: emerging knowledge to guide future dietetic advice toward individualized health'. *Journal of the American Dietetic Association*, **105**, 1425–1432.

Givens DI (2005) 'The role of animal nutrition in improving the nutritive value of animal-derived foods in relation to chronic disease'. *Proceedings of the Nutrition Society*, **64**, 395–402.

Grinder-Pedersen L, Rasmussen S, Bügel S, Jørensen L, Dragsted L, Gundersen V and Sandstrom B (2003) 'Effects of diets based on foods from conventional versus organic production on intake and excretion of flavonoids and markers of antioxidant defense in humans'. *Journal of Agricultural and Food Chemistry*, **51**, 5671–5676.

Haglund A, Johansson L, Garedal L and Dlouhy J (1997) 'Sensory quality of tomatoes cultivated with ecological fertilizing systems'. *Swedish Journal of Agricultural Research*, **27**, 135–145.

Haglund A, Johansson L and Dahlstedt L (1998) 'Sensory evaluation of wholemeal bread from ecologically and conventionally grown wheat'. *Journal of Cereal Science*, **27**, 199–207.

Haglund A, Johansson L, Berglund L and Dahlstedt L (1999) 'Sensory evaluation of carrots from ecological and conventional growing systems'. *Food Quality and Preference*, **10**, 23–29.

Hajšlová J, Schulzová V, Slanina P, Janné K, Hellenäs KE and Andersson C (2005) 'Quality of organically and conventionally grown potatoes: Four-year study of micronutrients, metals, secondary metabolites, enzymic browning and organoleptic properties'. *Food Additives and Contaminants*, **22**, 514–534.

Hansson I, Hamilton C, Ekman T and Forslund K (2000) 'Carcass quality in certified organic production compared with conventional livestock production'. *Journal of Veterinary Medicine Series B-Infectious Diseases and Veterinary Public Health*, **47**, 111–120.

Heeb A, Lundergårdh B, Ericsson T and Savage G (2005) 'Nitrogen form affects yield and taste of tomatoes'. *Journal of the Science of Food and Agriculture*, **85**, 1405–1414.

Henderson L, Gregory J and Irving K (2003) *The National Diet and Nutrition Survey: Adults Aged 19 to 64 Years. Energy, protein, carbohydrate, fat and alcohol intake.* The Stationery Office, London.

Hodges R and Scofield A (1983) 'Effect of agricultural practices on the health of plants and animals: a review'. In Lockeretz W, *Environmentally Sound Agriculture*, Praeger Scientific, New York, 3–34.

Jensen TK, Giwercman A, Carlsen E, Scheike T and Skakkebaek NE (1996) 'Semen quality among members of organic food associations in Zealand, Denmark'. *Lancet*, **347**, 1844–1844.

Johansson L, Haglund A, Berglund L, Lea P and Risvik E (1999) 'Preference for tomatoes, affected by sensory attributes and information about growth conditions'. *Food Quality and Preference*, **10**, 289–298.

Juhler RK, Larsen SB, Meyer O, Jensen ND, Spanó M, Giwercman A and Bonde JP (1999) 'Human semen quality in relation to dietary pesticide exposure and organic diet'. *Archives of Environmental Contamination and Toxicology*, **37**, 415–423.

Kant AK (1996) 'Indexes of overall diet quality: a review'. *Journal of the American Dietetic Association*, **96**, 785–791.

Larsen SB, Spano M, Giwercman A and Bonde JP (1999) 'Semen quality and sex hormones among organic and traditional Danish farmers'. *Occupational and Environmental Medicine*, **56**, 139–144.

Linder M (1973) 'A review of the evidence for food quality differences in relation to fertilization of the soil with organic or mineral fertilisers'. *Biodynamics*, **107**, 1–11.

Magkos F, Arvaniti F and Zampelas A (2003) 'Putting the safety of organic food into perspective'. *Nutrition Research Reviews*, **16**, 211–221.

McCarrrison R (1926) 'The effect of manurial conditions on the nutritive and vitamin values of millet and wheat'. *Indian Journal of Medical Research*, **14**, 351–378.

Millet S, Cox E, Buyse J, Goddeeris BM and Janssens GPJ (2005) 'Immunocompetence of fattening pigs fed organic versus conventional diets in organic versus conventional housing'. *The Veterinary Journal*, **169**, 293–299.

Plochberger K (1989) 'Feeding experiments – a criterion for quality estimation of biologically and conventionally produced foods'. *Agriculture Ecosystems and Environment*, **27**, 419–428.

Prentice A and Jebb S (1995) 'Obesity in Britain: glutony or sloth'. *British Medical Journal*, **311**, 437–439.
Reiter H, Ertel H, Wendt H, Pies Prufer J, Barth L, Schroder H, Catel W, Dost F and Schneunert A (1938) 'Nutritional studies on the effects of vegetables produced with and without fertilisers'. *Ernährung (Liepzig)*, **3**, 53–69.
Roesch M, Doherr MG and Blum JW (2005) 'Performance of dairy cows on Swiss farms with organic and integrated production'. *Journal of Dairy Science*, **88**, 2462–2475.
Rowlands M and Wilkinson B (1930) 'Vitamin B content of grass reeds in relation to manures'. *Biochemical Journal*, **24**, 199–204.
Schuphan W (1972) 'Effects of application of inorganic and organic manures on market quality and on biological value of agricultural products'. *Qualitas Plantarum et Materiae Vegetabiles*, **21**, 381–398.
Schutz HC and Lorenz O (1976) 'Consumer preferences for vegetables grown under "commercial" and "organic" conditions'. *Journal of Food Science*, **40**, 70–73.
Scott P, Greaves J and Scott M (1960) 'Reproduction in laboratory animals as a measure of the value of some natural and processed foods'. *Journal of Reproduction and Fertility*, **1**, 130–138.
Shepherd R, Magnusson M and Sjoden PO (2005) 'Determinants of consumer behavior related to organic foods'. *Ambio*, **34**, 352–359.
Thybo AK, Molgaard JP and Kidmose U (2002) 'Effect of different organic growing conditions on quality of cooked potatoes'. *Journal of the Science of Food and Agriculture*, **82**, 12–18.
Torjusen H, Lieblein G, Wandel M and Francis CA (2001) 'Food system orientation and quality perception among consumers and producers of organic food in Hedmark County, Norway'. *Food Quality and Preference*, **12**, 207–216.
Velimirov A, Plochberger K, Huspeka U and Schott W (1992) 'The influence of biologically and conventionally cultivated food on the fertility of rats'. *Biological Agriculture and Horticulture*, **8**, 325–337.
Vogtmann H (1988) 'From healthy soil to healthy food: an analysis of the quality of food produced under contrasting agricultural systems'. *Nutrition and Health*, **6**, 21–35.
Von Alvensleben R and Meier T (1990) 'The influence of origin and variety on consumer perecption'. *Acta Horticulturae*, **259**, 151–161.
Wendt H (1943) 'Long term studies in humans on the effect of vegetables and potatoes produced with different fertilisers'. *Ernährung (Liepzig)*, **8**, 281–295.
de Whalley CV, Rankin SM, Hoult JRS, Jessup W and Leake DS (1990) 'Flavonoids inhibit the oxidative modification of low density lipoproteins by macrophages'. *Biochemical Pharmacology*, **39**, 1743–1750.
Williams C (2002) 'Nutritional quality of organic food: shades of grey or shades of green?' *Proceedings of the Nutrition Society*, **61**, 19–24.
Williams PRD and Hammitt JK (2001) 'Perceived risks of conventional and organic produce: pesticides, pathogens and natural toxins'. *Risk Analysis*, **21**, 319–330.
Woese K, Lange D, Boess C and Bogl KW (1997) 'A comparison of organically and conventionally grown foods – Results of a review of the relevant literature'. *Journal of the Science of Food and Agriculture*, **74**, 281–293.
Worthington V (1999) 'Nutrition and biodynamics: evidence for the nutritional superiority of organic crops'. *Biodynamics*, **224**. http://www.biodynamics.com/biodynamicsarticles/worth.html
Wszelaki AL, Delwiche JF, Walker SD, Liggett RE, Scheerens JC and Kleinhenz MD (2005) 'Sensory quality and mineral and glycoalkaloid concentrations in organically and conventionally grown redskin potatoes *(Solanum tuberosum)*'. *Journal of the Science of Food and Agriculture*, **85**, 720–726.

4

Quality assurance, inspection and certification of organic foods

Bo van Elzakker, Agro Eco Consultancy BV, The Netherlands and Jochen Neuendorff, Gesellschaft für Ressourcenschutz mbH, Germany

4.1 Introduction to quality assurance in organic foods

The history of quality assurance in organic foods is very short. Whereas first standards or guidelines for organic farming emerged in the first half of the last century, labelling became common only in the 1970s. Organic agriculture evolved from various schools of thought, each with their own (spiritual, social, cultural) motivation. These include biodynamic farming as taught by Rudolf Steiner, organic farming as promoted by Lady Eve Balfour, bio-organic farming according to Müller and Rusch, the biological farming methods of Lemaire-Boucher, Bob Rodale's regenerative farming and natural farming if one adhered to Fukuoka (Lampkin, 2002). Production standards, often only a few pages, were guidance documents to ensure that those within a group of adherents applied the same rules for growing organic crops, keeping organic animals or processing organic foods.

In 1976, a number of these organic agriculture movements came together and the International Federation of Organic Agriculture Movements (IFOAM) was founded. As the interest in organic products increased and international markets came into existence, the need for one international reference standard arose. The first basic standard was formulated by the Technical Committee of IFOAM in 1980 (Schmidt and Haccius, 1998). Subsequently, more organic farming groups started to formulate standards for organic agriculture, applicable to their specific situation. As the standards were somewhat different, the question arose 'is your product as organic as my organic?'

Owing to increasing international trade, labelling came into being, bringing with it false labelling and fraud. From the early 1980s onwards, the organic movement was plagued by cases of fraudulent labelling. At that time, farm advisors doubled as 'inspectors', with little attention paid to inspection methodology or documentation. However, the need for an independent, third party inspection and certification was recognised and within a few years was put in place. To facilitate the growth of the international market and to address the issue of 'my organic is more organic then your organic', IFOAM started a programme of evaluating the standards and inspection practices of its members. This has since become the IFOAM Accreditation Programme, implemented by the International Organic Accreditation Service, a non-profit international accreditation body, founded by IFOAM. By the end of the 1980s there was a system of private standard setting, third-party inspection and certification in place and a service to evaluate certifiers against the one international reference standard. This industry self-regulation however was not able to address fraud properly. Fraudulent operators could be identified and persons blacklisted, but without a law they could not be prosecuted. The private organic movement needed regulations to be able to address this issue.

Stories about fraud spurred some national authorities into action. In 1985, France was the first European country to have its own national organic legislation. This triggered the formulation of a EU-wide regulation that came into force in 1993, the Regulation (EEC) No. 2092/91. Much time has since been spent in further describing organic farming, notably animal husbandry, crop production and, to a lesser extent, organic processing. By December 2004, this regulation had been complemented and modified by 57 amendments and supplementary regulations. The result is an unwieldy legislative text that is open to different interpretations and needs authorisation of exceptions. The European Commission has called for an overall revision, with revised regulations taking full effect within the EU on 1 January 2009.

4.2 The regulation

Article 5 regulating the labelling of organic products is the central element in the EEC Regulation No. 2092/91 (Schmidt and Haccius, 1998). The regulation applies only if agricultural products are labelled with references to organic farming. Article 5 also refers to the production rules and the inspection system. The production rules guarantee a farming system that is particularly environmentally friendly, under which the use of synthetic chemical fertilisers and pesticides is prohibited. Appropriate animal husbandry is an essential requirement for organic animal production. When organic products are processed, only a very limited number of conventional ingredients and processing aids are permitted.

The inspection system is described in Articles 8 and 9 of the Regulation. The member states can choose between using the inspection authorities to apply the inspection system or using private inspection bodies do it, or a combination of both. Private inspection bodies must be authorised and supervised by competent authorities. A purely state-administered system is in place in only two EU member states (Denmark and Finland). The majority of the EU member states have opted for a system incorporating private inspection bodies subject to state supervision. These private inspection bodies are obliged, under Article 9 (11) of the EEC Regulation No. 2092/91, to meet the requirements of the quality management norm EN 45011 (General requirements for bodies operating product certification systems). They are then allowed to certify products derived from organic farming once the requirements of EEC Regulation No. 2092/91 are met. Annex III of the EEC Regulation No. 2092/91 contains detailed 'minimum inspection requirements' for agricultural operations as well as processing companies, importers and companies that are producing animal feedstuffs. Its weakness in focussing on particular risks has been criticised in the past (Darnhofer and Vogl, 2002; Heinonen, 2001).

Exactly how inspection is to be done is not clear. Whereas there was some good collaboration and an effort in the early days within IFOAM, the inspection bodies now focus on what their national competent authority requires. The result is that inspection regimes in the various EU member states vary, although the extent of this variation is not known.

The application of EN 45011 led to a strong reduction in the number of certification bodies in several countries, like France. In some EU countries the inspection bodies have to be accredited by the national accreditation body, in others they do not (Michaud *et al.*, 2004).

For over ten years, the organic movement diligently cooperated with the authorities to make the regulation a success. What few realise is that the regulation has completely taken away the role of standard setting from the original main stakeholders. The private organic sector has become an advisor, sometimes appreciated, sometimes neglected. Within a period of 20 years, it has come close to losing the ownership of the term 'organic'. The private sector, represented by IFOAM, is trying to maintain private ownership by maintaining its landmark basic standards for organic production and processing, and accreditation criteria. Various organic groups maintain their private standards over and above the regulation, in order to maintain some hold on what happens to the standard. Most of these are accredited by IFOAM.

As the regulation is a labelling law, there was an expectation that it would resolve the issue of labelling. An EU wide logo has been developed, but each country also has its own system, sometimes with various organic logos that represent more or less the same, on sometimes different, things. However, all these different labels are subject to the EEC regulation, thus ensuring flexibility, but also transparency in the organic marketplace.

4.3 Responsibilities

With the mandate to define organic farming being transferred from the private sector to the public sector, the sense of responsibility for the product's organic quality has also been removed from the private sector. Operators hold the inspection bodies responsible for organic quality. As the competent authorities approve and supervise the inspection bodies, the authorities are second in the line of responsibility. This is aggravated by the fact that, when a buyer has doubts about the organic status of a product, he/she cannot question the validity of the certificate; product cannot be refused because it comes from country X or is certified by inspection body Y.

It is only through learning hard lessons that operators have started to realise again that the organic integrity of the product is their affair as well and have begun to seek refuge in additional private standards and/or an in-house quality assurance system. Others have invested in longer-term, personal relationships with their suppliers and buyers, to be more sure of the product's quality, regardless of what certificate comes with the product. Threatened by liability claims, some certification bodies maintain that the certificate that accompanies the goods does not guarantee that the product is organic, it only says that, based on the information available, the certification body had to conclude that it is organic. If it turns out not to be organic, the organisation that provided the information is at fault, not the certification body (SKAL letter, 2000). There is thus some trend towards responsibility for product integrity returning to where it primarily belongs, the private sector organic operators themselves.

4.4 Quality assurance

Until now, Annex III of the EEC Regulation 2092/91 requires an equally intensive evaluation of all operators. Throughout the organic world, operators, small or big, complicated or simple, are thus subjected to an annual inspection visit by the certification body. The advantage is that every operator is treated equally. It adds considerably to consumer confidence, to be able to say that all operators are visited once a year by an independent third party. On the other hand, the annual inspection approach does not reflect the potential for real risks, which vary according to the type of operator. That one annual inspection visit might not be the most appropriate way to deal with quality assurance was, until recently, not given much thought. That the larger or more complex units, or those dealing with both organic and conventional produce, might need more frequent inspection is obvious even to laymen.

Organic product certification is a process certification, that is the nature of the product is defined by its production process. By the end of the 1990s, process certification was still a very new concept, certainly in agriculture. The inspector has to issue qualitative statements rather than report on quantitative measurements. There has thus been a drive to become more exact. Early approaches centred on making sure that operators knew what they were doing, identifying possible problems (for example with pests) and making sure that sufficient guarantees were in place so that prohibited substances were not needed. The focus was on the farming system, with processing inspections often done by persons trained in agriculture.

In the USA and among some European private inspection bodies, the concept of identifying organic critical control points was developed, normally abbreviated to OCCPs (GFRS, 2003; Riddle and Ford, 2000). Together with the operator, the inspector analyses the product flow and identifies the critical points where the organic quality can be at risk. The various OCCPs are prioritised and measures to reduce or eliminate the risk are checked. The certification body can request corrective actions from the operator to put measures in place. The use of OCCPs comes close to the concept of hazard analysis critical control points, or HACCP, which is dealt with in Section 4.7. The current legislative tendency is however, to describe in painful detail what organic farming is (e.g. the area needed (m^2) for hens' access to the outdoors) and the inspection must, according to the opinion of most competent authorities, go through the regulation line by line. How else could the inspector make sure that the operation complies with all requirements? There is, in practice, less and less space either for investigation or for risk assessments, as more time is needed to fill in the forms.

Despite the fact that the operation of the certification body has been standardised by the application of EN 45011, there is still variation in the local interpretation of the EEC regulation and in how inspection is done. In practice, one sees a mixture of approaches, depending on the certification body and the opinion of the competent authority in charge of the system. The situation is worse when this authority is staffed by persons who do not have any experience or much understanding of organic agriculture, which is not uncommon among the (sometimes frequently changing) staff of the competent authorities in the various member states.

Twenty years ago, nobody ever expected that the organic market would become so globalised, especially as for many the concept of organic includes the principle of regionality. Most of what are dubbed the main consumer countries import around 50% of what they consume. At the same time they export a large part of their production. The consumer expectation that organic products will come from a local farmer is not very realistic. The focus on primary production, especially on animal husbandry, and the ongoing discussion about approved inputs and ingredient lists, did not allow the issue of quality assurance in international trade to be sufficiently addressed. Quality assurance in organic farming should be far more concerned with international trade and multi-ingredient processing in often mixed (organic and conventional) processing units.

4.5 Private, additional certifications

Apart from the legal requirements of EEC regulation, suppliers of organic products, depending on the final market destination, often have to fulfil additional private standard requirements for market labels which are historically owned by organic farmers associations. Within Europe, the Soil Association in the UK and KRAV in Sweden are strong labels in their national marketplace. In Germany, most organic farmers are members of an association, for example Bioland, Naturland or Demeter, while this is quite uncommon for processors, importers and traders. These private logos differentiate the market, suggest to the final consumer a link with a certain group of organic farmers and give the organic product a unique identity. As most of these certification bodies are accredited by IFOAM, which has stricter accreditation criteria, it also provides for a certain additional organic integrity guarantee.

A scheme that was initiated by a Dutch organic importer, Eosta BV, deserves special mention. In its Nature & More scheme, products are scored in terms of compliance with organic, health and environmental standards. This is done as an add-on to the organic certification process. By entering a code that is on the product label, the consumer can see how the product he or she purchased scores. It is a communication tool that links the farmers to the consumers.

Fair trade is a true add-on to organic certification. It guarantees that small-scale producers in developing countries receive a price for their product which covers the real production costs. NGOs and members of the Fair Trade Labelling Organisation (FLO) based in Bonn, Germany, developed the concept. It was originally targeted at smallholders, but owing to lack of sufficient volumes in some product categories, there are also standards for larger operators using hired labour. Certification to the FLO standard is done by FLOCert only. FLOCert normally subcontracts individuals and sometimes organic inspection bodies to conduct the necessary inspections. Being certified by FLO means that the product can carry, for example, the Max Havelaar symbol, alongside the organic logo. To satisfy the increasing market demand, but also for ideological reasons, there are some organic certification bodies that have developed their own social scheme, more integrated with their organic standards, like the Soil Association in the UK, Naturland in Germany and ICEA in Italy.

Last but not least, supermarkets have introduced their own private certification schemes to ensure suppliers' compliance with good agricultural practices (GAP), good manufacturing practices (GMP), food and feed hygiene and other legal requirements. Examples of these standards are EurepGAP for farmers, the requirements of the British Retail Consortium (BRC) or the International Food Standard (IFS) for food processing, and GMP+ for animal feed. Some buyers insist on HACCP and/or ISO 9001 certification. While these standards were originally designed for conventional production, organic operators must also increasingly demonstrate compliance with the above requirements by providing the corresponding certificate alongside the organic one.

4.6 Quality assurance to ensure quality and safety of organic and 'low input' foods

Traditionally, there is a conviction in the organic sector, from farmers to consumers, that organic foods are healthier. They are produced in a more natural way, without the use of synthetic pesticides and fertilisers and processing is done using a minimum of (synthetic) additives. Investigations by official food surveillance agencies and laboratories in Germany showed that organic fruits and vegetables contained only very small amounts of pesticide, around 0.002 ppm, while conventional fruits and vegetables contained around 0.4 to 0.5 ppm (CVUA, 2004; Stolz *et al.*, 2005). The CVUA, a governmental service with laboratories in various German States, also analysed GMOs in conventional and organic food samples and did not find GMO contamination in the organic products. Nevertheless, statements about low levels of pesticide and GMO contamination in organic foods should be made with caution; we live in a contaminated world and organic producers and processors cannot guarantee that their products are residue-free.

Organic foods are subject to the same standards for pesticide residues (EC 91/414) as conventional foodstuffs. Some groups within the organic sector would like to see lower acceptable limits for organic foods. Before the EU regulation came into force, the German organic wholesalers association, BNN, had a rule that acceptable pesticide levels in organic product should not be more than 10% of what was allowed by law for conventional foods. There are some voices that want to re-institute this standard.

The issue of what the organic sector should do about GMO contamination is even more contentious. The authorities have proposed a threshold, whereas some organic advocates insist on a complete ban on any GMO contamination, largely because of consumers' expectations that organic foods will be GMO free (Anon, 2001).

The fact that food quality involves more than assurances about pesticide and GMO residues is only recently becoming clear, because of the demands of large retailers. Organic fruit and vegetable producers who deliver to the larger supermarkets are asked to comply with the Eurep GAP protocol, organic food processors with the BRC and IFS standards, and traders of organic feed with the Dutch GMP+. All these standards focus on good practices in primary and secondary production, feed and finally, food safety. This means that suppliers have to install an internal quality assurance system in their company. Analogous to the requirement in the organic regulation, the certification bodies offering these inspections must be accredited to EN 45011. In some cases the national accreditation body is required to accredit them. This causes some problems in the international arena where not every country has a recognised national accreditation body.

Organic producers have to comply with an increasing number of certifications apart from the organic one. Some of the above schemes can be combined with organic certification, either when it is done by one body offering different certifications (e.g. SKAL in The Netherlands) or through cooperation between different bodies (e.g. Quavera Alliance, a network of Austrian, Canadian and German certifiers). Too often, these requirements are fulfilled in completely separate certification processes, requiring that inspection is done by inspectors from different certification bodies, through multiple site visits, insisting on specific administrative procedures.

For a long time there has been private, in-house quality research, especially in the biodynamic sector and by some larger organic traders and processors. These vary from crystallisation techniques to broad screen residue analyses of raw materials. These in-house quality management schemes are not standard; they depend on how seriously the companies take their responsibility towards their clients and/or want to avoid surprises by the food authorities.

4.7 Risk assessment in organic quality assurance

Recent developments in organic inspection and certification focus on the development of a risk-oriented approach, in addition to the annual inspection visit. When inspecting the operators, the amount of time spent must be flexible and should be more orientated to goals and risks. Sanctions must be effective and strongly discourage violations.

4.7.1 Organic critical points

For large specialised holdings and for processing companies and import companies, it makes sense to prepare diagrams of the flow of the product(s) and identify the 'organic critical points' of the holding or company. The critical points are those points in the handling of the product where its organic quality can be put at risk. For that reason the company must take measures to ensure that the potential risks are reduced or eliminated. The implementation of these measures must be monitored and documented. In the case of non-compliance, corrective measures must be established.

On organic farms, the use of prohibited substances is 'not tolerable'. Feed for organic animals therefore must comply with the relevant provisions of the EEC regulation. In processing companies, errors that cannot be tolerated include, for example, using conventional ingredients or employing processing aids such as ionising radiation in the processing of organic foods. In situations where parallel production is taking place, inadequate cleaning measures that do not prevent commingling of organic with conventional products cannot be tolerated. In an import company, traces of prohibited pesticides in agricultural products being imported from third countries are not acceptable.

In order to identify the critical points during the process, every stage of the process must be analysed. The precautionary measures at the critical points must be designed in such a way that they ensure that the organic integrity is subjected to the least possible risk during the production process.

In the case of parallel production in processing operations, the precautionary measure may be an adequate cleansing process that has been demonstrated to be effective. In the case of the importers, a measure to be taken may be the introduction of an internal plan to take samples to be tested for possible traces of pesticides.

During the inspection, the inspector should proceed with particular emphasis on the critical points. This system guarantees that the internal quality assurance is oriented to the risks associated with the particular type of production. It is a preventive approach. The employees who identify with the 'organic objective', who are involved in the process and assume responsibility for it, are more careful, because they understand the sense of 'quality assurance' and do not feel that it is an additional burden, impractical, bothersome, a mere formality and bureaucratic.

4.7.2 Classification of the risks as the basis for external inspection

Irregularities that can compromise the integrity of organic products can occur at any step in the chain, from farm production to sale to the final consumer. The risk of irregularities increases with:

- increasingly similar demands for quality for conventional and organic products
- the more sophisticated production technology required for organic production
- the shorter shelf-life of the product, or greater turnover of products
- a larger price differential between conventional and organic products
- the greater availability and/or easier application of prohibited technologies and ingredients (in situations of parallel production), and
- a decreasing liability of the supplier.

Certification bodies can assign the agricultural holdings and companies inspected into risk categories. The following classification criteria can be applied: the structure and complexity of the company; production of risk products; the risk of commingling; the company's importance in the market; the risk of liability; previous sanctions imposed; and the existence of an internal quality assurance system. The kind and frequency of inspections to be carried out by the certification body can be deduced from the risk category. In some cases the inspection body might carry out an annual inspection of the entire unit as an unannounced inspection. This would be appropriate in those companies in which the turnover of perishable products is high (e.g. certain specialised holdings such as those of laying hens or large-scale mushroom production, certain processing companies such as catering services, restaurants and cafeterias, or importers of organic fruits and vegetables). Other operations such as grassland farms can continue to be inspected with prior notice.

The frequency of the inspections carried out by the certification bodies should also be set based on the risk category. For example, it makes sense to inspect an organic fruit and vegetable wholesaler with a significant share of the market several times a year. As part of the inspections, special emphasis should be placed on the critical points. To complement this, spot check inspections can be done to monitor certain higher-risk steps in the process (e.g. sufficient separation of the lots in the case of parallel production) and to verify compliance with the conditions imposed by the inspection body.

4.7.3 Cross checks

Cross checks at different levels of the production chain make it possible to verify certificates and the flow of goods. They provide an opportunity to analyse the data prepared by the inspected company so that monitoring outside of the 'company logic' can be done. A good initiative is under way in Austria, Germany and Luxemburg. In an effort to reduce the falsification of certificates, all certificates issued by the participating inspection bodies to operators can be verified through the BioC.info database (http://www.bioc.info). Participation in BioC is voluntary. It is in the German language. The authenticity of the certificates of inspection bodies that do not participate in this system can be verified with the help of cross checks between inspection bodies by using more formalised procedures (i.e. sending a fax with a request to confirm).

Cross checks help the inspector to calculate and check the flow of organic products for the operator inspected, the so-called input–output reconciliation. The intention is to prevent the holdings from selling more organic products than they can produce in their own fields and to prevent the processing companies and importers from selling more organic products than would be normally be produced using the organic raw materials and ingredients they have bought. Additional cross checks can be done by sharing information gathered during the inspection process with other inspection bodies. In this system, the inspection body certifying the purchaser of an organic lot consults the inspection body that has certified the supplier, on a random basis, to verify if it is plausible that a certain consignment has come from that particular supplier and if the data on the documentation from both the purchaser and the supplier are identical.

4.8 Outlook

To be accepted by the public and the operators, the implementation of the EEC regulation should be in clear accordance with the four principal objectives of this regulation: protection of the consumer, fair competition, transparency in the marketplace and encouragement of organic farming. Standards for organic production must be simple and transparent. While the call for simplicity becomes more and more popular, authorities must still be convinced that this is the best approach. Member states should learn to accept that interpretation in other European countries can be different, as their realities are different. The inspection system, which has been a model for other certification schemes, must be further improved. Risk-orientated procedures that efficiently tackle the risk of fraud must be developed. In future, organic certification will be combined with other inspection regimes. Food security issues and fair trade are good examples of this integration which should be encouraged.

4.9 Sources of further information and advice

www.ifoam.org the website of the International Federation of Organic Agriculture Movements, Bonn, Germany

http://europa.eu.int/eur-lex/en/consleg/main/1991/en_1991R2092_index.html, the consolidated version of the EC regulation 2092/91

http://www.unctad.org/trade_env/itf-organic, the website of the International Taskforce on Harmonization and Equivalence in Organic Agriculture, an initiative of UNCTAD, IFOAM and FAO.

www.ioia.net, the website of the Independent Organic Inspectors Association.

http://www.soilassociation.org, website of the Soil Association in the UK, one of Europe's oldest organic farming associations.

http://www.krav.se/english.asp, the Swedish organic label

http://www.bioland.de/bioland/biolandqualitaet/standards.html about the main German organic farmers association

http://www.naturland.de/englisch/frame_defs/framedef.html, another German association for organic agriculture, with a social certification standard

http://www.demeter.net/ is the English language website of Demeter International

http://www.ioas.org/ website of the International Organic Accreditation Service, which accredits organic certification bodies with private standards and labels.

http://www.natureandmore.com the website of the labelling initiative that expresses varying degrees of compliance with organic, health and fair trading practices.

http://www.fairtrade.net/ the website of Fair Trade Labelling Organisations International (FLO), with a link to FLO-Cert, its certification branch

http://www.icea.info/Default.aspx?language=en-US the English language starting page of the Institute for Ethical and Environmental Certification in Italy

http://www.eurep.org/Languages/English/index_html, the home page for the European retailers working group's Good Agricultural Practices scheme

http://www.brc.org.uk/details04.asp?id=483&kcat=&kdata=1 on the new Global Food Standard of the British Retail Consortium

https://www.food-care.info/index.php? is the starting page for the International Food Standard, a co-operation between German and French retailer organisations

http://www.pdv.nl/index_eng.php?switch=1 is the starting page for the Good Manufacturing Practices+ standard for animal feed industries
http://www.n-bnn.de/, the purely German language website of the German organic processors and retail associations.
http://www.skal.com/ the website of the Dutch national organic inspection body and http://www.controlunion.com/certification/default.htm, the website of its international counterpart that also certifies to various other quality schemes.
http://www.quavera.org/home.htm is an international alliance between three certification bodies that provide various certifications including organic.

4.10 References

Anon (2001), *GMOs in Food Production: Evidence of Risks*. Briefing paper, Soil Association, Bristol.
CVUA Chemisches Landes- and Staatsliche Veterinäruntersuchungsamt (2004), *Bericht Öko-Monitoring*, http://www.untersuchungsaemter-bw.de/pdf/oekomonitoring2004.pdf
Darnhofer I and Vogl CR (2002), 'Certification and accreditation of organics in Austria: implementation, strengths and weaknesses', in *Ecolabels and the Greening of the Food Market*, Tufts University Boston.
GfRS Gesellschaft für Ressourcenschutz (2003), *Systematic Gap Analysis of the Control System Under Regulation (EEC) No. 2092/91 and Proposals for the Further Development of the Control System and Inspection Procedures in Organic Agriculture*, http://www.orgprints.org/2495/
Heinonen S (2001), 'The role of legislation', in *Proceedings of the European Conference Organic Food and Farming*, IFOAM Copenhagen.
Lampkin N (2002), *Organic Farming*, Old Pond Ipswich.
Michaud J, Wynen E and Bowen D (2004), *Harmonisation and Equivalence in Organic Agriculture,* Volume 1, UNCTAD, FAO & IFOAM, Geneva.
Riddle JA and Ford JE (2000), *International Organic Inspection Manual*, IFOAM/IOIA, Bonn.
Schmidt H and Haccius M (1998), *EG-Verordnung, Ökologischer Landbau,* C.F. Müller, Karlsruhe.
Stolz P, Weber A and Strube J (2005), *Analyses of Pesticide Residue Data of Food Samples from Organic and Conventional Agriculture,* http://www.orgprints.org/5399/

5

A new food quality concept based on life processes

Joke Bloksma, Martin Northolt, Machteld Huber, Geert-Jan van der Burgt and Lucy van de Vijver, Louis Bolk Instituut, The Netherlands

5.1 Introduction

5.1.1 Demand for a new quality concept

Consumers expect organic producers to provide healthy and tasty products. But which qualities enhance health and what is tasty? And how can all this be realised by crop or stock management?

In the conventional vision, product quality is mainly based on external, nutritive and sensory properties and is strongly directed by traders and trends. Besides tastiness and ripeness, organic consumers expect products to have properties such as 'vitality' and 'coherence', which are not easy to define and thus to explain and transfer. In the past, experimental parameters have been proposed to estimate 'vitality' and 'coherence', but they were neither scientifically validated nor related to a validated quality concept with relation to human health.

A quality concept which matches the expectations of the organic consumer with the organic view on agricultural production and human health was developed on the basis of two apple studies (Bloksma *et al.*, 2001, 2004) and a carrot study (Northolt *et al.*, 2004). The new quality concept is based on the life processes of growth and differentiation, and their integration. These life processes can be defined in plant physiological terms in order to link the concepts to generally accepted science. Growth and differentiation (including ripening) are familiar processes for organic producers. They are aware that effective management of these processes is necessary to obtain a crop with higher resistance (to stress, pests and diseases) and a product with better taste and keeping quality and which may also be better for human health.

Meanwhile, new questions have been raised. Is there indeed a relation between soil health, plant health and human health as expected in organic agriculture? Is the quality of genetically modified and hybrid varieties less 'coherent', and if so, is this a health concern? Do food crops with increased levels of vitamins or phenols enhance health? What do 'coherence' and 'ripeness' mean in terms of taste and consumer health? These questions are very topical, but they are based on vague notions of food quality. A new conceptual framework for these topical questions is needed, as well as better-defined concepts to operationalise these questions.

5.1.2 Long-term aims for research

The new quality concept was developed in cooperation with other research members of the international research association 'Organic Food Quality and Health' (FQH, for projects and partners see www.organicfqhresearch.org). The research association was established to promote research into the health effects of good quality organic food and to develop parameters for quality assessment. The research members distinguish four lines of research:

- the design and validation of a quality concept for organic produce;
- the validation of individual parameters to assess quality differences;
- the establishment of the relationship between quality and agricultural management practices;
- the study of the relationship between good quality food and animal and human health.

5.2 Description of the inner quality concept

5.2.1 Organic food is not by definition better quality food

Many organically grown products have won the acclaim of the best chefs. Growers know that good taste depends on moderate fertilisation and yield, careful ripening and freshness. Among conventional producers, the need to cut costs has, however, prompted concessions to be made with respect to ripening and freshness. In an effort to cut the cost price, organic agriculture, too, is moving in the direction of higher fertilisation, higher yields, earlier harvests, long trade chains and extended storage. Farmers, growers and traders are exploring the extent to which they can realise these economies without excessively compromising quality. This may explain the large variation in quality in both organic and conventional products. Organic products might score either better or worse on quality. In most cases, it is difficult to attribute these differences to specific cultivation factors. Successful cultivation factors are not always connected with organic regulations.

The further development of organic agriculture thus depends not only on animal friendly and environmentally friendly production methods, but also on the continued acclaim of the best chefs. The new quality concept, named the inner quality concept (IQC), is described below. In this concept, quality is related to the life processes of growth and differentiation and their integration.

5.2.2 Purpose of the quality concept

The inner quality concept (IQC) provides a framework:

1. To link product properties to farm management during production. Organic growers manage the plant's life processes to optimise quality in positive terms of taste, keeping quality and supposed healthfulness. This contrasts with the often negative emphasis of food safety standards (no residues, no microbes, etc.) in industrial agriculture.
2. To verify the assumption of organic agricultural communities that healthy food needs to be ripe and coherent (coherence is defined as a high degree of organisation in the plant).

5.2.3 Motive: comparing leaf lettuce grown in high-tech and organic conditions

In 1990, a large-scale lettuce grower applied for organic certification. This company grew lettuce in a closed hydro culture system (no nutrient leaching), with full climate control (no disruptions in growth), sterile conditions and insect netting. Pesticides were not necessary. In the eyes of the company's directors, this was a first-class 'organic' approach. In the eyes of long-standing organic growers, however, such cultivation practices could never result in top quality lettuce.

The Louis Bolk Instituut conducted an experiment to compare this modern cultivation system with biodynamic cultivation. The purpose of the experiment was further to develop organic standards and quality standards (Lammerts van Bueren and Hospers, 1991). Leaf lettuce of the same variety was sown on the same date in the two different growing situations.

Table 5.1 presents the differences in cultivation, crop properties and quality of the harvested product. Figure 5.1, depicting the development series of both crops, shows that the high-tech lettuce completed its growth twice as quickly as the biodynamic lettuce. The high growth rate resulted in a young, fresh, vulnerable lettuce with a watery taste, high nitrate content and low levels of vitamin C. The slow growing lettuce was firm, with a full taste, better keeping quality, low nitrate content and high levels of vitamin C. After harvesting, a selection of plants from each type of crop was planted in an organic greenhouse so that their further development could also be studied. The fast grown lettuce grew enormously, producing leaves, until it started to rot. The slow-growing lettuce flowered and formed seed following the natural pattern of development.

The fast grown lettuce had lost its capacity to flower and form seed after harvest, or in other words, it may have lost an aspect of differentiation. It

Table 5.1 Comparison of the cultivation and quality of leaf lettuce grown under high-tech (intensive) and biodynamic conditions (Lammerts van Bueren and Hospers, 1991)

	Corgrow: high-tech	De Vijfsprong: biodynamic
Cultivation		
Cultivation	Automatic, continuous	Rotation
Substrate	Plastic pot with peat in water	Peat block in greenhouse soil
Manure	Water with minerals	60 tonne composted manure/ha
Light	Illumination 20 hours/day	Natural
Temperature	Heated continuously	Only during raising
Pesticides	None	None
Prevention of diseases	Disinfected, insect netting	Rotation, resistance
Weed control	Sterile substrate	1 × hoeing, 1 × hand weeding
Various treatments	None	4 × biodynamic preparations
Crop		
Start of growth	Rapidly	Gradually
Cultivation period	5 weeks	10 weeks
Lettuce head	Luxurious open	Compact
Leaf colour	Bright green	Light green
Leaf form	Long narrow	Round
Leaf	Vulnerable	Firm
Weight in grams	110 (2 plants/pot)	136 (1 plant)
Quality		
Taste	Fresh watery	soft sweet, like lettuce
Storability	Wilts easily	stays crisp longer
Nitrate (mg/l)	2837 (very high)	687 (moderate)
Sugar (% total soluble solids)	2.1 (low)	2.5 (good)
Dry matter (%)	4.2 (low)	5.3 (good)
Vitamin C	Low	Moderate
Ripeness	Little	Good
Rot in self-disintegration test	Much	Little
Characteristics	Young and vulnerable	Full ripe and strong
Quality	Watery	Firm
Cultivation dependent on	Technique, energy, nutrients	Weather, labour, manure

was also high in nitrate, and low in sugar, vitamin C and dry weight. The question that was not answered in this study was what the consequences of this might be for the crop's nutritive quality.

The lettuce example demonstrated the importance of balanced growth in the growing stages of the crop in order to acquire mature quality properties. We had to work out the balance and tried to do this using the optimal proportion between two major life processes of 'mass growth' and 'differentiation in form and substances'.

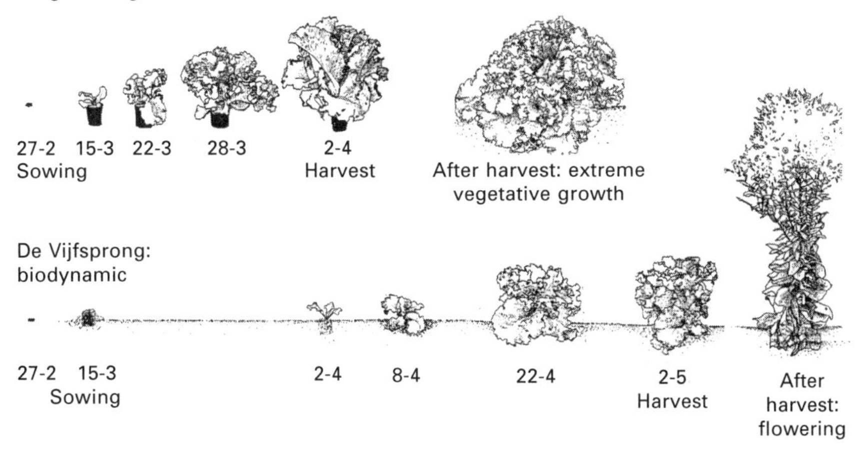

Fig. 5.1 Comparison of crop development of leaf lettuce (Lammerts van Bueren and Hospers 1991). Dates are presented as numbers ('27-2' means 27 February).

5.2.4 Life processes of growth and differentiation in plants

Growth can be defined as the production of organic matter by increase in size or volume. This process involves the uptake of water, carbon dioxide and minerals. In plants, growth is made possible by the process of photosynthesis, which produces the sugars (as primary components) from which compounds such as starch, cellulose, amino acids and proteins are derived.

Differentiation can be defined as the process of specialisation in terms of shape and function. An example is cell differentiation in plants, animals and humans: a young cell, which is initially multifunctional, gradually acquires one specific function and shape. Specialisation is a refinement that is expressed in terms of shape, scent and colour. For example, fruits ripen, leaves change colour in the autumn, the growth of a shoot ends in a terminal bud and seeds become dormant. The primary components are converted into secondary components such as phenols, vitamins, aromas, wax, and so on. Thus 'differentiation' in this context has a broader meaning than only the 'formation of a new plant organ'.

Since these two processes, growth and differentiation, occur simultaneously in living organisms, they cannot be separated. But as agricultural practice shows, it is nevertheless useful to *distinguish* between them and to see them in their relative proportions. For example, Fig. 5.2 shows that growth processes dominate the first formed leaves and that differentiation processes gradually become more pronounced in later formed leaves.

Growth does not take place at equal rates in the different organs of a plant. In carrot, leaves grow vigorously in July while the root grows thicker in August and September. In the case of apple, shoots, leaves, roots, flowers,

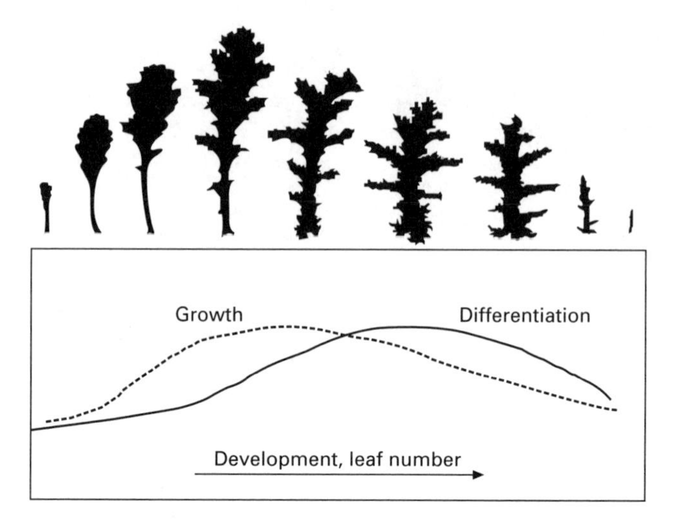

Fig. 5.2 Growth and differentiation expressed in the leaf form during development of groundsel. The leaves of groundsel (*Senecio vulgaris*) are presented in a consecutive line-up from the first leaf grown on the stalk, far left, to the last leaf, far right. The first leaf is small and undifferentiated. This is followed by a stage of vigorous growth of both petiole and leaf blade. The leaves at the centre have attained their maximum size and the differentiation of their shape is becoming apparent. Differentiation gradually takes over and the leaves become smaller and more toothed. The influence of growth has given way to the influence of differentiation and the last leaf 'ripens' into a narrow, pointed shape. It is time for flowering to start.

flower buds and fruits each realise their own maximum growth rate at different periods in the year (Bloksma *et al.*, 2004). This implies that growth and differentiation should be considered per organ and in relation to time.

5.2.5 Balance of growth and differentiation to obtain maximum quality

The need for some balance between growth and differentiation can be recognised by looking at situations of imbalance, as in the next examples:

- Plants grown in shade are tall and weak and lack the fine structured forms developed by light (may indicate too little differentiation).
- A lettuce crop that grows rapidly owing to excessive fertiliser often lacks taste and may be susceptible to disease (may indicate too much growth).
- Emergency flowering in drought reduces vigour (may indicate too little growth).
- Aphids suck growing substances and this results in dry, mummified fruits (may indicate too little growth).

We define 'balance in life processes' as the development of crops with a moderate resistance to stress and diseases and with aromatic and firm food products. Some authors describe this as 'maturing' (Rosenfeld, 1998; Strube and Stolz, 2004) or 'maximum natural development for vegetable growth, flowering and fruiting'. Biodynamic growers express this as 'coherence' and 'plant specificity' (Koepf *et al.*, 1976).

Defining optimum food quality as a balance in life processes is not a new concept. This mode of thought has long held sway among biodynamic researchers (e.g. Schuphan, 1961; Klett, 1968; Pettersson, 1970; Koepf *et al.*, 1976; Kunz, 1999; Bauer, 1999) and in plant physiology (e.g. Herms and Mattson, 1992; Lerdau *et al.*, 1994; Galston, 1994). The way we use this idea, however, is new: the life processes are used to form a framework for a coherent quality concept including various quality properties. In addition the quality properties of the final product can be linked to crop properties and to management tools.

Another new aspect of this concept is that it does not include a single optimum quality. For example, some consumers like green, firm, juicy apples, and others prefer blushed, sweet, aromatic apples. So, there is some freedom to choose a more growth-related or a more differentiation-related optimal balance when managing the crop.

5.2.6 The inner quality concept

In Table 5.2 we present the 'inner quality concept' for a general crop, which is a generalisation of the IQC developed for apple (Bloksma *et al.*, 2004) and for carrot (Northolt *et al.*, 2004). It is described more popularly in a brochure by Bloksma and Huber (2002).

The IQC has three components, in horizontal rows, which describe the life processes of growth and differentiation, and their integration. The vertical columns give the four different descriptions for crop management, the physiological life processes, the properties of the crop and the properties of the food product.

5.2.7 Two meanings of 'vitality'

The term 'vital products' is attractive for a concept dealing with life processes. In any event, it has considerable commercial appeal. Nevertheless we omitted 'vitality' from the first quality concept because of its ambiguity. Some, mainly in conventional agriculture, use 'vital' for products with an emphasis on growth processes: the fresh, green, growing, young and lively product (vitality meaning 1 in Table 5.2). Others use 'vital' for products which were grown with a balance between growth and differentiation and with optimum self-regulatory properties, coherence and self-realisation (vitality meaning 2 in Table 5.2). Initially, we gave the concept the name vital quality concept, but because of these different confusing meanings of 'vital', the concept was renamed inner quality concept (IQC).

Table 5.2 The inner quality concept for a general crop

	Crop management in communication with grower	Life processes grower, plant physiologist	Properties of crop grower	Properties of product grower, consumer, retailer	
Growth	• No limits to nutrients, light and water (fertilisation, breaking and watering) • Extra CO_2 (in greenhouses) • Warmth (crop ridges, greenhouses) • More space (wider plant distance, defruiting, weeding)	• Production of mass: forming cells, tissues, organs • Expansion • Absorption of water and nutrients • Photosynthesis → primary metabolites • Maintenance of basic metabolism	• Big mass, high yield • Large, dark-green leaves • Grows until harvest • Many fungal diseases and sucking insects • Strong seeds and flower buds	• Firm, tart, crisp, crunchy, juicy • High glucose, starch, nitrate, amino acid, protein, etc • Low dry matter, vulnerable, short storable • High initial value in luminescence (biophotons) • Perradation, fullness, expansion in biocrystallisation pictures	Vitality (meaning 1)
Differentiation	• Light (no shade, pruning, clean windows) • A little growth stress (limited water, limited fertilisation, root pruning) • Binding down young apple twigs • Ethylene hormone	• Refining, ordering • Ripening of all organs • Replenish reserves • Secondary metabolites • Induction of generative organs	• Differentiated refined forms (fine leaf serrations, cork, colour) • Order, symmetry • Growth is completed (final bud) • Many flower buds and seeds • Biting insects	• Form and colour is completed (stumpy carrot, yellow ground colour in apple) • High dry matter • High secondary metabolites (phenols, vitamins, tannin, resin, wax, aromas) • Bitterness • Hyperbolic decay in luminescence (biophotons) • Structure in biocrystallisation	
Integration of growth and differentiation	• Optimal proportion in stimulating growth and differentiation (species and stage typical) • Appropriate varieties • Disease preventive soil • Diversity of agro-ecosystem • Biodynamic preparations?	• Maturing • Enough primary → enough secondary metabolites • Self-regulation	• Species and stage-specific full-grown and ripe • Resistance against pest, diseases and stress • Crop: wound healing after damage • Many fertile generative organs	• Attractive (coloured, glossy) • Optimum taste (juicy, crispy and aromatic) • Optimal nutrient composition • High ratio protein/total N • Species typical in spectral range luminescence • Coherence in biocrystallisation	Coherence, vitality (meaning 2)

5.2.8 Experimental parameters mainly focus on coherence

Several experimental parameters are proposed to assess the coherence aspects of products:

- biocrystallisations or copper chloride crystallisation (Engquist, 1970; Busscher *et al.*, 2004),
- luminescence or biophotones (Popp *et al.*, 1981; Popp and Li, 1993; Strube, 2003),
- physiological amino acid status (Wistinghausen 1975; Stolz, 2003),
- electrochemical measurements (Kollath, 1978; Staller, 2003),
- capillary rising pictures (Tingstad, 2002; Skjerbaek *et al.*, 2005).

The parameters, mentioned however, have not been clearly interpreted in terms of quality, i.e. what aspect of quality is measured? The IQC provides a framework for further investigations to demonstrate the meaning of these parameters. In the apple and carrot experiments of the Louis Bolk Instituut, we demonstrated some experimental parameters in food products grown with different balances between growth and differentiation. Biocrystallisation pictures in particular, are able to show both the life processes of growth and differentiation, and their integration. In future we expect to find the key to work out the relevance of 'coherence' for health in the balance (the integration) between the life processes.

5.2.9 Perspective to enlarge the concept to animals and humans

We see some similarities between the major life processes in plants (growth and differentiation) and the major life processes in animals and humans (proliferation and differentiation). We expect in future to relate this concept to animal production and to human health, to be able to cross the bridge from soil to plant to animal and finally human health. For example, the development in medicine of 'differentiation therapy' in which vitamin A-derivates are used to treat human cancer cells *in vitro* (De Luca *et al.*, 1995). Cancer is defined by too much uncontrolled growth of cells without enough differentiation. Using treatment with vitamin A-derivatives – a product of differentiation processes in the plant – undifferentiated cancer cells change into differentiated more healthy ones.

5.3 Method for validation of the inner quality concept

5.3.1 Validating a new concept

There is a risk in using circular reasoning when introducing a new overall quality concept (such as the IQC) that contains new aspects *and* new parameters. It is difficult to introduce an unknown aspect such as integration, to associate it with crop management measures and to measure it with experimental parameters such as luminescence and biocrystallisations. A methodological

Table 5.3 Validation route for the inner quality concept

	Completed (+) or partly (+/–)
1. Development of quality concept for organic products	
a. Based on life processes (growth, differentiation and integration of both)	+
b. Relating processes and properties (see Table 5.2)	+
c. Making processes measurable by parameters	+
d. Relating to holistic health concept by physicians and dieticians	+/–
2. Testing face validity	
a. Life processes recognised by workers in the field (e.g. farmers)	+
b. Life processes recognised by specialists (e.g. physiologists)	+/–
3. Testing content validity of concept	
a. Is concept consistent in itself?	+
b. Is concept consistent with current theories?	+/–
4. Testing predictive validity of concept Is concept consistent with existing empirical data?	+/–
5. Testing reliability of established parameters	
a. Good correlation between parameters for the same item?	+/–
b. Same results by different observers and laboratories?	+/–
6. Responsivity to change Do parameters discriminate sufficiently?	+/–
7. Development of a new parameter	
a. Parameter compared with established parameter in controlled field study	+/–
b. If no established parameter, the parameter is based on logical reasoning (here on physiological theories)	+/–

foothold can be achieved by simultaneously working on the theory of the new concept, and executing experiments to evaluate the concept (Streiner and Norman, 2001). In Table 5.3 we present the course of validation followed for the inner quality concept.

The first step was to validate the new quality concept as described at the beginning of this chapter (step 1). Over the years, we discussed and improved the quality concept presented in Table 5.2 with many colleagues in science and the field (step 2). Growth and differentiation are recognised by plant physiologists, for example in the 'growth-differentiation-balance-hypothesis' or GDBH (Herms *et al.*, 1992; Lerdau *et al.*, 1994). According to the GDBH, growth is necessary for primary metabolism and differentiation for secondary metabolism. Unfortunately, plant physiological theory on integration processes such as self-regulation is still underdeveloped. In other words, this most interesting aspect of the content validity (step 3) is not yet completed.

Table 5.4 Overview of the FQH experiments by Louis Bolk Instituut and the presumed effects of each factor on the life processes

Crop	Harvest	Series in	Growth	Differentiation	Integration
Lettuce[1]	1991	High-tech versus organic	↑↑	↓?	↓?
Apple[2]	2000	Bearing (= yield, 5×)	↓↓		
		Sunlight (3×)	↑	↑↑	↑
		Ripening (5 harvest dates)	↑	↑↑	
		Post-harvest ageing (5×)	↓↓		↓
Apple[3]	2002	Nutrients (5×)	↑↑	↓	
	(2001–2003)	Compost/ commercial organic fertilisers (2×)			↑?
		Biodynamic preparations/ none (2×)			↑?
Carrot[4]	2003	Nutrients (3×)	↑↑	↓	
		Sunlight (3×, shade nets)	↑	↑↑	↑
		Ripening (3 harvest dates)	↑	↑↑	

[1]Lammerts van Bueren and Hospers 1991. [2]Bloksma *et al.*, 2001. [3]Bloksma *et al.*, 2004. [4]Northolt *et al.*, 2004.

5.3.2 Controlled series with extremes in growth and differentiation

In order to find the range of and the optimum balance between growth and differentiation during cultivation, it is necessary to grow crops under extreme conditions of growth and differentiation and to demonstrate the consequences of one or the other for the harvested product. It is also necessary to know how the balance between the life processes can be managed during cultivation.

To this end, we designed experiments in which conditions were varied in order to induce the extremes of growth or differentiation into the crop. In Table 5.4 we present an overview of the experiments with the presumed effects of varying the cultivation factors on the life processes. By comparing the results with our expectations, we largely completed step 4 of the validation course for apple and carrot.

5.3.3 Experimental parameters for quality

In order to understand the significance of experimental parameters such as luminescence and biocrystallisation, we assessed them for products grown in controlled conditions. We also correlated the results with as many established parameters as possible (step 7).

Experimental parameters may correlate with established parameters (*convergent validity*), such as resistance to diseases and evaluation of taste (step 7a). In that case, the experimental parameters offer no added value and the parameter which is least expensive to implement will be chosen in the future. On the other hand, there may be no correlation which might suggest that a new aspect of quality may have been found. A new aspect such as integration might be derived from plant physiology theories by logical reasoning (for example, self-regulation may be derived from growth and differentiation). This method of argumentation is known as *construct validity* (step 7b).

Individual parameters must be validated by the laboratories that produce them. The biocrystallisation method and the luminescence method have been validated for selected crops almost up to steps 5 and 6 (Busscher *et al.*, 2004; Kahl *et al.*, 2003). More correlations with products grown in controlled conditions and evaluations with regard to human health are necessary in order to understand the significance of these methods.

5.4 Experiments to validate the inner quality concept

5.4.1 Apple, between growth and differentiation

We chose to develop the quality concept using apple because we had already gained a considerable amount of knowledge about the inner quality aspects of apples in previous research. In addition, the aspects of growth and differentiation are commonly used by apple growers. They are familiar with 'vigour and bearing' and take measures to regulate these.

Method

In Fig. 5.3, we present results from two apple studies in which an orchard was divided into several different plots, so that one cultivation factor could be varied in gradual degrees from too little to too much. The cultivation factors were: bearing level (35, 75, 100, 125, 140 fruits per tree by hand thinning), sunlight exposure (three positions in the tree), ripening (five harvest dates with 7-day intervals), post-harvest ageing (1, 4, 8, 12 days on the shelf after three months cold storage), nutrient level (0, 40, 80, 120, 160 kg N/ha supplied by commercial organic fertilisers, farmyard compost or commercial fertilisers) and, finally, biodynamic preparations (present/absent).

In both studies the apples were grown in the biodynamic orchard 'Boomgaard ter Linde' in the south-western part of the Netherlands. We used full grown apple trees of the Dutch variety Elstar on dwarf rootstock M9, 2460 trees/ha, grown on limey humus sea loam with trickle irrigation. The experiment was replicated four times.

We measured many properties of the soil, the growing trees and the apples after harvest and after storage. The results were studied in relation to the management factors and mutual correlations were computed. This procedure enabled us to contribute to the validation of the IQC (step 3, Table 5.3), the validation of new parameters (steps 5 and 6) and the evaluation of orchard management. Details of the methods and results of the apple series are described elsewhere (Bloksma *et al.*, 2001, 2004). Here, we mention only some of the results that are relevant to the IQC.

Parameters for growth
The parameters for tree growth are fruit-bearing (with equal shoot growth and bearing in previous year), shoot growth (with equal bearing), leaf size or colour (with equal shoot growth), nitrogen content in bud, nitrogen and magnesium contents in leaf and scab infestation. The parameters for fruit growth are fruit size or weight (with same bearing), firmness (with equal bearing and shoot growth), acidity, nitrogen content, amino acid content, protein content, tart and crisp taste, growth score on crystallisation pictures, initial luminescence and susceptibility to fruit rot.

Parameters for differentiation
Parameters for tree differentiation are autumn colours and bud formation (with equal bearing). Parameters for fruit differentiation are degree and hue of blush, yellow ground (background) colour, shape of fruit, sheen, starch conversion, differentiation score on crystallisation pictures and luminescence.

Parameters for growth or differentiation depending on limiting factor
Both growth and differentiation are important for most plant processes. We recognised that many parameters can be expressions of either growth or differentiation, depending on which is the limiting factor for the crop. For example, when fertilisation is taken as the limiting factor for bud formation, then bud formation is a growth parameter. If the limiting factor is sunlight exposure, bud formation acts as a differentiation parameter in a light exposure series.

Parameters for integration
The results found in the literature and from these experiments enabled us to select parameters which may give an indication of the degree of integration. These are resistance to diseases and pests, overall taste, phenols, ratio of proteins to free amino acids (physiological amino acid status), integration score on crystallisation pictures and species-typical colour ratio in spectral-range luminescence.

The secondary metabolites in the plant, such as phenols, vitamins, aromas and colouring agents, are integration parameters. The formation of the raw material (assimilates, primary metabolites) is a growth process; the subsequent conversion to secondary metabolites is a differentiation process. The integration of both processes is necessary for high content of secondary metabolites. A correlation with either growth or differentiation is found depending on the time of year that measurements are taken and the limiting factors in the production system. We found that the experimental series for apple were not suitable for evaluating the degree of integration with the required degree of certainty. Too many assumptions remained unproven regarding the effects of biodynamic preparations in promoting integration.

Evaluating crop management for apple quality
The biocrystallisation pictures of apple juice from progressively riper fruits showed increasing openness towards the periphery. Looking at ripening as

'an opening of gestures' also allowed us to recognise the successive transition of solids into gaseous substances in conventional content analyses: firm, sour fruits with high starch and phenols changed into tasty, juicy fruits with vaporous, aromatic substances. A picture-creating method is useful in getting a physical image of a process and increases overall insight. In the future, this expensive method might be replaced by the cheaper conventional analyses, but for now its expense is more than offset by its usefulness in revealing the process.

With increasing sunlight exposure, the spectral range luminescence that is typical for apple increased. In addition, the free amino acid content decreased, the ratio of pure/raw protein increased and the ratio of sugar/nitrogen also increased (Fig. 5.3). All these parameters are indications of the same general conclusion, that the product is 'more completed' in the sun. Also, higher contents of phenols and vitamin C indicate that a sun-ripened product is more resistant to decay and is thus more integrated. In the shade, more building substances (amino acids and nitrate) accumulate, as the plant waits for sugar to assimilate in the tissue. High quantities of free amino acids and nitrate in fruit and vegetables are known to be undesirable for human health and keeping quality. We found that shade and over-fertilisation had a similar effect on the balance between life processes: the accent shifted from differentiation to growth. Both can be explained by carbon/nitrogen plant physiology. The crop can be improved both by providing more light and using less fertiliser.

As expected, higher fertilisation stimulated growth characteristics and caused a decline in differentiation characteristics. There was a longer period of shoot growth and there were more shoots, but there were also more fungal infections, darker and larger leaves, higher nitrogen levels in the bud, leaf and fruit, and more and stronger blossom formation for the following year. Higher levels of fertilisation resulted in larger apples which were less firm and slightly less tart, with less blush, a lower phenol content and a greater susceptibility to fruit rot. After two to three years it had become clear that the 0 kg N/ha and 40 kg N/ha regimes were too low and inadequate (the trees had a non-bearing year) and that fertilising with 160 kg N/ha was too much (more fruit rot, less tasty). For this orchard, a fertilisation regime of about 100 kg N/ha best achieved the two-fold objective of regular yield and optimal inner quality. This is what we call the optimal apple-specific balance between growth and differentiation.

5.4.2 Carrot, between growth and differentiation

Carrot has a more straightforward physiology than a perennial crop such as apple. It might therefore be easier to study the integration aspects of growth and differentiation in this crop. In addition, some organic carrots are famous for their carroty taste which is often lacking in conventional carrots.

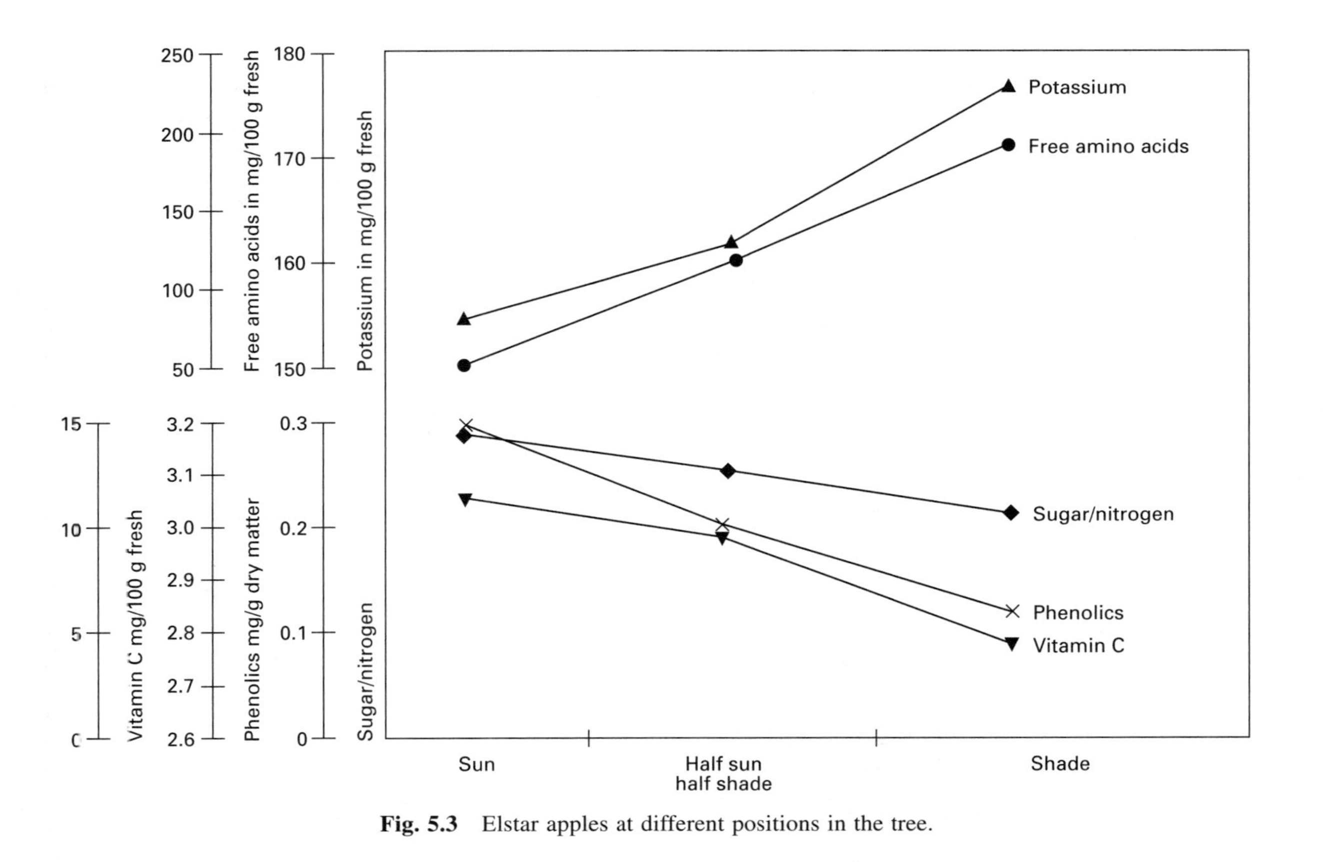

Fig. 5.3 Elstar apples at different positions in the tree.

Method
In Table 5.4, we present the carrot study, in which a carrot crop was divided into several series which gradually varied: three nutrient levels, three sunlight exposure levels, ripening (three harvest dates). The carrots were grown on the Dutch biodynamic mixed farm 'Warmonderhof', on reclaimed, well-drained sea clay. The field was moderately fertile and the crop followed onions in the rotation. We used the open-pollinated cultivar Rodelika (Bingenheimer Saatgut AG). We divided a large carrot field into four replicate blocks. The carrots grew on ridges. The nutrient levels were 0, 100, 200 kgN/ha, applied in the form of blood and feather meal pellets. Unfortunately, it was a warm and wet season, so that the nutrient level was not a true limiting factor. The light levels 100%, 85% and 52% light were realised by using shade nets once the young plants were established. Plants were evaluated on six different dates.

We measured many properties of the soil, the growing crop and the carrots, after harvest and after storage. The results were studied in relation to the management factors, and mutual correlations were computed. This procedure enabled us to contribute to the validation of the IQC, the validation of the new parameters and the evaluation of carrot management. Details of the methods and results of the carrot experiments are described elsewhere (Northolt *et al.*, 2004). As with apple, we only mention some of the results here that are relevant to the IQC.

Parameters for growth
Parameters for crop growth are weight of leaves and leaf colour. Parameters for carrot growth are root weight, monosaccharide content, nitrate content, emission of spectral range luminescence and rot in storage test (trend).

Parameters for differentiation
Parameters for crop differentiation are fine forms in leaves (trend), colouration of the leaves (hypothesis, but here too much growth for autumn colouration). Parameters for carrot differentiation are root form (from pointed to stump, Fig. 5.4), dry matter and emission ratio in spectral range luminescence.

Parameters for integration of growth and differentiation
Parameters for crop integration are incidence of pests and diseases (hypothesis, because no spontaneous pests and diseases occurred). Parameters for carrot integration are carotene content, orange colour of root (trend), saccharose content, total sensory appreciation and carrot taste.

Evaluating crop management for carrot quality
The design of this carrot experiment, with only three levels of fertilisation, light and ripeness, did not reveal the desired optimum levels. A series with at least five levels will give a better chance of finding significant results. The experiment did clearly show, however, that the best quality resulted from a

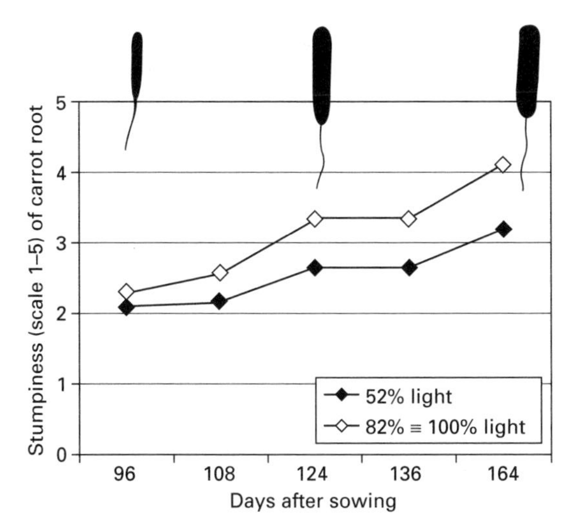

Fig. 5.4 Form of the carrot root in relation to ripeness and light exposure (Northolt *et al.*, 2004).

combination of the lowest level of fertiliser, the highest degree of light and the latest harvest date. Carrots grown in these conditions had the best taste, lowest nitrate content and best keeping quality. All three factors shift the balance in favour of differentiation. The type of soil on which this experiment took place is highly conducive to growth; all management measures taken here should aim to reinforce the differentiation process.

5.5 Progress made in the validation of the concept

The IQC based on life processes offers good prospects as a tool for improving the production of high quality crops and measuring the health effects of these products in the future. Growth and differentiation are well distinguished. The integration aspect is still the weakest part of the concept and needs to be developed further, see Part 3 of Table 5.2.

Parameters are especially useful for the IQC when they express the three aspects of the concept: growth, differentiation and integration. Such parameters might involve crop observations (e.g. a test of resistance to stress, diseases and pests), content analyses dealing with sugar/nitrogen ratios, physiological amino acid status and secondary metabolites (phenols), biocrystallisation pictures and spectral-range luminescence. The new holistic parameters have a secondary purpose: they also provide scientists with a new perspective from which to study life processes. In the future, these parameters might be replaced by the cheaper content analyses.

Another holistic parameter, the electrochemical measurements (pH, redox potential, resistance) yielded little to no effect in these apple and carrot studies.

5.5.1 Recommendations for future research

Further research is necessary to complete the validation of the quality concept. Our suggestions are listed below:

- The integration aspect and associated management measures need to be translated in plant physiological terms by experiments with some physiologically well-documented crops.
- Experiments should be carried out with other crops than apple and carrot. Such experiments should fulfil the following requirements:
    - a simple crop with few organs;
    - serial implementation of management measures, with at least five levels of one variable from too low to too high in growers' opinion, while other factors remain constant;
    - diseases and pests should not be prevented (and may even be intentionally introduced) after the young stage, as degree of resistance is an important integration parameter.
- The relationship between the quality concept and human or animal health needs to be established from a holistic health perspective. The Louis Bolk Instituut has already started animal feeding experiments. Such research can only be done if foodstuffs are available which have been produced using well-balanced processes, in the sense described above, and which have thus achieved an optimum level of quality (defined hereunder).

5.6 Perspective for farmers, traders and consumers

Elsewhere in this book, experiments are described which compare organically grown products with conventionally grown products. The next steps will be to interpret these differences in terms of their effects on human health and finding ways of improving crop quality in the production phase. This will require a coherent concept of food quality, a concept in which food quality is more than the sum of exterior characteristics some specific health components and the absence of harmful contaminants. Moreover, we need a concept that connects the different phases of plant growth to properties of the harvested product and to human or animal health. The IQC can meet these requirements.

5.6.1 Traders and consumers should recognise quality

Consumers of organic products do not only attach value to organic cultivation methods, but also to the freshness, taste, ripeness and keeping quality of the product. Together, these attributes justify the higher price of organic products. In Table 5.1 (right column) traders and consumers considered the characteristics of the products with an emphasis on growing processes (e.g. big, firm, juicy, storable) or an emphasis on differentiation processes (e.g. ripeness, aroma, colour). Optimum quality is a balance of characteristics that falls somewhere in between. Consumers have some leeway in choosing for emphasis on one or the other. Traders can use Table 5.1 to find the corresponding crop properties and can ask growers to keep this in mind in stimulating the corresponding life processes.

5.6.2 Quality-driven cultivation

In Table 5.1 (left column), growers will find the management tools to influence life processes. Fertilisation is the most common method of stimulating plant growth (especially using nitrogen), but this may also be achieved by breaking the soil, to promote mineralisation and watering. In greenhouses, extra carbon dioxide can be added. Growth is also enhanced by a warm growing site, more space between plants, defruiting and weeding.

Management tools for stimulating differentiation are also available. The most important of these is to maximise light: shade must be minimised by pruning trees, thinning and cleaning greenhouse windows. Differentiation can also be stimulated by taking measures to limit growth. Mild forms of growth limitation include restricted water supply, limited fertilisation and root pruning. In apple production, binding down young twigs is a trick to initiate flower bud formation (differentiation). In tomato growing, artificial ethylene hormone is applied.

The grower thus has different tools by which to achieve an optimum balance and integration between growth and differentiation (specific for the plant species and the development stage). Moreover, we expect more integration by using appropriate varieties, soil that helps prevent soil diseases, by encouraging diversity of the agro-ecosystem and, probably, by applying biodynamic preparations.

5.6.3 Relation to human health

Since growth, differentiation and their integration are universal processes in all living systems, it is not unthinkable that human health might be favourably influenced by the use of food products which balance and integrate between growth and differentiation. Does a ripe, balanced food product enhance the self-regulatory capacities of the consumer? These are exciting questions for future research.

5.7 References

Bauer D (1999), 'Die Fähigkeit zu reifen', *Lebendige Erde*, (3), 6.

Bloksma J and Huber M (2002), *Life Processes in Crops: On growth & Differentiation*, publication FQH-02, Driebergen, Louis Bolk Instituut, downloadable from www.louisbolk.nl.

Bloksma J, Northolt M and Huber M (2001), *Parameters for Apple Quality and an Outline for a New Quality Concept*, part 1 and 2, publication FQH-02, Driebergen, Louis Bolk Instituut, downloadable from www.louisbolk.nl.

Bloksma J, Northolt M, Huber M, Jansonius P J and Zanen M (2004), *Parameters for Apple Quality and the Development of the Inner Quality Concept 2001–2003*, publication FQH-03, Louis Bolk Instituut, Driebergen, downloadable from www.louisbolk.nl.

Busscher N, Kahl J, Huber M, Andersen J O, Mergardt G, Doesburg P, Paulsen M, Kretschmer S, de Weerd A and Meier-Ploeger A (2004), *Validation and Standardization of the Biocrystallization Method: Development of a Complementary Test to Assess Qualitative Features of Agricultural and Food Products*, Triangle report Nr.1, University Kassel, Louis Bolk Instituut and Biodynamic Research Association Denmark.

De Luca L M, Darwiche N, Jones C S and Scita G (1995), 'Retinoids in differentiation and neoplasia', *Am Science and Medicine*, **2** (4), 28–37.

Engquist M (1970), *Gestaltungskräfte des Lebendigen, die Kupferchlorid-Kristallisation, eine Methode zur Erfassung biologischer Veränderungen pflanzlicher Substanzen*, Verlag Vittorio Klostermann, Frankfurt am Main.

Galston A W (1994), *Life Processes of Plants*, The Scientific American Library, New York.

Herms D A and Mattson W J (1992), 'The dilemma of plants: to grow or defend', *Quart Rev Biol*, **67**, 283–335.

Kahl J, Busscher N and Meier-Ploeger A (2003), *Abschlussbericht Projektnummer 02OE170 Ganzheitliche Untersuchungsmethoden und Prüfung der Qualität ökologischer Lebensmittel: Stand der Entwikkelung und Validierung*, Fachgebiet Okologische Lebensmittelqualität und Ernährungskultur Universität Kassel, Kassel.

Klett M (1968), *Untersuchungen über Licht- und Schattenqualität in relation zum Anbau und test von Kieselpräparaten zur Qualitätshebung*, IBDF Darmstadt.

Koepf H, Pettersson B D and Schaumann W (1976), *Biologisch-dynamische Landwirtschaft*, Eugen Ulmer, Stuttgart.

Kollath W (1978), *Regulatoren des Lebens-Vom Wesen der Redox-Systeme*, 2. Auflage, Haug-Verlag, Heidelberg.

Kunz P (1999), 'Reife, sorten, qualität', *Lebendige Erde*, (1), 34–36.

Lammerts van Bueren E M and Hospers M (1991), *Technologisch groen versus biologisch groen – een onderzoek naar de kwaliteit van industrieel en biologisch-dynamisch geteelde pluksla*, Louis Bolk Instituut, Driebergen.

Lerdau M, Litvak M and Monson R (1994), 'Plant chemical defence: monoterpenes and the growth – differentiation balance hypothesis', *Trends Ecol Evol*, **9**, 58–61.

Meier-Ploeger A and Vogtmann H (1988), 'Lebensmittelqualität – ganzheitliche Methoden und Konzepte'. *Alternative Konzepte 66*, C.F. Müller, Karlsruhe.

Northolt M, van der Burgt G J, Buisman T and Vanden Bogaerde A (2004), *Parameters for Carrot Quality and the Development of the Inner Quality Concept*, publication FQH 04, Louis Bolk Instituut, Driebergen,

Pettersson B (1970), 'Die Eindwirkung von Standort, Düngung und wachstumbeeinflussende Stoffen auf die Qualitätseigenschaften von Speisekartoffeln' *Lebendige Erde*, (3), 4.

Popp F A and Li K H (1993), 'Hyberbolic relaxation as a sufficient condition of a fully coherent ergodic field', *Int J Theoretical Physics*, **32**, 1573–1583.

Popp F A, Ruth B, Bahr W, Böhm J, Grass P, Grolig G, Rattenmeyer M, Schmidt H G and Wulle P (1981), 'Emission of visible and ultraviolet radiation by active biological systems', *Collective Phenomena*, **3**, 187–214.

Rosenfeld H J (1998), 'Maturity and development of the carrot (*Daucus carota L.*) root', *Gartenbauwissenschaft*, **63** (2), 87–94.

Schuphan W (1961), *Zur Qualität der Nahrungsplantzen*, BLV-Verlaggesellschaft, München.

Skjerbaek K, Zalecka A, Kahl J, Huber M and Doesburg P (2005), *Development and Characterization of the Capillary Dynamolysis Method for Food Quality Analysis*, Triangle report Nr.2, University Kassel, Louis Bolk Instituut and Biodynamic Research Association Denmark.

Staller B (2003), 'Elektrochemische messungen', in Kahl J, Busscher N and Meier-Ploeger A, Abschlussbericht Projektnummer 02OE170 *Ganzheitliche Untersuchungsmethoden und Prüfung der Qualität ökologischer Lebensmittel: Stand der Entwikkelung und Validierung*, Fachgebiet Okologische Lebensmittelqualität und Ernährungskultur, Universität Kassel, Kassel, 203–238.

Stolz P (2003), 'Physiologischer aminosäurestatus', in Kahl J, Busscher N and Meier-Ploeger A, Abschlussbericht Projektnummer 02OE170 *Ganzheitliche Untersuchungsmethoden und Prüfung der Qualität ökologischer Lebensmittel: Stand der Entwikkelung und Validierung*, Fachgebiet Okologische Lebensmittelqualität und Ernährungskultur, Universität Kassel, Kassel, 158–202.

Streiner D L and Norman G R (2001), *Health Measurement Scales*, Oxford Medical Publications, Oxford.

Strube J (2003), 'Fluoreszenz-anregungsspektroskopie', in Kahl J, Busscher N and Meier-Ploeger A, Abschlussbericht Projektnummer 02OE170 *Ganzheitliche Untersuchungsmethoden und Prüfung der Qualität ökologischer Lebensmittel: Stand der Entwikkelung und Validierung*, Fachgebiet Okologische Lebensmittelqualität und Ernährungskultur, Universität Kassel, Kassel, 61–157.

Strube J and Stolz P (2004), *Lebensmittel vermitteln Leben – Lebensmittelqualität in erweiterter Sicht*, Kwalis Qualitätsforschung Fulda GmbH, Dipperz.

Tingstad A (2002), *Quality and Method, Rising Pictures in Evaluation of Food Quality*, Gads Forlag, Copenhagen.

Wistinghausen E von (1975), 'Die Qualität von Möhren, Rote Bete und Weizen in Beziehung zu ihren Standortverhältnissen und Bodenbedingungen', *Lebendige Erde*, (3) 4.

Wistinghausen E von (1979), *Was ist Qualität, wie Entsteht sie und wie ist sie Nachzuweisen?*, Verlag Lebendige Erde, Darmstadt.

6

Food consumers and organic agriculture[1]

Elizabeth Oughton and Christopher Ritson, Newcastle University, UK

6.1 Introduction

This chapter focuses on describing our current understanding of why some people choose to consume organic food and others do not. This can provide important additional insights into food consumer behaviour in general. Thus the purpose of this chapter is not to describe 'the organic consumer' *per se*, but to explore which, and to what extent, factors underlying food choice influence consumption of organic products. A number of features of organic products (known in some European countries also as ecological or biological products) suggest that they have the potential to provide a valuable case study for food choice.

First, although historically organic production has been associated with fruit and vegetables, today it is possible to buy an organic version of virtually any food product, from milk to wine, eggs to bread, and baby foods to chocolate. The market share of organic beef, milk and sheep is as high as the share for organic fruit and vegetables in 15 European Union (EU) countries (Hamm and Gronefeld, 2004). Equally, there is always an alternative – usually described in the literature dealing with organic foods as 'conventional' or non-organic – product available. So although it is quite possible that an organic consumer may, for example, be more likely to be a vegetarian, or perhaps less likely to purchase processed products, neither 'conventional' nor 'organic' constrains the product choice range available in many EU countries. Organic consumption therefore resembles a real world 'food choice laboratory' in which in almost all cases the choice to purchase the organic

[1]This chapter is adapted from Ritson and Oughton, 2006.

product will be a consequence of the perceived attributes of organic versus the conventional alternative (and nothing else), although of course, occasionally the organic purchase may be an 'accident' or a consequence of temporary availability.

The second interesting feature of an organic product from the perspective of consumer food choice is, paradoxically, the lack of choice. Lampkin and Measures (2001) describe organic farming as:

> an approach to agriculture where the aim is to create integrated, humane, environmentally and economically sustainable agricultural production systems. Maximum reliance is placed on locally or farm-derived, renewable resources and the management of self-regulating ecological and biological processes and interactions in order to provide acceptable levels of crop, livestock and human nutrition, protection from pests and diseases, and an appropriate return to the human and other resources employed. Reliance on external inputs, whether chemical or organic, is reduced as far as possible.

All this is backed up by a complex set of rules relating to farm production, and to some extent food processing, under the umbrella of two EU regulations, 2092/91 for plants and 1804/99 for animals. National certifications bodies implement and monitor the regulations, sometimes imposing 'stronger' rules. In some countries there is a single body/organic label, in others there are competing bodies/labels. However, even in countries where more then one 'certification label' is available, choice with respect to individual food products is usually limited to one 'label' in most shops. This is mainly due to the still relatively small proportion of organic (compared to non-organic) foods in the overall turnover of retail outlets. Thus an organic product comprises a set of attributes and the consumer buys a prescribed package and is not in a position to select their own preferred combination of organic product attributes.

The third feature of the organic product category of interest, from the perspective of food choice, is that many of the product attributes which distinguish the organic product from the conventional alternative can only be contributed by the primary producer, rather than the food manufacturer, distributor, or retailer (although, of course, there must be conformity along the food chain with organic regulations). This provides a link between consumer food choice and farmer decision making which is lacking in much of the modern conventional food system. Moreover, most of these product characteristics which distinguish organic from conventional are 'credence attributes' (Ritson and Mai, 1998), that is, not attributes that can be identified before purchase ('search') or ascertained after consumption ('experience'), but ones which require 'belief'. This therefore involves trust on the part of the consumer in the behaviour of the primary producer and other actors in the organic food chain and the mechanism for achieving that trust will be an important factor in food choice.

Finally, organic consumption represents an excellent real world empirical base for understanding aspects of food choice because it is a dynamic market.

Real world purchase data can help to explain the factors underlying food choice in one of two ways:

- cross-sectionally, that is, what are the differences between consumers which lead one to choose to purchase a specific product and the other not, or
- time-related, that is, what has changed about an individual to lead them to purchase a specific product this year, but not last year.

To provide useful data, the latter requires change over time. The limitation of this type of analysis is that, as yet, there is no systematic collection of organic market data, which makes it difficult to generalise across countries. The situation in Europe is being addressed by the activities of EISfOM. See for example, Recke *et al.* (2004). The current chapter therefore draws on a number of different studies and data collections in order to illustrate the arguments put forward.

A notable feature of the market for organic products over the past 10 to 15 years has been rapid growth worldwide with global sales reaching €23.5 billion in 2004. Of this total, Europe and North America account for 96% of organic product sales and the market in western Europe alone accounts for €11.6 billion. Sales in other regions similarly continue to grow, although from a very small base (Willer and Yussefi, 2005; Sahota, 2006). Rapid growth has continued in US markets although market growth in Europe has slowed from 8% in 2002 to approximately 5% in 2003 (Richter and Padel, 2005). The slower growth in Europe in general masks considerable national differences: Spain, the UK and many of the new EU accession countries have shown annual increases of more than 10% per annum, whereas there has been slower growth in Denmark, Germany, Austria and Switzerland which had higher growth rates in the 1980s and early 1990s). In terms of the absolute market size, Germany is the largest market in Europe with sales of organic food over €3.6 billion in 2004, followed by Great Britain €1.6 billion, France €1.5 billion, Italy €1.4 billion and Switzerland €0.74 billion. (Figures for Germany and UK are for 2004 and for the remaining countries 2003.) However, average consumer expenditure in Switzerland is highest in Europe at €105 per consumer per year, which is double the second highest annual consumer expenditure (Denmark at €51 per year). Per capita consumption of organic food in the USA is closer to the UK, France or Netherlands at €30 per head per year (Richter and Padel, 2005). As Richter *et al.* (2006) note, these comparisons should be treated with caution as there is as yet no standard method of data collection for organic food statistics and differences will occur between methods of collection in different European countries.

Across Europe the relative significance of organic foods within different food groups varies. The most recent comprehensive analysis of these patterns is provided by Hamm and Gronefeld (2004). In the EU as a whole, the most significant share of organics (by volume) in total consumption is for cereals

(1.8%), beef (1.6%), eggs (1.3%), vegetables (1.3%), fruit and nuts (1.3%) and milk and milk products (1.2%). These relatively low European averages mask some high shares within countries: for example, in Denmark 8.4% of cereals, 8.8% of potatoes, 8.8% of eggs and 10% of milk and milk products consumed are organic. Similarly, in Switzerland, organic sales by volume account for 8.9% of cereals, 7.3% of oilseeds, 3.7% of milk and milk products and 3.5% of vegetables. There is also evidence for the increasingly rapid growth of prepared organic foods. For example, data collected by Mintel in Britain show very large increases in prepared organic foods and baby and infant foods between 1998 and 2003 (Mintel, 2003). In 2004 the value of the UK baby food market was estimated at 148 million of which 43% (equivalent to £63 million) was accounted for by organic baby foods (Soil Association, 2005).

Against this brief background description of organic markets, we now look more carefully at the factors accounting for these patterns of consumption of organic foods.

6.2 The expanding organic market: consumer led or producer driven?

It is tempting to assume that the rapid growth in consumption of a particular kind of food must represent incontrovertible evidence of a fundamental change in consumer attitudes, perceptions and beliefs about that kind of food, and that an end to that growth in turn indicates a stabilising of those changing attitudes. However, a number of studies suggest that the organic market is more complicated than this.

In simple market economics, a distinction can be made between a demand led and supply driven growth in market size. Changing consumer attitudes and perceptions towards the food product could indeed lead to market growth. Changing tastes, lifestyles, meal patterns, environmental or health concerns, but also income growth, can all lead consumers to purchase more of the product, causing a shortage at existing price levels. Production would then be expected to respond to market signals (higher prices–more profitable) so that more is produced and can be consumed. But an increase in consumption can also be supply driven. The usual reason for this is that a new, cost reducing, technology is adopted which makes production more profitable at existing prices. Supply increases, prices fall and consumption increases. In the case of organic agriculture, there are two further and rather peculiar reasons why the increase in consumption could be at least partially supply driven.

The first is what one might term 'producer values'. The early producers of organic products did so almost exclusively because 'they believed in it'. They were often not particularly market orientated or market aware and had to develop new and alternative marketing methods and supply chain structures.

This often limited efficiency at many stages of the supply chain and increased cost. (See various papers in the special edition of *Sociologia Ruralis* **41**(1) on 'Politics, ideology and practice of organic farming', and Kaltoft (1999).)

Second, conversion of land to organic production has been subject to substantial government incentives. When the government subsidises the production of anything, more is produced and more consumed (Ritson, 1977). Subsidies reduce the unit cost of production. Producers find it profitable to expand at prevailing prices and new producers are attracted into the market. Supply to the market increases, causing market prices to fall. This in turn stimulates an increase in consumption. In the case of organic production a critical issue is the impact of subsidies on the profitability of organic production relative to conventional agriculture. Where subsidies push profit margins well above conventional production, which appears to have been the case in many countries for milk and red meat (Hamm and Gronefeld, 2004), then a substantial supply response results. The increase in production may also lead to increased efficiency of processing and distribution and reduced 'waste' at the retail stage which contribute to increased consumer participation in the organic market.

It is possible in principle for a market to grow owing to the interaction of independent demand-led and supply-driven factors which, if balanced, allow consumption to increase at constant prices (or in the case of the organic market, what are usually referred to as price premiums over 'conventional' produce, see later). In addition, there is a particularly interesting way in which supply-driven market growth can lead to an increase in consumption. If a particular product simply becomes more available, consumers become more aware of its existence, qualities and attributes, and more is purchased without the incentive of cheaper prices or any fundamental change in consumer attitudes, tastes and preferences for the product. Increased availability has also involved a greater organic product range so that the introduction of a complementary organic product may increase the consumption of another.

Ritson (1993) argues that there was an element of this in the rapid growth in consumption of farmed salmon during the late 1980s and early 1990s. The new, sophisticated, technology of salmon farming reduced costs. Salmon farming was profitable at prevailing prices, attracted substantial investment and supply increased rapidly. But this coincided with a period in which the product itself was an ideal match for a demand trend towards more 'healthy' and convenient products. In addition, though, a product that had previously been mostly associated with restaurant meals became increasingly available in supermarkets, product awareness increased and more was purchased. Marketing specialists at the time indicated that the most important reason for increased consumption was 'availability'. The conditions were therefore in place for a rapid expansion of the market at constant prices. Quite suddenly, though, the 'supply drive' overtook the demand pull and prices collapsed, that is, the continual increase in supply could only be converted into an equivalent growth in consumption by falling prices.

Many of the same forces appear to have been present in the case of the recent rapid rise in the consumption of organic food. To oversimplify a little, it is probably the case that the growth in the organic market in most European countries will have displayed an early, supply-driven period, followed by a demand-led phase, with supply growth again taking over more recently. The substitution of a supply-driven growth by a demand-led one is illustrated by the Danish experience. In the early 1990s in Denmark, many farmers who converted to organic did so because of the personal values that they held about nature and the environment. By the end of the decade this had been replaced by the market-orientated lure of the organic price premiums prevailing, with many quoting 'higher incomes' as their reason for conversion as well as the presence of an institutional framework developed to support organic farming (Kaltoft and Risgaard, 2005).

Similarly, the rapid growth in the UK at the end of the 20th century appears to have been demand led. The best evidence of this is the degree to which the UK was reliant on imports for its organic supplies (See Table 6.1). Domestic production was failing to meet the growth in demand and high price premiums were pulling in imports. The UK House of Commons Agriculture Committee observed: 'The expansion in organic production is racing to keep up with the growth in customer demand' (House of Commons, 2001). At the same time, however, a German academic commented: '...the market growth during the last years has not primarily been driven by the demand side, but was mainly caused by activities on the supply side.' (Alvenslaben, 2001).

The main explanation for this was that generous government incentives for conversion to organic were stimulating the increase in production in some countries. This was evidenced by the proportion of organic production which was sold into conventional marketing channels (that is, not sold to consumers as being organically produced). This is illustrated in Table 6.2. It should be noted though that the gap between the production and consumption of organic food may also reflect a failure in the institutions governing the processing, transport and retailing of organics, that is structural failures in the link between producers and consumers.

It is important to point out here the significant characteristic of organically produced food that distinguishes it from conventionally produced food and

Table 6.1 UK self sufficiency (%)

Product	2000	2001	Product	2000	2001
Beef	77	60	Cereals	19	28
Sheep	97	94	Potatoes	63	66
Pork	66	34	Vegetables	40	55
Poultry	46	67	Fruit	16	4
Eggs	93	90	Milk	80	97

Source: based on data in Hamm and Gronefeld (2004).

Table 6.2 Share of organic production sold as organic in the EU: (2001)

Animal products (%)		Vegetable products (%)	
Milk	68	Cereals	93
Beef	69	Oilseeds	91
Sheep	54	Olive oil	73
Pork	94	Potatoes	96
Poultry	99	Vegetables	95
Eggs	97	Fruit	84
		Wine	61

Source: based on data in Hamm and Gronefeld (2004).

which raises complications for market analysis. The European Action Plan for Organic Food and Farming (SEC2004 739) argues that the environmental benefits of organic farming are important public goods and should therefore be financed by public means:

> Both roles of organic farming contribute to the income for farmers… In order to achieve the objectives of consumers, producers and the general public, organic farming should develop a balanced approach to these societal roles. It should offer a fair and long-term support for public goods, and at the same time foster the development of a stable market (European Commission, 2004).

The Action Plan notes, for example, that in Sweden farmers are encouraged to produce organic for its public good attributes even though they sell into a conventional food chain. The private benefits are reaped by consumers who have organic foods available to them, but private benefits should be subject to market rules. Given that any organic product embodies both these benefits, analysis of the market becomes very complex indeed.

In summary, it is clear that a substantial part of the increase in consumption of organic products has been demand led, the consequence of a positive shift in consumer attitudes to organically produced food. But part has also been supply driven, with consumer reaction to more competitive prices and increasing availability the main vehicle for increasing consumption. It is to these two features of organic consumer behaviour that we now turn.

6.3 Factors influencing organic purchase

In the previous section we examined the development of the European market for organic agricultural products and the inter-relationship between supply and demand factors within the market. In this section we draw on current work being carried out on organic and low input food supply chains in Europe (http://www.qlif.org/). We examine more closely the consumer of organic foods and ask: what factors influence consumers to choose organic rather than conventional foods? This is: '... a potentially complex task in which many different aspects might need to be considered. Health, environmental concern, ethics, authenticity and taste, and concerns about the relations between people and nature are examples of broad themes that recur in the literature' (Torjusen *et al.*, 2004).

There are a large number of studies of European organic consumers but it is difficult to generalise the findings across countries or to untangle the complex interconnections because of the different methodologies and conceptual models that have been used. As Ritson and Kuznesof (2006) note in their study of food consumption and alternative production technologies, a number of models of food choice have emerged drawing on contributions from different academic disciplines and including: economic factors, sensory aspects of eating, perceptions relating to health, nutrition and well-being, lifestyle factors and beliefs about production technologies. Approaches to the study of food consumers can be split broadly into those taking a cognitive or behavioural approach and those with a more socially or culturally determined view of behaviour (see Torjusen *et al.*, 2004).

Cognitive or behavioural approaches depend on psychological models that explore the consumers' knowledge and perceptions of the characteristics of the food in relation to the needs that they are trying to satisfy through their purchase. Within this approach, differing emphases are placed upon the consumers' values, beliefs and attitudes, their intentions to act and their actions.

Social and cultural studies of organic food, on the other hand, emphasise the many symbolic meanings of food and the activities surrounding its purchase, preparation and consumption. Both approaches show consumers concerned with quality and safety aspects of organic food but these concerns are constructed in different ways. As Midmore *et al.* (2005) note:

> From the point of view of the organic consumer, 'organic' implies 'quality' in itself, and support for organic agriculture and 'safe' food-processing techniques. The use of wholesome, unadulterated ingredients contributes not only to the individual good, in terms of healthy eating, but also to broader environmental and social goals, which benefit the community as a whole through fundamentally sustainable and 'caring' production methods.

Just as in the case of conventional foods, differences in organic food choices are found according to socioeconomic and demographic factors. Other significant factors affecting patterns of organic food choice include whether consumers are traditional, heavy, medium or light consumers of organic foods. There is also an interesting pattern emerging that shows that consumer participation in organic food markets changes across the life cycle, for example families with children in the 15–20 age group living at home having lower consumption than those with younger children.

The proportion of organic foods sold through different supply chains and retail channels differs between countries and according to the overall level of participation in organic food. For example, over 80% of organic food purchased in Britain, Denmark, Finland and Sweden is from supermarkets, whereas in Belgium, Germany, Spain, Greece, Portugal and Norway the majority of organic food sales is through organic or whole food shops or through direct sales from farmers. Also, in the UK and some other European countries, most 'medium' and 'light' users make purchases mainly from the large supermarket chains, whereas 'heavy' users of organic food frequently buy at least a proportion of their organic foods though alternative, small scale retail channels (Hamm and Gronefeld, 2004). Table 6.3 provides an overview of the reasons that consumers have given for purchasing organic foods across a number of European countries.

Clearly, the table does show some general patterns emerging across the eight countries. (For details of consumer organic preferences based on laddering interviews and focus groups across a range of product categories, see Zanoli *et al.*, 2004 and for an additional review of Denmark, Italy, UK and Hungary, see Torjusen *et al.* 2004). In all of the countries studied, health, either for self and/or family, appears as an important factor for consumers. Health benefits may be associated with the idea of (a) fewer additives and chemicals in the food that it is produced 'naturally' and (b) healthy eating, which in turn helps to avoid health problems. Furthermore, there is a strong association between health, well-being and quality of life in general. Health associations derive not only from what is absent from the food, but also the belief that it contains higher nutrient values. This view is particularly true for fruit, vegetable and cereal products, which it is believed contain more vitamins and minerals, and contribute to a more wholesome meal.

The 'health' attribute of organic food is not just a reflection of a positive 'pull' feature of organic products, but reflects a 'push' from the conventional food market. Consumers have become anxious, following food scares, such as those associated with BSE and *Salmonella*, and uncertain about the effects of novel technologies such as genetic engineering of food, thus increasingly seeking 'safe' food. Purchasing organic food is perceived as one of the ways of dealing with the anxieties associated with conventional food production and processing systems (Alvenslaben, 2001).

This feature of the organic food consumer – the belief that organic products lack the negative credence attributes associated with conventional agriculture and food production, is illustrated by Tables 6.4 and 6.5. In Germany 2000 consumers were asked what they most associated with bio (organic) products. The various responses are shown in Table 6.4; the responses are ranked according to the most frequently mentioned association.

Second, in a survey of 1000 British consumers, respondents were asked 'how worried' they were about a series of potential food safety issues previously identified from focus groups as things which concerned consumers about food consumption. In Table 6.5 the 'worries' are ranked according to the percentage of respondents who said they were either 'very' or 'extremely' worried about the food safety issue. For example, 60% were very or extremely

Table 6.3 Reasons why consumers purchase organic foods

Country	Reasons for buying organic food			Reasons for not buying organic food
Austria	Own health (improvement, avoidance of risk)	Responsibility for children	Contribution to regional development	Price, habit, mistrust and lack of motivation, poor availability and product range
Switzerland	Better taste	Health, especially for mothers and people with illness	Altruistic motives; environment, animal husbandry, remuneration of farmers	Price, low perception of difference between organic and conventional production, mistrust in organic standards
Germany	Own or children's health (avoidance of harmful ingredients)	Support of organic shops and farmers in their aspiration	Occasional consumers mention taste as a buying motive more often than regular consumers	Price, poor availability, shopping habit doubts about quality, lack of interest, taste
Denmark	Motives reflect a lifestyle choice: environmental protection	Own health	Support of, and contribution to, a better world	Poor quality, no perceptible difference between organic and conventional food
Finland	Motives reflect a lifestyle choice: environmental protection	Health (products are pure and contain no residues)	Conscience	Price conscious consumer affected by unreliable quality
France	Healthy nutrition (healthy, nutritious, unaltered)	Taste	Respect for the living world	Lack of information, large number of different labels
Italy	Health (safety)	Taste		Availability, lack of trust in standards, product quality, price for regular users
UK	Own health (no chemicals, purity)	Local farming and fair trade	Environmental protection	Product related (price, appearance, availability, quality, variety, taste) information about the product (confusion, habit, trust, information)

Adapted from Zanoli *et al.* (2004), various sources.

Table 6.4 Ideas that German consumers associate with the stimulus 'bio-products'

Association			
1.	Without chemicals	10.	Expensive
2.	Natural products	11.	No pesticides
3.	Without artificial fertiliser	12.	Controlled farming
4.	Biological' farming'	13.	Not containing noxious agents
5.	Healthy	14.	Not genetically modified
6.	'Ecological' farming	15.	Natural manure
7.	Caring animal husbandry	16.	Free range animals
8.	Not sprayed	17.	Negative associations
9.	Environmentally friendly		

Source: based on data in Alvenslaben, 2001.

Table 6.5 UK public concerns about food

Concern	
1.	Use of hormones in animal production
2.	Use of antibiotics in food production
3.	Use of pesticides in food production
4.	Animal welfare standards in food production
5.	Eating genetically modified food
6.	Safety of meat products produced by intensive farming methods
7.	Use of additives in food
8.	Quality of food using intensive farming methods
9.	Conflicting information on food safety
10.	Lack of information about food from government
11.	Hygiene standards in the food industry
12.	Hygiene standards in restaurants and take-aways
13.	Being able to afford good quality food
14.	Amount of fat in your diet
15.	Information about what foods are good for you keeps changing
16.	Knowing what to do when there is a food scare
17.	Getting food poisoning
18.	Hygiene standards in your home

Source: Miles *et al.* (2004).

worried about the use of hormones, but only 12% about hygiene standards in the home. The striking observation is that many of the features of food consumption which seem to cause most concern to consumers – pesticides, hormones, antibiotics, additives, intensive farming and poor animal welfare – represent negative characteristics thought to be absent from organic products (without chemicals, without artificial fertilisers, no pesticides, not sprayed, caring animal husbandry).

A second frequently mentioned characteristic of organic food is the taste. Taste is a sensory or organoleptic attribute of food that may be experienced directly by the consumer and may be compared directly to the physical aspects of other foods. Controlled blind sensory comparisons of conventional and organic foods of organic foods have shown no consistent difference for a range of product categories in the taste of organic and conventional foods (Fillion and Arazi, 2002). However, specific foods may have a measurably better or different taste when organic and this is thought to be due to differences in the agronomic and/or the processing methods used. As Midmore *et al.*, (2005) point out, these physical factors can be viewed as an effect of the organic production process as well as hedonistic, or pleasure giving, characteristics (see Johansson *et al.*, 1999). Taste is very subjective and positive feelings about taste tend to be linked to the authenticity of the organic product. Consumers describe the taste as being 'real' or 'genuine' and as in the case of health there is an association with 'naturalness'. However, taste is also given as a reason for non-purchase of organic foods where there may be no discernable difference in taste between conventional and organic foods or where the freshness or 'look' of the food suggests that it will not taste good.

A range of ethical issues are given for the consumption of organic foods including animal welfare – natural rearing, humane slaughter techniques, protection of the environment, fair trade, local production and the reduction of food miles, as well as broader economic and social impacts. An interesting aspect of this 'ethical' group of issues is the number of characteristics that are not (i) regulated through European Organic Standards and/or (ii) assured through the organic certification system for organic production and processing, but are nevertheless associated with organic by the consumer. These factors include, for example, fair trade, food miles, small-scale production, origin labels, regional images and so on. These issues are complicated by the number of different standards to which organic foods are produced and labels under which organic foods are marketed – some have additional rules or restrictions to address additional characteristics (e.g. with respect to animal welfare or fair trade), thus being more inclusive of broader social values than other labels. Further systematic work is required on the ways in which the differing cultural values across Europe may relate to different ethical concerns.

Tables 6.3 and 6.4 illustrate clearly that many of the reasons that consumers offer for purchasing organic foods are 'non-sensory' positive credence characteristics. Even after eating the product, the consumer may not be certain that it was organic. Thus the degree of knowledge on the part of the consumer, and the amount of information on the production and processing techniques that the food has undergone, play an important role in the decision to purchase. It is for this reason that the certification and labelling of food plays such an important part in organic food choice. It is not surprising then that 'trust' or 'lack of trust' is mentioned frequently as a feature of the purchasing decision. Trust illustrates well the differences that cognitive/ behavioural and socio/cultural theories offer in understanding and analysing organic food choices. Whereas in the former, lack of trust is seen as being something that may be remedied by education and the provision of information, in the latter it may be regarded as a positive communication within the development of the food system (see Kjærns 2003 quoted in Torjusen, 2004). Many of the credence characteristics associated with the positive decision to purchase organic food are not required by the formal EU regulations governing its production and processing. For example, 'small scale' and 'local' are not organic attributes, but are valued by organic consumers and associated by them with organic agriculture.

Whereas many of the positive characteristics associated with the decision to buy organic foods are credence characteristics, many of those given for not purchasing are more directly 'experience' characteristics (Zanoli, 2004; Torjusen *et al.*, 2004). For example, price, poor availability, limited product range, too many different labels, poor taste. The implications of this are discussed in more detail below.

The attributes that consumers attach to organic food can be split into those with private use values, such as health, taste and freshness, and public use values, for example, animal welfare, environmental conservation. Private use values benefit the individual, whereas public good values are shared by everyone and consumption by any one person does not exclude consumption by another. Historically, the decision to buy organic produce has been associated more strongly with public use values, in particular a concern for the environment. However, more recently, as consumers have become more concerned with food safety and health, the significance of private values has increased. It is interesting to note therefore that all the reasons given for *not* buying organic described above are private use values. The public good attributes of organic foods are not mentioned. Is it then the private good aspects of organic foods that fail to satisfy consumer needs?

Work being carried out in Denmark is revealing an interesting and dynamic relationship between the values that consumers hold with respect to organic foods and their purchasing patterns (Wier *et al.*, 2005). Denmark has the highest consumption per capita of organic food in the EU and government support has emphasised the public good aspects of organic farming. Wier *et al.* show that public good attributes are widely acknowledged by the Danish respondents in the study and that over one-quarter would be willing to pay extra taxes to support the future of organic farming. However, the study of household purchasing behaviour shows that, although public-good values appear to be a pre-requisite for purchasing organic foods, it is private good attributes that determine the actual degree to which purchases take place.

The paradox generated by these findings is further illustrated by a comparative study of Denmark and Britain, a country with the fastest growth of organic food consumption in Europe (Weir *et al.*, 2005). Demand in both markets is shown to be sustained primarily by the private good attributes, that is health and safety, of organic foods. However, in both countries, much of the organic food is produced and handled in concentrated and industrialised sectors characteristic of the conventional food systems that consumers are trying to avoid. These results indicate that there is still much to learn about the development of the organic food market and what it is about it that affects consumer choice.

6.4 The price premium

One of the most fundamental and durable things that we know about food consumers (Ritson and Petrovici, 2001) is that, in almost all cases, when price falls (and nothing else changes) more is purchased. The increase in consumption will be a combination of existing consumers increasing their frequency/amount of consumption and lower prices inducing new consumers of the product. Almost all of the increase in consumption will involve substitution. In the case of organic produce, most of the substitution is likely to be between the organic food product and a similar 'conventionally produced' one. This in turn has led to much of the debate over organic prices being characterised in the form of the 'organic price premium' – the percentage excess of the organic product price over the conventionally produced alternative. Table 6.6 provides organic price premium estimates from different sources for selected products in selected EU member states. It is tempting to conclude that these, often substantial, organic price premiums provide us with monetary estimates of the value consumers attach to the attributes of organic products relative to conventionally produced ones. Up to a point, this is correct, but subject to a number of important qualifications.

First, the price premium may reflect attribute differences other than those specifically associated with organic production. It is clearly difficult in many cases to establish a conventional benchmark when a range of quality exists for the conventional product, for example, for chicken and eggs, there are

Table 6.6 Consumer price premiums (%) for organic products in selected EU member states (2002)

Product	Denmark	UK	France	Germany	Italy	EU (15) average
Bread	13	34	106	25	38	41
Potatoes	56	128	140	83	101	94
Tomatoes	85	118	126	110	68	102
Apples	36	53	41	98	53	75
Milk	19	38	64	42	117	50
Yoghurt	8	8	85	25	63	37
Eggs	17	56	51	53	121	54
Chicken	91	138	64	181	107	129
Steak	46	70	56	74	70	59

Source: based on data in Hamm and Gronefeld (2004).

value, standard, corn fed, free range, farm assured and extra quality products. Hill and Lynchehaum (2002) claim that 'the primary reason that organic is more expensive is simply because it is good quality food. Organic is not expensive when compared to other quality foods'. They are perhaps oversimplifying the reason for an organic price premium, but this view does underline the importance of taking into account the range of quality available from conventional agriculture.

Second, the pricing policy adopted by retail stores may incorporate elements of consumer value additional to those tied to organic production. In a study of retail store pricing policies for organic fruit and vegetables in France, Germany, Spain and the UK, La Via and Nucifora (2002) found that the organic price premium had a basic component of about 40% which appeared to reflect the higher production and marketing costs of organic products. Beyond that, there could be an additional mark-up of up to another 40%, reflecting store location and additional services and display information associated with the organic sections of the store.

Third, the price premiums are unstable over time and very sensitive to changes in the balance of supply and demand in a dynamic market. As indicated earlier, the time taken for farms to convert to organic can lead the growth in demand to outstrip supply and price premiums to rise. Equally, when produce from the converted land comes onto the market, supply can overshoot demand and price premiums are squeezed (and some organic produce goes into conventional marketing channels). In a more stable market one would expect the price premium to reflect the extra organic production and marketing costs and the fact that supermarket pricing policies may take higher margins at the high cost/low turnover end of a product range (e.g. organic chickens).

Clearly these fluctuations in price premiums do not in themselves indicate major shifts in the value attached by consumers to organic attributes. Rather consumers with lower valuations are varying their levels of consumption, or coming in and out of the market, in response to varying price premiums. The growth in organic consumption has typically been associated with the view that it indicated 'extra' consumers joining the existing market. However, analysis of panel data in Denmark (Wier *et al.*, 2005) clearly shows existing consumers 'dropping out' of organic consumption, as well as new consumers joining. Thus, what the price premium indicates is the lower boundary of consumer valuation, at the prevailing levels of consumption. It does not tell us the extent to which 'committed consumers' value organic attributes – what they would be 'willing to pay'. There are three ways in which to explore this issue: analysis of the impact of changing prices on purchases; 'stated preference' interviews; and 'choice experiments'. In the case of the first of these – price analysis – statistical estimates can be made of the degree to which consumption rises or falls in response to a price increase or decrease. Ideally, this requires substantial price changes to have been experienced, but it is possible to infer what would be likely to happen at prices above those actually experienced. Broadly speaking, the less sensitive is the level of organic consumption in response to a price increase, then the more this indicates the presence of 'committed consumers', valuing organic attributes at more than the prevailing price premium.

Wier and Smed (2000) use data for 2000 Danish households in 1997–98 to estimate the response of organic consumption to price changes. Their results suggest that the demand for organic products is much more sensitive to price than the demand for conventional food. More formally, the price elasticity of demand – the percentage change in consumption for a given percentage change in price – is greater for organic products.

This is to be expected if variations in price premiums cause consumers to switch between organic and conventional consumption. For example, the price premium could increase either because of a rise in organic prices or a fall in conventional prices. If this increase in premiums leads to, say, 10% less organic consumption, then the corresponding increase in conventional consumption will be much lower when expressed as a percentage of total non-organic consumption. Wier and Smed estimated that a decrease of 20% in the organic price premium would increase the market share of organic consumption for dairy products from 10 to 15%, for bread and cereals from 5 to 7% and for meat from 1 to 2%.

Another way of exploring the sensitivity of organic consumption to organic prices, relative to benchmark conventionally produced product prices, is by the concept of cross-price elasticity of demand – the percentage change in purchases of one product relative to a given percentage change in the price of another product. If food products are substitutes for each other, that is consumers are willing to switch consumption in response to changes in relative prices, then one would expect a rise in the price of one product to lead to an increase in the consumption of the substitute product. This is shown in Table 6.7 for organically and conventionally produced dairy products in Denmark.

The estimates are intuitively very credible. They suggest, for example, that a 10% increase in the price of organic dairy products (with no change in conventional prices) would decrease organic consumption by 22.7% and increase consumption of conventional products by 1.3%. However, an increase

Table 6.7 Own and cross price elasticities of demand for organic and conventionally produced dairy products in Denmark

	Price	
Quantity	Organic	Conventional
Organic	–2.27	1.27
Conventional	0.13	–1.13

Source: Wier *et al.* (2001).

in the price of conventional products by 10% (with no change in organic prices and thus narrowing the price premium) would decrease the consumption of conventional products by 11.3% and increase organic consumption by 12.7%. Again, the fact that organic consumption seems to be much more sensitive to changes in conventional prices, than conventional consumption is to organic prices, reflects the much lower market share (about 10% in this case) for organic dairy products.

Table 6.8 provides a more comprehensive set of estimates of own-price elasticities of demand for organic products, compared to the conventional equivalent, this time for the UK. In all cases the response of consumption of organic products to change in price is about double that for conventional produce.

The second way of attempting to estimate the extent to which consumers value organic attributes, in the sense of being willing to pay a price premium, is simply to ask them. Wier and Calverley (2002) report on a range of such studies and a synthesis of these results for the studies carried out in Sweden, Norway, Denmark and the UK, are presented in Table 6.9.

Table 6.8 Own price elasticities of demand for organic and conventional products in the UK

	Conventional	Organic
Dairy	–0.57	–1.14
Milk	–0.76	–1.54
Eggs	–0.26	–0.52
Cheese	–0.34	–0.67
Meat	–0.95	–1.91
Beef	–1.64	–3.28
Lamb	–0.52	–1.05
Pork	–1.87	–3.74
Chicken	–1.37	–2.75
Vegetables	–0.31	–0.62
Processed vegetables	–0.67	–1.34
Potatoes	–0.21	–0.43
Green vegetables	–0.47	–0.94
Fruit	–0.21	–0.43
Bananas	–1.31	–2.62
Apples	–0.49	–0.97
Citrus fruits	–0.46	–0.92

Source: ADAS (2004).

Table 6.9 European consumers willingness to pay for organic food

Price premium (%)	5–10	10–20	20–30	30–40	40–50	50–60
Proportion of consumers willing to buy (%)	45–80	20–50	10–25	5–20	3–18	3–15

Source: based on data collected by Wier and Calverley (2002).

The two notable features of the table are, first, the substantial proportion of consumers willing to purchase organic at low price premiums – in general much lower premiums than those shown as prevailing in Table 6.6. Second is that the proportion indicating a willingness to pay seems to level off, once premiums exceed 30%, suggesting a core of 'committed' organic consumers willing to pay the higher price premiums that tend to apply. Wier and Calverley also note that studies in the Netherlands and Germany indicated higher proportions of consumers willing to pay high price premiums.

The strength of the 'willingness to pay' approach is that it allows insight into potential consumer behaviour lying outside the range of price premiums provided by market data. There are, however, serious doubts over the reliability of individual consumers' own estimates of how they would behave in different market circumstances, and the willingness to pay technique (sometimes described as 'contingent valuation' or 'stated preference') is more commonly used to attempt to value goods for which markets do not exist, in particular environmental goods.

Some experts argue that, for products which do possess markets, a more reliable method for predicting consumer behaviour outside the range of observed market prices is by 'choice experiments', in which consumers use real money and real products under laboratory conditions. Soler and Gil (2002) used an experimental auction market to attempt to elicit consumer willingness to buy organic olive oil. They did this first on its own, and then, as a reference point, provided information relating to prevailing prices of conventional olive oil. They found that only 5% of consumers were willing to pay the prevailing organic price premium in Spain, but that up to 70% were willing to pay some premium.

6.5 Conclusions

The past 10 to 15 years have seen rapid growth in the consumption of organic products in Europe and North America. Not all of this consumption growth can however be attributed to a fundamental shift in consumer attitudes towards organic products. Part of the growth in consumption has been supply driven by government support and because of the environmental goals that lead some producers to convert to organic production. Many studies of organic consumption indicate that health reasons underpin much of the consumer motivation to purchase organic. The 'health' attribute of organic food is not just a reflection of positive 'pull' factors such as perceived higher nutrient value, but also 'push' factors associated with the absence of negative associations of conventionally produced food.

For markets to function efficiently, it is assumed that buyers and sellers have complete information. However, the organic food market shows not just that consumers lack information, but in some cases that they have misconceptions about certain positive or negative characteristics of organic foods. This may be particularly true for 'occasional' consumers who will be significant in the further development of the organic market. Many of the positive attributes associated with organic foods are strongly linked to attributes associated with other 'alternative' production and food handling systems, for example, fair trade, small scale, local production, low travel miles and so on. The reasons that consumers give for purchasing organic foods reflect their beliefs about the characteristics of organic foods. There does not appear to be a great level of knowledge of the organic principles and standards and certification systems used to control production and processing. Consumers need to be better informed about the production and processing methods used in organic production. Further evidence for sensory and composition differences between conventionally produced and organic foods and their demonstrated and potential impact on the environment, human health, animal welfare and sustainability of food production are required. Until this knowledge is produced and available to consumers, it is difficult to be sure about the longevity of current perceptions, attitudes and demand levels.

Similarly, price analysis suggests that organic consumption appears to be very sensitive to changes in the price premium over conventional produce with, however, a core of 'committed consumers' willing to pay the substantial premiums which usually prevail. However, there is a much larger pool of potential consumers at more modest price premiums. The analysis of the organic food market is complicated, however, by the fact that organic foods embody both public and private goods and these are intimately bound together. Whereas it may be argued that private goods should be subject to market institutions, the situation is more complex for public goods. Subsidy of the public good – the environmental benefits of organic farming – is thus also a subsidy of the private good. Moreover, emerging evidence seems to suggest that, although consumers list the environment as a significant and positive attribute of organic food, their market behaviour seems to indicate that it is the private good attributes – health and taste – that determine purchase. Public good characteristics may form a necessary but not sufficient condition for private action.

6.6 References

ADAS (2004) *Evidence Assessment to Inform the Review of the Organic Farming Scheme – November 2003,* http://statistics.defra.gov.uk/esg/evaluation/ofs/

Alvensleben R von (2001) 'Beliefs associated with agricultural production methods', in Frewer L, Risuik E and Schiferstein H, *Food People and Society*, Springer Verlag, Berlin-Heidelberg.

European Commission (2004) *European Action Plan for Organic Food and Farming*, Commission Staff Working Document SEC (2004) 739.

Fillion, L and Arazi S (2002) 'Does organic food taste better? A claim substantiation approach', *Nutrition and Food Science*, **32** (4), 153–157.

Hamm, U and Gronefeld F (2004) *The European Market for Organic Food: Revised and*

Updated Analysis, *Organic Marketing Initiatives and Rural Development Vol.5,* The University of Wales, Aberystwyth.
Hill H and Lynchehaum F (2002) 'Organic milk: attitudes and consumption patterns', *British Food Journal*, **104**, (7), 526–542.
House of Commons (2001) *Select Committee Report on Organic Farming*, HMSO, London.
Johansson L, Haglund Å, Berglund L, Lea P and Risvik E (1999) 'Preference for tomatoes, affected by sensory attributes and information about growth conditions', *Food Quality and Preference*, **10**, 289–298.
Kaltoft P (1999) 'Values about nature in organic farming practice and knowledge', *Sociologia Ruralis*, **39** (1), 39–53.
Kaltoft P and Risgaard M-L (2005) 'Has organic farming modernized itself out of business? – Reverting to conventional methods in Denmark', in Reed M and Holt G, *Sociological Perspectives of Organic Agriculture: From Pioneer to Policy*, chapter 8. CABI, Wallingford, UK.
Lampkin W and Measures M (2001) *Organic Farm Management Handbook,* University of Wales, Aberystwyth.
La Via G and Nucifora AMD (2002) 'The determinants of the price mark-up for organic fruit and vegetables in the European Union', *British Food Journal*, **104** (3/4/5), 319–336.
Midmore P, Naspetti S, Sherwood A-M, Vairo D, Wier M and Zanoli R (2005) *Consumer Attitudes to Quality and Safety of Organic and Low Input Foods: a review*, September. EU Integrated Project 506358, Quality Low Input Food, p 67.
Miles S, Brennan M, Kuznesof S, Ness M, Ritson C and Frewer L J (2004), 'Public worry about specific food safety issues', *British Food Journal*, **106**, (1) 9–22.
Mintel (2003) *Organic Foods – UK– November 2003*, Mintel International Group Ltd.
Michelsen J. (2001) 'Recent development and political acceptance of organic farming in Europe', *Sociologia Ruralis*, **41** (1) on Politics, Ideology and Practice of Organic Farming, 3–20.
Padel S (2001) 'Conversion to organic farming: a typical example of the diffusion of an innovation?' *Sociologia Ruralis*, **41** (1) on Politics, Ideology and Practice of Organic Farming, 40–61.
Recke G, Willer H, Lampkin N and Vaughan A (2004) 'Development of a European information system for organic markets – improving the scope and quality of statistical data', *Proceedings of the 1st EISfOM* (European Information System for Organic Markets) Seminar, Berlin, Germany, 26–27 April 2004.
Reed M. (2001) 'Fight the future! How the contemporary campaigns of the UK organic movement have arisen from their composting of the past', *Sociologia Ruralis*, **41** (1) on Politics, Ideology and Practice of Organic Farming, 131–145.
Richter T and Padel R (2005) 'The European market for organic foods', in Willer H and Yussefi M, *The World of Organic Agriculture Statistics and Emerging Trends 2005*, IFOAM, Berlin.
Richter T, Padel S, Lowman S and Jansen B (2006) 'The European market for organic food 2004/2005', in Willer H and Yussefi M, *The World of Organic Agriculture: Statistics and Emerging Trends 2006*, IFOAM, Germany.
Ritson C. (1977) *Agricultural Economics Principles and Policy,* Collins, London.
Ritson C (1993) *The Behaviour of the Farmed Salmon Market in Europe: A Review*, Centre for Rural Economy, University of Newcastle upon Tyne.
Ritson C and Kuznesov S (2006) 'Food consumption, risk perception and alternative production technologies', in Eilenbery J and Hokkanen HMT, *An Ecological and Societal Approach to Biological Control*, Springer, Dordrecht Chapter 3.
Ritson C and Mai LW (1998) 'The economics of food safety', *Nutrition and Food Science*, (5) 253–259.
Ritson C and Oughton, E (2006) 'Food consumers and organic agriculture', in Frewer L and Van Trijp H *Understanding Consumers of Food Products*, Woodhead, Cambridge, 254–272.

Ritson C and Petrovici D (2001) 'The economics of food choice: is price important?' in Frewer L, Risuik E and Schifferstein H *Food People and Society*, Springer, Dordrecht.

Sahota A (2006) 'Overview of the global market for organic food and drink', in Willer H and Yussefi M *The World of Organic Agriculture: Statistics and Emerging Trends 2006*, IFOAM, Germany.

Soil Association (2005) *Organic Market Report 2005*, Bristol.

Soler F and Gil JM (2002) 'Consumers' acceptability of organic food in Spain: results from an experimental auction market', *British Food Journal*, **104**, (8), 670–687.

Torjusen H, Sangstad L, O'Doherty Jensen K and Kjærnes U (2004) *European Consumers' Conceptions of Organic Food: A Review of Available Research*, Professional Report no. 4-2004, National Institute for Consumer Research, Oslo, Norway.

Wier M and Calverley C (2002) 'Market potential for organic foods in Europe', *British Food Journal*, **104**, (1) 45–62.

Wier M and Smed S (2000) Reported in Wier and Calverley, 'Modeling demand for organic foods', paper presented at *The 13th International Scientific IFOAM Conference*, Basel, Switzerland.

Wier M Hansen and LG Smed S (2001) 'Explaining demand for organic foods', paper presented at *11th Annual EAERE Conference*, Southampton, UK, June 2001.

Wier M, Anderson LM and Millock K (2005) 'Information provision, consumer perceptions and values – the case of organic foods', forthcoming in Russell C and Karup S, *Environmental Information and Consumer Behaviour*, New Horizons in Environmental Economics Series, Edward Elgar, Cheltenham, UK.

Wier M, O'Doherty Jensen K, Andersen LM, Millock K and Rosenkuist L 'The character of demand in mature organic food markets', *Food Policy* (submitted).

Willer H and Yussefi M (eds.) (2005) *The World of Organic Agriculture Statistics and Emerging Trends 2005*, IFOAM, Bonn.

Zanoli R (ed) (2004) 'The European consumer and organic food', *Organic Marketing Initiatives and Rural Development, Vol 4,* University of Wales, Aberystwyth.

Part II

Organic livestock foods

7

Effects of organic and conventional feeding regimes and husbandry methods on the quality of milk and dairy products

Richard F. Weller, Christina L. Marley and Jon M. Moorby
Institute of Grassland and Environmental Research, UK

7.1 Introduction

This chapter outlines the main factors affecting the quality of milk that is produced for both the liquid and processing markets. These include the main quality parameters in the milk, including fatty acid composition, protein components, vitamins, somatic cell counts and their effect on the processing quality of the milk. The nutritional quality of milk is influenced by many factors and the inter-relationship between the type of management system, quality of the diet, breed of cow and both the nutritional quality and processing properties of milk is discussed in the chapter. The production of quality milk for both liquid consumption and processing is dependent on achieving the optimal balance between the available feed sources for inclusion in the diet and the genetic potential and requirements of the cow, with the objective of continually producing quality milk throughout the lactation. However, compared with both conventional intensive and low-input systems, where the composition of the diets is not restricted by the contribution of concentrates to the total ration, the organic standards for milk production require the feeding of high-forage diets with a minimum forage content of 60–90%. Therefore, there is less flexibility when organic diets are formulated and the quality of the diets is dependent on producing high-quality forage that meets a major part of the cow's nutrient requirements, particularly during the critical early and mid-lactation periods.

7.2 Quality parameters in dairy products

7.2.1 Fatty acid composition

The fatty acid composition of milk is an important factor that affects the physical properties during processing and has also been linked to health factors in the human population. Beneficial effects of milk for human consumption have been associated with the *n*-3 series of polyunsaturated fatty acid (PUFA) content and conjugated linoleic acids (CLA), including the prevention of carcinogenesis, a reduced incidence of heart disease and benefits for the immune system (Tricon *et al.*, 2004). As both grasses and legumes contain significant levels of PUFAs, the feeding of high-forage diets in both organic and conventional low-input systems has the potential to enhance the value of milk as an essential component in the human diet. When Dewhurst *et al.* (2003) fed dairy cows either grass or legume silages, they reported not only improved intakes and milk yields from the legume silages (lucerne, red clover, white clover), but also higher levels of PUFAs in milk, particularly α-linolenic acid, with the highest concentrations in the milk from cows fed red clover silage.

On many conventional and organic farms a high proportion of the annual milk production is from low-cost systems based on grazed rather than conserved forages. This leads to potential benefits as a number of studies have shown that changing from the feeding of diets based on conserved forages to the grazing of fresh herbage increases the CLA concentration in the milk (Chilliard *et al.*, 2001). Conserving crops as hay or silage leads to a reduction in the CLA content of forages and any changes in the CLA content of the conserved forages fed during the winter period of housing are likely to be smaller than the changes that have been found in grazed herbage during the growing season. On farms in a suitable location for growing forage maize, the ensiled crop provides a valuable energy feed for the early-lactation period and also has a higher CLA content than grass silages. The linoleic acid content of milk from cows fed maize silage is higher compared with the concentrations found in the milk from cows fed grass silage diets, with the total PUFA concentrations being similar. The CLA content of milk is also influenced by the breed of cow, with milk from Jersey cows having a lower concentration than milk from either Friesian or Holstein cows. Conjugated linoleic acids remain stable in dairy products and therefore the concentration in these products is directly correlated to the content found in raw milk.

7.2.2 Protein

The protein content of milk is primarily influenced by the breed of cow, the stage of lactation, type of diet being fed and the health status of the cow, and is important in processing because the protein (and specifically casein) content of milk determines its cheese yield. Milk provides a highly digestible source of protein for a large proportion of the world's population, either as raw milk or processed into dairy products. In addition to this basic nutrition, milk contains various compounds with physiological functions, ranging from immunoglobulins which are important in newborn mammals, to a number of bioactive molecules that are 'encrypted' within milk proteins and released during digestion and/or milk processing (Shah, 2000; Steijns, 2001). Casein phosphopeptide fragments, formed from the cleavage of casein by digestive enzymes, may complex with calcium and enhance its absorption from the small intestine and may also help reduce tooth decay (Aimutis, 2004). Casokinins, derived from caseins, and lactokinins, derived from whey proteins, are potent inhibitors of angiotension-I-converting enzyme, which is involved in blood pressure regulation. Human studies have shown that high blood pressure can be reduced by consuming milk protein hydrolysates and fermented dairy products (FitzGerald *et al.*, 2004). Opioid peptides (e.g. β-casomorphins, from β-casein) are peptide fragments of casein and have opium-like properties, which can slow gut motility and have analgesic and sedative effects in the bloodstream. Some casein-derived peptides (e.g. casoxins, from κ-casein) behave as opioid antagonists. Milk also contains small quantities of other proteins and polypeptides, such as immunoglobulins and lactoferrin, which have antimicrobial properties. The importance of the immunoglobulin content of colostrum in the transfer of disease immunity to newborn mammals is well known and immunoglobulins concentrated from bovine milk may have beneficial effects on human health, particularly with respect to intestinal infections.

Many conventional systems that aim to maximise lactation yields, feed diets with a high proportion of concentrates and with protein contents above those found on many low-input and organic farms. Milk protein concentration tends to decrease in the post-calving period and the lowest values are often recorded in the 35–70 days after calving followed by a steady increase during the remainder of the lactation. True milk protein constitutes about 95% of the total milk nitrogen and includes caseins (α, β, κ and γ), whey proteins, serum albumin and immunoglobulins. Animal breed has an important influence on milk protein concentration, with cows of the Holstein and Shorthorn dairy breeds often producing milk with total protein concentrations of <3.30% compared with >3.40% in the milk produced by Brown Swiss, Meuse Rhine Issel and Montbeliarde cows and >3.50% in the milk from Guernsey and Jersey cows. The proportion of casein in the total milk protein ranges from 76% to 86% (DePeters and Cant, 1992). Cows with mastitis produce milk with a lower total casein content, and with β- and α_{s1}-casein both decreased, but with increased proportions of κ-casein and other unidentified proteins (DePeters and Cant, 1992). As parity (the number of times a cow has calved) increases, the casein content of the milk decreases. While both milk protein and casein concentrations are influenced by changes in the energy density of the diet, the ratio of the different types of casein remains unaffected.

The concentration of milk protein is controlled by the rate of milk protein synthesis (milk protein yield) in relation to milk yield (milk volume). The rate of milk protein synthesis depends on the supply of protein precursors (amino acids) to the udder, while milk volume is largely dependent on the production of lactose and therefore the supply of glucose to the udder. Milk crude protein (nitrogen content × 6.38) concentration was found not to be related to diet protein concentration (National Research Council, 2001), although milk protein yield was found to be significantly related to feed dry matter intake (which influences milk yield) and the dietary concentrations of rumen degradable protein and rumen bypass protein. Milk protein concentration is frequently seen to increase as the intake of energy increases, although the form of the energy intake is important. Carbohydrates, particularly non-structural carbohydrates such as sugars, starches and fructans, can improve the growth rate of rumen microbes and the efficiency of use of nitrogen in the rumen, leading to higher yields of microbial protein from the rumen and hence an increase in the supply of amino acids to the dairy cow. Fats and oils also supply energy, although a fat content of greater than about 8% of feed dry matter can depress rumen fermentation and therefore reduce milk yield. Dietary fats can improve milk yield, but do not contribute to the synthesis of milk protein, and therefore increasing the proportion of fat in the dairy cow's diet can lead to increased milk yields and decreased milk protein concentrations through a dilution effect.

Ideally, milk protein concentration should be increased by enabling an increase in milk protein synthesis without decreasing milk yield or without increasing it at a rate that allows any extra protein production to be diluted. Forage legumes, which are widely used in all types of dairy production systems, tend to have a higher intake potential than grasses and are used successfully both fresh (grazed) and ensiled. Cows tend to eat more red clover silage than grass silage and therefore yield more milk (Thomas *et al.*, 1985), and cows offered lucerne silage can produce more milk and milk protein than those offered red clover silage (Broderick *et al.*, 2000). Although there has been a lot of recent interest in red clover silage, white clover has long been included as part of grass pastures used for grazing and ensiling. Rapid breakdown of the physical structure of white clover silage can lead to higher rates of intake, increased protein flows to the animal from the rumen and increased milk production. Polyphenol oxidase is an enzyme present in many plants and is found at relatively high concentrations in red clover. In the silo and also possibly in the rumen, this enzyme can help to reduce the breakdown of plant proteins by rumen microorganisms and increase the supply of plant proteins to the duodenum of the animal where they are digested and absorbed as amino acids.

Urea is the major fraction in the non-protein nitrogen content of milk and high values are indicators of either excess protein or energy deficiency in the diet. The urea concentration also influences the processing quality of milk and high values can increase the coagulation time and reduce curd firmness during cheese production. In conventional systems, high urea concentrations are often recorded in the milk during the spring period owing to the high rates of nitrogen fertiliser that are applied to ryegrass-dominant pastures.

7.2.3 Vitamins and minerals

Milk is an excellent source of calcium, phosphorus, riboflavin (vitamin B2), thiamine (vitamin B1) and vitamin B12, and a valuable source of folate, niacin, magnesium and zinc (Food Standards Agency, 2002). In particular, dairy products are an important source of calcium, which is vital for maintaining optimal bone health in humans (Prentice, 2004). The vitamins and minerals it provides are all bioavailable (i.e. available for absorption and use by the body) and thus milk consumption in humans increases the chances of achieving nutritional recommendations for daily vitamins and mineral intake (Bellew *et al.*, 2000).

The vitamin content in the dairy cow's diet is likely to vary throughout the year with differences recorded between different farming areas. A number of studies have shown grazed herbage has a higher content of both vitamins A and E than conserved forage, with both vitamin A and β-carotene declining in the herbage as the plants mature. In a review of a number of studies, Lucas *et al.* (2005), concluded that there are variations between geographical locations in the concentrations of both the fat-soluble (vitamins A, D E and β-carotene) and water-soluble fractions (vitamin B and minerals) that are found in dairy products. They suggested that these differences are likely to be due to a number of factors, including the milk being produced from different breeds, the effect of stage of lactation and the feeding of a range of diets. For example, milk from Guernsey cows had a higher β-carotene concentration than milk from Holstein cows.

In a comprehensive review, Underwood and Suttle (1999) reported that the total mineral concentration of milk did not change with increasing parity; however, there was a decline in mineral content as the lactation progressed. They concluded that the mineral concentration in milk varies, depending on the stage of lactation of the animal, and declines as the lactation progresses. Kirchgessner *et al.* (1967) reported that the effects of the diet on the composition of milk vary greatly for different minerals. Dietary deficiencies of calcium, phosphorus, sodium and iron are reflected in diminished yield of milk, but not in the concentrations of these minerals in the secreted milk. In contrast, when copper and iodine deficiencies occur there can be a marked fall in the quantity secreted in the milk. Studies on the iodine, molybdenum and selenium concentrations in milk found the level was correlated to the selenium content in the diet.

There have been relatively few studies comparing the mineral and vitamin composition of milk from different dairy systems. Toledo *et al.* (2002) compared milk samples from organic and conventional dairy systems in Sweden, for iodine and selenium, and found that levels of selenium but not iodine were lower in organic milk. Studies in Denmark compared the vitamin E concentrations in milk from organic and conventional systems and found that, despite less synthetic vitamin E being used in organic systems, the concentrations of vitamin E were higher in organic milk than in conventional (Nielsen *et al.*, 2004). Overall, reports comparing the content of various minerals and vitamins in milk from organic and conventional farming systems state that results comparing these systems must be treated with caution owing to the effects of factors (genetic variation and scales of production) which often differ between these systems (Woese *et al.*, 1997; Magkos *et al.*, 2003).

Milk from clinically mastitis infected, sub-clinically mastitis infected and healthy cows has been shown to differ in sodium, potassium, calcium, magnesium and phosphorus concentrations (El Zubeir *et al.*, 2005). The study reported that milk from cows infected with sub-clinical mastitis had significantly lower potassium and higher sodium and phosphorus concentrations. Similarly, the milk from cows with the clinical form of the disease had a higher sodium and lower potassium, magnesium and calcium concentration.

7.2.4 Somatic cell counts

Both clinical and sub-clinical mastitis are major health problems that occur regularly in many dairy herds, leading to increased somatic cell counts. Mastitis control is a more challenging problem in systems where the routine use of antibiotics is restricted and other alternative treatments are only partially effective. The occurrence of mastitis leads to an increase in cell counts, with the raised level often maintained when the cause of sub-clinical mastitis is *Staphylococcus aureus*. The somatic cell count of milk is a key factor in relation to both the processing properties and stability of dairy products, with leucocytes accounting for >90% of the total cells. High cell counts also reduce the financial viability of a dairy system as increasing cell counts are negatively correlated with milk yield, with the decline in production commencing when cells increase to >100,000/ml (Renau, 1986). In both conventional and organic systems the cell counts also change during lactation with the highest values recorded at both the beginning and end of lactation and the increasing lactation number of the dairy cow positively correlated to higher cell counts. Good dairy hygiene and the culling of cows with persistently high cell counts are essential practices in the maintenance of acceptable cell count levels in the bulk tank.

The increase in pH when the level of cell counts in the milk is raised affects not only the processing quality of the milk but also the stability of the dairy products (Munro *et al.*, 1984). In cheese production, high cell count milk leads to lower cheese yields, longer coagulation times, weaker curds and longer ripening times. As stated earlier, both the total casein and ratios between the individual casein types also change as the cell count increases. Butter from milk with high cell counts deteriorates faster when stored at 10°C, acid development of cream is slower and the flavour quality reduced, while in cultured products (e.g. yoghurt) the effect is correlated to the type of mastitis infection with *Staphylococcus aureus* having a particularly negative effect. Other milk components that change as the cell counts in the milk increase, include higher free fatty acid, whey protein, sodium and chlorine concentrations and a decrease in the calcium, potassium and lactose concentrations.

Based on treatment records, cases of mastitis in conventional dairy systems have been found to be lower than in organic systems in Germany (Sundrum, 2001) and the UK (Hovi and Roderick, 2000) but higher in Denmark (Bennedsgaard *et al.*, 2003), Norway (Ebbesvik and Loes, 1994) and Sweden (Hamilton *et al.*, 2002). Studies comparing somatic cell counts have shown that organic systems had lower counts than conventional systems in Sweden (Toledo *et al.*, 2002) and Denmark (Vaarst and Enevoldsen, 1994), had similar counts in Norway (Hardeng and Edge, 2001) and higher counts in the UK (Weller and Cooper, 1996).

7.2.5 Sensory quality

Sensory quality can be defined as texture, flavour (taste), aroma and visual aspect. The sensory properties of milk are highly influenced by its fat content (Phillips *et al.*, 1995a). As a result, research has examined the effects of various food additives on sensory quality when used as a substitute for fat in milk (Philips *et al.*, 1995b). Frøst *et al.* (2001) showed that a combination of thickener, whitener and cream aroma in 0.1% fat milk was successful in mimicking the sensory quality of 1.3% fat milk. With the interest in the production of milk enriched with *cis*-9, *trans*-11 CLAs, owing to their relevance to human health (Tricon *et al.*, 2004), recent research has examined the effects of CLA on the sensory quality of dairy products and found that it is possible to produce CLA-enriched dairy products with acceptable sensory characteristics (Jones *et al.*, 2005).

Milk from cows affected with mastitis alters the sensory quality of raw milk and cheese (Munro *et al.*, 1984). Sensory defects are reported as increased rancidity and bitterness, factors which are consistent with higher levels of lipolysis and proteolysis (Ma *et al.*, 2000).

Studies into the protected denomination of origin have led to a greater understanding of the impact of different livestock systems and pastures on the sensory qualities of milk and dairy products (Buchin *et al.*, 1999). Differences in sensory quality can be attributed to the volatile flavouring compounds (Bosset and Jeangros, 2000), the microbes present in the raw milk (Mariani and Battistotti, 1999) or the effects of one on the other (i.e. studies have shown that terpenoids modified the metabolic activity of microbes during cheese ripening (Bugaud *et al.*, 2002).

7.2.6 Shelf life and processing quality

Shelf life can be defined as the period of time that a product can be kept under practical storage conditions and still retain acceptable quality, as determined by sensory, chemical and microbiological analysis. Current consumption and marketing patterns require that most dairy processors manufacture milk that has a shelf life of 14 days or more (Duyvesteyn *et al.*, 2001). Whilst early microbial contamination of milk is affected by farm conditions and has a considerable affect on milk processing quality (Bertoni *et al.*, 2005), in general, the shelf life of milk is affected by contamination by spoilage bacteria that have the ability to grow at refrigerated temperatures (Elliot *et al.*, 1974). However, poor quality raw milk will result in defective products. Milk from cows with mastitis, which has high somatic cell counts (SCC), affects the shelf life of pasteurised fluid milk, mostly owing to an increase in free fatty acids (i.e. lipolysis) and casein hydrolysis (Ma *et al.*, 2000). High SCC milk also has adverse effects on cheese production, by decreasing cheese yield (Barbano *et al.*, 1991), reducing curd firmness and increasing casein and fat loss in whey (Politis *et al.*, 1988a,b).

The restricted shelf life of liquid milk continues to be a problem that is often more influenced by the type of milk being sold rather than the pasteurisation technique. The shelf life of processed milk is determined primarily by the quality of the raw milk from the dairy herd. Increasing cell counts in the milk and a higher concentration of free fatty acids, contribute to rancidity in both liquid milk and milk products. Janzen (1972) reported that the 0–14 day shelf life of pasteurised milk is influenced by the somatic cell concentration in the raw milk and found that after 14 days any observed changes in the flavour and stability of the milk were attributable to microbial activity during storage.

Legumes are an essential component of organic diets and Bertilsson *et al.* (2002) compared the effects of feeding red clover, white clover, lucerne and grass silages to dairy cows. They concluded that feeding legumes, particularly red clover, had a negative effect on the organoleptic quality of milk and Al-Mabruk *et al.* (2004) also found increased oxidative deterioration of milk produced from cows fed red clover silage. As the majority of organic diets are based on grass/legume mixtures, with legumes unlikely to provide more than 40% of the total annual forage input, the negative effect of the legumes on the stability of organic milk is likely to be lower compared with changes in the milk from cows fed 100% legume diets. Studies on the processing quality of milk from different breeds of dairy cows have shown that the processing qualities (clotting time, curd firmness) were better in milk from Jersey, Montbeliarde and Tarentaise cows compared with milk from either Friesian or Holstein cows. Furthermore, there is anecodotal evidence that milk from cows on herb-rich pastures has increased keeping qualities compared to milk from cows fed on predominantly ryegrass swards and some research indicates that milk from cows fed a basal diet including herbs had a higher anti-oxidative activity than milk from cows fed the basal diet only (Uegaki *et al.*, 2001).

Research is under way to investigate the potential to develop the natural antimicrobial system present in milk, such as the lysozyme, lactoferrin, immunoglobulins and vitamin binding proteins (Bertoni *et al.*, 2005) and, in particular, the lactoperoxidase system (Seifu *et al.*, 2005). This method activates a reaction between thiocyanate and hydrogen peroxide, which is catalysed by the enzyme lactoperoxidase and leads to the formation of antimicrobial compounds (de Wit and Hooydonk, 1996). The application of this method to extend the shelf life of milk will need to overcome the variation in lactoperoxidase activity in milk observed between individual animals and the many factors shown to affect it, including diet and stage of lactation (Fonteh *et al.*, 2002).

7.3 Factors affecting the nutritional quality of liquid milk and milk products

7.3.1 Production systems

The type of system that is established for the production of milk has a major influence on both the quality and quantity of milk produced by the dairy cow. Conventional systems have the flexibility to adjust both the quantity and type of concentrates being fed to match any adverse or beneficial changes in the availability and quality of forage grown on the farm. Although the standards for organic dairy production require the feeding of high-forage diets, there are different management options available for the farmer to meet the standards. Milk production costs are lowest when cows produce the major portion of their total annual yield from grazed grass/legume herbage, with the optimum production from herbage achieved by ensuring cows reach peak milk production when the grazing season commences. Cows calving in mid-summer when both the availability and quality of the herbage is lower produce either lower lactation yields per cow or require the herbage to be supplemented with concentrate feeds or high-quality forage to maintain both milk quality and yield.

In wetter western areas, where the winter period of housing can be up to six months, conserved silages are a major component of the diet. Here the challenge is to conserve high-quality forage to meet the cow's nutrient requirements, while ensuring sufficient forage is available for the winter period, with quality a key factor for any cows that are fed diets based on conserved forages during the first 15 weeks of lactation. As forages are generally markedly more variable in quality than concentrate feeds, intensive dairy systems feeding a high quantity of concentrates per cow (e.g. 2+ tonnes) have the potential to more readily meet the cow's nutrient requirements and maintain milk protein levels than low input and organic farms where concentrate inputs are limited by either the management strategy or organic feed standards. The stage of lactation has an influence on the quality of milk produced by the dairy cow and the month to month variation in milk quality can be minimised when the cows in a herd calve evenly throughout the year rather than in a specific 8–12 week period each year. On units where the milk is both produced and processed, the calving season is critical to meet the requirements for both a level milk volume each month and milk with minimal variations in quality.

7.3.2 Forage to concentrate ratios

Increasing the proportion of concentrates in the dairy cow's diet tends to raise milk yields and improve milk protein concentrations, particularly when feeding starchy concentrates. At the same time, however, milk fat concentrations may decline as the fermentation of fibre is reduced. Conversely, in both low-input and organic dairy systems, in which the aim is to maximise milk production from forage, a low proportion of concentrates is used in the diets, which is why organic dairy farms tend to have lower milk yields than conventional equivalents (Roesch *et al.*, 2005). These low concentrate diets are often inadequate to meet the cow's feed energy requirement (Knaus *et al.*, 2001) and can lead to lower milk protein concentrations that are below the standards required for both the liquid and processing markets, and a greater proportion of milk samples with a high fat to protein ratio that indicate sub-clinical ketosis.

In many intensive conventional systems, the feeding of high-concentrate diets ensures that the energy density of the diet is maintained. In high-forage systems, farms in a suitable location growing fodder beet or forage maize can increase the energy density of the diet. Other high energy feeds include cereal grains, sugar beet pulp and molasses. However, in the future the latter two feeds will be not be available to organic farmers owing to changes in the standards for organic production. Compared with many conventional herds, the level of protein in the diets of both low-input and organic herds may be more variable during the year, owing to the greater proportion of forage in the total diet. High-forage diets based on conserved forage (e.g. haylage, silage), which are fed during the winter period of housing, are frequently low in energy, leading to lower protein concentrations in the milk. These diets may also be deficient in protein. Conversely in many organic systems during the spring to autumn period, when minimal concentrate supplements are fed, the grazing of grass/clover swards with a high white clover content can lead to excess protein in the diet (Weller and Cooper, 2001) and markedly higher milk protein and urea values than those found in the milk from cows on winter diets. Excess protein in the diet is also likely to occur during the summer months on low-input dairy farms where purchased nitrogen fertiliser is being replaced by nitrogen fixed by legumes.

High concentrate rations can reduce milk fat concentrations, with the effect being more marked with ground, rather than cracked or rolled, cereal grains, the feeding of starch rather than fibre-based concentrates and also the feeding of concentrates only once or twice daily rather than evenly throughout the day via total mixed rations of both forage and concentrate feeds (Doreau *et al.*, 1999). The production of forage for grazing from grass/white clover swards, where the nitrogen for plant growth is primarily from N-fixation by the clover plants rather than from fertiliser applications, may have a negative effect on milk fat content. The lactating dairy cow requires 30–50% fibre in the diet (Chamberlain and Wilkinson, 1998). While the fibre content of the grass is >40%, the white clover plants are low in fibre, potentially leading to low-fat milk that fails to meet the requirements of either the liquid milk or processing markets.

7.3.3 Protein, mineral and vitamin supplements

Feeding supplementary proteins is a widely used technique for balancing rations in conventional dairy production systems, particularly during the winter months when conserved feeds are used. These typically have lower protein concentrations than the fresh feeds from which they are made and although supplementation with energy-rich concentrates can improve the efficiency of utilisation of available forage proteins, a basal forage diet with a low crude protein concentration may need to be supplemented. Soya bean meal and high protein by-products such as maize gluten meal and oil seed meals (sunflower seed, rapeseed, linseed) are commonly used as ingredients in conventional dairy rations, either as part of a pelleted concentrate feed, or as a purchased single ingredient in an on-farm mix. The options available for low-input and organic farmers include purchasing limited quantities (and if organic, at a high unit cost) or growing their own high-protein crops. However, most conventional protein crops are not viable in the parts of the UK predominantly used for dairying, although alternative crops such as lupins, field beans and peas have much potential, either as whole-crop forage for mixing with lower protein forages, or as a grain crop for mixing as a concentrate supplement. Legume crops have the additional benefit of improving soil fertility for subsequent crops.

While most conventional dairy farms feed purchased mineral supplements to the cows to ensure that adequate minerals (or in some cases excess minerals) are included in the diet, many organic dairy farms rely on reduced mineral inputs from only seaweed meal and the available minerals from multi-species leys. Unlike conventional systems, organic dairy production is based on forage from leys with mixed species rather than a monoculture of grass and these leys have the potential to provide a greater source of minerals for the dairy cow. Weller and Bowling (2002) reported that, compared with perennial ryegrass, white clover, chicory and plantain have respectively higher mineral contents by 124%, 134% and 128% for calcium, 26%, 128% and 66% for magnesium and 140%, 817% and –11% for sodium.

As organic dairy cows graze leys with mixed plant species, there are likely to be differences in the mineral concentration of the herbage consumed by the cow and this may influence the mineral concentration in the milk. For example, animals show a preference for different types and parts of plants, which can both vary widely in mineral concentration (Underwood and Suttle, 1999). Both of these factors will be influenced by the diversity of plant species, stocking density and quantity of total herbage available to the animal. As organic dairy herds graze multi-species leys and often produce lower lactation yields than many conventional herds with higher concentrate inputs, the need for vitamin supplementation should be minimal, providing the diet includes quality grazing species and forages for the winter period are conserved with minimal losses during the fermentation (e.g. silage) and drying (e.g. hay) processes. Cobalt deficiency results in a decrease in the concentration of vitamin B12 in the milk owing to the inability of the rumen microorganisms to synthesise sufficient B12 while cobalt remains inadequate (Underwood and Suttle, 1999). The relationship between vitamin E and selenium is important and a deficiency in either can influence both the incidence of retained placenta and the reproductive performance of the dairy herd (Hogan *et al.*, 1993; Underwood and Suttle, 1999).

Vitamin E supplementation has been shown to have positive effects on milk quality in a conventional commercial dairy herd. Milk obtained from cows that received supplemental vitamin E had lower somatic cell counts and plasmin concentrations than corresponding values in milk obtained from control cows. The reduction in plasmin as a result of vitamin E supplementation is very beneficial to the dairy industry because plasmin reduces the cheese-yielding capacity of milk, affects the coagulating properties of milk and its overall ability to withstand processing during cheesemaking (Politis *et al.*, 2004).

In conventional systems, feeds are supplemented with minerals and vitamins. Studies indicate that supplementing the winter feed of cows leads to enhanced levels of Ca, P, Mg, Na, K, and particularly Cu, in the milk of cows receiving a metal amino acid chelate and vitamins (Strusinska *et al.*, 2004).

The requirements of dairy cattle for B-vitamins, determined almost half a century ago, concluded that a ruminant animal does not require an exogenous supply of B-vitamins because its rumen microflora should synthesise enough of these compounds to avoid deficiency. Since then, dairy cows have greatly increased their average milk and milk component yields. More recent studies have shown that B-vitamin supply in dairy cows is increased by supplementation, although losses in the rumen are extensive (Santschi *et al.*, 2005). Whilst there are few reports of B-vitamin supplementation affecting milk quality, supplemental biotin has been shown to directly improve milk yield (Majee *et al.*, 2003).

7.3.4 Livestock breeds and husbandry

The trend towards the Holstein breed in the UK over the last 20 years has led to problems in relation to milk quality. While these cows have the potential to produce high yields by the greater partitioning of nutrients into milk, their efficiency in converting metabolisable feed energy into milk has not increased

(Gordon *et al.*, 2000). Therefore, to sustain higher yields these cows need to achieve high intakes of both energy-rich forages and significant quantities of concentrate feeds, if their potential is to be met and the problems of negative energy balance and low milk proteins are to be avoided.

Many dairy herds with Holstein cows produce high volumes of milk per cow but compared with many other breeds the milk has a lower solids concentration. When milk with low solids content is sold for processing, the costs of production may increase owing to a requirement to reduce the water fraction in the raw milk. During the early lactation period, the milk protein concentration in the milk from some Holstein cows may be below the minimum level required for processing unless the diet includes a significant proportion of concentrate feeds. The problem is less acute in herds where the cows calve throughout the year as there are comparatively small variations in the monthly milk protein values that are usually attributable to changes in the forage component of the diet (e.g. changing from grazed herbage to silage). However, for farms block-calving a herd within an 8–12 week period each year there is the potential problem of unacceptably low protein concentrations during the early post-calving period. Other breeds have the potential to produce better milk protein concentrations from high-forage diets and in a number of herds either a proportion of the Holstein cows are being crossed with a breed that has better milk quality (e.g. Brown Swiss, Jersey, Meuse Rhine Issel) or a nucleus of cows from these breeds is being included in the herd to improve the milk quality profile in the bulk tank. The breed of cow being managed within a system is particularly critical in low-input and organic systems as the flexibility to improve the energy density of the diet is more limited.

Studies into the effect of frequency of milking on milk quality compared once-daily and twice-daily milking and showed that milk yield was significantly reduced and milk fat and protein concentrations were increased, with once-daily milking compared with twice-daily. Furthermore, casein concentrations in the milks were similar, but once-daily milk had higher whey protein content (O'Brien *et al.*, 2002).

A study of 28 dairy farms in The Netherlands found that the introduction of an automatic milking system led to a reduction in milk quality compared to conventional farms, owing to an increase in the total bacterial plate count and of free fatty acids (Klungel *et al.*, 2000).

7.3.5 Effect of processing methods

The way in which raw milk is handled or processed into other dairy products can influence its nutritional and sensory quality. During milking itself, certain milking machines, especially those with overhead lines, can cause mechanical shearing of milk fat globules. This leads to the release of triglycerides, usually protected from lipolysis by being encapsulated inside fat globules and leads to off-flavours of the milk (Deeth, 1993). Prior to reaching the consumer, raw milk is separated, homogenised, pasteurised and packaged or even sterilised, concentrated or treated with ultra-high temperatures. All of these processes can alter the nutritional or sensory quality of milk. For example, homogenisation has the effect of breaking down the fat globules to sizes so small that the milk fat can no longer form a cream layer and this improves the organoleptic and physical properties of the milk. Typically, homogenisation would induce lipolysis but this process is usually done after an extra pasteurisation step that inactivates the endogenous lipases. High pressure treatment only has limited effects on covalent bonds and, thus, it is expected that this process only marginally affects the nutritional characteristics of milk but its influence on the organoleptic properties of milk are well recognised. Thermal treatment can affect the appearance, taste, nutritional value and processing characteristics of milk (Lewis, 2003). Furthermore, research has shown that the packaging used for milk can affect its quality, with lower losses of vitamin A and riboflavin occurring when milk is packaged in coloured plastic packaging or paperboard cartons compared with clear plastic (Moyssiadi *et al.*, 2004).

7.3.6 Economic effects of changing systems and management practices to improve milk quality

Changing a dairy system to improve milk quality will often have both benefits and disadvantages for the viability of the dairy enterprise. Changing the calving pattern of a herd from autumn to the spring period has the potential to reduce concentrate inputs, improve milk protein concentrations and reduce production costs, by increasing the proportion of milk that is produced from grazed herbage rather than conserved forage. However, the potential disadvantages for many farmers will include the difficulty of sustaining consistent herbage availability during the grazing season, the decline in herbage quality as the season progresses and price penalties imposed by the milk buyer as a result of excess milk being produced in the spring period.

Manipulating milk quality by increasing the proportion of concentrates in the diet to overcome short-term problems (e.g. low milk protein values) may also significantly increase feed costs. Conversely, changing to high-forage diets may enhance milk quality components but also leads to a reduction in the stocking density, milk income and margins per hectare, when less concentrates are fed and the area required for producing forage increases. Introducing new crops into the cropping plan to improve the energy or protein of the diets may lead to an improvement in milk quality but also affect both the management and financial viability of the farm when other factors are accounted for, including the costs of production per unit of energy or protein and requirements for extra labour and specialist machinery for growing and harvesting the crop.

7.4 Procedures for implementing methods to improve the nutritional quality of milk products

Many of the current payments for milk production are based on achieving a level milk production profile throughout the year, with increased payments for producing milk when the total volume is below the market requirement for both liquid consumption and processing. These variable payments are often effective in preventing the majority of dairy farmers, both conventional and organic, from producing the primary part of the annual total volume during the grazing season when the cost of feed and other costs is lower than at other times of the year. Changing to producing milk, for example in a system based on herb-rich pastures or all forage diets to enhance specific nutritional benefits in the milk, may need a high premium to persuade dairy farmers of the financial viability of the enterprise. Costs of production may be markedly higher compared with their current system and therefore not financially viable.

Better methods of technology transfer to dairy farms are required to ensure more rapid progress is made in implementing the results from laboratory trials on methods for improving the nutritional quality of milk. For the majority of farmers, both the fat and protein concentrations of the milk, hygiene band and the level of somatic cell counts are the only factors that are monitored on a regular basis and usually only because of their impact on the price paid for the milk. Better communications between researchers working in a laboratory environment and dairy farmers is required, including the testing and implementation of new techniques and procedures to improve the nutritional quality of milk in relation to the human diet, and the role of dairy products in the total diet. In the future, the price structure for milk is likely to be based on a wider range of quality factors than is currently required by the milk buyer.

7.5 Future trends and priority areas for research and development

The need to reduce the environmental impact of agriculture, including dairy production, requires changes in the management strategy and methods of production of dairy systems. Dairy farms need to be managed as 'whole' systems with the environmental impact of the system, sustainability and efficiency of production per unit of dairy product, and animal welfare standards, all being important factors. In the future, systems will need to be more sustainable and less reliant on the importation of purchased concentrate feeds. Many plant species can be grown on the farm to produce either forage or combinable grain crops with the potential to improve the energy or protein content of the diet. However, many of these species have not been evaluated fully or subjected to breeding programmes to improve their key traits and more research is required to exploit fully the potential of these plants. Globalisation and the removal of trade barriers indicate that the milk price paid to dairy farmers will continue to be low irrespective of whether the milk is destined for the liquid or processing markets. This trend suggests that dairy farmers may have to consider new strategies to ensure their businesses remain viable in the future. One option is to minimise costs by establishing larger dairy units where milk volume and low production costs are the key factors with the milk sold and marketed away from the farm. An alternative option that is increasingly being undertaken is to both produce and process the milk within the same business, with the marketing of the dairy products aimed at niche markets that will often be within the same geographical location as the dairy farm. Niche markets require high-quality products and further research is required to improve the shelf life of dairy products, enhance flavour and establish a closer link between the producer and consumer. This improves the understanding of the 'whole system' and boosts confidence in the benefits dairy products can provide in the human diet.

7.6 References

Aimutis W R, (2004), 'Bioactive properties of milk proteins with particular focus on anticariogenesis', *Journal of Nutrition*, **134**, 989S–995S.

Al-Mabruk R M, Beck N F G and Dewhurst R J (2004), 'Effects of silage species and supplemental vitamin E on the oxidative stability of milk', *Journal of Dairy Science*, **87**, 406–412.

Barbano D M, Rasmussen R R and Lynch J M (1991), 'Influence of milk somatic cell count and milk age on cheese yield', *Journal of Dairy Science*, **74**, 369–388.

Bellew C, Kuesters S and Gillespie C (2000), 'Beverage choices affect adequacy of children's nutrient intakes', *Archives of Pediatrics and Adolescent Medicine*, **154**, 1148–1152.

Bennedsgaard T W, Thamsborg S M, Vaarst M and Enevoldsen C (2003), 'Eleven years of organic production in Denmark: herd health and production related to time of conversion and compared to conventional production', *Livestock Production Science*, **80**, 121–131.

Bertilsson J, Dewhurst R J and Tuori M (2002), 'Effects of legume silages on feed intake, milk production and nitrogen efficiency', in Wilkins R J and Paul C, *Legume Silage for Animal Production–LEGSIL, Proceedings of an International Workshop*, 2001, Braunschweig, Germany, 8–9 July 2001. (Landbauforschung Volkenrode, Special Issue 234), 39–45.

Bertoni G, Calamari L, Maianti M G and Battistotti B (2005), 'Milk for Protected Denomination of Origin (PDO) cheeses: I. The main required features', in Hocquette J H and Gigli S, *Indicators of Milk and Beef Quality*, EAAP Pub. 112, Wageningen Academic Publishers, The Netherlands.

Bosset J O and Jeangros B (2000), 'Comparison of some highland and lowland Gruyere type cheeses from Switzerland: a study of their potential PDO-characteristics', in Gagnaux D and Poffet J R, *Livestock Farming Systems. Integrating Animal Science Advances into the Search for Sustainability, Proceedings of the Fifth International Symposium on Livestock Farming Systems*, Posieux (Fribourg), Switzerland, 19–20 August 1999, 337–339.

Broderick G A, Walgenbach R P and Sterrenburg, E (2000), 'Performance of lactating dairy cows fed alfalfa or red clover silage as the sole forage', *Journal of Dairy Science*, **83**, 1543–1551.

Buchin S, Martin D, Dupont D, Bornard A and Achilleos C (1999), 'Influence of the composition of Alpine highland pasture on the chemical, rheological and sensory properties of cheese', *Journal of Dairy Research*, **66**, 579–588.

Bugaud C, Buchin S, Hauwuy A and Coulon J B (2002), 'Flavour and texture of cheeses according to grazing type: the Abundance cheese', *Productions Animales*, **15**, 31–36.

Chamberlain A T and Wilkinson J M (1998), *Feeding the Dairy Cow*, Chalcombe Publications Lincoln, UK.

Chilliard Y, Ferlay A and Doreau M (2001), 'Effects of different types of forages, animal fat or marine oils in cow's diet on milk fat secretion and composition, especially conjugated linoleic acid (CLA) and polyunsaturated fatty acids', *Livestock Production Science*, **70**, 31–48.

Deeth H C, (1993), 'Lipase activity and its effect on milk quality', *Australian Journal of Dairy Technology*, **48**, 96–98.

DePeters E J and Cant J P (1992), 'Nutritional factors influencing the nitrogen composition of bovine milk: A review', *Journal of Dairy Science*, **75**, 2043–2070.

Dewhurst R J, Fisher W J, Tweed J K S and Wilkins R J (2003), 'Comparison of grass and legume silages for milk production. 1. Production responses with different levels of concentrates', *Journal of Dairy Science*, **86**, 2598–2611.

Doreau M, Chilliard Y, Rulquin H and Demeyer D I (1999), 'Manipulation of milk fat in dairy cows', in Garnsworthy P and Wiseman J, *Recent Advances in Animal Nutrition 1999*, Nottingham University Press Nottingham, UK, 81–109.

Duyvesteyn W S, Shimoni E and Labuza T P (2001), 'Determination of the end of shelf life for milk using Weibull Hazard Method', *Food Science and Technology*, **34**, 143–148.

Ebbesvik M and Loes A K (1994), 'Organic dairy production in Norway – feeding, health, fodder production, nutrient balance and economy – results from the 30-farm-project 1989–1992', in Granstedt A and Koistinen R, *Converting to Organic Agriculture Seminar*, Scandinavian Association of Agricultural Scientists, **93**, 35–43.

Elliot J A, Emmons D B and Yates A R (1974), 'The influence of the bacterial quality of milk on the properties of dairy products: A review', *Journal of the Institute of Canadian Science and Technology*, **7**, 32–39.

El Zubeir I E M, ElOwni O A O and Mohamed G E (2005), 'Effect of mastitis on macro-minerals of bovine milk and blood serum in Sudan', *Journal of the South African Veterinary Association*, **76**, 22–25.

FitzGerald R J, Murray B A and Walsh D J (2004), 'Hypotensive peptides from milk proteins,' *Journal of Nutrition*, **134**, 980S–988S.

Fonteh F A, Grandison A S and Lewis M J (2002), 'Variations of lactoperoxidase activity and thiocyanate content in cows' and goats' milk throughout lactation', *Journal of Dairy Research*, **69**, 401–409.

Food Standards Agency, (2002), *McCance and Widdowson's the Composition of Foods*, 6th summary edition, Royal Society of Chemistry, Cambridge.

Frøst M B, Dijksterhuis G and Matens M (2001), 'Sensory perception of fat in milk', *Food Quality and Preference*, **12**, 327–336.

Gordon F J, Ferris C P, Patterson D C and Mayne C S (2000), 'A comparison of two grassland-based systems for autumn-calving dairy cows of high genetic merit', *Grass and Forage Science*, **55**, 83–96.

Hamilton C, Hansson I, Ekman T, Emanuelsson U and Forslund K (2002), 'Health in cows, calves and young stock on 26 organic dairy herds in Sweden', *Veterinary Record*, **150**, 503–508.

Hardeng F and Edge V L (2001), 'Mastitis, ketosis and milk fever in 31 organic and 93 conventional Norwegian dairy herds', *Journal of Dairy Science*, **84**, 2673–2679.

Hogan J S, Weiss W P and Smith K L (1993), 'Role of vitamin E and selenium in host defense against mastitis', *Journal of Dairy Science*, **76**, 2795–2803.

Hovi M and Roderick S (2000), 'Mastitis and mastitis control strategies in organic milk', *Cattle Practice*, **8**, 259–264.

Janzen J J (1972), 'The effect of somatic cell concentration in the raw milk on the shelf-life of the processed product', *Journal of Milk and Food Technology*, **35**, 112–114.

Jones E L, Shingfield K J, Kohen C, Jones A K, Lupoli B, Grandison A S, Beever D E, Williams C M, Calder P C and Yaqoob P (2005), 'Chemical, physical, and sensory properties of dairy products enriched with conjugated linoleic acid', *Journal of Dairy Science*, **88**, 2923–2937.

Kirchgessner M, Friesecke H and Koch G (1967), *Nutrition and the Composition of Milk*, Lockward, London.

Klungel G H, Slaghuis B A and Hogeveen H (2000), 'The effect of the introduction of automatic milking systems on milk quality', *Journal of Dairy Science*, **83**, 1998–2003.

Knaus W F, Steinwidder A and Zollitsch W (2001), 'Energy and protein balance inorganic dairy cow nutrition – model calculations based on EU regulations', in Hovi M and Baars T, *Proceedings of the 4th NAHWOA Workshop,* Wageningen, The Netherlands. University of Reading, Reading, UK.

Lewis M J (2003), 'Improvements in the pasteurisation and sterilisation of milk', in *Dairy Processing – Improving quality*, Gerrit Smith, Woodhead Publishing, Cambridge, UK 81–103.

Lucas A, Coulon J B, Grolier P, Martin B and Rock E (2005), 'Nutritional quality of dairy products and human health', in Hocquette J H and Gigli S, *Indicators of Milk and Beef Quality*, EAAP Pub. 112, Wageningen Academic Publishers, The Netherlands.

Ma Y, Ryan C, Barbano D M, Galton D M, Rudan M A and Boor K J (2000), 'Effects of somatic cell count on quality and shelf-life of pasteurized milk fluid', *Journal of Dairy Science*, **83**, 264–274.

Magkos F, Arvaniti F and Zampelas A (2003), 'Organic food: nutritious food or food for thought? A review of the evidence', *International Journal of Food Sciences and Nutrition*, **54**, 357–371.

Majee D N, Schwab E C, Bertics S J, Seymour W M and Shaver R D (2003), 'Lactation performance by dairy cows fed supplemental biotin and a B-vitamin blend', *Journal of Dairy Science*, **86**, 2106–2112.

Mariani P and Battistotti B (1999), 'Milk quality for cheesemaking', *Proceedings of the ASPA XIII Congress*, Piacenza, June 21–24, 499–516.

Moyssiadi T, Badeka A, Kondyli E, Vakirtzi T, Savvaidis I and Kontominas M G (2004), 'Effect of light transmittance and oxygen permeability of various packaging materials on keeping quality of low fat pasteurized milk: chemical and sensorial aspects', *International Dairy Journal*, **14**, 429–436.

Munro G L, Grieve P A and Kitchen B J (1984),'Effects of mastitis on milk yield, milk composition, processing properties and yield and quality of milk products', *Australian Journal of Dairy Technology*, **39**, 7–16.

National Research Council, (2001), *Nutrient Requirements of Dairy Cattle*, 7th revised edition, National Academy Press, Washington, DC.

Nielsen J H, Lund-Nielsen T and Skibsted L (2004), 'Higher antioxidant content in organic milk than in conventional milk due to feeding strategy', *Newsletter from Danish Research Centre for Organic Farming*, September, No. 3.

O'Brien B, Ryan G, Meaney WJ, McDonagh D and Kelly A (2002), 'Effect of frequency of milking on yield, composition and processing quality of milk' *Journal of Dairy Research*, **69**, 367–374.

Phillips L G, McGiff M L, Barbano D M and Lawless H T (1995a), 'The influence of fat on the sensory properties, viscosity, and color of lowfat milk', *Journal of Dairy Science*, **78**, 1258–1266.

Phillips L G, McGiff M L, Barbano D M and Lawless H T (1995b), 'The influence of nonfat dry milk on the sensory properties, viscosity, and color of low-fat milks', *Journal of Dairy Science*, **78**, 2113–2118.

Politis I and Ng-Kwai-Hang K F (1988a), 'Effect of somatic cell counts and milk composition on cheese composition and coagulating properties of milk', *Journal of Dairy Science*, **71**, 1711–1719.

Politis I and Ng-Kwai-Hang K F (1988b), 'Association between somatic cell counts of milk and cheese yielding capacity', *Journal of Dairy Science*, **71**, 1720–1727.

Politis I, Bizelis I, Tsiaras A and Baldi A (2004), 'Effect of vitamin E supplementation on neutrophil function, milk composition and plasmin activity in dairy cows in a commercial herd', *Journal of Dairy Research*, **71**, 273–278.

Prentice A, (2004), 'Diet, nutrition and the prevention of osteoporosis'. *Public Health Nutrition*, **7**, 227–243.

Renau J K, (1986), 'Effective use of dairy herd improvement somatic cell counts in mastitis control', *Journal of Dairy Science*, **69**, 1708–1720.

Roesch, M, Doherr M G and Blum J W (2005), 'Performance of dairy cows on Swiss farms with organic and integrated production,' *Journal of Dairy Science*, **88**, 2462–2475.

Santschi D E, Berthiaume R, Matte J J, Mustafa A F and Girard C L (2005), 'Fate of supplementary B-vitamins in the gastrointestinal tract of dairy cows', *Journal of Dairy Science*, **88**, 2043–2054.

Seifu E, Buys E M and Donkin E F (2005), 'Significance of the lactoperoxidase system in the dairy industry and its potential applications: a review', *Trends in Food Science and Technology*, **16**, 137–154.

Shah N P (2000), 'Effects of milk-derived bioactives: a review', *British Journal of Nutrition*, **84**, S3–S10.

Steijns J M, (2001), 'Milk ingredients as nutraceuticals', *International Journal of Dairy Technology*, **54**, 81–88.

Strusinska D, Mierzejewska J and Skok A (2004), 'Concentration of mineral components, beta-carotene, vitamins A and E in cow colostrum and milk when using mineral-vitamin supplements', *Medycyna Weterynaryjna*, **60**, 202–206.

Sundrum A, (2001), 'Organic livestock farming: a critical review', *Livestock Production Science*, **67**, 207–215.

Thomas C, Aston K and Daley S R (1985), 'Milk production from silage. 3. A comparison of red clover with grass silage', *Animal Production*, **41**, 21–31.

Toledo P, Andren A and Bjorck L (2002), 'Composition of raw milk from sustainable production systems', *International Dairy Journal*, **12**, 75–80.

Tricon S, Burdge G C, Kew S, Banerjee T, Russell J J, Jones E L, Grimble R F, Williams C M, Calder P C and Yaqoob P (2004), 'Effects of *cis*-9, *trans*-11 and *trans*-10, *cis*-12 conjugated linoleic acid on immune function in healthy humans', *American Journal of Clinical Nutrition*, **80**, 1626–1633.

Uegaki R, Ando S, Ishida, M, Takada O, Shiokura K and Kohchi Y (2001), 'Antioxidative activity of milk from cows fed herbs', *Journal of the Japanese Society for Bioscience Biotechnology and Agrochemistry*, **75**, 669–671.

Underwood E J and Suttle N F (1999), *The Mineral Nutrition of Livestock*, CABI Publishing Wallingford, UK.

Vaarst M and Enevoldsen C (1994), 'Patterns of clinical mastitis manifestations in Danish organic dairy herds', *Journal of Dairy Research*, **64**, 23–37.

Weller R F and Bowling P J (2002), 'The yield and quality of plant species grown in mixed organic swards', in Kyriazakis I and Zervas G, *Organic Meat and Milk from Ruminants*, Wageningen Academic Publishers, The Netherlands 177–180.

Weller R F and Cooper A (1996), 'The health status of dairy herds converting from conventional to organic dairy farming', *The Veterinary Record*, **139**, 141–142.

Weller R F and Cooper A (2001), 'Seasonal changes in crude protein concentration of

white clover/perennial ryegrass swards grown without fertiliser N in an organic farming system', *Grass and Forage Science*, **56**, 92–95.
de Wit J N and van Hooydonk A C M (1996), 'Structure, functions and applications and lactoperoxidase in natural antimicrobial systems', *Netherlands Milk and Dairy Journal*, **50**, 227–244.
Woese K, Lange D, Boess C and Bogl K W (1997), 'A comparison of organically and conventionally grown foods – results of a review of the relevant literature', *Journal of the Science of Food and Agriculture*, **74**, 281–293.

8

Effects of organic husbandry methods and feeding regimes on poultry quality

Helen Hirt and Esther Zeltner, FiBL, Switzerland and Carlo Leifert, Newcastle University, UK

8.1 Introduction

Over the last 40 years conventional poultry (and in particular chicken) production has undergone the most extreme level of intensification of all major livestock production systems. This was driven by a more rapid increase in demand for poultry meat (which was perceived by consumers as leaner and healthier) and to a lesser extent eggs, compared to other livestock products (e.g. red meat, milk), which often saw unchanged or decreasing consumption patterns (Frazer, 1996; Tilman *et al.*, 2002).

Intensification was mainly focused on achieving higher growth rates and feed to meat/egg conversion rates and was associated with (i) increasingly complex hybrid breeding systems, (ii) separate breeding systems for egg (laying hens) and meat production resulting in males from laying hen production having to be destroyed, (iii) a switch from at least partially outdoor to caged and indoor production systems with extremely high stocking densities, (iv) use of high energy and protein concentrate based diets without access to green plant materials and (v) prophylactic use of antibiotic growth regulators and veterinary medicines (Mench, 1992; Siegel, 1993; Frazer, 1996; Martrenchar *et al.*, 1999; Mellor and Stafford, 2001; Tilman *et al.*, 2002; Keppler *et al.*, 2003; see also Chapter 11).

However, as the negative animal welfare impact associated with caged and other intensive indoor production systems became apparent, animal welfare became a primary driver for consumer demand and product differentiation in the poultry sector. As a result, both organic and conventional 'free range' products have achieved significant market shares and price premiums compared to products from intensive indoor production in most European countries (Hamm *et al.*, 2002).

More recently, a range of other quality and safety issues have been recognised by consumers and now influence poultry meat and egg buying patterns and behaviour. Most importantly this includes (i) the routine use of antibiotics as growth promoters and curative medicines and the potential for development of transferable antibiotic resistance, (ii) risk associated with enteric pathogen (e.g. *Salmonella* and *Campylobacter*) and toxin (e.g. dioxin) contaminants of poultry products, (iii) the environmental impact of poultry production and (iv) the sensory and nutritional quality of eggs and poultry meat (Menzi *et al.*, 1997; Hamm *et al.*, 2002; Rodenburg *et al.*, 2004; Horsted *et al.*, 2005).

The basis for organic production in Europe is EU Regulation 2092/91 (CONSLEG: 1991R2092 – 01/05/2004). Table 8.1 lists the general principles underlying organic poultry standards and Table 8.2 lists the minimum areas, indoors and outdoors, and other characteristics of housing for different poultry species and types of production.

There is still significant controversy within the organic poultry sector/ industry about the level of 'extensification' that should be achieved and there are significant differences in standards (especially with respect to flock sizes/stocking densities, level access to the outside and pasture and feeding regimes) used by different organic sector bodies.

Thus, the level of intensification in conventional (including free-range) production systems and in particular the lack of poultry breeds suitable for organic systems, results in significant challenges for farmers converting to organic production and organic sector bodies who aim to develop and improve organic production systems and standards in tune with the underlying 'principles' and ideals of organic farming (see also Chapter 2.1).

This chapter will therefore focus on reviewing the current knowledge about quality and safety parameters (including animal welfare) associated by consumers with organic and other low input systems. Separate sections will focus on describing (i) nutritional and sensory quality differences between organic, other free range and conventional poultry products, (ii) animal welfare related benefits and challenges, (iii) poultry health management and relative risk from foodborne diseases, (iv) risks associated with the relative levels of veterinary medicine, (v) toxic chemicals and heavy metals and (vi) maintaining quality during processing and novel methods for quality assessments of eggs and poultry meat.

The issue of anthelmintic use in poultry production is covered in Chapter 12 and quality assurance systems based on hazard analysis by critical control point (HACCP) systems are described in detail by van Elzakker *et al.* (2004) and in Chapters 22 and 23 and are therefore not addressed here.

8.2 Sensory and nutritional quality

There are relatively few scientific studies which have compared the sensory and nutritional quality of eggs and poultry meat from organic, 'low input',

Table 8.1 General principles for organic poultry production according to the EU regulation

Subject	EU regulation	Notice
Origin of the animals	In the choice of breeds or strains, account must be taken of the capacity of animals to adapt to local conditions, their vitality and their resistance to disease. In addition, breeds or strains of animals shall be selected to avoid specific diseases or health problems associated with some breeds or strains used in intensive production Preference is to be given to indigenous breeds and strains. Poultry for meat production must be less than three days old.	See Section 8.2.1
Conversion	10 weeks for poultry for meat production, brought in before they are three days old 6 weeks in the case of poultry for egg production.	
Feed	Roughage, fresh or dried fodder, or silage must be added to the daily ration.	See Section 8.2.4
Disease prevention and veterinary treatment	The withdrawal period between the last administration of an allopathic veterinary medicinal product to an animal under normal conditions of use, and the production of organically produced foodstuffs from such animals, is to be twice the legal withdrawal period or, in a case in which this period is not specified, 48 hours.	See Sections 8.2.5 and 8.2.6
Minimum age at slaughter	81 days for chickens 150 days for capons 49 days for Peking ducks 70 days for female Muscovy ducks 84 days for male Muscovy ducks 92 days for Mallard ducks 94 days for guineafowl 140 days for turkeys and roasting geese Where producers do not apply these minimum slaughter ages, they must use slow-growing strains.	See Section 8.2.1
Free range areas and livestock housing	Poultry must be reared in open-range conditions and cannot be kept in cages. Water fowl must have access to a stream, pond or lake whenever the weather conditions permit in order to respect animal welfare requirements or hygienic conditions. Buildings for all poultry must meet the following minimum conditions: • at least one third shall be solid, that is, not of slatted or of grid construction and covered with a litter material such as straw, wood shavings, sand or turf; • in poultry houses for laying hens, a sufficiently large part of the floor area available to the hens must be available for the collection of bird droppings; • they must have perches of a size and number	See Sections 8.2.2 and 8.2.3

Table 8.1 Continued

Subject	EU regulation	Notice
	commensurate with the size of the group and of the birds as laid down in Annex VIII; • they must have exit/entry pop-holes of a size adequate for the birds, and these pop-holes must have a combined length of at least 4 m per 100 m² area of the house available to the birds; • each poultry house must not contain more than: – 4800 chickens, – 3000 laying hens, – 5200 guinea fowl, – 4000 female Muscovy or Peking ducks or 3200 male Muscovy or Peking ducks or other ducks, – 2500 capons, geese or turkeys; • the total usable area of poultry houses for meat production on any single production unit, must not exceed 1600 m². In the case of laying hens natural light may be supplemented by artificial means to provide a maximum of 16 hours light per day with a continuous nocturnal rest period without artificial light of at least eight hours. Poultry must have access to an open-air run whenever the weather conditions permit and, whenever possible, must have such access for at least one-third of their life. These open-air runs must be mainly covered with vegetation and provided with protective facilities, and permit animals to have easy access to adequate numbers of drinking and feeding troughs. For health reasons, buildings must be emptied of livestock between each batch of poultry reared. The buildings and fittings are to be cleaned and disinfected during this time. In addition, when the rearing of each batch of poultry has been completed, runs must be left empty to allow vegetation to grow back and for health reasons. Member States will establish the period in which runs must be empty and they will communicate their decision to the Commission and the other Member States. These requirements shall not apply to small numbers of poultry which are not kept in runs and which are free to roam throughout the day.	See Table 8.2

'free range' and/or conventional productions systems. In a recent review of the literature on quality characteristics of all organic livestock commodities, Kouba (2003) concluded that there was no evidence for consistent differences in flavour or nutritional qualities between organic and conventional livestock products, but that organic products contain lower levels of residues from veterinary drugs and pesticides. He also reported that there is no clear evidence for differences in mycotoxin contamination and risks from enteric pathogen contamination between organic and conventional livestock foods.

Table 8.2 Minimum areas indoors and outdoors and other characteristics of housing in the different poultry species and types of production according to the EU regulation

	Indoor area (net area available to animals)			Outdoor area (m^2 of area available in rotation/head)
	No animals/m^2	cm perch/ animal	Nest	
Laying hens	6	18	8 laying hens per nest or in case of common nest 120 cm^2/bird	4, provided that the limit of 170 kg of N/ha/year is not exceeded
Fattening poultry (in fixed housing)	10 with a maximum of 21 kg liveweight/m^2	20 (for guinea fowl only)		4 broilers and guinea fowl, 4,5 ducks, 10 turkey, 15 geese In all the species mentioned above the limit of 170 kg of N/ha/year is not to be exceeded
Fattening poultry in mobile housing	16 (*) in mobile poultry houses with a maximum of 30 kg liveweight/m^2			2,5, provided that the limit of 170 kg of N/ha/year is not exceeded

*Only in the case of mobile houses not exceeding 150 m^2 floor space which remain open at night.

However, a series of investigations in Italy reported higher breast and drumstick percentages and lower levels of abdominal fat in the carcasses of organically compared to conventionally produced chickens. Also muscle tissue from organic production systems had a lower pH, water-holding capacity and cooking loss, better lightness and shear values, and increased content of Fe. Finally, they reported higher sensory quality, omega-3 fatty acid and α-tocopherol content and oxidative stability in meat from organic systems (Castellini *et al.*, 2002). Increased dietary intake of omega-3 fatty acids has been linked to lower rates of cardiovascular diseases and certain cancers (see Chapter 7 and Butler *et al.*, 2006 for further details). However, levels taken up with poultry meat (breast muscle tissue only contains around 1% of fat, Jahan *et al.*, 2004) are relatively low, compared to dietary intake associated with other livestock products (e.g. milk, butter and cheese).

A study carried out in Ireland which also compared oxidative stability and α-tocopherol levels in organic, free-range and conventional breast meat burgers from France, Ireland and the UK showed significant and greater variability between different samples from the same production system, than between production systems. They concluded that oxidative stability and α-tocopherol levels in poultry meat are mainly related to the diets used (Lawlor *et al.*, 2003).

Effects of dietary composition on product quality are well known and both over and under supply and/or consumption of protein are shown to affect product quality. For example, if low temperatures stimulate high feed intake or, in contrast, hot weather results in reduced feed intake, growth and carcass quality characteristics are negatively affected (Walker and Gordon, 2001). Increasing dietary content of essential amino acids and protein was shown to increase egg weight and decrease shell percentage and albumen quality (Hammershøj and Kjaer, 1999). With phase feeding, the size of eggs can be affected and a larger number of marketable eggs may be achieved. However, three or more changes of diet during the laying period was also shown to increase feather pecking (Green *et al.*, 2000; see also Section 8.3 below). To improve egg shell quality, it is also important to have optimal calcium content and a balanced proportion of sodium and chloride.

Omega-3 fatty acid and α-tocopherol levels in eggs were shown to increase if hens are reared in 'free range' conditions which permit foraging on pasture by chickens (Lopez-Bote *et al.*, 1998). In an earlier study, eggs from free range production systems were shown to also contain significantly higher levels of folic acid and vitamin B12 than eggs from battery and indoor production systems, but this was not linked to access to and foraging on pasture (Tolan, 1974). Access to outdoor roaming areas, but not the level of access to forage/green pasture is compulsory under EU organic standards. However, access to pasture is prescribed by 'higher level poultry standards' developed by some European sector bodies (e.g. the Soil Association). Differences in grass uptake between organic production units used in studies could explain the variability in α-tocopherol and omega-3 fatty acid content observed in chicken meat from different organic systems in the study by Lawlor *et al.* (2003).

The chicken genotype chosen was recently shown to have a significant effect on the total lipid, α-tocopherol and carotenoid content and total antioxidant activity in organically reared chicken, with the faster growing 'Ross' genotype showing lower antioxidant content and antioxidant activity, but higher levels of intramuscular fat (Castellini *et al.*, 2005). However, when the fat composition was examined, the slower growing 'Kabir' genotype was found to have a significantly higher proportion of mono- and polyunsaturated fatty acids compared with the fast growing Ross genotype, which had higher levels of saturated fatty acids. When eaten within 48 hours of slaughter the sensory quality of meat from Kabir was also assessed to be better by taste panels, while taste panels did not detect significant differences between genotypes when chickens were stored for longer periods of time (Castellini *et al.*, 2005). Similar results with respect to the sensory quality of chicken grown under organic productions systems were reported by Horsted *et al.* (2005), who reported that fast growing commercial hybrids produced less tender meat consistency with age, whereas the opposite was found for slower growing pure breeds. These data strongly indicate that (i) genotype choice is an important sensory and nutritional quality-determining factor in organic chicken production, (ii) that slower growing genotypes may be better suited to produce quality meat characteristics than fast growing hybrid strains developed for intensive farming systems and (iii) that the use of different genotypes could have contributed to the variability in α-tocopherol and omega-3 content observed in chicken meat from different organic systems in the study by Lawlor *et al.* (2003).

A more recent study (Jahan *et al.*, 2004, 2006) based on a supermarket survey of chicken breast fillets, also indicated an overall lower fat content in organic breast meat, but reported a lower omega-3 and omega-6 content and higher cholesterol content for organic and rare breed chicken compared with conventional chicken. However, the level of replication was insufficient to allow statistically sound quality comparisons of meat between production systems. The authors collected three packets of fresh organic breast meat, fresh conventional breast meat and frozen conventional breast meat from two supermarkets on one occasion only. As a result, it was likely that the three packets collected in the same retail outlet were from the same farm, resulting in breast meat from only two organic and four conventional farms being compared. However, despite its design faults the study may indicate that the more 'intensive', minimum EU-standard-based chicken production systems (which often use greater flock sizes, fast growing conventional breeds and do not provide access to pasture) that supply the supermarket supply chains in the UK may not produce meat with higher nutritional quality. It is therefore essential to carry out a properly designed survey in which meat quality in organic systems working to different standards (EU minimum standard and the higher Soil Association standard) is compared with that of conventionally produced chicken.

Yolk colour plays a role in consumer acceptance, but the preferred colour varies in different countries. Yolk colour has no relation to nutrient content, flavour or freshness, but is often enhanced in conventional production systems by addition of synthetic pigments to the animal feed. In organic production, synthetic yolk pigments are prohibited and this normally results in paler yolks, but may also lead to greater variability in yolk colour intensity. In many European countries paler yolk colour is perceived by consumers as being associated with 'less natural' production systems, an issue that clearly needs to be addressed by improved consumer information.

8.3 Animal welfare related quality parameters

Organic poultry production systems used in Europe are very variable (Zollitsch, 2003). However, the Sustaining Animal Health and Food Safety in Organic Farming (SAFO) Working Group on Organic Poultry Production described the following common trends/background conditions in organic production systems (Zollitsch, 2003):

- traditionally small flock sizes; however, over the last decade flocks of layers and broilers have become larger (500 to several thousand birds per flock) mainly owing to large-scale conventional producers having converted to organic farming practice;
- increasing premiums for organic poultry products have attracted an increasing number of producers into the industry who have mainly commercial objectives and often push for minimum restrictions being imposed by organic farming standards (especially with respect to stocking rates, flock sizes, diet composition and access to outdoors and especially foraging on pasture);
- changing market conditions, vertical market integration and the need to adapt new marketing strategies; with higher proportions of eggs and poultry meat being retailed through multiple retailers;
- higher price premiums for organic broilers than for organic eggs;
- increased competition between organic and other 'free-range' egg production systems.

The trend towards larger flock sizes, while often perceived as undesirable from an animal welfare point of view, coincided with more 'professional' practices being introduced into the organic poultry industry, especially with respect to food safety and HACCP based quality assurance systems. However, it did result in a decrease in time spent on individual animals and their welfare.

The SAFO Poultry Working Group also identified the need to address the following main welfare-related challenges in many parts of the organic poultry industry (Zollitsch, 2003):

- feather pecking and cannibalism;
- disposal of spent laying hens;
- nutritional needs of poultry, particularly essential amino acids;
- finding optimum animal welfare systems within the framework of existing economic constraints;
- linking animal welfare and product quality within consumer perceptions of organic poultry production;.
- defining 'slow-growing birds' within organic standards.

Feather pecking and cannibalism are a major problem in conventional production systems and can be reduced by extensifying production but remain a problem in organic production in alternative housing systems. The disposal of spent laying hens is a problem in both organic and conventional systems. It has become a problem because of changing eating habits and because the market for stewing chicken has virtually disappeared in many European countries and an alternative use of spent hens not yet being established.

The provision of sufficient supplies of essential amino acids for poultry is a specific organic farming issue and results from (i) an insufficient supply of suitable protein crops (especially organically grown soya) for the expanding organic poultry sector and (ii) the prohibition (or restrictions) on the use of synthetic amino acid supplements. If chicken are raised on diets with insufficient levels of essential amino acids, their growth rate decreases and susceptibility to diseases increases. High growth rate chicken breeds are particularly affected by insufficient and imbalanced supplies of essential amino acids (Zollitsch, 2003).

The need to define and breed an appropriate 'slow-growing bird' is also a specific demand of the organic farming industry. It is thought by many to be the most important development issue in organic poultry production, because it would potentially result in (i) improved feed conversion efficiency over the longer lifespan specified under organic poultry standards (Table 8.1), (ii) better adaptation to outdoor rearing systems (see below), (iii) greater resistance to lower or imbalanced supplies of essential amino acids and (iv) reduced negative animal physiological and welfare problems associated with high growth rate hybrids. The main challenges described above are discussed in separate sections below.

8.3.1 Breeding to improve animal welfare

Commercial poultry breeding focuses on either egg production or meat yield. The combination of both criteria is difficult as they were shown to be genetically negatively correlated. This has resulted in completely separate hybrid breeding programmes for laying hens and fattening poultry. Table 8.3 gives an overview of the performance features and criteria for laying hens, fattening poultry and parental stock currently used in poultry breeding. Most of the criteria focus on yield increases and cross/hybrid breeding has generated enormous yield increases. But this strategy has also resulted in some constitutional problems.

With high performance laying hen genotypes, feather pecking and possibly cannibalism are the main reasons for injuries and deaths in alternative husbandry. This is a severe problem for the laying hens and can result in significant economic loss for the producers. However, reducing egg yield, feather pecking and cannibalism are thought to have no direct effect on egg quality. There are currently few slow growing laying hen genotypes available.

In broilers of modern fast growing, meat production genotypes, most deaths are due to sudden heart failure and parasite infections (see also Chapter 12). Also such genotypes suffer from leg disorders (perosis, tibial dyschondroplasia, etc.) and muscle diseases to such an extent that they have difficulty in walking normally. They are thought to suffer severe pain and consequently spend much longer lying down. This in turn may lead to serious breast blisters.

Muscle diseases and breast blisters may affect slaughter performance negatively. The poultry breeding industries provide a range of slow-growing strains (see Section 8.6 for further information), some of which show fewer health problems. The choice of type depends on the desired weight (country-specific) or the length of fattening period and the type of production. As

Table 8.3 Performance features and criteria for laying hens, fattening poultry and parental stock (after Merkblätter der Geflügelhaltung, 2004)

Laying hen	Egg production capacity: age at point of lay, laying peak, persistence Egg weight and grading eggs: egg weight in grams, proportion of grades important if differences between prices Entire production of eggs: eggs per initial hen, egg weight in kilograms important if the egg revenue is calculated in kilograms Egg quality: egg shell quality (egg shell strength, proportion of soft-shell eggs), interior quality (egg white quality: Haugh-Units), proportion of yolk, yolk collar, proportion of blood spots Feed consumption: per hen and day, per egg, per egg weight in kilograms Mortality: particularly dependent on resistance to disease and disposition to cannibalism Further criteria: nest appetite and acceptance, disposition to cannibalism and feather pecking, feather quality, proportion of broody hens, unwanted behaviour in general (nervousness, flying in the hen run)
Fattening poultry	Fattening performance (important for fattener): weight gain, feed conversion, mortality, constitution (more leg problems and heart/circulatory failure in lines with fast growing rate), productivity index (EBI: European broiler index) Slaughter performance (important for butcher): slaughter yield, balanced carcass weight, composition of carcass Meat quality (important for consumer): measurable or sensory meat quality criteria (appearance, colour, tenderness, juiciness, flavour, roast and boil capacity, biochemical values (e.g. pH-value)), and nutritional value
Parental stock	Percentage of eggs capable of brooding Fertility rate Hatching rate Number of chicks per hen Feed consumption per chick Chicks quality: weight at hatch, balance and weight development of the chicks

slow-growing animals get older, the meat becomes denser, darker, more flavourful and lower in fat content (see Section 8.2 above).

Most organic poultry production is currently based on the same breeding stock as conventional farming and the principal problems (see above) are the same. The discussion group at the fourth NAHWOA Workshop (Network for Animal Health and Welfare in Organic Agriculture) identified the following main problems (Hirt *et al.*, 2001):

- feather pecking/cannibalism in laying hens;
- health problems in fast-growing fattening birds and
- endoparasites in free-range systems.

The group further recognised that some of the problems identified were likely to be related to the high production levels and very specific feeding regimes the birds were bred for. However, organic poultry production conditions (especially husbandry and feeding regimes) are more variable. For example, feeding regimes are based on energy and protein crops suitable for the agronomic and climatic conditions on individual farms and therefore differ greatly from enterprise to enterprise. Also the proportion of on-farm produced feed that will be required to be used in poultry production is likely to increase in the future, further increasing variability. There is therefore a need for 'robust' slower growing breeds that are flexible and can adapt to a wider range of husbandry and feeding conditions (Hirt *et al.*, 2001).

The discussion group agreed that poultry breeds and breeding programmes for organic farming should have the following characteristics (Hirt *et al.*, 2001):

- The animals should show normal social behaviour in large groups (no or low levels of feather pecking and cannibalism).
- Fattening poultry should grow slowly (at least 81 days before slaughter).
- Facilities common in organic farming, such as free ranging and perches, should be used in breeding facilities.
- Birds should be able to adapt to regional characteristics, for instance the local climate.
- Birds should be able to cope with different feed compositions.
- Birds should show resistance to diseases and parasites, even if selecting for a certain trait means poorer performance in another area.

Sørensen (2001) discusses the demand for breeding strategies that focus on genetic adaptation to the organic environment. Regarding access to breeding material for laying hens for organic egg production he concludes the following:

- The breeding material available is genetically adapted to cage systems.
- This cage-adapted material has shortcomings in terms of behaviour in larger flocks. This manifests itself in an excessive tendency for feather pecking and cannibalism, and, independent of flock size, many mislaid eggs.
- Experiments have shown that feather pecking, cannibalism and nesting behaviour can be improved by selection.
- A search for alternative breeds/hybrids has not yet produced the ideal 'organic hen'.
- It is possible that laying birds could be adapted to thrive on diets with lower crude proteins, which would allow production based on 'home-grown' crops, even in the northern parts of Europe.

Sørensen (2001) proposes two strategies that should be initiated as soon as possible: (i) a continuous search for breeds or lines that are suitable for organic egg production and (ii) starting a breeding programme for laying hens that takes into consideration the specific requirements of organic egg and meat production. He also states that there has been and still are small national projects dealing with item (i), but little has been done to address item (ii), and that there is an urgent need to start the relevant breeding programmes before organic poultry production is faced with a crisis when Regulation EC 1804/99 enforces sourcing of organically bred stock (Sørensen, 2001).

Koene (2001) gives a broad view of animal welfare and genetics in organic farming of layers, especially concerning cannibalism in the Netherlands and states that one of the most promising potential solutions to welfare issues is to adapt the animal to its ecological environment by genetic selection. He also points to the need to educate consumers about the link between product qualities and welfare parameters by noting:

> Although it is found that white layers may perform better in organic conditions concerning feather pecking … and cannibalism …, such strains are not used in the Netherlands. This is due to the marketing facts: the consumer learns that brown eggs are laid in animal-friendly strawyard systems, and that white eggs come from battery cages! So the choice of strain is for marketing/consumers, not for the layers welfare! (Koene, 2001).

8.3.2 Husbandry practices to improve animal welfare

Moulting

Other animal welfare concerns related to laying hen production are that laying hens are used for just one year and the males of the laying hen strain are killed after hatching. The suggestion to produce 'dual purpose' breeds/ breeding programmes is widely considered as unrealistic or non-commercially viable, but it should be stressed that relatively little breeding effort has focused on this approach. In the studies that were carried out, own cross-breeding did not achieve economically viable egg production and commercial lines recommended as dual-purpose strains fell short of the aims, with behavioural disorders being very frequent (Hirt, 2001). Furthermore, cooperation with breeding companies turned out to be difficult as they felt the market potential did not justify significant investment in 'dual purpose breed' focused breeding efforts (Hirt, 2001).

An approach for prolonging the productive life of laying hens is moulting. From an ethical point of view, a longer life for laying hens is desirable, not only because the hens could be used for a longer period, but also because fewer one day-old male chicks would have to be killed. Naturally, birds have to lose their plumage, or moult, occasionally. During this time, they hardly take in any food and they stop laying. This laying interval presents a recuperation period for the egg-producing organs of the hen and the plumage, resulting in better plumage condition and better egg quality. In large conventional flocks of laying hens, moulting has to be induced via hormonal treatments (a practice not permitted under organic farming standards) to make it simultaneous among all the hens and prevent welfare problems related to social instability or even cannibalism in the flock. A method of inducing moulting through food and light restriction succeeded in improving egg quality and plumage condition, making it suitable for organic farming (Zeltner and Hirt, 2005a). The desired positive effects of a moult – improving egg white and shell quality, increasing egg production and improving plumage condition – were achieved using this method. The better egg-shell strength is supported by the findings of Al-Batshan *et al.* (1994), who detected an increased shell thickness after moulting. For ethical purposes, it is important to establish a trade-off between sustainable use of resources and stress on the animals.

Housing systems

Housing systems in organic farming should allow poultry access to a free range area where they find a wider range of feed stuff, including green forages, insects and worms, and are able to express their natural social behaviour. Traditionally, poultry on organic farms were kept in small flocks and housing systems with very low stocking densities. However, over the last two decades, the number of larger commercial organic poultry farms with higher stocking densities and larger flock sizes has increased. Nowadays, large organic poultry farms are showing the same trends with regard to stocking densities and housing systems as conventional farms, except that farm sizes are still smaller and maximum flock sizes are limited by organic farming standards (see Table 8.1).

Higher stocking densities and flock sizes on organic poultry farms were made possible by the development of new housing systems. In the 1980s, research was carried out in conventional farming to develop alternative systems to battery cages, when these became of increasing concern to consumers. These new systems aimed to provide conditions essential for normal poultry behaviour, including daylight, sufficient space per bird, appropriate litter, perches and egg-laying areas with suitable nesting facilities. These new housing systems provided significant advantages with regard to animal welfare. For example, access to perches, from not later than the age of four weeks, decreases the prevalence of floor eggs and the occurrence of cloacal cannibalism and feather pecking (Gunnarsson *et al.*, 1999).

However, it was also realised that the poultry farmers who worked with larger stocks in these new housing systems need to have more experience than those running rural poultry housing systems or battery cages. In 'free range' and organic laying hen husbandry systems, close attention must be paid to litter management. Hauser and Fölsch (2002) found the lowest dust concentrations in battery cages. The significantly lower dust concentration in battery farming is due to the lack of a scratching floor and dust bath. Optimum litter management can reduce the dust content in aviary systems (Carlucci *et al.*, 1994).

Hauser and Fölsch (2002) also found that there are no differences in the microbiological quality of eggs from four different farming systems (Table

Table 8.4 Microorganisms per eggshell surfaces (Hauser and Fölsch, 2002)

	Free range	Two levels	Aviary	Battery
TMO $\log_{10}$	7.43	6.71	7.0	6.3

TMO = total number of microorganisms.

8.4). In their investigation, 7 of 16 units had littered nests. Littered nests are closest to the natural nest established by wild fowl (the closest relative to domesticated chicken), which scratch a small cavity on the floor, sit in it and use loose material from the surrounding area to build a nest.

Keppler *et al.* (2001) found that there is less cannibalism in littered nests. This is probably due to the fact that hens in the standard nests, where the eggs disappear immediately after laying, are more restless, which may lead to pecks directed to the cloaca. With littered nests, there are also fewer misplaced eggs and the eggs seem to be cleaner (Keppler *et al.*, 2001). However, the hygiene levels tend to be higher in non-littered nests, as there could be problems with broken eggs in nest litter and/or with ectoparasites (see Chapter 12).

Good insulation for the poultry house is not only important to prevent frozen water in the cold season, but also to keep temperatures down on hot days in summer. High temperatures reduce the well-being of the hens and reduce egg shell quality, resulting in more broken eggs (Yahav *et al.*, 2000).

Bad weather runs

A bad weather run (or veranda or winter garden) joined onto to the poultry house should be part of housing systems used in most organic production systems. It typically has a concrete floor, a roof and litter. This run is used instead of the wider roaming area, when weather conditions are poor, for instance when it is raining or the ground is very damp. A bad weather run protects the health of the hens and the physical condition of the hen run itself, which can deteriorate if it is used in wet weather. The bad weather run gives the hens access to fresh air and enables dust bathing behaviour in a dry area in daylight, even on wet days, contributing greatly to their welfare.

Flock and range management

Special attention must be paid to the rearing period as the aim is to produce a flock that remains calm and healthy during its entire lifespan. Trouble during the rearing period may lead to serious problems in subsequent phases of life, therefore good management is essential during this formative period.

For example, early access to a littered floor instead of the plastic grid typical in conventional aviary systems leads to increased foraging (which was shown to increase the nutritional quality of eggs; see Section 8.2), less feather pecking and better tail feather condition during the remainder of the rearing period (Huber-Eicher and Sebö, 2001; Blokhuis and Arkes, 1984). Also, if there is no loose litter by the end of laying, the risk of feather pecking was shown to increase (Green *et al.*, 2000).

Another risk factor for feather pecking occurs when only a small proportion of the flock is using the outdoor area (Green *et al.*, 2000). If a large proportion of the flock goes outdoors, then the stocking density within the house will be reduced. High stocking density is associated with the occurrence of feather pecking (Huber-Eicher and Audigé, 1999). Keppler *et al.*, (2003) found more chickens with small featherless areas, more injuries and cannibalism in groups with higher stocking density. Stocking density is a major management factor for poultry welfare and health, and in turkeys, more hip and foot lesions are found at higher densities (Martrenchar *et al.*, 1999).

During the rearing period, the light regime plays an important role in managing animal welfare issues such as feather pecking. Management practices regarding light intensity should match the particular hen genotype, as not all strains show the same response to different light intensities with regard to egg production (Renema *et al.*, 2001). Light may even have an influence in the brooder house, as prenatal light exposure leads to more feather pecking in laying hen chicks (Riedstra and Groothuis, 2004). However, light intensity during rearing and age at access to the range does not appear to affect feather pecking behaviour in free-range laying hens (Kjaer and Sørensen, 2002).

As described above, litter management practices may affect animal welfare in housing systems used for organic poultry production; however, accurate litter management is essential to avoid problems. Special attention must be paid to the effect of litter on air quality. The basic principle for a healthy flock is to have a poultry house without draughts, with dry litter and with a bad weather run. A litter-drying system effectively keeps the dry matter content of the litter at a high level and minimises the degradation of nitrogenous components into ammonia (Groot Koerkamp *et al.*, 1998). Higher activity in a littered and structured housing system may lead to higher dust and ammonia concentrations in the air. Berk *et al.* (2001) found a tendency to higher dust concentrations in structured pens for turkeys, but not a higher ammonia concentration.

The uneven use of the hen run is one of the main problems in free-range husbandry. In flocks of free-range poultry, generally only a small proportion of the flock is outside at any one time. Indeed, frequent use of the hen run is advantageous as the risk of feather pecking is reduced when more hens use the range area (Nicol *et al.*, 2003). Hence special attention should be paid to ensuring good and frequent use of the range area. Hirt *et al.* (2000) have shown that the percentage of hens in the range area decreases with increasing flock size and that the hens in the range mostly stay near the poultry house. Keeling *et al.* (1988) mentioned three hypotheses to explain the small proportion of the flock using the free range: (i) little necessity or motivation to go out as most of the hens' basic requirements are met inside the poultry house, (ii) fear of their surroundings and predators (e.g birds of prey) and (iii) the gregarious nature of the hens, which results in their tending to flock together and therefore remain inside the house, or for those that go outside to be drawn back inside. An important agronomic consequence of uneven distribution of the hens in the range area is an uneven distribution of nutrients. For example, Menzi *et al.* (1997) found a nutrient and heavy metal overload in the parts of the run that were used most frequently.

The forage/grassland areas recommended for hen runs should be supplemented with structures that result in a more even distribution of the flock (Zeltner and Hirt, 2003). Bubier and Bradshaw (1998) recommend planting of vegetation features such as shrubs or trees, which break up the barren expanse of a grassy field and therefore promote a better distribution of the hens in the hen run. This may also reduce the hens' fear of their surroundings and predators by providing areas of 'shelter' (Keeling *et al.*, 1988). The quality and variety of such structures influence the use of the hen run more than the number of structures. This is probably due to individual preferences of hens with regard to the structures to which they are attracted, or their need for different structures in different circumstances. Providing a variety of structures was shown to encourage different types of behaviour and improves the overall use of the hen run (Zeltner and Hirt, 2005a, b).

From an ecological point of view, another important factor for hen run management is to reduce nutrient load, especially in the parts of the hen run close to the poultry houses. Green plants take up some of the accumulated nutrients and prevent leaching. Dense vegetation slows the accumulation of mud, which is a reservoir for parasites and diseases and prevents water and wind erosion, maintaining soil fertility. It also encourages the presence of earthworms and insects. Good hen run management helps maintain the turf. By alternating use of the hen run (Maurer and Hirt 2000), and/or the use of mobile poultry houses which can be relocated frequently, the turf quality can be significantly improved (Fürmetz *et al.*, 2005).

There are basically two methods of rotational system. Either the rotation is done approximately every fortnight, depending on how highly the hen run is used and how quickly the plants grow, or the change is done annually after the flock is replaced. The first method is better for ensuring a high turf quality.

8.3.3 Forage and feeding regimes to improve animal welfare

The largest component of any organic poultry diet will be cereal, and one of the most important problems limiting an expansion of organic poultry production is the limited availability of organic cereals and the relatively slow expansion of organic cereal growing in Europe. The availability of suitable protein sources for poultry, especially sources which provide adequate levels and compositions of essential amino acids, is also a primary concern. The fact that synthetic amino acids or vitamins are not allowed in organic poultry feed may result in a nutritional imbalance, leading to increased disease susceptibility. It has been shown that amino acid contents are lower in organic than in conventional diets (Zollitsch and Baumung, 2004). To obtain a 100% organic diet, it is therefore important to provide enough organically grown feed components with a suitable protein composition. The uncertainty of obtaining non-GMO soya, especially from North America, makes it difficult to provide poultry with sufficient amino acids. It should be pointed out that it is easier to supply adequate essential amino acids to broilers than to laying hens, as the slow-growing broiler breeds used in organic farming have lower requirements for proteins than the conventional ones (Zollitsch and Baumung, 2004).

Oily fishmeal is allowed in organic rations and it has an even higher essential amino acid content than full-fat soya. However, its use in poultry rations is limited partly by cost, restrictions on the source of the fishmeal imposed by organic standards, the fact that some customers demand birds that are fed on a vegetable-based diet and concerns about fishy taints to the product (Walker and Gordon, 2001).

Apart from the negative effects on animal health and reduced growth rates, an inadequate supply of amino acids also results in other welfare problems. For example, feather pecking seems to be more severe if the methionine supply is low (Zollitsch and Baumung, 2004; Kjaer and Sørensen, 2002). Prolonged storage of feed in fodder silos may impair the quality of the feed, especially in higher summer temperatures. Under bad storage conditions, vitamins, pigments and other bio-active feed components are degraded and the feed may oxidise. Additionally, condensation water in fodder silos is a medium for microbes, which produce toxins. Kijlstra *et al.* (2003) estimate that at least 20% of the food crops grown in the EU and used for food and feeds contain measurable amounts of mycotoxins.

Water is an important factor in the nutrition of poultry and should be provided in sufficient amounts and quality. As a medium for vaccines, medications and other supplements, water also plays an important role. Although hens are able to drink from nipples without any problems, they prefer to drink from open water surfaces, as this matches their normal drinking behaviour. In some organic farming standards/regimes, the use of nipples is forbidden. However, open water surfaces may pose hygiene risks, as the water often becomes polluted.

8.4 Poultry health management and risk from foodborne diseases

Poultry health is an important animal welfare related quality parameter and food safety parameter (especially with respect to risks from foodborne enteric pathogens) and this is increasingly recognised by consumers. Despite the introduction of common EU regulations for organic production, poultry systems and their management across Europe have remained highly diverse. Not surprisingly, therefore, health and mortality rates related to disease vary between countries in Europe. For example, in the Netherlands, the mean mortality in organic laying hens is 11% (0–21%), caused mainly by infectious diseases such as *E. coli*, infectious bronchitis, coccidiosis and brachyspira (Fiks *et al.*, 2002). In Switzerland, mean mortality in organic layer flocks is 8% (3–25%; Bio Suisse, 2006).

The main health problems in broilers differ from those of laying hens. Bestman and Maurer (2006) state that diseases with a long pre-patent or incubation period (e.g. ascarids), do not usually break out (even in the relatively long-living) organic broilers (see also Chapter 12), but are a problem in laying hen production systems. Preventing common diseases affecting young animals, for example diarrhoea, is therefore the main priority in broiler production. Organic broilers live long enough to exhibit specific or, more frequently, unspecific, clinical symptoms of Marek's disease, against which they should be vaccinated (Bestman and Maurer, 2006). Vaccination (Paracox) is also recommended, and usually applied on organic farms, against coccidiosis, which otherwise is a major health problem (Früh *et al.*, 2003).

8.4.1 Parasite infections

The risk of parasite infections may be greater in free-range systems. A survey in Switzerland showed that laying hens with access to a free-range area had an incidence of 75% of intestinal worm eggs in the faeces, whereas flocks with no access to free-range had an incidence of 42.5% (Häne, 1998). A bright yolk may be an indicator of intestinal worms. However, the faeces should be analysed for parasites before treatment to decide if the well-being of the poultry is at risk. Approaches used to reduce the use of anthelmintics, such as breeding for parasite resistance, preventive management, biological control and phytotherapy, are pointed out in Chapter 12. Control of red mites in poultry is becoming more and more important for organic as well as for conventional egg producers, owing to increased resistance to treatment. The use of mechanically acting substances like silicas may become a preferred alternative treatment (see Maurer and Perler, 2006).

Predation (e.g. by foxes and birds of prey) may also be a serious cause of mortality in free-range husbandry. Foxes can be kept away with a solid fence, which can be supplemented by an electric fence. Birds of prey are not a big problem for poultry houses that are closed at night and free of holes. However, keeping birds of prey such as the hawks away is not easy, especially for farms surrounded by forest or in a landscape with a lot of trees. In such situations, low structures should be provided as hiding places for the hens.

8.4.2 Enteric pathogen infection in poultry and transfer risk to the human food chain

The main poultry health problems that may affect safety of poultry products are *Salmonella* and *Campylobacter* infections (Ogden *et al.*, 2004a,b; see also Chapters 13,19, 20 and 23). While the incidence of *Samonella* infections in humans has decreased (especially in Northern European countries), *Campylobacter* infections associated with the consumption of poultry meat and eggs has increased and recent epidemiological studies indicate that it is now responsible for the majority of cases of sporadic gastroenteritis in humans (Lee and Newell, 2005). The epidemiology of *Salmonella* and *Campylobacter* in poultry and the human food chain will not be discussed in detail here and the reader is instead referred to a recent UK Food Standard Agency report (Allen and Newell, 2005). However, some of the most important issues arising are summarised below.

Present scientific evidence indicates that transmission of enteric pathogens (particularly *Campylobacter*) is mainly horizontal, while vertical transmission from parent to progeny via the egg is considered much less likely. Studies have shown that that the majority of *Salmonella* and *Campylobacter* strains do not cause disease symptoms in poultry and that between 40% and 80% of chicken flocks are *Campylobacter* positive.

However, there is more limited information on the level of infection from environmental sources, especially transmission by external vectors such as small ruminants and wild birds, which are known to carry both *Salmonella* and *Campylobacter* (Allen and Newell, 2005; Lee and Newell, 2005). Clearly, if the level of infection via vectors is significant, this would affect mainly free-range systems (Allen and Newell, 2005; FSA, 2006). There are a limited number of studies comparing the prevalence of enteric pathogens among free-range organic, extensive conventional and indoor intensive conventional flocks. In one relatively extensive study in Denmark (in which 160 broiler flocks were tested) *Campylobacter* could be isolated from 100% of organic broiler flocks, but only from 37% of conventional indoor broiler flocks and 50% of extensive indoor broiler flocks. The study also reported that antibiotic resistance was scarce among *Campylobacter* isolates from all rearing systems (Heuer *et al.*, 2001). However, recent studies into the epidemiology of *Campylobacter* in UK free-range chicken production systems indicate that infection occurs early in the production cycle, before animals are exposed to the outside (Newell, personnel communication).

Approaches such as vaccination are not yet feasible and the currently available HACCP systems to prevent or minimise the impact of enteric pathogen contamination of eggs and in particular poultry meat rely on optimising control at several critical control points (CCPs) in the poultry food supply chain. This includes minimising pathogen inocula during primary production, reducing contamination levels during processing and, most importantly, an emphasis on proper heat treatment during the preparation of poultry products (in particular meat) (Allen and Newell, 2005; FSA, 2006).

8.4.3 Avian influenza

The risk of contracting avian influenza from free-range poultry is currently much lower than the risks for the diseases mentioned above, but widely discussed in relation to contact with wild birds (FSA, 2006). However, since vector transmission (especially by wild birds) is important, free-range systems are thought to be at greater risk. However, the temporary outdoor ban during the migration periods of wild birds may influence the health and well-being of poultry.

8.5 Veterinary medicine use and residues

According to EU Regulation CONSLEG: 1991R2092–01/05/2004, the withdrawal period for the use of allopathic veterinary medical products in organic production is twice the legal withdrawal period. It is assumed that this lowers or even prevents residues in the product, but there is, to our knowledge, no information on the relative levels of residues in eggs and poultry meat from organic and conventional systems in the literature.

The regulation also stipulates that a maximum of three courses of treatment with chemically synthesised allopathic veterinary medical products or antibiotics within one year (or no more than one course of treatment if the productive life cycle is less than one year) is acceptable. These regulations are designed to encourage the use of preventive management and alternative treatments for the control of parasites and diseases. Vaccinations, veterinary medicine treatments for parasites and any compulsory eradication schemes established by Member States are exempt from the treatment maximums, in order to ensure animal welfare.

8.6 Toxic chemicals and heavy metals

Organic agriculture aims to eliminate residues from production systems, rather than meeting maximum permissible values. Owing to the risks of contamination being introduced onto farms from external sources (e.g. with feed, via air pollution and/or irrigation water) and with greater accuracy of pollution measurement/detection systems, this aim appears to be more and more difficult to achieve.

8.6.1 Dioxins

Dioxins recently hit the headlines as a contamination issue in organic farming. Dioxins are formed during incomplete combustion and are regarded as some of the most toxic substances known. Dioxin residues in eggs may result either from contaminated soil (air pollution) or contaminated feed. Kijlstra (2005) discusses in detail the role of organic and free-range poultry production systems on dioxin levels in eggs. He states that the most likely source of dioxins in eggs is directly (soil uptake) or indirectly (uptake worms/insects) related to soil dioxin levels. He also concludes that the marked differences between European countries may be related to historical dioxin emissions, which are lower in countries such as Sweden and Ireland than in more densely populated countries, such as The Netherlands. So dioxin residues in eggs are not primarily a problem of organic agriculture or free-range husbandry, but related to heavy historical environmental burdens. Decontamination of large surfaces is complex and expensive. Dioxin contamination in feed should be minimised through exact regulations and strict supervision (see Chapter 23).

8.6.2 Heavy metals

Another issue associated with poultry production is the potential for heavy metal contamination of the environment, in particular the soil. Schumacher (2005) explains that for cadmium (Cd) and lead (Pb), the main source is deposition (and not farming) but that for copper (Cu) and zinc (Zn) the main source is animal manure. Conventional pig feeds are highly supplemented with Cu and Zn, hence high levels of these metals occur in pig manure. The main source of metals in cattle manure is (partly illegal) prophylactic foot baths used to prevent certain types of lameness (the bathing liquids are often disposed of in the slurry tank after use) (Schumacher, 2005). He suggests that the most practical solution is to adjust the supplements in pig and poultry feeds and bedding materials, and to improve practices in the use and disposal of cattle foot baths (Schumacher, 2005). The lower stocking densities prescribed also limit the amount of heavy metals recycled to land as fertiliser.

8.7 Maintaining quality during processing

Apart from the primary production methods used, harvesting, transport and processing of poultry products may also affect product quality and safety. To maintain egg and meat quality throughout the food supply chain, specific methods and quality assurance practices need to be adapted. Since harvest, transport and processing methods for poultry products do not differ greatly between organic and conventional systems, they are not described in detail here and the reader is referred to appropriate textbooks and guides (e.g. Merkblätter der Geflügelhaltung, 2004).

However, the most important issues/recommendations are given below:

- Egg collection:
  - Housing: attention should be paid to maintaining a good microclimate, minimising floor eggs, collecting eggs daily and ensuring dust- and odour-free interim storage at temperatures below 20°C.
  - Storage: up to 20 days at a maximum temperature of 5°C, few fluctuations of temperature, continuous cold chain, humidity 50–85%, clean packing material, good air circulation.

  - Carriage: minimum agitation.
- Egg products (pasteurised whole egg, egg white and egg yolk):
  - Demand has increased enormously owing to ease of use and hygiene.
  - Never use cracked eggs, use only correctly stored eggs (no waste recycling!).
  - Ensure maximal hygiene (pasteurisation, filling and storage).
- Boiled eggs:
  - Use only eggs with good egg shell quality (from hens up to 40 weeks of age).
  - Eggs should not be fresh, but must have been correctly stored (1–2 weeks, for ease of peeling).
  - Colouring acts as a preservative (seals porous egg shell).
  - Only food colouring should be used (redwood or onionskin have minimum sealing properties so eggs are poorly preserved).
- Poultry meat:
  - Carcass quality depends on correct housing, management, transport and slaughtering.
  - Meat quality is most affected by the length of the fattening period, meat from older animals is tighter, darker, more aromatic and has lower water-holding capacity.

8.8 Alternative assessment systems for organic food quality

Because of the difficulty in (i) identifying non-compliances with organic standards (e.g. mineral N and P fertiliser use) in organic food production and (ii) assessing the 'overall quality' levels of organic foods, a range of alternative food assessment systems has been developed (see also Chapters 2 and 5). Biophoton measurement is used as a supplement to conventional food quality parameters. In this method, food is illuminated by light of different wavelengths and, using optical parameters such as light storage capacity and the kinetics of light re-emission, a quality declaration can be made for the food (Lambing, 1992). Köhler (2000) has investigated the influence of housing, feeding and illumination on the delayed luminescence of hen's eggs and found that natural light and forage intake result in different biophoton measurements. However, it should be stressed that the physiological and food quality implications of differences detected by biophoton measurements are not clear.

8.9 Acknowledgements

The authors gratefully acknowledge financial support from the EU under the IP QualityLowInputFood (506358) and from TESCO (UK).

8.10 Sources of further information and advice

Further information can be found on the following internet sites:

General information
Proceedings from the NAHWOA Workshops (Network for Animal Health and Welfare in Organic Agriculture): http://www.veeru.reading.ac.uk/organic/proceedings.htm (7.21.2005)
Proceedings from the SAFO Workshops (Sustaining Animal Health and Food Safety in Organic Farming): http://www.safonetwork.org/ (7.21.2005)
http://orgprints.org/view/subjects/9poultry.html (8.31.2005)

International standards and regulations
Homepage IFOAM (International Federation of Organic Agriculture Movements): http://www.ifoam.org/ (7.21.2005)

Breeding companies for laying hens:
Lohmann-Tierzucht (Germany): http://www.ltz.de/ (8.25.2005)
H&N and Hy-Line International (Germany): http://www.hy-line.com/ (8.25.2005)
ISA Poultry (France): http://www.isapoultry.com/ (8.25.2005)
Hendrix Poultry Breeders (The Netherlands): www.hendrix-poultry.nl (8.25.2005)
Bàbolna TETRA (Hungary): http://www.babolnatetra.com/ (8.25.2005)
Dominant (Czech Republic): www.dominant-cz.cz (8.25.2005)

Breeding companies for broilers
Aviagen (UK, USA): http://www.aviagen.com/ (8.25.2005)
Hubbard (France): http://www.hubbardbreeders.com
Cobb-Vantress (UK): http://www.cobb-vantress.com/ (8.25.2005)
Hybro (The Netherlands): http://www.hybrobreeders.com/ (8.25.2005)
Sasso (France): http://www.sasso.fr/ (8.25.2005)

8.11 References

Al-Batshan H.A., Scheideler S.E., Black B.L., Garlich J.D. and Anderson K.E. (1994). 'Duodenal calcium uptake, femur ash, and eggshell quality decline with age and increase following molt', *Poultry Science*, **73**, 1590–1596.

Allen V.M and Newell D.G. (2005). Food Standard Agency Report on project MS0004. http://www.food.gov.uk

Berk J., Hinz T. and Linke S. (2001). *Strukturierung der Haltungsumwelt und ihr Einfluss auf Tierverhalten, Luftqualität und Emissionen*. Tagungsband der 15. IGN-Tagung Tierschutz und Nutztierhaltung, 4–6 Oktober 2001, Halle, 136–140.

Bestman M and And Maurer V. (2006). Health and welfare in organic poultry in Europe: state of the art and future challenges. http://www.louisbolk.org/downloads/1769.pdf

Bio Suisse (2006). Calculation of production costs (internal document).

Blokhuis H.J. and Arkes J.G. (1984). 'Some observations on the development of feather pecking in poultry', *Applied Animal Behaviour Science*, **12**, 145–157.

Bubier N.E. and Bradshaw R.H. (1998). 'Movement of flocks of laying hens in and out of the hen house in four free-range systems', *British Poultry Science*, **39**, 5–18.

Butler G., Stergiadis S., Eyre M. and Leifert C. (2006). 'Effect of production system and geographic location on milk quality parameters', *Aspects of Applied Biology*, **80**, 189–193.

Carlucci L., Hauser R.H. and Fölsch D.W. (1994). 'Preventative measures in health care in hen houses', in *Biological Basis of Sustainable Animal Production*, Proceedings of 4th Zodiac Symposium, Wageningen NL, 224–231.

Castellini C., Mugnai C. and Dal Bosco A. (2002). 'Effect of organic production system on broiler carcass and meat quality'. *Meat Science*, **60**, 219–225.

Castellini C., Dal Bosco A., Mugnai C. and Pedrazzoli M. (2005). 'Comparison of two chicken genotypes organically reared: oxidative stability and other qualitative traits of the meat'. *Italian Journal of Animal Science*, **5**, 29–42.

CONSLEG: 1991R2092 – 01/05/2004: Council Regulation (EEC) No 2092/91 of 24 June 1991 on organic production of agricultural products and indications referring thereto on agricultural products and foodstuffs.

Fiks T.G.C.M., Reuvekamp B.F.J. and Landman W.J.M. (2002). *Pluimveehouderij*, **33**(2), 10–11.

Frazer A.F. (1996). *Farm Animal Behaviour and Welfare,* CABI Publishing, New York.

Früh B., Hirt H., Hossle I., Maurer V. and Richter T. (2003). *Pouletmast in Biolandbau*, FiBL, Frick, Switzerland, 16 pp.

FSA (2006) Avian flu – updates. http://www.food.gov.uk

Fürmetz A., Keppler C., Knierim U., Deerberg F. and Hess J. (2005). 'Legehennen in einem mobilen Sallsystem – Auslaufnutzung und Flächenzustand', in *Ende der Nische – Beiträge zur 8. Wissenschaftstagung Ökologischer Landbau*. Kassel, 1–4 März 2005, Hess J. and Rahmann G. (eds), Kassel University Press GmbH, Kassel.

Green L.E., Lewis K., Kimpton A. and Nicol C.J. (2000). 'Cross-sectional study of the prevalence of feather pecking in laying hens in alternative systems and its associations with management and disease', *Veterinary Record*, **147**, 233–238.

Groot Koerkamp P.W.G., Speelman L. and Metz J.H.M. (1998). 'Litter composition and ammonia emission in aviary houses for laying hens. Part 1: Performance of a litter drying system'. *Journal of Agricultural Engineering Research*, **70**, 375–382.

Gunnarsson S., Keeling L.J. and Svedberg J. (1999). 'Effect of rearing factors on the prevalence of floor eggs, cloacal cannibalism and feather pecking in commercial flocks of loose housed laying hens', *British Poultry Science*, **40**, 12–18.

Hamm U., Gronefeld F. and Halpin D. (2002). *Analysis of the European Market for Organic Food*, University of Wales, Aberwystwyth.

Hammershøj M. and Kjaer J.B. (1999). 'Phase feeding for laying hens: effect of protein and essential amino acids on egg quality and production', *Acta Agriculturae Scandinavica, Section A – Animal Science*, **49**, 31–41.

Häne M. (1998). Legehennenhaltung in der Schweiz 1998 – Schlussbericht zum Forschungsproject 2.97.1 des Bundesamts für Veterinärwesen, Zentrum für tiergerechte Haltung von Geflügel und Kaninchen, CH-3052 Zollikofen, CH.

Hauser R.H. and Fölsch D.W. (2002). 'How does the housing system affect the hygienic quality of eggs?' *Archiv für Geflügelkunde, Sonderheft*, **66**, 143.

Heuer O.E., Pedersen K., Andersen J.S. and Madsen M. (2001). 'Prevalence and antimicrobial susceptibility of thermophilic Campylobacter in organic and conventional broiler flocks', *Letters in Applied Microbiology*, **33**, 269–274.

Hirt H. (2001). 'Dual purpose poultry – A challenge or an illusion?' *Ecology and Farming*, **27**, 20–21.

Hirt H., Hördegen P. and Zeltner E. (2000). 'Laying hen husbandry: group size and use of hen-runs', in Alföldi T. Lockeretz W. and Niggli U. *Proceedings 13th International IFOAM Scientific Conference*, Basel, 363.

Hirt H., Bestmann M., Nauta W., Philipps L. and Spoolder H. (2001). 'Breeding for health and welfare', in *Breeding and Feeding for Animal Health and Welfare in Organic Livestock Systems*, Proceeding of the Fourth NAHWOA Workshop Wageningen, 24–27 March 2001. Hovi M. and Baars T. (eds), University of Reading, UK.

Horsted K., Henning, J. and Hermansen J.E. (2005). 'Growth and sensory characteristics of organically reared broilers differeing in strain, sex and age at slaughter, *Acta Agriculturae Scandinavica, Section A – Animal Science*, **55**, 149–157.

Huber-Eicher B. and Audigé L. (1999). 'Analysis of risk factors for the occurrence of feather pecking in laying hen growers', *British Poultry Science*, **40**, 599–604.

Huber-Eicher B. and Sebö F. (2001). 'Reducing feather pecking when raising laying hen chicks in aviary systems', *Applied Animal Behaviour Science*, **73**, 59–68.

Jahan K., Paterson A. and Spickett C.M. (2004). 'Fatty acid composition, antioxidants and lipid oxidation in chicken breasts from different production regimes', *International Journal of Food Science & Technology*, **39**, 443–453.

Jahan K. Paterson A. and Spickett C.M. (2006). 'Relationship between flavour, lipid composition and antioxidants in organic, free-range and conventional chicken breasts from modelling', *International Journal of Food Science & Technology*, **57**, 229–243.

Keeling L.J., Hughes B.O. and Dun P. (1988). 'Performance of free-range laying hens in a polythene house and their behaviour on range', *Farm Building Progress*, **94**, 21–28.

Keppler C., Lange K., Weiland I. and Fölsch D.W. (2001). 'Die Bedeutung des natürlichen Nestsuch- und Eiablageverhaltens von Legehennen für Eiproduktion und Tierschutz', in Tagungsbeiträge der 15. IGN-Tagung, 4–6. Oktober, Halle, 130–135.

Keppler C., Lange K. and Fölsch D.W. (2003). 'Influence of breed and stocking density of laying hens in improved rearing systems on behaviour, feather condition and injuries', in *Current Research in Applied Ethology 2002*, KTBL-Schrift 418, KTBL, Darmstadt, 19–29.

Kijlstra A. (2005). 'The role of organic and free range poultry production system on the dioxin levels in eggs', in Hovi, M. Jan Zastawny and S. Padel *Proceeding of the 3rd SAFO Workshop*, 16–18 September 2004, Falenty, Poland. University of Reading, UK, 83–90.

Kijlstra A., Groot M., van der Roest J., Kasteel D. and Eijck I. (2003). *Analysis of Black Holes in our Knowledge Concerning Animal Health in the Organic Food Production Chain*, Bericht, Animal Sciences Group, Wageningen UR.

Kjaer J.B. and Sørensen P. (2002). 'Feather pecking and cannibalism in free range laying hens as affected by genotype, dietary level of methionine + cystine, light intensity during rearing and age at first access to the range area', *Applied Animal Behaviour Science*, **76**, 21–39.

Koene P. (2001). 'Animal welfare and genetics in organic farming of layers: the example of cannibalism', in Hovi M. and Baars T. (eds). *Breeding and Feeding for Animal Health and Welfare in Organic Livestock Systems*, Proceeding of the Fourth NAHWOA Workshop Wageningen, 24–27 March 2001. University of Reading, UK.

Köhler B., (2000). *Der Einfluß von Haltung, Fütterung und Beleuchtung auf die Biophotonenemission (delayed luminescence) von Hühnereiern*, Univ. Kassel (GhK), Witzenhausen, Diss. agr. (Schriftenreihe KWALIS, Bd. 1, Fulda).

Kouba M. (2003). 'Quality of organic animal products', *Livestock Production Science*, **80**, 33–40.

Lambing K. (1992). 'Biophoton measurement as a supplement to the conventional consideration of food quality', in Popp F.A., Li K.H., Gu Q. *Recent Advances in Biophoton Research and its Applications*, Technology Center Kaiserslautern, International Institute of Biophysics, Germany, World Scientific.

Lawlor J.B., Sheehan E.M., Delahunty P.A., Morrissey P.A. and Kerry J.P. (2003). 'Oxidative stability od cooked chicken breast burgers obtained from organic, free-range and conventional reared animals', *International Journal of Poultry Science*, **2**, 398–403.

Lee M.D. and Newell D.G. (2005). *Campylobacter* in poultry: filling an ecological niche, *Avian Diseases*, **50**, 1–9.

Lopez-Bote C.J., Sanz Arias R., Rey A.I, Castano A., Isabel B. and Thos J. (1998). 'Effect of free-range feeding on omega-3 fatty acids and alpha tocopherol content and oxidative stability of eggs', *Animal Feed Science and Technology*, **72**, 33–40.

Martrenchar A., Huonnic D., Cotte J.P., Boilletot E. and Morisse J.P. (1999). 'Influence of stocking density on behavioural, health and productivity traits of turkeys in large flocks', *British Poultry Science*, **40**, 323–331.

Maurer V. and Hirt H. (2000). 'Laying hen husbandry: effects of run management on turf quality', in Alföldi T., Lockeretz W. and Niggli. U, *Proceedings 13th International IFOAM Scientific Conference*, Basel, 368.

Maurer V. and Perler E. (2006). *Silicas for Control of the Poultry Red Mite* Dermanyssus gallinae. Presentation at Joint Organic Congress, Odense, Denmark, May 30–31, 2006.

Mellor D.J. and Stafford K.J. (2001). 'Integrating practical, regulatory and ethical strategies for enhancing farm animal welfare', *Australian Veterinary Journal*, **79**, 762–768.

Mench J.A. (1992). 'The welfare of poultry in modern production', *Poultry Science Review*, **4**: 107–128.

Menzi H., Shariatmadari H., Meierhans D. and Wiedmer H. (1997). 'Nähr- und Schadstoffbelastung von Geflügelausläufen', *Agrarforschung*, **4**, 361–364.

Merkblätter der Geflügelhaltung (2004). Aviforum, CH-3052 Zollikofen.

Nicol C.J., Pötzsch C., Lewis K. and Green L.E. (2003). 'Matched concurrent case-control study of risk factors for feather pecking in hens on free-range commercial farms in the UK', *British Poultry Science*, **44**, 515–523.

Ogden I., Rosa E., Wyss G. and Brandt K. (2004a). *Safety and Contamination, Information to Consumers regarding Control of Quality and Safety in Organic Production Chains*, Technical leaflet, Research Institute of Organic Agriculture FiBL, CH-5070 Frick, Switzerland.

Ogden I.D., Lück L., Wyss G.S. and Brandt K. (2004b). *Production of Eggs, Control of Quality and Safety in Organic Production Chains*, Technical leaflet, Research Institute of Organic Agriculture FiBL, CH-5070 Frick, Switzerland.

Renema R.A., Robinson F.E., Feddes J.J., Fasenko G.M. and Zuidhoft M.J. (2001). 'Effects of light intensity from photostimulation in four strains of commercial egg layers: 2. Egg production parameters', *Poultry Science*, **80**, 1121–1131.

Riedstra B. and Groothuis T.G.G. (2004). 'Prenatal light exposure affects early feather pecking behaviour in the domestic chick', *Animal Behaviour,* **67**, 1037–1042.

Rodenburg T.B., Van der Hulst-van Arkel, M.C. and Kwakkel R.P. (2004). *Campylobacter* and *Salmonella* infections on organic broiler farms, *Netherlands Journal of Agricultural Science*, **52**, 101–108.

Schumacher U. (2005). 'Reduction of heavy metal input – a task also for organic animal husbandry', in, Hovi M., Zastawny J. and Padel S. *Proceeding of the 3rd SAFO Workshop*, 16–18 September 2004, Falenty, Poland. University of Reading, 91–92.

Siegel, P.B. (1993). 'Behavior-genetic analyses and poultry hushandry', *Poultry Science*, **72**, 1–6.

Sørensen P. (2001). 'Breeding strategies in poultry for genetic adaptation to the organic environment', in, Hovi M. and Baars T., *Breeding and Feeding for Animal Health and Welfare in Organic Livestock Systems*. Proceeding of the Fourth NAHWOA Workshop Wageningen, 24–27 March 2001. University of Reading.

Tilman D., Cassman K.G., Matson P.A., Naylor R. and Polasky S. (2002). 'Agricultural sustainability and intensive production practices', *Nature*, **418**, 671–677.

Tolan, A. (1974). 'Studies on the composition of food: the chemical composition of eggs produced under battery, deep litter and free-range conditions', *British Journal of Nutrition*, **31**, 185–189.

van Elzakker B., Torjusen H., O'Doherty Jensen K. and Brandt K. (2004). *Authenticity and Fraud, Information to Consumers regarding Control of Quality and Safety in Organic Production Chains*, Technical leaflet Research Institute of Organic Agriculture FiBL, CH–5070 Frick, Switzerland.

Walker A. and Gordon S. (2001). 'Nutrition issues in organic poultry systems', in Hovi M. and Baars T., *Breeding and Feeding for Animal Health and Welfare in Organic Livestock Systems*, Proceeding of the Fourth NAHWOA Workshop Wageningen, 24–27 March 2001. University of Reading.

Yahav S., Shinder D., Razpakovski V., Rusal M. and Bar A. (2000). 'Lack of response of laying hens to relative humidity at high ambient temperature', *British Poultry Science*, **41**, 660–663.

Zeltner E. and Hirt H. (2003). 'Effect of artificial structuring on the use of laying hen runs in a free-range system', *British Poultry Science*, **44**, 533–537.

Zeltner E. and Hirt H. (2005a). 'Ethological investigation on moulting laying hens in organic farming', *Animal Science Papers and Reports 23*, Supplement **1**, 175–179.

Zeltner E. and Hirt H. (2005b). 'Präferenz der Hühner für Menge und Variation von Strukturen', in *Aktuelle Arbeiten zur artgemässen Tierhaltung 2004*, KTBL–Schrift, **437**, 204–208.

Zollitsch W. (2003). 'Working group report: Poultry production: constraints and recommendations for enhancing health, welfare and food safety', in Hovi M., Martini A. and Padel S., *Socio-Economic Aspects of Animal Health and Food Safety in Organic Farming System*, Proceeding of the 1st SAFO Workshop 5–7 September 2003, Florence, Italy. University of Reading.

Zollitsch W. and Baumung R. (2004). 'Protein supply for organic poultry: options and shortcomings', in Hovi M. Sundrum A. and Padel S. *Diversity of Livestock Systems and Definition of Animal Welfare,* Proceeding of the 2nd SAFO Workshop 25–27 March 2004, Witzenhausen, Germany. University of Reading.

9

Quality in organic, low-input and conventional pig production

Albert Sundrum, University of Kassel, Germany

9.1 Introduction

From various viewpoints, the issue of intrinsic and extrinsic pork quality is a complex and multifactorial concept. It encompasses both objective carcass and organoleptic measures and subjective perceptions about the product and production methods (Edwards and Casabianca, 1997). Apart from the various factors that influence the different quality traits and their synergetic or antagonistic impact, variability within the traits aggravates reasonable and valid conclusions drawn from previously available data. In the following, consumer perceptions and the differences in framework conditions for pig production in relation to high- and low-input, as well as organic production systems, are described. Moreover, the factors relevant for quality assessment and the main factors that affect product and process quality are elucidated. Finally, the potentials and disadvantages of different production systems in relation to the various quality traits are discussed. Owing to the complexity of the production and growth process of pork, this chapter can only consider the most relevant aspects of the different systems and try to structure the rough lines, while the 'fine tuning' remains a challenge for each farmer in his or her own individual farm system.

9.2 Perception of quality

It has been said that 'quality, like beauty, is in the eyes of the beholder'. Thus, quality has different meanings for different people, and the interests of one particular group may conflict with those of others. The attributes included in the concept of quality primarily depend on who is making the definition. Typical actors participating in the evaluation of food quality are food producers, government officials, marketing people and consumer groups.

In general, quality is assessed by quantifiable traits that are more or less related to specific attributes of the product and the production process. Moreover, the assessment depends on the information delivered by the sensory organs. Information is filtered and evaluated by the brain depending on the specific information provided but also on the concept of understanding that already exists in the cerebral cortex (Singer, 2000). A mental representation of a sensory event can shape neural processes that underlie the formulation of the actual sensory experience. Thus, the subjective sensory experience is shaped by interactions between expectations and incoming sensory information.

Sensory stimuli provide great variability as does the discernment of different individuals' sensory organs (the beholder) and the interpretation of the information by the cerebral cortex by the means of the existing concept of understanding. Thus, there is a huge variability of quality pictures in the 'eyes of the beholder'. This makes it very difficult to deal with the quality issue in general and with the quality of pork in particular. Owing to the comprehensive and subjective characteristics of quality, it is reasonable to understand why scientists encounter difficulties when their research involves dealing with qualitative dimensions. On the other hand, the image and previous experience of the quality of specific foods become more and more decisive in the purchase situation (Grunert *et al.*, 2004). Process-related qualities of a food product are almost exclusively credence characteristics, since the consumer is seldom able to evaluate whether the food product has the promised process qualities. Thus, food image and reputation, whether based on facts or anecdotal, are increasingly important in food marketing. Currently, meat has a negative image for the public both because of the food scandals of past years and the anecdotal statement that present-day meat is fat and not proper from a nutritional viewpoint (Kubberod *et al.*, 2002). In relation to fresh meat, some of the most crucial quality characteristics demanded by the European society of today and tomorrow are outlined in Table 9.1.

Pork is the product of a very complex process. All the various characteristics of pork quality cannot be assessed directly in each carcass because these measurements and assessments would be too expensive. Therefore, previous scientific quality assessment of meat is primarily an indirect approach based on a few easily detectable quantitative traits and on the prescription of minimal standards in relation to the product in terms of size or composition and in relation to the production process. The prescriptions and the exclusion criteria vary between countries or between labelling programmes. The most encompassing prescriptions are enshrined in the EC regulation on organic livestock production (EEC No. 2092/91). Owing to this approach, extreme deviations in quality traits and deleterious effects are prevented. However, there is still space left within these framework conditions for huge variability in pork quality.

Table 9.1 Characteristics of fresh meat demanded in Europe today (Andersen *et al.*, 2005)

Characteristics	Individual elements	Examples	Characteristics	Individual elements	Examples
Safety	Pathogens	*Salmonella*	Wholesomeness	Welfare aspects	
	Residues	Dioxin, PCB		Outdoor rearing	
	Contaminants			Ethics	
Eating quality	Appearance			Sustainability	
	Flavour		Convenience	Organic farming	
	Tenderness			'Fresh and appealing' for days	
	Juiciness		Technological quality	Water holding capacity (WHC)	
Healthiness	Lean				
	Lipid content/composition	PUFA, CLA		pH-value	
	Vitamins	Vitamin B, E		Protein content and its characteristics	
	Minerals	Iron		Lipid content and its characteristics	
Traceability					
Experiences	Origin			Anti-oxidative status	
	History			WHC	
Diversity			Price		

Although basic scientific research deals with variation and its sources, the results of research work are mainly described and compared in terms of mean values supplemented with information about whether specific factors have a significant impact or not. Most research studies focus on individual factors in isolation and there are limited data in the literature on the interaction of a number of factors, particularly in relation to on-farm production practice. Consequently, the meaningfulness of previous results is often limited and often does not allow general conclusions to be drawn. As the relevance of the various factors changes between different production systems it is even more difficult to assess the ranking position of each factor within each production system in relation to the variation of product and process quality traits.

Taking into consideration the complexity of the production process and the variability of the product, a large gap becomes obvious between the huge variability in product and process quality and the low variability in these of which the consumer is aware when buying pork. In the past, labelling programmes tried to address this huge variability by establishing different production lines and thereby providing greater differentiation between pork products. However, labelling programmes or food marketing in general put an emphasis on some very specific traits in the first place while neglecting or ignoring others. Based merely on single characteristics to convey images to the consumer, food marketing ignores the existing and production related variability and tries to sell a more or less uniform image of the product. The market is full of simplified but mostly inaccurate images that have found their way into the 'eyes' of the 'beholders', making it even more difficult to bring quality assessment into line with reality.

9.3 Framework conditions of pig production

In pig production systems, the use of external resources like bought-in feedstuffs and synthetic chemicals (mineral fertilizers, pesticides, synthetic amino acids, growth promoters etc.) has become standard practice. However, consumers and producers alike are becoming increasingly concerned about the amount of chemicals used in conventional agriculture and are looking for options that reduce reliance on synthetic chemicals. As a result, interest in organic and low-input agriculture is growing steadily. Although there is a huge variation in production methods, they are characterised by specific framework conditions and by the use of production tools. As pork consists, to a high degree, of protein, the protein source in the feeding rations is of great importance. The main aspects in relation to quality production are discussed below.

9.3.1 Conventional production systems

Soybean meal is used worldwide in conventional pig production, as the most important protein source. In addition, there are high value proteins from the food industry, such as brewers, yeast, potato protein and maize gluten, while grain legumes and extracted rape meal are used only in small quantities. In recent years, the use of synthetic amino acids has markedly increased as a tool to increase the supply of essential amino acids.

In general, genotypes used in conventional production are characterised by a high capacity for protein accretion. This is achieved by crossbreeding different strains, mostly belonging to the Landrace, Large White, Duroc, and/or Hampshire breeds. Additionally, boars of the Pietrain breed are used to sire because of their famous effect on lean meat percentage.

Conventional pig production has developed with the use of larger production units, which have become increasingly specialised and produce more or less independently of the agricultural area. Moreover, a shift has taken place to concertrate pig production in centres in Denmark, Brittany in France, Flanders in Belgium, Lower Saxony and Westphalia in Germany as well as Catalonia in Spain and the eastern parts of the Po valley in Italy. However, from a global perspective, pig production in Europe has to face competition from growing export-oriented, low-cost pork producing regions such as North America, Brazil and potentially several Asian countries.

9.3.2 Low-input production systems

Low-input farming systems operate in-between conventional and organic systems, trying on the one hand to minimise the use of external resources and synthetic tools and on the other hand to implement basic principles such as in organic systems. Owing to a lack of a clear definition, low-input systems of husbandry methods are difficult to characterise. The type and amount of input especially vary between farm units to a high degree. In comparison to industrialised systems, the use of bought-in feedstuffs, chemicals like growth promoters, prophylactic use of antibiotics, and so on is reduced.

In recent years, a large number of labelling programmes have been developed which can be summarised under the topic of 'low-input' systems. They often include some prescription in relation to housing methods and with respect to the use of chemicals or bought-in feedstuffs. In low-input systems a wide range of genotypes is used. While production is mainly based on those types that are also used in conventional production, some special breeds are used in labelling programmes. Iberian pig production in La Dehesa in Spain (Lopez-Bote, 1998) and Bisaro pig production in Portugal (Santos e Silva *et al.*, 2000) are examples of unconventional production systems with specific and highly demanded characteristics. These production systems have served as models to introduce new production systems, for example free-range systems. At present, only limited information is available regarding the influence of these systems on eating quality and some of the data available often involve eating quality problems rather than improvements (Jakobsen and Hermansen, 2001). Owing to the huge variation within this production method, the following does not especially refer to this production method.

9.3.3 Organic farming systems

Based on production guidelines, organic livestock farming has set itself goals of establishing environmentally friendly production, sustaining animals in good health, achieving high animal welfare standards and producing products of high quality. The idea and concept of organic livestock production was developed to provide an alternative to the progressive intensification in conventional animal production. The concept refers to the whole farm as the base and starting point of a comprehensive system. Realisation of a system-oriented approach usually requires a complete re-organisation of the farm, in which plant requirements have to be tailored to the concerns of animal husbandry and the extent and direction of animal husbandry adapted to home-grown feedstuffs. The aim is to achieve sustainable production of animal products principally through precautionary and avoidance strategies.

While the energy supply does not provoke severe difficulties, the supply of essential amino acids is a challenge under the framework conditions of organic agriculture. The possibilities and limitations of protein supply in organic pig production have been recently reviewed (Sundrum *et al.*, 2005). As soybean meal from conventional processing and synthetic amino acids are banned, grain legumes represent the main protein source. The availability and digestibility of essential amino acids and the type and quantity of anti-nutritive factors (ANFs) are most relevant with respect to the maximum inclusion rate of home grown corn legumes. Moreover, the replacement of conventional by organically produced commercial feed (like soybean meal) comes to the fore in order to meet the protein requirements under organic conditions. In contrast to conventional production, determining factors for the use of feedstuffs are not only the price, but in particular the availability of home grown feedstuffs. Every farm has its own optimum standards with respect to marginal utility for the use of bought-in feedstuffs.

The genotypes used in organic farming are often the same crossbreeds as in conventional production but also encompass a wide range of rare breeds characterised by specific traits. Sometimes conservation of rare breeds is the objective in different farm units.

Organic systems vary between different European countries, depending on the prevalent conventional pig production systems. In Spain, the traditional Dehesa systems have lent themselves to organic conversion fairly easily (Trujillo and Mata, 2000). Similarly in the UK, the prevalent outdoor breeding units have been relatively easy to convert to organic production, where farm size and crop rotations have provided adequate space. In western Europe and in Scandinavia, more intensive systems have been converted to organic production by running the breeding stock outside during lactation and allowing pigs after weaning access to outdoor pens (Vermeer *et al.*, 2000; Baumgartner *et al.*, 2002). In Germany, organic pig production is primarily limited to small units where pigs are kept indoors with access to an open yard (Sundrum and Ebke, 2004), whereas outdoor fattening is the exception rather than

Table 9.2 Differences in priorities between conventional and organic production systems

Conventional	Organic
(i) Minimising production costs	(i) System-orientated production, based on land use and use of organic feedstuffs
(ii) Maximising productivity of farm animals	(ii) Maximising efficiency within the whole farm system
(iii) Maximising performance (carcass yield, lean meat percentage)	(iii) Optimising product and process quality (animal health and welfare, environmentally friendly production, naturalness)
(iv) Optimising single quality traits	(iv) Reducing production costs

the rule. In contrast, outdoor fattening has developed into a standard production method for organic pigs in Scandinavia (Jakobsen and Hermansen, 2001).

In 2002, there were estimated to be 553,000 organic pigs within the EU (Olmos and Lampkin, 2005). The major producing countries are Germany, Denmark, France and the UK. While total livestock numbers increased in the years that followed, especially in the new member states, the number of pigs declined in most of the main pig production countries between 2002 and 2003 to a total number of about 472,000 pigs. Although increase and decline in demand for organic pork cannot be extrapolated into the future, from previous knowledge several challenges can be identified that must be faced in organic pig production with regard to the future development.

9.3.4 Difference between organic and conventional livestock production

In organic livestock production the objectives of a land-based system, the avoidance of specific production methods, and the priority of quality production rather than maximising production are of overriding importance. Therefore, a main feature of organic production is dealing with limited availability of resources, while maximisation of protein accretion is of subordinate importance. Differences in priorities between conventional and organic livestock production and a comparison of hierarchy related to objectives are presented in Table 9.2. With regard to the different objectives, different priorities, and different framework conditions, it has to be taken into consideration that organic and conventional livestock production belong to completely different farm systems. Therefore, the traditional approach which reduces agricultural problems to the level of single production traits is not directly comparable and compatible with the organic approach. General conclusions derived from conventional production systems are not directly compatible and therefore do not have the same validity in organic livestock production.

9.4 Consumer perception

Consumers are interested in many aspects related to the quality of food, such as appearance, freshness, taste, nutritional value and food safety. Different consumers show different preferences and subjective perceptions in relation to different features. People in different countries rank quality criteria differently. Price is an extremely visible attribute of pork related to quality by the notion of value. When purchasing pork products, customers try to assess the other quality characteristics, that is, the degree to which the need and expectations have been met in relation to the price given. A survey in Scotland indicated that price and product appearance are the primary meat selection criteria, the latter being used as a predictor of eating quality (McEachern and Schröder, 2002). Consumers are also becoming interested in foods produced according to ethical aspects of animal welfare principles and the aspect of environmental pollution. However, the main focus of interest is expected to vary to a high degree between groups and individuals.

Quality evaluation of fresh meat by consumers consists of two phases: (i) prior to the actual purchase and (ii) while eating the meat (Glitsch, 2000). Concerning the first stage, place of purchase and colour play a major role, while the price is considered to be less significant. After purchase, flavour is one of the most significant quality characteristics. According to Andersen (2000), appearance often has become the customer's only consideration when evaluating the quality of pork. Consumers look for colour and fluid-retaining characteristics in the hope that these will indicate the potential eating quality. However, consumer preferences vary to a great degree. While some groups of consumers exhibit the greatest preference for dark, firm and dry (DFD) chops, others prefer normal appearing chops and a relevant number of consumers even show a preference for pale, soft and exudative (PSE) chops (Jeremiah, 1994). Compared with highly marbled chops, leaner chops are darker and have a more acceptable appearance and are more likely to be purchased. However, they are also less tender, juicy, oily and flavoursome, indicating a disparity between purchase intent based on visual evaluation and quality attributes of the cooked product (Brewer *et al.*, 2001).

An increase in intramuscular fat (IMF) level is associated with an increase in visual perception of fat and a corresponding decrease in the willingness of consumers to purchase the meat, when expressed before testing (Fernandez *et al.*, 1999). On the other hand, the perception of texture and taste was enhanced with an increased IMF level. The authors conclude that the positive effect of increased IMF holds true as long as it is not associated with an increase in the level of visible intermuscular fat. However, consumers also show a huge variability in their eating preference for non-visible IMF. In an hedonic test panel nearly one-third of the consumers behaved indifferently towards an increase in IMF, for a minority preference of pork decreased with an increase in IMF, while the majority preferred the pork with the highest IMF (Sundrum and Acosta Aragon, 2005).

Alternatively produced products have in common that their unique selling proposition is not directly visible to the consumer. Only additional information will identify the nature of the origin or the production process of these foods (Oude Ophuis, 1993). Within a sensory evaluation of 'free range pork' under different conditions of experience and awareness, labelling and prior experience of the product have favourable influence on the sensory evaluation of 'free range pork' for a number of attributes. The author assumes that contextual elements are very important in the sensory evaluation of fresh foods.

A number of food scandals have stimulated consumer concerns about the safety and quality of food. The 'convenience-driven' consumer segment especially is characterised by being concerned with meat quality and safety (Raewell, 1999). Sometimes the concept of animal welfare is mixed with items of food safety and sensory quality. Some consumers appear to delegate responsibility for ethical issues in meat production to the meat retailer or the government, as consumers do not seem to wish to be reminded about issues connected with the animal when choosing meat (Bernués *et al.*, 2003). Knowledge of production systems often appears of little consequence in terms of any pork market potentials, as consumer groups often freely remark that there is no link between the negative images of production methods and their purchase behaviour (Ngapo *et al.*, 2003). According to the authors, consumer groups are clearly confused and mistrust the limited information available at the point of purchase. According to Von Alvensleben (2003), consumers often are in a conflict situation where it is difficult for them to decide whether it is worthwhile or should they be willing to pay a premium price for a premium product. In order to avoid dissonances and incoherencies when confronted with inconsistent information and uncertainty, consumers often ignore or suppress specific aspects that do not fit into their world view, or justify their doubts in relation to the credibility of the information, or deny being involved.

In several countries, the first attempts have been made to use the information obtained throughout the whole production chain in different marketing strategies to evoke positive associations with the product. The change from confinement to more free-range systems has been one of the tools used to sell stories and improve marketing of meats. The promotion of such meat production systems has been based on anecdotal information rather than on real facts regarding the overall quality obtained (Andersen *et al.*, 2005). Introduction of old breeds, which have not been exposed to genetic selection for decades, has become another tool to improve the image of meat (Santos e Silva *et al.*, 2000; Hollo *et al.*, 2003). This approach has allowed the introduction of the term 'original' to show potential in successful marketing of foods. Introduction of the wholesomeness concept in meat production, most often represented by organic production, is mainly due to a wish to re-establish positive meat sector images, for example, meat safety and animal welfare aspects (Verbeke and Viaene, 2000). Consumers make a whole range of positive inferences from the label 'organic' and these refer not only to concern for the environment and health, but also to animal welfare and better taste (Bech-Larsen and Grunert, 1998; Bruhn, 2002). However, positive interferences do not necessarily lead to a purchase if consumers do not think that the trade-off between give and get components is sufficiently favourable.

9.5 Product quality

The modern use of 'quality' raises problems: how to measure and evaluate quality, what is measured and which scale has to be used? It is commonly acknowledged that meat quality is a difficult characteristic to assess, as many different aspects, both objective and subjective, make up the overall trait (Hofmann, 1994). Plenty of literature is available describing quality of pork quality either by using a single criterion for the product or by enumerating various traits in the production process. Without a fairly precise idea of what is measured and how the measurement is performed, a term such as 'quality assurance' makes no sense (Andersen, 2000). The customers of fresh pork are both the meat-processing industry and the consumers, who buy 65–80% and 20–35%, respectively, of the pork produced (Andersen, 2000). Currently, the market does not distinguish between technological qualities for processing and those for direct consumption. However, in the future there may be reasons to provide different pathways for technological and eating quality, which are therefore elucidated separately.

9.5.1 Technological quality

The term 'technological quality' embraces all attributes relevant to the processing of fresh pork, including general meat processing and specific processing, such as for catering.

Food safety. From the perspective of technological quality, safety is the major issue, as control of safety during processing and distribution of meat products according to legislation calls for high safety demands. Additionally, safety refers to the shelf life of meat products.

Uniformity. The term uniformity includes muscle size, lean meat percentage, fat content of the carcass and pH-value or marbling pattern as characteristics of the meat composition. Uniformity is crucial especially in brand production (Branscheid, 1996) and becomes increasingly important as a consequence of ongoing automation at the slaughter line and in the meat-processing industry. The objective to obtain meat cuts and products that fit into the 'convenience-driven' consumer segment, give further rise to greater uniformity in simple handling and preparation procedures.

Water-holding capacity (WHC). WHC includes the ability of fresh pork to retain water and bind extra water. The higher the WHC, the more valuable the pork will be for use in highly processed products. Drip loss is an ongoing process involving the transfer of water from myofibrils to the extracellular space affected by structural features at several levels of organisation within muscle tissue (Bertram *et al.*, 2002). The physical features of muscle that affect, or result from, fluid loss also influence pork colour. Increased extracellular fluid results in a pale product owing to greater light reflectance and scatter. Many of the structural features of meat that affect colour and water-holding capacity are dictated by early post-mortem events. According to Schäfer *et al.* (2002), early post-mortem pH and temperature are sufficient to account for almost 90% of the variation in drip loss from pork. Markedly higher drip losses in pork with low pH-values occur especially after frozen storage (Person *et al.*, 2005). Even though higher pH results in better WHC of pork, it also results in inferior colour, flavour and microbial stability of the product.

Lipid characteristics. If the lipid becomes too unsaturated, the meat is not suitable for, for example, sausage production. Furthermore, products become oxidative unstable, accelerating rancidity problems, especially in many preheated catering products with an increased incidence of the development in warmed-over-flavour. Therefore, the anti-oxidative status of pork, for example content of vitamin, is an important technological quality criterion.

9.5.2 Eating quality

The image of fresh pork is to a high degree determined by its sensory quality (palatability, tenderness, juiciness) and its appearance (especially colour, visible fat content and WHC) at purchase in the shop (Schifferstein and Ophius, 1988; Becker *et al.*, 2000).

Appearance. Even though appearance does not have much to do with the eating quality of pork, it is becoming increasingly important in determining the consumer's purchase of pork. A bright red colour is perceived by consumers as indicative of freshness, while consumers discriminate against meat that has turned brown in colour. The rate of discoloration of fresh meat is related to both the rate of pigment oxidation, oxygen consumption and the effectiveness of the metmyoglobin enzymatic reducing systems (Ledward, 1991). Moreover, drip loss and visible fat play an important role with regard to attractiveness.

Tenderness. The contractile state of the muscle after rigor mortis is a major factor in meat tenderness, which is affected by post-mortem conditions creating differences in tenderness. Ageing of fresh pork can be used to improve tenderness. The process is based on a continuous weakening of the structural elements by different endogenous muscle peptidases along with an improved palatability (Taylor *et al.*, 1995).

Juiciness. Juiciness of pork is associated with the amount of moisture present in the cooked product and the amount of IMF. Increase in ultimate pH of the pork is associated with increased moisture retention in the heated product. In contrast, low pH after slaughter, which gives rise to PSE pork, is closely related to low moisture content of cooked pork. Also the distribution of water within muscles and the WHC affect juiciness and tenderness (Offer and Knight, 1988).

Meat flavour. Meat flavour represents a large number of compounds formed during heating of the product. The flavour development mainly depends on constituents in the fresh meat, for example, fat composition, peptides, glycogen concentration, vitamin content and the heat treatment of the product. Increasing temperatures increase flavour development. According to Andersen (2000), the intensity of pork flavour has decreased during recent years, most probably as a result of the production of pork with a minimal content of intramuscular fat.

Intramuscular fat (IMF). The importance of the IMF content has been defined by many authors as the flavour carrier (Schwörer and Morel, 1987; Affentranger *et al.*, 1996). Only the finely distributed fat in the muscle, recognisable in higher contents as marbling, makes a taste differentiation between animal varieties possible, whereas low-fat muscle meat is almost neutral in taste. Moreover, IMF is decisive in relation to the capacity to maintain juiciness and, when cooked, by tenderness and a typical aroma (Fischer, 2001). While the IMF optimal for taste is between 2.5% and 3% (Fernandez *et al.*, 1999), modern slaughter pigs currently show an average IMF content of only 1% (Köhler *et al.*, 1999). Higher pH-value and lipid content tend to decrease mechanical resistance and hardness of meat (Monin *et al.*, 1999). According to Warkup and Kempster (1991) a combination of high growth rate and high IMF is beneficial to eating quality whereas low growth rate and low IMF produce tougher and drier meat.

9.5.3 Factors affecting pork quality

Factors of significance for pork quality have been in the focus of pig science for decades. However, most of the present knowledge is based on studies investigating the influence of a single or at the most two factors. However, there are limited data in the literature concerning the understanding of how production and slaughter factors interact in relation to pork quality, particularly in relation to on-farm production practices (Rosenvold and Andersen, 2003). In the following, the most relevant factors are outlined.

Genetics

Genetics are known to have a marked influence on both production and quality traits. The capacity for protein accretion is determined principally by the genotype. However, the dominant effect of the first-limiting amino acids (AA) for protein accretion cannot be affected by the characteristics of the animals (Susenbeth *et al.*, 1999). Through breeding for high meat content and low feed expenditure, the extent of the utilisation of AAs for protein accretion has not changed (Susenbeth, 2002). Thus, the superiority of animals with a high genetic capacity for protein accretion becomes manifest only in combination with a corresponding protein supply. Numerous reports on the development of PSE pork have focussed on major gene effects. Advances in the understanding of calcium release from the sarcoplasmic channel led to the discovery of alterations in amino acid sequence (HAL-1843 mutation) in the calcium release channel protein, also called the ryanodine receptor (RYR1; Fujii *et al.*, 1991). The n allele of the HAL gene affects numerous pork texture traits in a way that can be considered as detrimental to acceptance (Monin *et al.*, 1999). The meat is less rough, more cohesive, harder, more fibrous, less granular, more elastic and less easy to swallow. Also the mechanical resistance for raw and cooked meat is higher in comparison to meat from pigs without the HAL-1843 mutation.

The Napole gene is another genetic abnormality that can lead to inferior pork quality. The dominant RN^- allele results in higher than normal muscle glycogen stores and in extended post-mortem pH decline that leads to pork with a lower than normal ultimate meat pH, reduced water-holding capacity and greater cooking losses (LeRoy *et al.*, 2000; Jonsäll *et al.*, 2001). Monin and Sellier (1985) referred to this condition as the Hampshire effect, owing to its prevalence in the Hampshire breed.

Despite an abundance of research describing PSE pork characteristics and a reduction in the frequency of major genes with known deleterious effects on pork quality, Cassens (2000) concluded that little progress has been made in reducing the incidence of PSE pork. This highlights the complexity of the PSE pork problem, which is not only restricted to the alleles of genes involved. Murray and Johnson (1998) reported that, in a population of more than one thousand pigs, 90% of the PSE condition was caused by factors other than the HAL-1843 gene. Oeckel *et al.* (2001) conclude from the results of a consumer panel that the elimination of the HAL gene from the pig population does not guarantee a better palatability of pork. If the main gene defects (HAL and RN genes) can be excluded, genetics, in the view of De Vries *et al.* (2000) and Tribout and Bidanel (2000), contributes only by a proportion of less than 30% to the total variation of meat quality criteria.

With the exception of the above-mentioned major genes, the heritability of most attributes referring to the quality of pork is low, except for that of IMF content (reviewed by Rosenvold and Andersen, 2003). Increased IMF is associated with improved eating quality. Although the heritability of IMF and fat tissue is high, the genetic correlation between them is very low (Wood, 1990), which suggests that selection for high IMF in lean carcasses should be possible. However, an increase of IMF requires incorporating generally fatter breeds (e.g. Duroc) and is expected to lead to decreased income caused by poorer grading (Glodek, 1996). On the other hand, a successful implementation of a breeding programme focussing on IMF content in pork has been reported from Switzerland (Schwörer *et al.*, 2000).

In the past, there has been considerable interest in the use of genotypes (especially Duroc) with high levels of IMF in an attempt to improve eating quality. There have been various evaluations of the influence of the Duroc breed on eating quality with variable results. While several studies have found no significant advantage for the Duroc breed for eating quality (McGloughlin *et al.*, 1988; Martel *et al.*, 1988; Edwards *et al.*, 1992), other studies highlight the benefit of this breed in several but not all quality traits (MLC, 1992; Ellis *et al.*, 1996; Sundrum and Acosta Aragon, 2005). However, within the current carcass grading and payment system, final products of stress resistant sire lines or those producing a high IMF content (e.g. Duroc) are not favoured because of higher back-fat and lower lean meat (Glodek, 1996; Laube, 2000). Thus, products with a high pork quality have the lowest market price.

Sex

The difference in sensory quality between females and castrated males is not consistent over several studies. Enfält *et al.* (1997) and Jonsäll *et al.* (2001) found that loins from castrated males scored higher for tenderness and juiciness than loins from females. In contrast, Jonsäll *et al.* (2000) found no sensory effect of sex on loins and the same working group detected in a further investigation that loin from gilts scored higher for juiciness and lower for off-flavour than loin from castrated males. Obviously, the effect of sex on sensorial quality is of minor relevance and can be overruled by other effects.

Feeding strategies

Pig producers mainly try to approach maximal rates of lean tissue deposition and carcass index values by providing diets formulated to meet all of the known requirements. In the growing period, protein accretion increases as the supply of limiting amino acids increases (Heger *et al.*, 2002). The dose–effect ratio can be subdivided into the nutrition-dependent phase, which is substantially linear, and the plateau phase, which is independent of nutrition supply and whose maximum depends on features of the animals, primarily characterised by the genotype (Susenbeth, 2002).

Feeding is an important tool, as many dietary components are readily transferred from the feed to the fat tissues. Thus, the fatty acid (Wood and Enser, 1997) and vitamin composition (e.g. vitamin E) in the diet (Buckley *et al.*, 1995) has a direct influence on the quality of the meat. Extensive experimental evidence exists for feed induced optimisation of the fatty acid profile, that is, an optimal ratio between saturated, monounsaturated and polyunsaturated fatty acids, to meet the dietary recommendations of humans in their intake of pork (Jakobsen, 1999). The large amount of unsaturated fatty acids are responsible for softer back-fat observed for pigs fed corresponding feedstuffs. The 'softness' of pork can be a problem for further processing (e.g. sausage production), especially when pigs have been fed too high a content of polyunsaturated fatty acids (Jorgensen *et al.*, 1996; Warnants *et al.*, 1998).

Increased awareness by consumers of the link between an increased intake of omega-3 fatty acids and a reduction in the risk of cardiovascular disease, has led to an increased demand for products with a higher content of these unsaturated fatty acids. Leskanich *et al.* (1997) increased the omega-3 fatty acid content of pig fat, without effect on the sensory attributes of the meat, by feeding 2% rapeseed oil and 1% fish oil to pigs during the finishing phase. However, increasing the content of these fatty acids in the pig feed can lead to a problem with rancidity, as the unsaturated fatty acids become oxidised. Sensory analysis data have provided evidence of a higher detection rate of off-flavours in processed pork derived from pigs fed diets containing low-erucic acid rapeseed oil, owing to the presence of oxidation products of linolenic acid (Wood and Enser, 1997).

Oxidatively stable raw materials are necessary to obtain a profitable shelf life of the products (Sheard *et al.*, 2000). When fed above requirement levels, vitamin E increases the oxidative stability in fresh pork and pork products considerably (Jensen *et al.*, 1998). Selenium is also involved in reducing lipid oxidation; there is no evidence, however, that supplying additional selenium above the requirements improves pork quality (NRC, 1998).

By feeding diets containing the optimal protein:energy ratio, an increase in protein accretion together with a reduction in carcass fat content can be realised. Several authors highlight the fact that the limit of leanness that will still allow a palatable commodity to be marketed has already been exceeded to a large degree (Honikel, 1996; Fischer, 2001). Negative selection in the back-fat thickness of the animals decreased the IMF content of the meat considerably (Rosner *et al.*, 2003).

According to Essen-Gustavsson *et al.* (1994), Castell *et al.* (1994) and Sundrum *et al.* (2000), the IMF content in the musculus longissimus dorsi can be clearly increased by a longer ratio of energy : crude protein (lysine) content in the diet without increasing fat area and back-fat thickness. Nutritional effects on intramuscular fat characteristics and content are clearly greater than genetic effects (Cameron *et al.*, 2000; Sundrum and Acosta Aragon, 2005). Furthermore, an increase of 30 days in the age of pigs slaughtered at 110 kg body weight greatly influences the carcass and the muscle chemical composition, including the IMF in *M. longissimus* (Lebret *et al.*, 2001).

In organic pig production, all animals must have access to roughage, primarily for welfare reasons. Increase of bacterially fermentable substances (BFS), enriched with roughage, reduces growth, protein retention and feed conversion efficiency (Kreuzer *et al.*, 1999). Furthermore, carcass yield is reduced with increasing BFS content of feed. Pigs with access to roughage and fed restrictively with a concentrate diet showed increased lean meat percentage and a decreased tenderness, while other eating quality attributes remained unchanged (Sather *et al.*, 1997; Danielsen *et al.*, 2000). Fatty acid composition of pork from free-range pigs with access to roughage has been reported to be more unsaturated compared to conventional production (Hansen *et al.*, 2000; Nilzén *et al.*, 2001). Moreover, recent studies showed that feeding with lupines has the potential to reduce the amount of skatole in back-fat of pigs (Hansen and Claudi-Magnussen, 2004). Muscle glycogen stores have been recognised to be decisive for meat quality. Muscle glycogen stores at the time of slaughter can be manipulated through feeding (Rosenvold *et al.*, 2001, 2002) and thus influence the rate of pH decline and possibly the technological pork quality.

Housing methods and management practice

Standardised production systems are of high priority for the modern pig industry, which is based on slaughter lines with rapid carcass flow and produces meat of known quality (high uniformity) compared to many of the alternative management systems or niche productions, which are emerging owing to customer demand. The intermediary mechanism between animal welfare, effect of stress and impaired organoleptic meat quality is situated at the level of the muscle cell metabolism. The more anaerobic metabolism can be avoided, the less risk for a deteriorated meat quality. Hence, risk factors impairing ethical and organoleptic features of fresh pork are really overlapping. However, studies comparing indoor and outdoor fattening are inconsistent. According to Edwards and Casabianca (1997) measures of carcass and meat quality have generally failed to demonstrate any superiority of products from outdoors. Several studies even showed a disadvantage of outdoor production in relation to the eating quality of pork compared to indoor production. In investigations by Enfält *et al.* (1997) and Jonsäll *et al.* (2000) loins from pigs reared outdoors were less juicy and had a lower intensity of acidulous taste than loins from pigs reared indoors. In contrast, Jonsäll *et al.* (2002), Stern *et al.* (2003) and Gentry *et al.* (2004) could find no differences in pork eating qualities of outdoor reared pigs compared to intensive rearing. According to Beattie *et al.* (2000) pigs from enriched environments produced pork with greater tenderness than pigs fattened in barren environments.

Animal health

It is well known that diseases in fattening pigs can lead to a considerable reduction in growth performance. Respiratory diseases cause the highest economic losses owing to reduced growth and higher production costs including costs for veterinarians and remedies (Miller and Dorn, 1990, Hammel and Blaha, 1993). The impact of animal diseases on quantitative carcass traits generally seems to be of minor importance (Doedt, 1997) and can even improve the evaluation by carcass grading because of a reduced fat and an increased lean meat content (Wittmann *et al.*, 1995). The literature provides inconsistent results on the impact of diseases on meat quality traits. While some investigators found a higher incidence of PSE meat in the carcass of pigs with respiratory diseases (Hammel and Blaha, 1993; Doedt, 1997), others did not found such a relationship (Wittmann *et al.*, 1995) or even assessed a reduced incidence of 'PSE meat' while the incidence of 'DFD meat' clearly increased (Schütte *et al.*, 1996). Obviously, the implications for quality traits vary between animals and between farms. They are not only related to the extend of the stress factor itself but are modified to a high degree by the specific physiological status and the specific circumstances. In general, pathological findings in the plucks of pigs are not the exception. They are often found to a high degree in pigs assessed at the abattoir (Doedt, 1997; Bostelmann, 2000; Sundrum and Ebke, 2004), indicating a serious need for action.

Procedures at slaughter and post-mortem
Pre-slaughter handling includes removal from the home cage, transport, mixing of unfamiliar pigs, loading and abattoir lairage. These handling practices can all induce stress caused by exposure to novel environments, interactions with humans and aggressive interactions with conspecifics. Pre-slaughter stress is both an animal welfare issue and can adversely affect the quality of pork (Warriss *et al.*, 1998). Animals that have fought during pre-slaughter handling show glycogen depletion in muscles leading to a progressively higher ultimate pH in the meat, which results in an increase in the incidence of DFD meat (Faucitano, 1998; Warriss *et al.*, 1998). Prolonged transport of pigs (8 h versus 0.5 h) has been found to improve tenderness caused by reduced glycolytic potential at the time of slaughter and subsequent higher ultimate pH (Leheska *et al.*, 2003). Van der Wal *et al.* (1997) found that a 3–4 h resting period before slaughter was optimal with respect to pork quality. Short-term stress immediately prior to stunning is shown to result in lower pH values and higher temperatures early post-mortem (Van der Wal *et al.*, 1997, 1999; Stoier *et al.*, 2001), which is generally accepted to reduce the WHC of pork. Finally, CO_2 stunning of pigs should be recommended when the aim is superior WHC, as it is generally accepted that this results in higher WHC compared with electrical stunning (Channon *et al.*, 2002).

9.5.4 Relationships between quantitative and qualitative traits

A literature review shows that a large number of traits (including terminal sire genotype, sex, slaughter weight, feeding and housing regime and slaughter house) have been found to have a more or less clear effect on carcass and meat quality in pigs and on the organoleptic properties of fresh pork. The fact that relatively few significant interactions for eating quality traits have been found in multi-factorial experiments (Ellis *et al.*, 1996; Sundrum and Acosta Aragon, 2005), means that it can be assumed that the effects of most of the factors are likely to be additive. There is reason to assume that the degree of influence is not only due to the factor itself but very much depends on the system context (context-variant) in which they are at work.

For a high variation in growth rate as well as in protein and fat accretion differences in living conditions, environmental temperature (LeBellego *et al.*, 2002), stocking rate (Edmonds *et al.*, 1998) and general stress situations (Whittemore *et al.*, 1988; Hyun *et al.*, 1998a; 1998b) are held responsible. In an on-farm investigation, Elbers *et al.* (1989) found, for example, a considerable variation in the digestibility of organic matter (77–84%). The variation between the farms was greater than within the farms.

Although uniformity is often emphasised to be of high importance for the pork industry, only few in the pork industry and in the primary production of pigs have realised or cared about the huge variations in most quality traits (Andersen, 2000). Trade value of carcasses is determined principally by the lean meat proportion, slaughter weight and cuts composition on the basis of price labels. In responding to market pressures, improved nutritional practices in combination with genetic selection has improved the leanness of pork considerably. With an increase of lean meat and an extreme reduction of overlay fat, the risk that meat deficiencies will appear increases (Harr, 1989; Lengerken, 1990). Meat research experts have been demanding for years that a muscle meat proportion of 55%, for reasons of meat quality, should not be exceeded (Schepper *et al.*, 1983; Honikel, 1996; Fischer, 2001).

Lean meat percentage and IMF content are correlated negatively (Köhler *et al.*, 1999; Baulain *et al.*, 2000; Sundrum *et al.*, 2000). With corresponding inclusion of the IMF in the breeding strategy, the meat proportion will consequently be reduced. Because breeding for high meat proportion in pigs leads to an increased formation of 'white fibres', extremely hypertrophied muscles are not only almost fat-free, but strikingly light. Pigs with a high lean meat proportion produce more 'PSE meat' (Doedt, 1997). In addition, a high proportion of ham in relation to the total carcass is negatively correlated with the appearance of 'PSE meat' as a consequence of the related muscle hypertrophy (Fewson *et al.*, 1987; Kallweit, 1989). A consideration of the ham form in the payment system thus leads to an increase in meat condition deficiencies (Doedt, 1997).

One-sided breeding for rapid growth and meat quantity performance makes osteochondrosis and stress myopathies in pigs more likely (Claus, 1996; Bickardt, 1998). The accelerated protein accretion puts a strain on a young immature skeleton. The illness is determined by cartilage degeneration with epiphyseolyses and bone proliferative reactions. Pigs with high lean meat percentage and high growth rates showed poor assessment values in respect of carpal and tarsal joints (Lundeheim, 1987; Huang *et al.*, 1995). According to Rauw *et al.* (1998), the characteristic antagonism between growth intensity and health impairment give rise to the supposition that the domestication of pigs has led to reduced ability to deal with environmentally determined stress.

9.6 Animal welfare issues

An increasing number of consumers accept good living conditions, appropriate to the needs of the animals, as a 'process quality', which even justifies premium prices for animal products (Badertscher-Fawaz *et al.*, 1998; Verbeke and Viane, 2000). There is a greater potential for optimal welfare in organic systems compared to intensive systems (Barton Gade, 2002). However, the realisation of minimal standards cannot be treated as equivalent to appropriate livestock housing conditions and high animal health and welfare status

(Sundrum, 1999). Minimal standards are primarily based on political decisions and a compromise between different interests that are not in all cases related directly to the animal welfare issue. The appropriateness of minimal standards cannot be judged by an absolute 'yes or no' decision, but requires a comparative assessment, ranking on a scale between very good and very poor. Furthermore, minimal standards do not include all relevant aspects of animal welfare. Thus, minimal standards represent only a small section of the interrelationship between farm animals and their living conditions. Besides the housing conditions, the quality of stockmanship and management, the feeding, climatic factors or the hygienic situation have an essential influence on animal health and welfare (Bergsten and Pettersson, 1992; Rushen and Passillé, 1992; Thielen and Kienzle, 1994). These factors are not part of the Council Directives or the EEC regulation owing to the difficulties in quantification and to the variation in time.

9.7 Environmental impact

The complex interactions between the different groups of substances, the constantly varying environment in which animals are reared and the considerable fluctuation found in the quantities of substances in circulation mean that any quantification of the emissions emanating from pig production must of necessity be only very approximate. Any assessment of the efficiency of individual measures or specific rearing systems can only be of limited scope and the resulting data can scarcely be considered valid, if the farm context is left out of consideration (Sundrum, 2002). In pig husbandry there are four principal feeding strategies to reduce nutrient losses (Flachowsky, 1993; Canh *et al.*, 1998):

- reduction of the protein content in the feed ration and thus the nitrogen input per animal, through a corresponding choice of protein carriers, differentiated feeding according to living mass and performance (phase feeding) and supplementation with industrial amino acids;
- increase of the animals' performance because the nitrogen emissions will thereby be lowered in terms of kilogram per growth increase or per piglet, if the farm conditions remain equal;
- displacement of nitrogen emission from the urine into excrement by raising the biologically fermentable substance in the ration of nitrogen in the hind gut through microbes;
- reduction of the pH-value in the urine and in faeces and thereby reduction the pH-value in the slurry by a targeted selection of feed components.

Reduction in nitrogen excretion is often set equal to a reduction in nitrogen emission. This is only true in those cases where the excess nitrogen in the commercial fertiliser can be caught and retained in the agricultural nutrient cycle. If the nutrients are, however, discharged from the cycle and emitted into the environment, this can mean correspondingly high environmental damage. A reduction of nitrogen input per animal is possible, especially by means of multi-phase feeding. The greater the use of nutrient-graduated feed mixtures, the higher the reduction potential with regard to nitrogen excretion. Substantial savings can thus be achieved if the phase feeding is accompanied by a simultaneous reduction of protein content (Kaiser *et al.*, 1998).

In organic animal husbandry the possibilities for adaptation of feed rations to the specific requirements by means of principal use of self-produced feed and avoidance of synthetic amino acids are considerably limited (Sundrum, 2002). However, the organic farmer has other possibilities and strategies available to reduce nutrient emissions. The main specifications concerning environmentally friendly production concern renunciation of pesticides and mineral nitrogen, the need to reduce the number of farm animals per area unit and minimisation of bought-in feedstuffs. Without these substitutes, organic farming must rely on efficient nutrient circulation within the farm to maintain soil fertility and high production. Reduction of pollution or energy consumption is reached by a systemic and causally related approach, while conventional strategies are often based on technical and management related measures (Kristensen and Halberg, 1997). The model calculations of Hermansen and Kristensen (1998) showed that mixed farm systems including cattle and pigs, as intended in organic agriculture, provide a clear potential for higher nutrient efficiency and an improved balance sheet compared to specialised farms because interactions between plant cultivation and livestock production can be better optimised.

9.8 Constraints and potentials for quality production

Scientific knowledge of how to produce pork with a high intrinsic and extrinsic quality has markedly increased during the last decades. On the other hand, it is obvious that knowledge alone is not sufficient to change the current unsatisfactory situation and implement an encompassing strategy to improve pork quality. The production of pork takes place within specific framework conditions characterised mainly by the availability of resources (nutrients, housing conditions, investments, labour time, management skills, etc) and by the potential to gain a high price on the market. As described above, framework conditions differ markedly between production systems. Each production system has to face specific constraints with regard to quality production. Some aspects of the different production systems are discussed below without claiming to be exhaustive.

9.8.1 Conventional production systems

A main objective in conventional pig production is to minimise production costs and to maximise those results that are directly rewarded by the market.

The main factors that affect productions costs are prices of feed, young stock and building. Furthermore, carcass yield, feed conversion and the economic gain per kilogram of pork have a high impact on the gross margin. Under the current framework conditions, carcasses with the highest lean meat gain the highest price. At the same time, the most efficient strategy is to reduce production costs as increase in protein accretion goes along with an increase in daily weight gain and an improvement in feed conversion (Susenbeth, 2002). Furthermore, a high proportion of lean meat provides the highest yield and profit for the processing industry. Owing to the antagonistic relationship between quantitative and qualitative traits, the current payment system conflicts with the production of pork with a high eating quality and consequently does not even prevent all the negative side effects of technological quality traits. For example, although an enormous amount of research has been directed towards the PSE problem, and the basis for, and understanding of, the syndrome is in hand, little progress in eliminating or minimising the problem has been recorded (Cassens, 2000). The author concludes that the reason for the little progress is due to the marketing system which does not reward quality by a premium, but rather pays for quantity. Moreover, farmers are punished by deductions if they do not fulfil the quantitative demands of the meat industry.

Focussing primarily on quantitative traits and on a reduction in production costs, there is little space for improvements in quality features. These are strived for in the first place by breeding techniques. However, breeding efforts to improve quality traits within the current framework conditions are limited because of the antagonistic relationship between quantitative and qualitative traits. Moreover, the limitations of quantitative traits become obvious. Thus, the genetic capacity for growth rate and lean meat gain has improved at a faster rate than has actually been achieved in practice. According to Boyd *et al.* (2000) the biggest constraint on the expression of growth potential appears not to be nutrition but immune stress and clinical and subclinical disease. Pigs that are capable of gaining 850–900 g/d (grams per day) in a good commercial environment often fail to achieve more than 700–750 g/d in many operations (Schinckel and De Lange, 1996; Holck *et al.*, 1997). Subclinical disease and immune challenge are generally associated with lower feed intake (reviewed by Johnson, 1997), while simultaneously the amino acid needs of challenged animals increases (Williams, 1998). Anabolic processes are interrupted and coincident catabolic processes are amplified (Spurlock, 1997). The net effect is a marked reduction in growth as the body places priority on defending against challenge. Any attempt to override the process through greater nutrient density seems to be futile.

Intensive pig production systems have shown severe negative impacts on animal health and welfare (reviewed by Borell *et al.*, 1997; Rauw *et al.*, 1998). Further constraints of intensive production are expected owing to the dramatic effects on environmental pollution caused by a high nutrient input and nutrient loss per land area (Erisman and Monteny, 1998; Van den Weghe, 1999). As if the limitations of the conventional approach in relation to quality production are not difficult enough, pig production in Europe is confronted by the challenge of an open global market. There is reason to assume that the present relatively cost-intensive pork production in EU countries has to face increasing challenges in order to compete with the increasing export oriented and low cost pork production in countries like Brazil or China. Currently, there seems to be no preparedness for the up and coming subsidy-free global market.

9.8.2 Low-input production systems

Pork production in low-input systems is highly diversified as the term 'low-input' is not clearly defined. Generally, the low-input is related to nutrients and chemicals. Because of production conditions that differ from conventional production, products are often offered as brand products. These systems included traditional southern European labelled products (e.g. 'Pata negra') a cured product with a high economic value from very extensive husbandry practices where genotype–environment interactions result in measurable effects on pork quality (Lopez-Bote, 1998). Also pork from meat labelling programmes have been shown to produce meat of higher quality compared with those without a labelling programme (Bostelmann, 2000). Obviously, the specific guidelines of meat labelling programmes in relation to breeding, transport and handling before slaughter are able to provide a positive influence on pork quality. On the other hand, pork from free range systems shows inconsistent results in relation to eating quality (Entfält *et al.*, 1997; Jonsäll *et al.*, 2002; Gentry *et al.*, 2004. In labelling programmes, single traits are often put into focus and highlighted with respect to a specific consumer clientele, while other traits are not different from conventional production. However, there is widespread evidence that consumers overestimate the predictiveness of such characteristics (Grunert *et al.*, 2004). Owing to various impact factors, the variation in the level of quality traits is expected to be huge, both between low-input systems and between production units. Therefore, conclusions in relation to the limitations and potential for quality production are difficult to draw and cannot be generalised but need clarifying in each production system.

The consumers' intrinsic desire for variety is recognised as an important characteristic that influences consumers' food choice behaviour and will become an ever-increasing factor in future food marketing owing to the increasing allowances for food consumption in most industrialised countries (Thiele and Weiss, 2003). Consequently, consumers' demand for food diversity might be one of the driving forces in the development of new concepts for fresh meat production in brand labelling programmes, that is, meats that differ in quality attributes, using all the possible quality control tools from conception to consumption.

9.8.3 Organic production systems

Organic pig production is clearly demarcated from the conventional production by the EEC regulation on organic agriculture as the basic standard. However, production conditions vary from outdoor pig production to indoor production and encompass huge differences between regions and seasons not only in relation to environmental conditions but especially in relation to nutrient supply and genotype. While in conventional production external resources, for example, high quality protein in the feed ration, are used to a high degree and only suboptimal living conditions for the animals are provided, the organic approach is restricted to the use of feed materials of organic origin whilst providing good living conditions for the animals. The holistic approach of organic farming is in line with the growing wish for improved traceability by the consumers, including the ever-rising consumer concern in relation to the introduction of genetically modified organisms in feed and food production. Owing to the far-reaching renunciation of external nutrient resources, organic pig production can claim to be environmentally friendly as a low input of nutrients results in low nutrient losses (Sundrum, 2002). Furthermore, the prescriptions concerning housing conditions and husbandry practice provide good conditions for the animal welfare issue. However, the framework conditions are not equivalent to a high animal health status (Sundrum, 1999). There is reason to assume that additional efforts have to be implemented to ensure a high health status and to ensure food safety. To manage a farm as a comprehensive system is a great challenge and requires highly qualified management skills. The mixed farm system, which is strived for in organic farming, contrasts with the specialisation in conventional farms and restricts the ability to meet the demands of a high quality production because of the widespread issues that have to be dealt with.

Because of the restricted availability of limited amino acids in organic livestock production, protein accretion capacity is limited. To optimise the use of limited resources, the farmer has to meet the challenge to adapt the level of amino acid supply to the protein accretion capacity of the animals to a high degree, as suboptimal supply reduces the performance, while excess supply of amino acids cannot increase performance further. Various measures are at the farmer's disposal to optimise the use of limited resources and to adapt the supply of limited amino acids to the growth process during the various stages of production (Millet *et al.*, 2004; Sundrum *et al.*, 2005). However, these measures are mainly characterised by increased effort, for example in terms of time input and expenditures. Thus, organic farmers in no way compete with the productivity and the production costs of conventional livestock production (Kempkens, 2003). When organic pork is offered, those choosing the organic variety expect it to be of better quality across all quality dimensions, including taste and tenderness. However, the quality experienced after preparing and eating organic pork provided varying results and often fell short of expectations (Grunert and Andersen, 2000; Jonsäll *et al.*, 2002; Olsson *et al.*, 2003). Thus, previous results show the pitfalls of positioning a product on process characteristics, which have little or unclear effects on those quality dimensions of the product that are accessible to consumer experience. Process characteristics may affect the formation of quality expectations more as general quality indicators than as single attributes.

On the other hand, reduced growth rates in organic pig production can be a good starting point for a better sensorial quality of pork owing to the antagonistic relationship between traits related to performance and those related to sensorial quality. However, the sensorial quality of meat does not occur automatically when extensifying the production process but needs special management skills to balance the various relevant factors in a comprehensive approach. Therefore, organic farming cannot *per se* claim to produce highly sensorial quality products but has the potential to do so if appropriate management tools are put into place. Instead of serving images that may be in the 'eye of the beholder', organic pig production faces the challenge of producing and offering pork products with clear and defined qualitative traits, thereby making the difference between a cheap no-name product and a premium product.

9.9 Conclusion

Quality is a multifactorial concept that includes a large number of factors which have to be taken into account and which are heterogeneous in outcome and in the perception of stakeholders and consumers. Currently, there is huge inconsistency between expectations and reality in relation to the level of different pork quality traits and the uniformity of products from different production systems.

Although often claimed by the pig industry, earlier pig production methods cannot claim to have been consumer driven. Independent of the production method, the criteria that are currently important to the farmer selling slaughter animals are not the same as those determining pork quality after slaughter. Although striving for a high lean meat carcass has clearly reduced the fat content, thereby pleasing a majority of consumers, this progress has become self-perpetuating as simultaneously the production costs decrease and economical gains increase. Thus, pig production is constrained by circumstances which cannot easily be changed despite the fact that the eating quality of pork has decreased considerably. While the difference between 1.0 g and 2.5 g fat in a 100 g chop is of negligible importance for the daily calorie intake of consumers, it is of great relevance for the eating quality and for the production costs.

Pig production today faces the difficult task of effectively meeting emerging consumer concerns while remaining competitive in its major target markets. Providing a high sensorial quality of pork and meeting consumer concerns about product safety and animal welfare are identified as key attention points for future livestock production. The relevance of these issues pertains to production efficiency and economic benefits and to the re-establishment of the meat sector image and consumer trust.

In current practice, the evaluation of a carcass according to the EUROP grading system is almost the only method used throughout Europe for quality forecasting, although it is a quantitative rather than a qualitative trait. The previous approach was very successful in producing lean meat on a low cost base. However, it had the disadvantage that the antagonistic relationship between quantitative and qualitative traits, as well as the negative side effects of maximisation of meatiness on animal health, have been easily neglected owing to a lack of data. Moreover, focussing only on quantitative traits and more or less ignoring quality traits does not diminish the huge variation between products in relation to quality traits. As a consequence, quality has been decreasing and commuted to a low level. The huge variation in intrinsic and extrinsic quality traits of pork cannot be explained by making a distinction between conventional, low-input or organic pig production. There are various traits in product quality where values overlap to a high degree, while the traits of process quality are more specific to one of the three production methods. However, differences in product and sensorial quality between production methods seem to be primarily due to variations within the production method rather than related to characteristics of that method. Standardised framework conditions for production (final crossbreds, feeding regime, housing conditions etc) are of great importance in relation to quantitative traits while their effect on qualitative traits remains an open question owing to lack of information. There is reason to assume that the variation in quality traits is primarily due to interactions between single factors. The sources of variation (availability of essential amino acids, feed intake, feed digestibility, protein accretion between genotypes, stress through crowding effects, lack of space allowance, deficits in air condition or challenges through pathogens, etc) vary to a great degree between farms. To reduce variation therefore requires the development of strategies that are more closely related to the specific farm situation.

The complexity of the quality issue provides difficulties for the perception of quality by stakeholders and consumers, giving each person the possibility of highlighting those aspects with which he or she is in favour and neglecting or even ignoring aspects which have not yet been met or are difficult to fulfil. Owing to different viewpoints, preferences and perceptions with regard to pork quality, the current situation is confused and lacks clarity. Every producer and every consumer has generated an image of what high quality pork should encompass. This image is based on individual interpretations and developed throughout life from tradition, learning, conscious thinking and reflection, social activities or from their own sensorial experiences. People seldom consider that their own concept of meaning in relation to quality is personal but take their own concept as a reference system to relate and integrate new information and new experiences. Each person has their own individual 'quality concept'. Currently, consumers have considerable difficulty in evaluating pork quality, resulting in uncertainty and dissatisfaction. It is apparent that little progress will be made until a firm resolution is made by all involved in the production chain. A first step towards resolving the current confused situation is direct measurement of quality traits with feedback to the farmers, so that they can react to this information. Simultaneously, it is important that above average quality pork is distinguished by adequate premium prices and it is important to inform the consumer of the distinguishing characteristics.

As it is too expensive to measure all the relevant quality traits, a start could be made with the most relevant such as pH value, conductivity and IMF which are particularly useful in relation to eating quality, pathological findings relevant to animal health and welfare status and for the assessment of environmental impacts via the nutrient balance sheets of pig-producing farms. Only information that is directly related to intrinsic and extrinsic quality traits will enable the consumer to decide what is the best fit to their specific preferences and perception. It can be expected that most consumers will be able to deal with cost–benefit relationships if transparency is clearly improved. Ever-increasing and more complex customer demands may turn to disappointment when quality production cannot live up to their expectations. Thus, there is ample room for the development of differentiated products, both in terms of improved intrinsic and extrinsic traits, providing a big challenge for the pig production and pork marketing industry.

Concurrent inclusion of multiple control trades must be general practice, if pork producing units are to be successful in the future. Production concepts taking several quality characteristics into consideration are most obviously handled in quality assurance schemes, which can guarantee the customer that the relevant characteristics have been included in the product. Only those production chains that are able to create a high correspondence between the different interests of stakeholders and consumers within the whole production chain and to provide consistent trustworthy strategies can expect to develop a sustainable production chain that can face the challenge of consumer expectations and is prepared for a subsidy-free global market and competition with the growing export-orientated, low-cost pork producing regions outside Europe.

9.10 References

Affentranger P, Gerwig C, Seewer G J, Schwörer D and Künzi N (1996), 'Growth and carcass characteristics as well as meat and fat quality of three types of pigs under different feeding regimens', *Livest Prod Sci*, **45**, 187–196.

Andersen H J (2000), 'What is pork quality', *EAAP-Publ*, **100**, 15–26.

Andersen H J, Oksbjerg N and Therkildsen M (2005), 'Potential quality control tools in the production of fresh pork, beef and lamb demanded by the European society', *Livest Prod Sci*, **94**, 105–124.

Badertscher-Fawaz R, Jörin R and Rieder P (1998), 'Einstellungen zu Tierschutzfragen: Wirkungen auf den Fleischkonsum', *Agrarwirtschaft*, **47**, 107–113.

Barton Gade P (2002), 'Welfare of animal production in intensive and organic systems with special reference to Danish organic pig production', *Meat Sci*, **62**, 353–358.

Baulain U, Köhler P, Kallweit E and Brade W (2000), 'Intramuscular fat content in some native German pig breeds', *EAAP-Publ*, **100**, 181–184.

Baumgartner J, Leeb T, Gruber T and Tiefenbacher R (2002), 'Pig health and health planning organic herds in Austria', *Proceedings of the 5th NAHOWA Workshop*, Rødding, Denmark, 11–13 November 126–131.

Beattie V E, Connell N E and Moss B W (2000), 'Influence of environmental enrichment on the behaviour, performance and meat quality of domestic pigs', *Livest Prod Sci*, **65**, 71–79.

Bech-Larsen T and Grunert K G (1998), 'Integrating the theory of planned behaviour with means-end chain theory – A study of possible improvements in predictive ability', in Andersson P, *Proceedings of the 27th EMAC Conference, Stockholm*, 20–23 May, 305–314.

Becker T, Benner E and Glitsch K (2000), 'Consumer perception of fresh meat quality in Germany', *Brit Food J*, **102**, 246–266.

Bergsten C and Pettersson B (1992), 'The cleanliness of cows tied in stalls and the health of their hooves as influenced by the use of electric trainers', *Prev Vet Med*, **13**, 229–238.

Bernués A, Olaizola A and Corcoran K (2003), 'Labelling information demanded by European consumers and relationships with purchasing motives, quality and safety of meat', *Meat Sci*, **65**, 1095–1106.

Bertram H C, Purslow P P and Andersen H J (2002), 'Relationship between meat structure, water mobility and distribution – A Low-Field NMR study', *J Agric Food Chem*, **50**, 824–829.

Bickardt K (1998), 'Belastungsmyopathie und Osteochondrose beim Schwein – Folge einer Züchtung auf Maximalleistung', *Tierärztl Umschau*, **53**, 129–134.

Boyd R D, Johnston M E and Castro G (2000), 'Feeding to achieve genetic potential', *Adv. in Pork Prod*, **11**, 97–115.

Borell von E, Broom D M, Csermely D, Dijkhuizen A A, Edwards S A, Jensen P, Madec F and Stamataris C (1997), *The Welfare of Intensively Kept Pigs*, Report of the Scientific Veterinary Committee of the EU Commission. Doc XXIV/B3/ScVC/0005/1997.

Bostelmann N (2000), *An Examination of the Influence of Marketing Organisations on Animal Health and Meat Quality of Fattening Pigs on the Basis of Collected Slaughter Check Results, pH-values and Meat Temperatures of the Ham*, Dissertation, University Berlin.

Branscheid W and Claus R (1989), 'Zur Qualität von Fleisch und Milch – Ansprüche der Verbraucher und Maßnahmen der Tierproduktion', *Ber Landwirtsch*, **74**, 103–117.

Brewer M S, Zhu L G and McKeith F K (2001), 'Marbling effects on quality characteristics of pork loin chops: consumer purchase intent, visual and sensory characteristics', *Meat Sci*, **59**, 153–163.

Bruhn M (2002), 'Warum kaufen Verbraucher Bioprodukte (nicht)?', *Ökologie und Landbau*, **121**, 15–18.

Buckley D J, Morrissey P A and Gray I I (1995), 'Influence of dietary vitamin E on the oxidative stability and quality of pig meat', *Anim Sci*, **73**, 3122–3130.

Cameron N D, Enser M, Nute G R, Whittington F M, Penman J C, Fisken A C, Perry A M and Wood J D (2000), 'Genotype with nutrition interaction on fatty acid composition of intramuscular fat and the relationship with flavour of pig meat', *Meat Sci*, **55**, 187–195.

Canh T T, Aarnink A J A, Verstegen M W A and Schrama J W (1998), 'Influence of dietary factors on the pH and ammonia emission of slurry from growing-finishing pigs', *J Anim Sci*, **76**, 1123–1130.

Cassens R G (2000), 'Historical perspectives and current aspects of pork meat quality in the USA', *Food Chem*, **69**, 357–363.

Castell A G, Cliplef R L, Paste-Flynn L M and Butler G (1994), 'Performance, carcass and pork characteristics of castrates and gilts self-fed diets differing in protein content and lysine: energy ratio', *Can J Anim Sci*, **74**, 519–528.

Channon H A, Payne A M and Warner R D (2002), 'Comparison of CO_2 stunning with manual electrical stunning (50 Hz) of pigs on carcass and meat quality', *Meat Sci*, **60**, 63–68.

Claus R (1996), 'Physiologische Grenzen der Leistungen beim Schwein', *Züchtungskunde*, **68**, 493–505.

Danielsen V, Hansen L L, Moller F, Bejerholm C and Nielsen S (2000), 'Production results and sensory meat quality of pigs feed different amounts of concentrate and ad lib. Clover-grass or clover grass silage', in Hermansen J E, Lund V and Thuen E, *Ecological Animal Husbandry in the Nordic Countries*, DARCOF Report, vol. 2, 79–86.

De Vries A G, Faucitano L, Sosnicki A and Plastow G S (2000), 'Influence of genetics on pork quality', *EAAP–Publ*, **100**, 27–35.

Doedt H (1997), *Qualitative und wirtschaftliche Aspekte der Schweinefleischproduktion unter Berücksichtigung von Handelswert und Gesundheitsstatus*, Dissertation, University of Kiel.

Edmonds M S, Arentson E and Mente G A (1998), 'Effect of protein levels and space allocations on performance of growing-finishing pigs', *J Anim Sci*, **76**, 814–821.

Edwards S A and Casabianca F (1997), 'Perception and reality of product quality from outdoor pig systems in Northern and Southern Europe', in Sorensen J T, *Livestock Farming Systems–more than food production,* Wageningen Pers, Wageningen, 145–156.

Edwards S A, Wood J D, Moncrieff C B and Porter S J (1992), 'Comparison of the Duroc and Large White as terminal sire breeds and their effect on pigmeat quality', *Anim Prod*, **54**, 289–297.

Elbers A R W, Den Hartog L A, Verstegen M W A and Zandstra T (1989), 'Between- and within-herd variation in the digestability of feed for growing-finishing pigs', *Livest Prod Sci*, **23**, 183–193.

Ellis M, Webb A J, Avery P I and Brown I (1996), 'The influence of terminal sire genotype, sex, slaughter weight, feeding regime and slaughter-house on growth performance and carcass and meat quality in pigs and on the organoleptic properties of fresh pork', *Anim Sci*, **62**, 521–530.

Enfält A-C, Lundström K, Hansson I, Johansen S and Nyström P E (1997), 'Comparison of non-carriers and heterozygous carriers of the RN allele for carcass composition, muscle distribution and technological meat quality in Hampshire-sired pigs', *Livest Prod Sci*, **47**, 221–22.

Erisman J W and Monteny G J (1998), 'Consequences of new scientific findings for future abatement of ammonia emissions', *Environ Pollution*, **101**, 275–282.

Essen-Gustafson B, Karlson A, Lundström K and Entfällt A-C (1994), 'Intramuscular fat and muscle fibre lipid contents in halothane-gene-free pigs fed high or low protein diets and its relation to meat quality', *Meat Sci*, **38**, 269–277.

Faucitano L (1998), 'Preslaughter stressors effects on pork: a review', *J Muscle Foods*, **9**, 293–303.

Fernandez X, Monin G, Talmant A, Mourot J and Lebret B (1999), 'Influence of intramuscular fat content on the quality of pig meat', *Meat Sci*, **53**, 67–72.

Fewson D, Branscheid W, Oster A, Sack E and Komender P (1987), 'Untersuchungen über die Beziehungen zwischen Klassifizierungskriterien und Merkmalen der Fleischbeschaffenheit', *Züchtungskunde*, **59**, 362–377.

Fischer K (2001), 'Bedingungen für die Produktion von Schweinefleisch guter sensorischer und technologischer Qualität', *Mitteilungsblatt BAFF Kulmbach*, **40**, 7–22.

Flachkowsky G (1993), 'Beiträge der Tierernährung zur Senkung der Umweltbelastung', *Lohmann-Information*, **4**, 1–9.

Fujii J, Otsu K, Zorzato F, De Leon S, Khanna V K, Weiler J E, O'Brien P J and MacLennan D H (1991), 'Identification of a mutation in the porcine ryanodine receptor associated with malignant hyperthermia', *Science*, **253**, 448–451.

Gentry J G, McGlone J J, Miller M F and Blanton J R (2004), 'Environmental effects on pig performance, meat quality, and muscle characteristics', *J Anim Sci*, **82**, 209–217.

Glitsch K (2000), 'Consumer perceptions of fresh meat quality: cross-national comparison', *Br Food J*, **102**, 177–194.

Glodek P (1996), 'The choice of sire lines determines the quality of final products in pigs', *Züchtungskunde*, **68**, 483–492.

Grunert K G and Andersen S (2000), 'Purchase decision, quality expectations and quality experience for organic pork', *9th Food Choice Conference*, Dublin, Trinity College, 28–31 July.

Grunert K G, Bredahl L and Brunso K (2004), 'Consumer perception of meat quality and implications for product development in the meat sector – a review', *Meat Sci*, **66**, 227–259.

Hammel von M-L, Blaha T (1993), 'Die Erfassung von pathologisch-anatomischen Organbefunden am Schlachthof', *Fleischwirtschaft*, **73**, 1427–1430.

Hansen L L and Claudi-Magnussen C (2004), 'Feeding with lupines reduces the amount of skatole in organic pigs', *DARCOFenews* (Newsletter from Danish Research Centre for Organic Farming) **4**, 1–2.

Hansen L L, Bejerholm C, Claudi-Magnussen C and Andersen H J (2000), 'Effects of organic feeding including roughage on pig performance, technological meat quality and the eating quality of the pork' in Alfôldig T, Lockeretz W and Niggli U, *Proceeding 13th International IFOAM Scientific Conference,* 288.

Harr G (1989), 'Qualitätsabweichung bei Schweinefleisch – Ursachen und Maßnahmen zur Verhinderung', *Fleischwirtschaft*, **69**, 1246–1248.

Heger J, Van Phung T and Krizova L (2002), 'Efficiency of amino acid utilization in the growing pig at suboptimal levels of intake: lysine, threonine, sulphur amino acids and tryptophan', *J Anim Physiol An N*, **86**, 153–165.

Hermansen J E and. Kristensen T (1998), *Research and Evaluation of Mixed Farming Systems for Ecological Animal Production in Denmark*, Workshop Proceedings: Mixed Farming Systems in Europe, Dronten, 97–101.

Hofmann K (1994), 'What is quality', *Meat Focus*, 73–82.

Holck J T, Schinckel A P, Coleman J L, Wilt V M, Thacker E L, Spurlock M E, Grant A L, Malven P L, Senn M K and Thacker B J (1997), 'Environmental effects on the growth of finisher pigs', *J Anim Sci*, **75**, 246 (abstr.)

Hollo G, Seregi J, Ender K, Nürnberg K, Wegner J, Seenger J and Repa L (2003), 'An evaluation of meat quality and fatty acid composition of *Mangalitsa hogs*', *Hung Agric Res*, **4**, 15–18.

Honikel K O (1996), 'Fleischqualitätskontrolle in der EU', *Agrarforschung*, **9**, 447–450.

Huang S Y, Tsou H L, Kan M T, Lin W K and CHI C S (1995), 'Genetic study on leg weakness and its relationship with economic traits in central tested boars in subtropical area', *Livest Prod Sci*, **44**, 53–59.

Hyun Y, Ellis M and Johnson R W (1998a), 'Effects of feeder type, space allowance, and mixing on the growth performance and feed intake pattern of growing pigs', *J Anim Sci*, **76**, 2771–2778.

Hyun Y, Ellis M, Riskowski G and Johnson R W (1998b), 'Growth performance of pigs subjected to multiple concurrent environmental stressors', *J Anim Sci*, **76**, 721–727.

Jakobsen K (1999), 'Dietary modifications of animal fats: status and future perspectives', Eur *J Lipid Sci Technol*, **101**, 475–483.

Jakobsen K and Hermansen J E (2001), 'Organic farming – a challenge to nutritionists', *J Anim Feed Sci*, **10**, 29–42.

Jensen C, Lauridsen C and Bertelsen G (1998), 'Dietary vitamin E: quality and storage stability of pork and poultry', *Trends Food Sci Technol* **9**, 62–72.

Jeremiah L E (1994), 'Consumer responses to pork loin chops with different degrees of muscle quality in two western Canadian cities', *Can J Anim Sci*, **74**, 425–432.

Johnson R W (1997), 'Inhibition of growth by pro-inflammatory cytokines: an integrated view', *J Anim Sci*, **75**, 1244–1255.

Jonsäll A, Johansson L and Lundström K (2000): 'Effects of red clover silage and RN genotype on sensory quality of prolonged frozen stored pork', *Food Qual and Pref*, **11**, 371–376.

Jonsäll A, Johansson L and Lundström K (2001), 'Sensory quality and cooking loss of ham muscle (*M. biceps femoris*) from pigs reared indoors and outdoors', *Meat Sci*, **57**, 245–250.

Jonsäll A, Johansson L, Lundström K, Andersson K H, Nilsen A N and Risvik E (2002), 'Effects of genotype and rearing system on sensory characteristics and preference for pork', *Food Qual and Pref*, **13**, 73–80.

Jorgensen H, Jensen S K and Eggum B O (1996), 'The influence of rapeseed oil on digestibility, energy metabolism and tissue fatty acid composition in pigs', *Acta Agric Scand A Anim Sci*, **45**, 65–75.

Kaiser S, Schlüter M and Van Den Weghe H F (1998), 'Auswirkungen eiweißreduzierter Multiphasenfütterung auf Ammoniakemissionen, Nährstoffbilanz und Wirtschaftlichkeit in einem einstreulosen Mastschweinestall', *KTBL-Arbeitspapier*, **259**, 66–90.

Kallweit E (1989), 'Empfehlungen zur Bewertung von Schweinehälften im Rahmen der Handelsklassenverordnung', *Züchtungskunde*, **61**, 251–252.

Kempkens K (2003), 'Ökologische Schweinehaltung – Der Einstieg muss gut kalkuliert sein', *DGS Magazin*, **5**, 38–42.

Köhler P, Hoppenbrock K-H, Adam F and Kallweit E (1999), 'Intramuskulärer Fettgehalt bei verschiedenen Schweineherkünften im Warentest', in Böhme H and Flachowsky G, *Aktuelle Aspekte bei der Erzeugung von Schweinefleisch*, Landbauforschung Völkenrode SH, **193**, 82–87.

Kreuzer M, Machmüller A, Hanneken H, Gerdemann M and Wittmann M (1999), 'BFS-Fermentable fibre in feeds for growing fattening pigs', *Züchtungskunde*, **71**, 306–322.

Kristensen E S and Halberg N (1997), 'A system approach for assessing sustainability in livestock farms', *Proceedings of the 4th Int. Liveststock Farming Systems Symposium*, Aug. 22–23, Foulum, Denmark, *EAAP-Publ*, **89**, 238–246.

Kubberod E, Ueland O, Tronstad A and Risvik E (2002), 'Attitudes towards meat and meat-eating among adolescents in Norway: a qualitative study', *Appetite*, **38**, 53–62.

Laube S (2000), *Die Eignung spezieller Schweinekreuzungen zur Qualitätsverbesserung von Markenschweinefleisch unter besonderer Berücksichtigung von MHS-Status, Hampshirefaktor und intramuskulärem Fettgehalt*, Dissertation University Göttingen.

Le Bellego L, Van Milgen J and Noblet J (2002), 'Effect of high ambient temperature on protein and lipid deposition and energy utilization in growing pigs', *Anim Sci*, **75**, 85–96.

Lebret B, Juin H, Noblet J and Bonneau M (2001), 'The effects of two methods of increasing age at slaughter on carcass and muscle traits and meat sensory quality in pigs', *Anim Sci*, **72**, 87–94.

Ledward D A (1991), 'Colour of raw and cooked meat', in Johnston D E, Knoght M K and Ledward D A, *The Chemistry of Muscle-based Foods,* Royal Society of Chemistry, Cambridge, 128–144.

Leheska J M, Wulf D M and Maddock R J (2003), 'Effects of fasting and transportation on pork quality development and extent of postmortem metabolism', *J Anim Sci*, **81**, 3194–3202.

Lengerken V G (1990), 'Einbeziehung von Methoden zur Reduzierung der Belastungsempfindlichkeit und von Fleischqualitätsmängeln in das Zuchtprogramm vom Schwein', *Tierzucht*, **44**, 465–467.

Le Roy P, Moreno C, Elsen J M, Caritez J C, Billon Y, Lagant H, Talmant A, Vernin P, Amigues Y, Sellier P and Monin G (2000), 'Interactive effects of the HAL and RN major genes on carcass quality traits in pigs: preliminary results', *EAAP-Publ*, **100**, 139–142.

Leskanich C O, Matthews K R, Warkup C C, Noble R C, and Hazzledine M (1997), 'The effect of dietary oil containing (n-3) fatty acids on the fatty acid, physiochemical, and organoleptic characteristics of pig meat and fat', *J Anim Sci*, **75**, 673–683.

Lopez-Bote C J (1998), 'Sustained utilization of the Iberian pig breed', Meat Sci **49**, 17–27.

Lundeheim N (1987), 'Genetic analysis of osteochondrosis and leg weakness in the Swedish pig progeny testing scheme', *Acta Agric Scand*, **37**, 159–173.

Martel J, Minvielle F and Poste L M (1988), 'Effects of crossbreeding and sex on carcass composition, cooking properties and sensory characteristics of pork', *J Anim Sci*, **66**, 41–46.

McEachern M G and Schröder M J (2002), 'The role of livestock production ethics in consumer values towards meat', *J Agric Environ Ethics*, **15**, 221–237.

McGloughlin P, Allen P, Tarrant P V and Joseph R L (1988), 'Growth and carcass quality of crossbred pigs sired by Duroc, Landrace and Large White boars', *Livest Prod Sci*, **18**, 275–288.

Miller G Y and Dorn R (1990), 'Costs of swine diseases to producers in Ohio', *Prev Vet Med*, **8**, 183–190.

Millet S, Hesta M, Seynaeve M, Ongenae E, De Smet S, Debraekeleer J and Janssens G P (2004), 'Performance, meat and carcass traits of fattening pigs with organic versus conventional housing and nutrition', *Livest Prod Sci*, **87**, 109–119.

MLC (Meat and Livestock Commision) (1992), *Stotfold Pig Development Unit Second Trial Results*, Meat and Livestock Commission, Milton Keynes.

Monin G and Sellier P (1985), 'Pork of low technological quality with a normal rate of muscle pH fall in the immediate post-mortem period: The case of the Hampshire breed', *Meat Sci*, **13**, 49–63.

Monin G, Larzul C, Le Roy P, Culioli J, Mourot J, Rousset-Akrim S, Talmant A, Touraille C and Sellier P (1999), 'Effects of the halothane genotype and slaughter weight on texture of pork', *J Anim Sci*, **77**, 408–415.

Murray A C and Johnson C P (1998), 'Impact of the halothane gene on muscle quality and pre-slaughter deaths in Western Canadian pigs', *Can J Anim Sci*, **78**, 543–548.

Ngapo T M, Dransfield E, Martin J F, Magnusson M, Bredahl L and Nute G R (2003), 'Consumer perceptions: pork and pig production. Insights from France, England, Sweden and Denmark', *Meat Sci*, **66**, 125–134.

Nilzén V, Babol J, Dutta P C, Lundeheim N, Enfalt A-C and Lundstrôm K (2001), 'Free range rearing of pigs with access to pasture grazing-effect on fatty acid composition and lipid oxidation products', *Meat Sci*, **58**, 267–275.

NRC – National Research Council (1998), *Nutrient Requirements of swine. Nutrient Requirements of Domestic Animals*, National Academy Press, Washington DC, USA.

Oeckel van M J, Warnants N, Boucque C V, Delputte P and Depuydt J (2001), 'The preference of the consumer for pork from homozygous or heterozygous halothane negative animals', *Meat Sci*, **58**, 247–251.

Offer G and Knight P (1988), 'The structural basis of water-holding in meat', in Lawrie R, *Developments in Meat Science,* Elsevier, London, 63–243.

Olmos S and Lampkin N (2005), *Statistical Report on the Development of Organic Farming in EU Member States and Switzerland for the Period 1997–2002 with Update for 2003*, Draft version of D5 of EU-CEE-OFP Organic Farming Policy, University of Wales, Aberystwyth.

Olsson V, Andersson K, Hansson I and Lundström K (2003), 'Differences in meat quality between organically and conventionally produced pigs', *Meat Sci*, **64**, 287–297.

Oude Ophuis P A M (1993), 'Sensory evaluation of 'free range' and regular pork meat under different conditions of experience and awareness', *Food Qual and Pref*, **5**, 173–178.

Person R C, McKenna D R, Ellebracht J W, Griffin D B, McKeith F K, Scanga J A, Beltk K E, Smith G C and Savell J W (2005), 'Benchmarking value in pork supply chain: processing and consumer characteristics of hams manufactured from different quality of raw materials', *Meat Sci*, **70**, 91–97.

Raewell H (1999), 'The wholesome quest', *World Ingred*, **58**, 60–61.

Rauw W M, Kanis E, Noordhuizen-Stassen E N and Grommers F J (1998), 'Undesirable side effects of selection for high production efficiency in farm animals: a review', *Livest Prod Sci*, **56**, 15–33.

Rosenvold, K and Andersen H J (2003), 'Factors of significance for pork quality – a review', *Meat Sci*, **64**, 219–237.

Rosenvold K, Petersen J S, Lrerke R N, Jensen S K, Therkildsen M, Karlsson A H, Moller R S and Andersen H J (2001), 'Muscle glycogen stores and meat quality as affected by strategic finishing feeding of slaughter pigs', *J Anim Sci*, **79**, 382–391.

Rosenvold K, Lrerke H N, Jensen S K, Karlsson A, Lundstrôm K and Andersen H J (2002), 'Manipulation of critical quality indicators and attributes in pork through vitamin E supplementation level, muscle glycogen reducing finishing feeding and pre-slaughter stress', *Meat Sci*, **62**, 485–496.

Rosner F, Von Lengerken G and Maak S (2003), 'The value of pig breedng herds in Germany and progress in improvement of meatiness and pork quality', *Anim Sci Papers and Reports*, **21**, 153–161.

Rushen J and de Passillé A M (1992), 'The scientific assessment of the impact of housing on animal welfare: A critical review', *Can J Sci*, **72**, 721–743.

Santos e Silva J, Ferreira-Cardoso J, Bernard A and Pires da Costa J S (2000), 'Conservation and development of the Bisaro pig. Characterisation and zootechnical evaluation of the breed for production and genetic management. Genetic and pork quality', *EAAP-Publ*, **100**, 85–92.

Sather A P, Jones S D, Schaefer A L, Colyn J and Robertson W M (1997), 'Feedlot performance, carcass composition and meat quality of free-range reared pigs', *Can J Anim Sci*, **77**, 225–232.

Schäfer A, Rosenvold K, Purslow P P, Andersen H J and Henckel P (2002), 'Physiological and structural events post mortem of importance for drip loss in pork', *Meat Sci*, **61**, 355–366.

Schepper J, Kallweit E. and Averdunk G (1983), 'Schweinefleischqualität', *Schweinezucht und Schweinemast*, **31**, 135–138.

Schifferstein H N J and Ophius P A M (1998), 'Health-related determinants of organic food consumption in the Netherlands', *Food Qual and Pref*, **9**, 119–133.

Schinckel A P and De Lange C F (1996), 'Characterization of growth parameters needed as inputs for pig growth models', *J Anim Sci*, **74**, 2021–2036.

Schütte A, Bork A, Mergens U, Pott U and Venthen S (1996), 'MHS-genetic, lean meat content and findings in lungs – the dominant factors in relation to meat quality?', *Landbauforschung Völkenrode*, **166**, 229–238.

Schwörer D and Morel P (1987), 'Verbesserung des Genusswertes von Schweinefleisch durch züchterische Bemühungen', *Der Kleinviehzüchter*, **35**, 1294–1304.

Schwörer D, Hofer A, Lorenz D and Rebsamen A (2000), 'Selection progress of intramuscular fat in Swiss pig production', *EAAP-Publ*, **100**, 69–72.

Sheard P R, Enser M, Wood J D, Nute G R, Gill B P and Richardson R I (2000), 'Shelf life and quality of pork and pork products with raised *n*–3 PUFA', *Meat Sci*, **55**, 213–221.

Singer W (2000), 'Phenomenal awareness and consciousness from a neurobiological perspective', in Metzinger T, *Neural Correlates of Consciousness*, Cambridge, 121–137.

Spurlock M E (1997), 'Regulation of metabolism and growth during immune challenge: An overview of cytokine function', *J Anim Sci*, **75**, 1773–1783.

Stern S, Heyer A, Andersson H K, Rydhmer L and Lundström K (2003), 'Production results and technological meat quality for pigs in indoor and outdoor rearing systems', *Acta Agric Scand, Sect A, Anim Sci*, **53**, 166–174.

Stoier S, Aaslyng M D, Olsen E V and Henckel P (2001), 'The effect of stress during lairage and stunning on muscle metabolism and drip loss in Danish pork', *Meat Sci*, **59**, 127–131.

Sundrum A (1999), 'EEC-Regulation on organic livestock production and their contribution to the animal welfare issue', *KTBL-Schrift*, **270**, 93–97.

Sundrum A (2002), 'Verfahrenstechnische und systemorientierte Strategien zur Emissionsminderung in der Nutztierhaltung im Vergleich', *Ber Ldw*, **80**, 556–570.

Sundrum A (2005), 'Carcass yield and meat quality of organic pig production', *Proceedings of the 4th SAFO-Workshop*, 17–19 March, Frick, Switzerland, 77–86.

Sundrum A and Ebke B (2004), 'Problems and challenges with certification of organic pigs', *Proceedings of the 2nd SAFO Workshop*, Witzenhausen, Germany, 25–27 March, 193–198.

Sundrum A and Acosta Aragon Y (2005), 'Nutritional strategies to improve the sensory quality and food safety of pork while improving production efficiency within organic framework conditions', Report of the EU-project, *Improving Quality and Safety and Reduction of Costs in the European Organic and 'Low Input' Supply Chains*, no. CT–2003 506358.

Sundrum A, Bütfering L, Henning M and Hoppenbrock K H (2000), 'Effects of on-farm diets for organic pig production on performance and carcass quality', *J Anim Sci*, **78**, 1199–1205.

Sundrum A, Schneider K and Richter U (2005), 'Possibilities and limitations of protein supply in organic poultry and pig production', Report of the EU-project, *Research to Support Revision of the EU Regulation on Organic Agriculture,* no. SSPE-CT-2004–502397.

Susenbeth A (2002), 'Anpassung der Aminosäurenversorgung an das Wachstumsvermögen von Schweinen', *Lohmann Information*, **1**, 21–24.

Susenbeth A, Dickel T, Diekenhorst A and Hohler D (1999), 'The effect of energy intake, genotype and body weight on protein retention in pigs when dietary lysine is the first-limiting factor', *J Anim Sci*, **77**, 2985–2989.

Taylor A A, Nute G R and Warkup C C (1995), 'The effect of chilling, electrical stimulation and conditioning on pork eating quality', *Meat Sci*, **39**, 339–347.

Thiele S and Weiss C (2003), 'Consumer demand for food diversity: evidence for Germany', *Food Policy*, **28**, 99–115.

Thielen C and Kienzle E (1994), 'Die Fütterung des 'Bioschweins' – eine Feldstudie', *Tierärztl Praxis*, **22**, 450–459.

Tribout T and Bidanel J P (2000), 'Genetic parameters of meat quality traits recorded on Large White and French Landrace station-tested pigs in France', *EAAP-Publ*, **100**, 37–41

Trujillo Garcia R and Mata C (2000), 'The Dehesa: an extensive livestock system in the Iberian Peninsula', in *Proceedings of the 2nd NAHWOA Workshop*, Cordoba, 8–11 January, 2000.

Van der Wal P G, Engel B and Hulsegge B (1997), 'Causes for variation in pork quality', *Meat Sci*, **46**, 319–327.

Van den Weghe H (1999), 'Environmental effects of livestock production systems', *Züchtungskunde*, **71**, 64–77.

Verbeke W A and Viaene J (2000), 'Ethical challenges for livestock production: meeting consumer concerns about meat safety and animal welfare', *J Agric Environ Ethics*, **12**, 141–151.

Vermeer H M, Altena H, Bestman M, Ellinger L, Cranen I, Spoolder H A M and Baars T (2000), 'Monitoring organic pig farms in The Netherlands', *Proceedings of the 51st Annual Meeting of the European Association of Animal Production*, The Hague, The Netherlands, 21–24 August, 211–224.

Von Alvensleben R (2003), 'Gesellschaft und Tierproduktion', in Lohde, E-J and Ellendorf F, *Perspektiven in der Tierproduktion,* Landbauforschung Völkenrode SH, 263, 15–21.

Warkup C C and Kempster A J (1991), 'A explanation of the variation in tenderness and juiciness of pig meat', *Anim Prod*, **52**, 599 (abstr.)

Warnants N, van Oeckel M J and Boucque C V (1998), 'Effect of incorporation of dietary polyunsaturated fatty acids in pork back-fat on the quality of salami', *Meat Sci*, **49**, 435–445.

Warriss P D, Brown S N, Edwards J E and Knowles T G (1998), 'Effect of lairage time on levels of stress and meat quality in pigs', *Anim Sci*, **66**, 255–261.

Whittemore C T, Tullis J B and, Emmans, G.C. (1988), 'Protein growth in pigs', *Anim Prod*, **46**, 437–445.

Williams N H (1998), 'Impact of immune system activation on pig growth and amino acid needs', in Wiseman J, Varley M A and Chadwick J P, *Progress in Pig Science*, Nottingham University Press, Loughborough (UK).

Wittmann M, Gerdemann M M, Scheeder M R, Hanneken H, Janecke D and Kreuzer M (1995), 'Zusammenhänge zwischen tierärztlichen Befunden und Schlachtkörper- bzw. Fleischqualität beim Schwein', *Fleischwirtschaft*, **75**, 492–495.

Wood J D (1990), 'Consequences for meat quality of reducing carcass fatness', in Wood J D and Fisher A V, *Reducing Fat in Meat Animals*, Elsevier Applied Science, London.

Wood J D and Enser M (1997), 'Factors influencing fatty acids in meat and the role of antioxidants in improving meat quality', *Br J Nutr*, **78**, 49–60.

10

Organic livestock husbandry methods and the microbiological safety of ruminant production systems

Francisco Diez-Gonzalez, University of Minnesota, USA

10.1 Introduction

10.1.1 Differences in feeding regimes and husbandry methods between organic, 'low input' and conventional ruminant production systems

The term 'conventional' is typically used to describe husbandry methods that employ legally accepted practices of animal production that do not follow specific guidelines. In general, conventional livestock farms use a variety of inputs and management practices. The meaning of 'organic' has evolved over the years, but according to the United Nations' Food and Agriculture Organization, organic agriculture is defined as a 'holistic production management system which promotes and enhances agroecosystem health, including biodiversity, biological cycles, and soil biological activity' (Codex Alimentarius Commission, 1999). In general, all organic standards agree that organic livestock production systems: (1) should use as few inputs as possible; (2) should avoid using a variety of synthetic substances such as antibiotics, hormones, genetically modified plant materials and animal by-products; and (3) should promote the welfare of animals. The concept of 'low input' ruminant production is not as well defined as organic, but it could be referred to as livestock operations that seek to utilize as few external feedstocks and veterinary inputs as possible.

Conventional livestock production systems can be very diverse and this diversity is influenced by economic, geographic, environmental and cultural factors. Conventional inputs for direct use in ruminant production include many types of plant feeds (i.e. forages, cereals, soybeans, etc.), industrial by-products (i.e. molasses, distiller's dried grain, meat bone meal, etc.), feed additives (i.e. antibiotics, growth promoters, bacterial cultures, etc.), drugs (i.e. hormones, antibiotics, etc.) and a variety of cleaning and sanitizing compounds (Church, 1991). Conventional management practices range from traditional grazing systems to intensive feedlot operations.

Organic livestock regulations differ from conventional ones because the use of antibiotics, hormones and growth promoters is not allowed. Also, most organic guidelines require the access to pasture or free range for animals. While there is still some disagreement about details among different national standards, organic livestock production in most countries is based on very similar principles. In general, low input ruminant operations are farms in which approximately 50% of the forage is harvested by livestock and they rely heavily on grazing (Weigel *et al.*, 1999). Organic systems are largely driven by a holistic philosophy, but the sole objective of some low input farms is an economic benefit for the producer.

10.1.2 Microbiological risks associated with ruminant production systems

A variety of foodborne pathogens are common inhabitants of the gastrointestinal tract of ruminants. From the animal's intestinal contents, pathogenic microorganisms can be disseminated to the environment via manure and directly to carcasses when the animals are slaughtered. Contaminated meat, water and other foods, such as fresh vegetables, can ultimately serve as vehicles of foodborne infections if consumed untreated. Frequent causes of gastrointestinal diseases have ultimately been linked to foods from animal origin and to fresh fruits and vegetables fertilized with animal manure. The most important pathogens responsible for these illnesses are *Salmonella*, *Campylobacter* and enterohemorrhagic *Escherichia coli* (EHEC) (Sivapalasingam *et al.*, 2004).

10.1.3 Incidence, epidemiology and public health implications of foodborne diseases linked to food products from ruminant livestock

The Centers for Disease Control (CDC) in the USA has estimated that *Campylobacter* and *Salmonella* are the top two bacterial foodborne pathogens as they are responsible for a total of 2 and 1.3 million foodborne infections every year (Mead *et al.*, 1999). Most of the cases of infection with *Campylobacter* and *Salmonella* appear to be due to eggs and poultry, but a significant number of cases are related to beef products (CDC, 2005; Jay, 2000). EHEC does not cause as many infections as the top two pathogens (approximately 100,000 in the USA), but EHEC is frequently linked to outbreaks of diarrhea caused by the consumption of contaminated ground beef (Mead *et al.*, 1999, Rangel *et al.*, 2005). Infections caused by EHEC are the major microbial public health concern related to ruminant food products.

10.2 Effect of forage to concentrate ratios on enteric pathogen prevalence and shedding

10.2.1 Types of ruminant feeds

The typical diets of livestock are formulated with a combination of ingredients to provide the necessary nutrients to sustain satisfactory weight gain, breeding and milk production (Church, 1991). Feed components can provide energy, protein, macronutrients (phosphorus, potassium, sodium, etc.) and micronutrients (vitamins and trace minerals). In most cattle and swine diets, the largest component supplies the energy to sustain animal productivity, accounting for more than 70% of the total dry matter composition of typical rations. Protein sources such as soybean meal are the second largest component in typical livestock diets.

The energy-supplying feed ingredients for weaned and adult animals are mostly plant materials such as forage and grains (Church, 1991). The plant materials fed to livestock are very diverse and this feed diversity is influenced by geographic location, availability of feeds, type of production system and stage in the animal lifecycle. In most of North America, grains are the predominant energy-providing feed during the finishing stage of beef cattle and swine, but in tropical countries molasses is sometimes the preferred ingredient owing to its availability. In beef cattle production, pasture often provides the main energy source for organic and low input farms and grains are typically used in conventional operations.

Energy-supplying feeds can be divided into two major categories: those that are largely composed of readily degradable carbohydrates and those that have high-fiber content. The term 'fiber' is typically defined as non-starch polysaccharides (NSP) (Grieshop *et al.*, 2001) NSP are a variety of substances that include pectins, hemicellulose, cellulose and lignin. Fibers are classified according to their solubility: neutral detergent soluble, acid detergent soluble and non-soluble fiber (Goering and Van Soest, 1970). Neutral detergent-soluble fiber includes pectins, glucans, mannan and inulin and the acid detergent-soluble NSP is largely hemicelullose. The acid-detergent insoluble fiber (ADF) components are mostly lignin and cellulose.

Ruminant feeding systems that feed high-fiber containing diets can be classified depending on whether the animal consumes living plants by grazing (pasture) or if the plants are harvested and preserved either by drying or ensiling. Pastures can consist of legumes and grasses. Some of the most important pastures used for animal production are alfalfa, timothy, Bahia grass, ryegrass, stargrass, Bermuda grass, and so on. Preserved high-fiber feeds include harvested dry grasses and legumes (hay), non-grain-containing cereal straw and silages. Some grasses and forages contain a fraction of soluble carbohydrates that are readily degradable, but their concentration is not sufficient to provide a significant source of energy for growth.

The rumen is a unique specialized organ that allows ruminants to obtain energy by degrading high-fiber plant materials via a complex microbial ecosystem that processes fiber for the benefit of the animal. Cellulose is the predominant type of fiber in grasses and forages and its conversion into absorbable volatile fatty acids (VFA) is a complex microbial process (Dehority, 2003). The hydrolysis of cellulose is catalyzed by cellulases produced by cellulolytic bacteria, which ferment the cellobiose units resulting from the cellulose hydrolysis. This fermentation leads to the production of VFA, which are absorbed by the animal and converted to glucose by the liver (Forsberg *et al.*, 1997). The slow cellulose degradation limits the utilization of fiber-rich feeds for high-productivity systems.

High-energy feeds (concentrates) are those that contain a large proportion of readily degradable carbohydrates. The two types of readily degradable carbohydrates in animal feeds are starch and sugars. High-starch feed ingredients include grains such as corn, barley, sorghum, wheat and oats and the most common high-sugar feed ingredient is molasses (Owens *et al.*, 1997). In cattle, starch and sugars are rapidly fermented in the rumen and this fast conversion of carbohydrates to volatile fatty acids supplies the animal with excess energy for growth. In mono-gastric animals such as pigs, starch and sugars are also readily digested and their monosaccharides quickly absorbed (Ewan, 2001).

Extracellular amylases produced by microorganisms in the rumen allow the rapid fermentation of starch granules (Huntington, 1997). The free maltose resulting from starch degradation is fermented by the bacterial species with the highest maltose affinity and is converted to VFA and lactic acid (Huntington and Reynolds, 1986). A significant fraction of starch (more than 20%) remains intact after rumen fermentation. This amount reaches the large intestine and can sustain an active microbial population (Huntington, 1997). In many regions of the world, the use of high-starch feeds has allowed highly productive animal systems, but the incidence of acidosis and liver abscesses in cattle that consume grains for long periods of time is very frequent (Nagaraja and Chengappa, 1998; Owens *et al.*, 1998). These health problems caused by grain feeding, indicate that high-starch feeds are not the natural feedstock for ruminant animals.

10.2.2 Effect of feed type on the gastrointestinal microbial ecosystem

The microbial populations of ruminants fed high-energy diets are markedly different from those fed high-fiber diets. It has long been recognized that forages and grasses promote the occurrence of celullolytic bacteria in the rumen that are necessary for digestion. Because the ability to hydrolyze starches is more widespread among bacteria, feeding grains typically promotes the growth of a more diverse bacterial population (Owens *et al.*, 1998). The more rapid fermentation of starches and sugars typically leads to higher concentrations of VFA and lower pH in grain- and molasses-fed cattle than in forage-fed animals (Hungate, 1966, Diez-Gonzalez *et al.*, 1998). Animal productivity can be markedly stimulated by such rapid carbohydrate fermentation.

The first evident difference between forage and concentrate-fed animals is the observation that forage-fed cattle can have a significantly lower total anaerobic bacterial population (Diez-Gonzalez *et al.*, 1998, Dehority and Orpin, 1988, Hungate, 1966). In animals fed grain, one of the predominant cellulolytic bacteria genera in the rumen was *Butyrivibrio* accounting for approximately 23% of the total anaerobic bacterial population (Dehority and Orpin, 1988). When animals were fed starch-rich feeds, the population of *Butyrivibrio* was reduced to less than 10% of the total and the numbers of amylolytic organisms (*Selenomonas*, *Peptostreptococcus*, *Streptococcus*, *Bacteroides*, *Lactobacillus*) and lactate utilizers (*Selenomonas*) increased. The rapid growth of lactic acid bacteria (*Streptococcus*, *Lactobacillus*) can cause an accumulation of lactic acid in grain fed animals that may lead to the development of rumen acidosis (Owens *et al.*, 1998). If feed differences can cause these marked changes in the overall population of the GI tract, it would then be expected that the animal diet also has a marked effect on the population of less predominant organisms such as foodborne pathogens.

10.2.3 Comparison of the effect of concentrates and forages on the natural prevalence of enteric pathogens in ruminants

Different methods have been utilized to determine the effect of ruminant feed on the prevalence of pathogenic microorganisms in their population. One of the most common strategies employed is to collect and analyse a large number of independent fecal samples obtained from as many animals as possible and subsequently perform microbiological analysis. By knowing the proportion of animals that are positive for a specific pathogen, inferences can be established related to management practices or types of feeding. In this section we will summarize the most important findings of prevalence studies of EHEC, *Salmonella* and *Campylobacter* in small and large ruminants.

Enterohemorrhagic Escherichia coli

In the last 20 years, an increased number of enterohemorrhagic diarrhea outbreaks have been caused by highly virulent *E. coli* strains. EHEC are capable of causing serious complications such as hemolytic uremic syndrome in children and vascular disease in elderly people. *E. coli* serotype O157:H7 has been the primary cause of outbreaks of EHEC, but other serotypes such as O26 and O111 are also a major public health concern. The Center for Disease Control has calculated that *E. coli* O157:H7 is responsible for more than 70,000 infections and 60 deaths, and other EHEC strains cause approximately 36,000 illnesses and 30 deaths every year in the USA (Mead *et al.*, 1999). In 1994 *E. coli* O157:H7 was legally declared an adulterant in ground beef in the USA (USDA/FSIS Directive 10,010.1, 2004). Since then the detection of serotype O157:H7 has resulted in recalls to the meat industry costing millions of dollars.

Different EHEC serotypes are often found in ruminant populations, but most of the research to elucidate the effect of feed type on EHEC prevalence has been based on serotype O157:H7. The preference for this pathogenic strain is due to its public health importance, its unique phenotypic characteristics that allow a relatively easy identification compared to other serotypes and the fact that ruminants are its most important natural reservoir. Shiga toxin-producing *E. coli* (STEC) are a larger group that includes EHEC and have also been found in ruminant populations in relatively large prevalence. Because the pathogenicity of most STEC has not been proven, we will focus our discussion on serotype O157:H7.

Initial cattle surveys reported that the prevalence of *E. coli* O157:H7 in cattle populations was less than 3% (Hancock *et al.*, 1997c; Garber *et al.*, 1995, Shere *et al.*, 1998), but more recent studies using immunomagnetic separation (IMS) techniques estimated that between 10% and 28% of cattle can be asymptomatic carriers (Elder *et al.*, 2000; Smith *et al.*, 2001; Sargeant *et al.*, 2003; Chapman *et al.*, 1997). The prevalence of *E. coli* O157:H7 in cattle population is markedly affected by seasonality. Several studies have observed that during the winter months the number of positive animals can be less than 5% (Mechie *et al.*, 1997). Because this microorganism does not sicken cattle, the only way to identify a carrier animal is by conducting microbiological tests.

The link between cattle diet and prevalence of *E. coli* O157:H7 has been investigated by longitudinal surveys of cattle. Garber *et al.* (1995) were the first to link management factors with O157:H7 prevalence and reported that grain-fed calves had a higher probability of being O157:H7 positive than calves that had been fed either ground hay or clover. In another study, the use of wheat fines was also linked to an increased likelihood of finding an O157-positive animal, but feeding alfalfa appeared to be related to decreased prevalence in dairy herds (Hancock *et al.*, 1997a). Barley was related to increased O157:H7 prevalence, but corn and wheat were not identified among risk factors (Dargatz *et al.*, 1997). Feeding corn silage appeared to be a practice that increased the chances of fecal shedding detection (Herriott *et al.*, 1998). From these initial studies, it appeared that high-energy feeds increased the prevalence of serotype O157:H7 in cattle populations.

The fecal shedding of serotype O157:H7 in cattle appears to be sporadic and transient, and this lack of consistent colonization might have a great impact on prevalence studies (Besser *et al.*, 1997; Mechie *et al.*, 1997; Sargeant *et al.*, 2000; Khaitsa *et al.*; 2003). Despite this potentially confounding factor for assessment of the effect of cattle diet on O157:H7 prevalence, a number of recent reports from different parts of the world have estimated consistently lower percentages of O157-positive samples in pasture-fed animals. In Argentina, an initial study on grazing herds detected no O157-positive cattle (Sanz *et al.*, 1998), but the same research group reported that 6.8% of grain-fed animals in feedlots shed this pathogenic bacteria in their feces (Padola *et al.*, 2004). In the United States, a study showed that approximately 83% of range beef calves had been exposed to *E. coli* O157:H7 before reaching the feedlots, but the fecal prevalence of this pathogen in the 15 herds tested was only 7.4% (Laegreid *et al.*, 1999).

In a recent study of range cattle, the overall prevalence of serotype O157:H7 from fecal samples was 1% and this parameter was only 0.6% in animals that were on cow–calf pasture (Renter *et al.*, 2003). New Zealand is a major cattle producing country that relies mostly on pasture feeding, but epidemiological evidence and cattle surveys suggested that the prevalence of *E. coli* O157:H7 appears to be extremely low in that country (Cook, personal communication). Other researchers reported only two O157-positive animals in a study that included 531 cattle in 55 dairy farms (Buncic and Avery, 1997). In studies conducted by New Zealand's Ministry of Agriculture and Forestry, *E. coli* O157:H7 was never detected from a total of 3000 bovine and 500 ovine carcasses (Cook, personal communication). Based on these reports, it is apparent that cattle fed pasture had a lower O157:H7 prevalence, but further work is needed to determine if the effect is due to the pasture feed itself or to the physical separation of the animals.

Several earlier cattle surveys, however, did not find a strong correlation between feed ingredients and prevalence of *E. coli* O157:H7 fecal shedding. The prevalence was almost the same in drylot and pasture herds, but the composition of the drylot feed was not reported (Hancock *et al.*, 1997b). In a survey of dairy cows, the same research group did not observe a strong association of corn silage and fecal shedding prevalence and none of the feeds investigated had a positive correlation (Garber *et al.*, 1999). The fecal prevalence did not appear to be different in beef cattle when compared with dairy cattle (1.8% vs. 1.66%) (Hancock *et al.*, 1997c; Herriott *et al.*, 1998). These studies, however, obtained their results using only culture methods which could have affected their final outcome.

In contrast to pasture feeding, the use of grain has been associated with increased risk of fecal shedding of *E. coli* O157:H7 by at least three separate studies. In a large-scale sampling of feedlots in 13 states, feeding barley was identified as a factor strongly related to an increased probability of shedding this pathogen (Dargatz *et al.*, 1997). In a report that involved dairy herds in Denmark, the number of O157-positive cows that were fed grains was twice the number of those that were not fed grains (Rugbjerg *et al.*, 2003). More recently, barley-fed cattle had a 2.4% O157-fecal prevalence, but corn-fed animals had only 1.3% (Berg *et al.*, 2004).

Based on the reportedly inhibitory effect of forage feeding and the stimulatory effect of grain feeding on prevalence of *E. coli* O157:H7, it could be hypothesized that the increased use of grain feeds combined with reduced grazing may have contributed to the emergence of this foodborne pathogen. The lack of forage and the increased amounts of grain feed might have provided a better environment for serotype O157:H7 to survive in the bovine GI tract. In addition, close confinement could have facilitated the spread from one animal to the other and eventually to the environment and to human food.

The reduction of fecal prevalence using pasture has been investigated by researchers from different countries. In Scotland, Ternent *et al.* (2001) indicated that the number of O157-positive animals was significantly reduced when cattle were turned out to grass after winter housing. A similar effect was observed in a Swedish study in which a group of calves that were initially O157-positive stopped shedding this pathogen when they were switched to pasture, while the control group that remained indoors receiving grains continued to shed serotype O157:H7 (Jonsson *et al.*, 2001). In two consecutive years in a cow–calf herd in Alberta, it was observed that the O157:H7 prevalence in cows and calves was reduced from approximately 20% to less than 2% after five weeks on native grass pasture (Gannon *et al.*, 2002).

One of the studies that showed the most conclusive evidence supporting the hypothesis that forage can reduce the fecal shedding of *E. coli* O157:H7, was conducted by USDA researchers (Keen *et al.*, 1999). In this report, an experiment that involved 200 feedlot cattle that had been consuming a finishing grain diet and had an average fecal prevalence of 54%, were divided into two equal groups. One of the groups continued receiving the same diet and the other group was switched to consume 100% alfalfa hay. After seven days, the fecal prevalence of the hay-fed cattle was only 14%, but the grain-fed cattle still shed the pathogenic *E. coli* at a 52% rate. This result clearly supported the idea that hay feeding could reduce the *E. coli* O157:H7 fecal prevalence.

Salmonella

Salmonella species are the only other bacteria whose fecal shedding prevalence has been studied extensively in bovines. The effect of cattle diet on *Salmonella* fecal prevalence has been studied by longitudinal studies that report management risks factors. Several researchers have found no association of the type of feed to the prevalence of this foodborne pathogen (Kabagambe *et al.*, 2000; Warnick *et al.*, 2003), but a couple of surveys found some relationship between prevalence and feeding type. Losinger *et al.* (1995) observed a 2.1% prevalence of *Salmonella*-positive samples from a total of 6861 fecal samples and reported that the use of hay feeding was linked to a decreased risk of fecal shedding. Another large-scale survey found a more than two-fold greater *Salmonella* prevalence in feedlot cattle when 4977 samples were analyzed (Losinger *et al.*, 1997). Feedlot cattle are typically fed larger amounts of grain than dairy cattle. This latter report indicated that the only two management practices related to a higher risk of fecal shedding were feeding tallow and feeding whole cottonseed, but found no association with any other type of feed.

Campylobacter

The presence of *Campylobacter* in cattle populations has not been studied as thoroughly as those of *Salmonella* and *E. coli* O157:H7, but a number of studies have detected this foodborne pathogen in beef and dairy cattle (Beach *et al.*, 2002; Wesley *et al.*, 2000). Feeding of alfalfa was identified as a risk factor for dairy cows, but the association between *Campylobacter* prevalence and this type of feed was not statistically significant (Wesley *et al.*, 2000). In a study that compared feedlot beef cattle to adult animals on pasture, the prevalence of *Campylobacter* was 13% and 25% in feedlots and no more than 2% in pasture animals (Beach *et al.*, 2002). In one of the most recent studies, the authors observed an association between *Campylobacter* shedding and feeding bromegrass hay, but that effect appeared to be confounded by the relatively low prevalence observed (less than 5%) (Berry *et al.*, 2006). A recent report that investigated the effect of diet on the presence of pathogenic microorganisms in animal manure also found no differences between grass feeding and other diets (Hutchison *et al.*, 2005).

10.2.4 Effect of different forage types (grass, clover, maize) on enteric pathogen persistence and shedding

Considerable attention from researchers has been devoted to comparing the effect of grain and forage feeding on the fecal shedding of *E. coli* O157:H7 since it was suggested that forage could be used to reduce carriage of this pathogen by cattle (Diez-Gonzalez *et al.*, 1998). However, there are very few studies that have compared the potential impact of different types of high-fiber feeds such as alfalfa, clover and silage. In one of the first reports that examined the association between *E. coli* O157:H7 prevalence and feed practices, Garber *et al.* (1999) reported that none of the farms in which dairy calves were given clover hay as their first forage had an *E. coli* O157-positive animal. In another longitudinal study among dairy herds, feeding corn silage was correlated with an increased fecal prevalence of *E. coli* O157:H7 in heifers (Herriott *et al.*, 1998). These contrasting results suggest that each forage type could have a different impact on fecal shedding of this particular pathogen and may explain the variability of effects observed by different research groups (Hovde *et al.*, 1999; Buchko *et al.*; 2000, Magnuson *et al.*, 2000).

In the first study that investigated the potential effect of forage type on the fecal shedding of *E. coli* O157:H7, Kudva *et al.* (1995) inoculated sheep with O157:H7 strains and compared alfalfa pellet feed to grazing on sagebrush-bunchgrass. The authors observed that alfalfa pellet-fed animals shed the pathogenic strains for longer periods of time than pasture-fed rams and suggested that nutritionally deficient and high-fiber diets could increase the bacterial count, but may cause clearing of serotype O157:H7 from the GI tract. The effect of fiber concentration on fecal shedding was also investigated using an inoculated sheep model by Lema *et al.* (2002). In that report, groups of lambs were fed seven different diets formulated with various amounts of fescue hay and cottonseed hull to contain ADF from 5% to 35%. After inoculation of lambs with *E. coli* O157:H7 strains, the population of this bacterium was significantly reduced in all six diets that had 10% or more forage compared with the diet that had no fescue hay or cottonseed hull.

The mechanism of the potential inhibitory effect of forages on the fecal shedding of serotype O157:H7 has not been elucidated, but some hypotheses have been explored. Duncan *et al.* (1998) have suggested that some plant metabolites can be inhibitory and even lethal for *E. coli* O157:H7. In pure cultures, the growth of this bacterium was inhibited by the coumarins esculetin, umbelliferone and scopoletin, and in mixed ruminal incubations esculin, a coumarin glycoside was also capable of inactivating this organism. More recently, when inoculated calves were given a daily dose of esculetin, the prevalence of *E. coli* O157:H7 in feces was reduced from 37% to 18% (Duncan *et al.*, 2004). An alternative explanation for the inhibitory forage effect is the lack of readily available carbon sources that *E. coli* O157:H7 could compete for when cellulose and lignin are the most important energy sources. However, this latter hypothesis has no experimental support, yet.

In addition to the above studies on *E. coli* O157:H7, there are very few studies on other foodborne pathogens that compare the effect of feeding different forages to ruminants. In one of these studies, Brownlie and Grau (1967) reported that *Salmonella* was rapidly eliminated from the intestine of heifers after inoculation when they were fed lucerne hay. Findings for fecal *E. coli* are also relevant for this discussion as a surrogate organism. Jordan and McEwen (1998) reported a reduction of 0.5 $\log_{10}$ CFU(colony forming units)/g in the count of fecal *E. coli* of beef cattle, after the animals were switched from a high-energy diet to a mixture of alfalfa hay and corn silage. Using relatively poor quality timothy hay, a greater reduction of 3 $\log_{10}$ CFU/g was reported by Diez-Gonzalez *et al.* (1998) when the diet of cows was changed from 90% cracked corn to forage. In another report, feeding a mixture of different forages resulted in a fecal *E. coli* count that was approximately 1.5 and 3.0 $\log_{10}$ CFU/g smaller than pasture feeding and than fasting, respectively (Gregory *et al.*, 2000).

10.3 Effect of livestock breed and husbandry (including veterinary antibiotic treatments) on the incidence of pathogens and antibiotic-resistant bacteria

The majority of studies that have tried to find an association between the cattle breed and the occurrence of foodborne pathogens have found a lack of correlation between these factors. Miyao *et al.* (1998) found no difference in the number of O157-positive cattle in Japanese Black and Holstein cows. In a study conducted in Denmark, the prevalence of *E. coli* O157:H7 did not appear to be influenced by the type of breed (Holstein Friesian, Jersey and Crossbred) (Nielsen *et al.*, 2002). In Turkey, Yilmaz *et al.* (2002) reported that Brown Swiss and Holstein cows had a greater prevalence, but their own data seems to indicate that there was no difference between these and another two breeds. Similar to *E. coli* O157:H7, the fecal prevalence of *Salmonella* and *Campylobacter* was not influenced by breed of beef cattle (Beach *et al.*, 2002). In the only study that has reported a breed-associated difference, Riley *et al.* (2003) indicated that Romosinuano cows had significantly lower fecal prevalence of serotype O157:H7 than Angus and Brahman animals, but no plausible explanation was provided.

One of the main differences between conventional and organic livestock operations is the use of antibiotics, ionophores and hormones intended to enhance growth and productivity. The potential link between these growth-promoting substances and the presence of foodborne pathogens has been suggested and some researchers have even argued that they have played a role in the emergence of some of these microorganisms. Feeding antibiotics at sub-therapeutic levels to ruminants is a common practice among conventional farms which is believed to be necessary to maintain high productivity. Significant evidence, however, indicates that this practice has led to the increase of antibiotic resistance in both commensal and pathogenic bacteria (McDermott *et al.*, 2002).

The best example of an emerging foodborne pathogen that probably originated from the use of antibiotics is *Salmonella* serovar *typhimurium* definitive type (DT) 104 (Glynn *et al.*, 1998). *Salmonella* DT104 is an animal and human pathogen that was first identified in cattle in the UK in 1984 (Hollinger *et al.*, 1998). Serovar DT104 is resistant to five or more different antibiotics and it has now been linked to cases of gastrointestinal disease in different parts of the world. In the UK, *Salmonella* DT104 has become so prevalent that the majority of all *S. typhimurium* isolates are now DT104. Dairy cattle appear to be its natural reservoir, but it is not clear if its emergence was due to the use of antibiotics in animals or in humans.

The use of antibiotics in ruminants and its effect in another emerging bacterium, *E. coli* O157:H7 has also been investigated. Several longitudinal surveys were not able to find an association between the use of antibiotics and the prevalence of *E. coli* O157:H7 (Garber *et al.*, 1999; Dargatz *et al.*, 1997). More recently, Dunn *et al.* (2004) reported that feeding oxytetracycline might have decreased the occurrence rate of this pathogen in beef calves. The opposite effect was reported by another study in which, artificially inoculated calves that received no oxytetracycline and neomycin in their milk replacer had a decreased risk of shedding serotype O157:H7 (Alali *et al.*, 2004). Feeding neomycin, however, has been investigated as a strategy to reduce prevalence of this pathogen in adult cattle populations. Elder *et al.* (2002) first reported that neomycin could reduce the fecal shedding to undetectable levels compared to the control groups. This effect was corroborated by Ransom *et al.* (2004) who were able to observe a fecal prevalence reduction from 45% to 0% after feedlot cattle were treated with neomycin. Despite this promising result, it is unlikely that neomycin could be adopted as a control strategy against *E. coli* O157:H7 because of concerns of cross-induction of antibiotic resistance.

Another group of growth promoting substances used by conventional livestock operations are the ionophores. Ionophores are different from traditional antibodies in their mode of action and are not typically used for humans. The most important ionophores currently fed to ruminants are monensin and lasalocid which are only effective against Gram-positive bacteria. Because Gram-negative bacteria are not inhibited, it was speculated that the emergence of *E. coli* O157:H7 could be due to the widespread utilization of ionophores (Buchanan and Doyle, 1997). A study in dairy cattle found an increased prevalence of this bacterium in heifers that had received ionophores in their diet (Herriott *et al.*, 1998). Two other studies that have used inoculated sheep and calves have, however, found no effect of ionophore feeding on the extent of fecal shedding of *E. coli* O157:H7 (Edrington *et al.*, 2003; Van Baale *et al.*, 2004).

The effect of feeding monensin on the fecal shedding of *Salmonella* by ruminants has received relatively little attention, but the published reports have contrasting results. Using sheep artificially inoculated with *Salmonella* strains, Edrington *et al.* (2003) found no difference in the fecal counts of the treatment and control groups. More recently in a cattle survey conducted in the midwest USA, a statistically significant association was found between increased prevalence of *Salmonella* and the lack of monensin in calf and heifer diets in dairy cattle (Fossler *et al.*, 2005). Further research needs to be conducted to determine if ionophores have an effect on gastrointestinal pathogen colonization in livestock populations.

10.4 Effect of stress on enteric pathogen shedding

Domestic ruminants used for meat and milk production are subjected to a variety of stresses during their life. The most important stresses related to a particular husbandry system are those caused by transport, weaning, drug application, diet changes, fasting, confinement and slaughterhouse practices. Animals are also subjected to other types of natural stresses such as heat, disease and reproduction. Because any type of stress has the potential to impact negatively weight and milk production, farmers seek to reduce its level in livestock. Conventional husbandry operations typically seek the reduction of stress levels in ruminants mostly when it affects animal productivity. In contrast, organic farmers are not only concerned about productivity, but they follow additional guidelines targeted to improve general animal welfare. Those required guidelines are intended to minimize stress in animals (NOP/USDA, 2000).

Fasting or feed withdrawal for no more than two days is commonly used in modern ruminant operations during transport and before slaughter (Church, 1991). The effect of fasting on the gastrointestinal microbial populations has been thoroughly studied. Brownlie and Grau (1967) published some of the first evidence that fasting caused a marked increase in the population of *Salmonella* in the rumen of cattle. A similar effect was observed by the same researchers in sheep (Grau *et al.*, 1969). The concentration of volatile fatty acids decreases rapidly during fasting and this decline is considered to be the factor that allows the growth of *Salmonella* and other Gram-negative organisms in the rumen of fasting animals (Wallace *et al.*, 1989).

Similar to *Salmonella*, the populations of commensal *E. coli* in cattle feces have been reported to increase after fasting periods of 1 to 2 days (Gregory *et al.*, 2000, Jordan and McEwen, 1998). Several studies have also provided evidence that fasting increases the fecal shedding of *E. coli* O157:H7. In the first study that reported this effect, Kudva *et al.* (1995) observed that fasting caused the re-incidence of fecal shedding in sheep. Fasting calves that had been inoculated with serotype O157:H7 strains shed from 1 to 2 $\log_{10}$ CFU/g more pathogenic cells than well-fed animals (Brown *et al.*, 1997). A number of reports, however, have indicated no effect or even a reduction of fecal shedding of serotype O157:H7 during periods of fasting.

In experimentally inoculated calves, Harmon *et al.* (1999) could not detect a correlation between numbers of *E. coli* O157:H7 and volatile fatty acid concentration of fasting animals and an increase of fecal shedding. Using naturally infected calves, Buchko *et al.* (2000) reported that a single fasting period did not increase fecal shedding, but when fasting was used in combination with diet shifting, the prevalence of this pathogenic bacterium sharply increased. In another larger cattle trial, Keen *et al.* (1999) indicated that feed withdrawal for 1 day reduced the fecal prevalence from 52% to 18%. These results suggest that the effect of fasting might be variable and might be affected by factors that need to be elucidated.

During the production cycle, ruminant animals are typically transported using vehicles at least once in their lifetime (Grandin, 1997). Current transportation methods in which animals are confined inside a moving vehicle are very stressful events that may have an impact on meat and milk production, as well as the populations of gastrointestinal microorganisms. The shedding pattern of *Salmonella* in swine populations during transport has been thoroughly studied and so far the data indicate that the bacterial count increases and *Salmonella* is spread to other organs (Hurd *et al.*, 2002). In ruminants, the effect of transport on foodborne pathogens does not appear to be very clear.

Among the studies that have investigated the effect of transport, the prevalence of *E. coli* O157:H7 in cattle after transport and outdoor temporary confinement (lairage) did not increase and remained slightly lower than at the farm (Minihan *et al.*, 2003). In a recent report that investigated the effect of transport duration on the prevalence of serotype O157 in calves that had been fed pasture, Bach *et al.* (2004) indicated that animals that were transported for approximately 13 hours were more susceptible to infection after arriving at the feedlot compared to those subjected to 3 hours transport. Similarly, the prevalence of *Salmonella* in feeder calves that were sent to a feedlot from three farms increased from 0% before transport to 1.5% after arriving at the feedlot and to 8% after 30 days in the feedyard (Corrier *et al.*, 1990). An even greater increase from 1 to 21% in fecal prevalence of *Salmonella* was observed in adult cattle after transportation (Beach *et al.*, 2002). In the same study, the prevalence of *Campylobacter* did not appear to be affected by transport.

Weaning is probably the most stressful stage during the first year of a ruminant animal's life. Because of the traumatic nature of weaning, it has been hypothesized that the gastrointestinal microbial populations would be altered and there are some reports in the literature that support this idea. In one of the first studies, Garber *et al.* (1995) reported that post-weaning calves had approximately three times greater risk of being *E. coli* O157-positive than pre-weaning animals. More recently, the prevalence of serotype O157:H7 was also increased from less than 1.5% one week before to more than 6% in the following two weeks after weaning (Gannon *et al.*, 2002). The findings of a longitudinal study conducted in Italy do not seem to support the above hypothesis, because a lack of association was found between pathogen prevalence and weaning (Conedera *et al.*, 2001).

According to organic guidelines, ruminants should have access to outdoors and to pasture (NOP/USDA, 2000) and they cannot be confined to the relatively small spaces of conventional feedlot operations. Confinement within a small area or pen with a large group of animals could also be a stressful situation that could influence the gastrointestinal pathogen population. There are some studies that have reported a decrease in the *E. coli* O157:H7 fecal prevalence of cattle that were moved to pasture from indoor pens, but it is not clear if this decline was the result of eating grass or free ranging (Gannon *et al.*, 2002; Jonsson *et al.*, 2001). In another study, all 174 calves that had been on pasture tested negative for serotype O157:H7 before being transported to a feedlot (Bach *et al.*, 2004). Beef cattle that were confined to a beef yard for 30 days had an increased *Salmonella* fecal prevalence (8%) compared with that at the time of their arrival (1.5%) (Corrier *et al.*, 1990). None of these studies however, measured physiological indicators of stress such as corticosteroids and the presence of pathogens.

10.5 Reducing enteric pathogen transfer risks in organic and 'low input' systems: outline of strategies

Organic agriculture is a production system that seeks to promote and enhance the health of agroecosystems by using few inputs, avoiding synthetic substances and promoting animal welfare (Codex Alimentarius Commission, 1999). While organic foods are expected to be safe, the production of safer foods is not one of the core priorities of organic food production. The question whether organic foods are more or less safe than their conventional counterparts is yet to be answered, but in recent years an increasing number of studies have been conducted to address it.

Because of the emphasis on the use of pasture for organic ruminant farms, it could be hypothesized that the prevalence of foodborne pathogens would be smaller among organic herds and flocks. However, the lack of antibiotic use could lead to increased incidence of animal infections such as those caused by *Salmonella* and enterotoxigenic *E. coli*, which could be transmitted to foods and the environment. On the other hand, the absence of antibiotics in organic herds would reduce the likelihood of antibiotic-resistant pathogens such as *Salmonella typhimurium* DT104. Some of those findings that have linked organic practices to increased safety are encouraging, but there is still too little evidence to advocate a particular strategy for prevention of foodborne disease.

10.5.1 Comparisons of organic versus conventional ruminant systems

Since 2003 an increased number of investigations have started to report comparative studies of organic and conventional livestock production systems. These reports have attempted to find differences in pathogen prevalence and antibiotic resistance incidence between organic and conventional farms. In a study that compared organic and conventional dairy farms in Wisconsin, the fecal prevalence of *Campylobacter* was very similar (27% vs. 29%, respectively) and the percentage of *Campylobacter* isolates that were resistant to ciprofloxacin, gentamicin, erythromycin and tetracycline was no different in both production systems (Sato *et al.*, 2004a). In a survey of dairy farms in Switzerland, no significant differences in the fecal prevalence of shiga toxin-producing *E. coli* and *E. coli* O157:H7 were found between organic and conventional farms (Kuhnert *et al.*, 2005). The fecal prevalence of *Salmonella* in dairy cows was almost the same (4.7% and 4.9%) in organic and conventional farms included in a survey of farms in Minnesota, Michigan, Wisconsin and New York (Fossler *et al.*, 2004). These results suggest that the prevalence of foodborne pathogens in dairy cattle is not markedly affected by the type of production system.

As an indirect measurement of the differences between organic and conventional livestock systems, some researchers have resorted to comparing samples of milk and meat. The presence of *Staphylococcus aureus* in bulk tank milk was investigated in organic and conventional organic farms in Wisconsin and Denmark (Sato *et al.*, 2004b). In Wisconsin, *S. aureus* was detected in 87% and 73% of organic and conventional farms, respectively, while in Denmark, this pathogen was detected in 50% and 85% of organic and conventional farms. *S. aureus* isolates from conventional farms in Wisconsin had an increased resistance to ciprofloxacin compared to isolates obtained from organic farms. LeJeune and Christie (2004) studied the antimicrobial susceptibility of *E. coli* isolated from meat produced conventionally and they reported that the proportion of isolates resistant to ceftiofur and chloramphenicol was from two to three times greater than the resistance of *E. coli* from meat produced with no antibiotics. Additional work is clearly needed to determine if the use of no antibiotics by organic and 'low input' farms results in fewer antimicrobial-resistant bacteria.

10.6 Future trends

The microbial safety of organically produced foods is a relatively new area of study. The interest of researchers and regulators in this novel field was not attracted until the demand and consumption of this type of foods reached a critical level. Because of this interest, several funding programs have been created since 2002 to support research, outreach and education on organic agriculture and a few of these programs have incorporated a focus on food safety. Based on the latest projections of demand growth of organic foods, it is estimated that both private and public resources will be devoted to study the likelihood of transmission of foodborne diseases by organic foods.

Some of the first published papers that focused on assessing the safety of organic foods of ruminant origin have yielded valuable information, but it is still too early to use that information for practical recommendations. Future research should involve a variety of approaches to understand the connection between husbandry practices and pathogen presence in animal populations. One of the areas that needs significant attention from researchers is the gastrointestinal ecology of foodborne pathogens. Most of the factors that determine whether a particular pathogen is capable of colonizing an animal remain to be elucidated. As an example, to date we do not understand the sporadic fecal shedding of *E. coli* O157:H7 and why this organism can suddenly appear in previously negative-tested animals. To advance the microbial ecology of pathogens, it is recommended that multi-disciplinary approaches should be taken that use *in vitro* models, artificially inoculated animals or naturally carrying ruminants.

In the coming years, we will probably continue to see publications of comparisons of organic versus conventional systems. This type of approach can be useful to identify potential associations between pathogen prevalence and specific agricultural practices. However, there will still be a need to conduct highly controlled farm experiments in which extensive analysis will identify the underlying mechanisms behind the effect of specific practices. In particular, current organic practices and their implications for pathogen prevalence need to be evaluated. Another type of research that we will be seeing more often is the systems approach in which many aspects of the production chain are studied.

Another important area of research is to investigate the potential application of some organically approved substances for pathogen control. As an example, in our laboratory we have identified that sodium carbonate is capable of reducing *E. coli* O157:H7 and *Salmonella* DT104 in cattle manure (Park and Diez-Gonzalez, 2003). The use of sodium carbonate is compatible with organic guidelines and its utilization for controlling pathogens in the farm environment could be promising. Similarly, the use of probiotic bacteria should be explored for the control of animal infections and prevalence of foodborne pathogens for organic, low input and conventional livestock.

10.7 Sources of further information and advice

There are several electronic and print sources of information on the topic of microbial safety of ruminant production systems. On organic livestock and food safety, the website: http://www.safonetwork.org of Sustainable Animal Health and Food Safety in Organic Farming (SAFO) is probably the best suited to the contents of this chapter. The SAFO website contains the proceedings of three recent conferences devoted to this topic. An interesting insight into the Risks of Organic vs. Conventional Foods is found at: www.misa.umn.edu/vd/risks.html by Riddle. The Proceedings of the First World Congress on Organic Foods provides several perspectives by presenters on general organic food safety: www.foodsafe.msu.edu/Documents/Conference%20Proceedings/Organics_Proceedings/Organics_Proceedings_MSU.pdf.

On organic livestock practices, in addition to the National Organic Rule referred to above, the *Organic Livestock Workbook* among other useful documents for organic farmers, is found at the National Center of Appropriate Technology's website: www.attra.org/livestock.html. The USDA's National Agricultural Library has prepared an exhaustive list of references on organic livestock production http://www.nal.usda.gov/afsic/AFSIC_pubs/srb0405.htm. On general organic agriculture the reader is also encouraged to visit the Organic Agriculture at FAO website: http://www.fao.org/organicag/default.htm.

A wide variety of valuable sources of information on the topic of animal diet and foodborne pathogens have been written. The reader is referred to other reviews by Callaway *et al.* (2003), Duncan *et al.* (2000) and Hancock *et al.* (2001) that cover the effect of forage. The topic specifically of cattle diet and *E. coli* O157:H7 fecal shedding/prevalence has been extensively discussed on the Internet. A recent Google search using the terms 'cattle diet' and 'O157' yielded 12,900 sites. One of those sites by Rena Orr (http://www.foodsafetynetwork.ca/ en/article-details.php?a=3&c=10&sc=79aid=271-0157-cattle.htm) has a very thorough discussion of the topic.

The recently published book on food safety related to animal products *Improving the Safety of Fresh Meat* edited by Sofos (Sofos, 2005) is a very valuable resource that covers different strategies of control and understanding the ecology of pathogens transmitted via meat. Another book that covers the topic of 'farm-to-fork' food safety with relevant information on livestock production and foodborne diseases is *Preharvest and Postharvest Food Safety: Contemporary Issues and Future Directions* edited by Beier (Beier *et al.*, 2004). On general information about foodborne pathogens, the reader is also invited to check some of the food microbiology textbooks recently authored by James Jay, Bibek Ray and Thomas Montville/KarlMatthews, from Springer, CRC Press and ASM Press.

10.8 References

Alali, W. Q., Sargeant, J. M., Nagaraja, T. G. and DeBey, B. M. (2004) *J. Anim. Sci.*, **82**, 2148–2152.

Bach, S. J., McAllister, T. A., Mears, G. J. and Schwartzkopf-Genswein, K. S. (2004) *J. Food Prot.*, **67**, 672–678.

Beach, J. C., Murano, E. A. and Acuff, G. R. (2002) *J. Food Prot.*, **65**, 1687–1693.

Beier, R., Pillai, S. D. and Phillips, T. D. (2004) *Preharvest and Postharvest Food Safety*, IFT Press/Blackwell, Ames, Iowa.

Berg, J., McAllister, T., Bach, S., Stilborn, R., Hancock, D. and LeJeune, J. (2004) *J. Food Prot.*, **67**, 666–671.

Berry, E. D., Wells, J. E., Archibeque, S. L., Ferrell, C. L., Freetly, H. C. and Miller, D. N. (2006) *J. Anim. Sci.*, **84**, 2523–2532.

Besser, T. E., Hancock, D. D., Pritchett, L. C., McRae, E. M., Rice, D. H. and Tarr, P. I. (1997) *J. Infect. Dis.*, **175**, 726–729.

Brown, C. A., Harmon, B. G., Zhao, T. and Doyle, M. P. (1997) *Appl. Environ. Microbiol.*, **63**, 27–32.

Brownlie, L. E. and Grau, F. H. (1967) *J. Gen. Microbiol.*, **46**, 125–134.

Buchanan, R. L. and Doyle, M. P. (1997) *Food Tech.*, **51**, 69–76.

Buchko, S. J., Holley, R. A., Olson, W. O., Gannon, V. P. J. and Veira, D. M. (2000) *Can. J. Anim. Sci.*, **80**, 741–744.

Buncic, S. and Avery, S. M. (1997) *N. Z. Vet. J.*, **45**, 45–48.

Callaway, T. R., Elder, R. O., Keen, J. E., Anderson, R. C. and Nisbet, D. J. (2003) *J. Dairy Sci.*, **86**, 852–860.

CDC (2005) US Foodborne Disease Outbreaks, Team Outbreak Response and Surveillance http://www.cdc.gov/foodborneoutbreaks/US_outb.htm.

Chapman, P. A., Siddons, C. A., Malo Cerdan, A. T. and Harkin, M. A. (1997) *Epidemiol. Infect.*, **119**, 245–250.

Church, D. C. (1991) *Livestock Feeds and Feeding*, Prentice Hall, Englewood Cliffs.

Codex Alimentarius Commission (1999) *Guidelines for the Production, Processing, Labelling and Marketing of Organically Produced Foods*, FAO/WHO, Rome.

Conedera, G., Chapman, P. A., Marangon, S., Tisato, E., Dalvit, P. and Zuin, A. (2001) *Int. J. Food Microbiol.*, **66**, 85–93.

Corrier, D. E., Purdy, C. W. and DeLoach, J. R. (1990) *Am. J. Vet. Res.*, **51**, 866–869.

Dargatz, D. A., Wells, S. J., Thomas, L. A., Hancock, D. D. and Garber, L. P. (1997) *J. Food Prot.*, **60**, 466–470.

Dehority, B. A. (2003) *Rumen Microbiology*, Nottingham University Press, Nottingham.

Dehority, B. A. and Orpin, C. G. (1988) *The Rumen Microbial Ecosystem*, Hobson, P. N. (ed.), Elsevier Applied Science, London.

Diez-Gonzalez, F., Callaway, T. R., Kizoulis, M. G. and Russell, J. B. (1998) *Science*, **281**, 1666–1668.

Duncan, S. H., Flint, H. J. and Stewart, C. S. (1998) *FEMS Microbiol. Lett.*, **164**, 283–288.

Duncan, S. H., Booth, I. R., Flint, H. J. and Stewart, C. S. (2000) *J. Appl. Microbiol.*, **88**, 157S–165S.

Duncan, S. H., Leitch, E. C., Stanley, K. N., Richardson, A. J., Laven, R. A., Flint, H. J. and Stewart, C. S. (2004) *Br. J. Nutr.*, **91**, 749–755.

Dunn, J. R., Keen, J. E., Del Vecchio, R., Wittum, T. E. and Thompson, R. A. (2004) *J. Food Prot.*, **67**, 2391–2396.

Edrington, T. S., Callaway, T. R., Bischoff, K. M., Genovese, K. J., Elder, R. O., Anderson, R. C. and Nisbet, D. J. (2003) *J. Anim. Sci.*, **81**, 553–560.

Elder, R. O., Keen, U. J. E., Siragusa, G. R., Barkocy-Gallagher, G. A., Koohmaraie, M. and Laegreid, W. W. (2000) *Proc. Natl. Acad. Sci.*, **97**, 2999–3003.

Elder, R. O., Keen, J. E., Wittum, T. E., Callaway, T. R., Edrington, T. S., Anderson, R. C. and Nisbet, D. J. (2002) *J. Anim. Sci.*, **80 (Suppl. 1)**, 151.

Ewan, R. C. (2001) *Swine Nutrition*, Lewis, A. J. and Southern, L. L. (eds), CRC Press, Boca Raton, FL.

Forsberg, C. W., Cheng, K.-J. and White, B. A. (1997) *Gastrointestinal Microbiology*, Vol. 1, Mackie, R. I. and White, B. A (eds), Chapman and Hall, New York.

Fossler, C. P., Wells, S. J., Kaneene, J. B., Ruegg, P. L., Warnick, L. D., Bender, J. B., Godden, S. M., Halbert, L. W., Campbell, A. M. and Zwald, A. M. (2004) *J. Am. Vet. Med. Assoc.*, **225**, 567–573.

Fossler, C. P., Wells, S. J., Kaneene, J. B., Ruegg, P. L., Warnick, L. D., Bender, J. B., Eberly, L. E., Godden, S. M. and Halbert, L. W. (2005) *Prev. Vet. Med.*, **70**, 257–277.

Gannon, V. P., Graham, T. A., King, R., Michel, P., Read, S., Ziebell, K. and Johnson, R. P. (2002) *Epidemiol. Infect.*, **129**, 163–172.

Garber, L., Wells, S., Schroeder-Tucker, L. and Ferris, K. (1999) *J. Food Protec.*, **62**, 307–312.

Garber, L. P., Wells, S. J., Hancock, D. D., Doyle, M. P., Tuttle, J., Shere, J. A. and Zhao, T. (1995) *JAVMA*, **207**, 46–49.

Glynn, M. K., Bopp, C., Dewitt, W., Dabney, P., Mokhtar, M. and Angulo, F. J. (1998) *N. Eng. J. Med.*, **338**, 1333–1338.

Goering, H. K. and Van Soest, P. J. (1970) *Forage Fiber Analysis*, Agricultural Research Service/United States Department of Agriculture, Washington, DC.

Grandin, T. (1997) *J. Anim. Sci.*, **75**, 249–257.

Grau, F. H., Brownlie, L. E. and Smith, M. G. (1969) *J. Appl. Bacteriol.*, **32**, 112–117.

Gregory, N. G., Jacobson, L. H., Nagle, T. A., Muirhead, R. W. and Leroux, G. J. (2000) *N. Z. J. Agric. Res.*, **43**, 351–361.

Grieshop, C. M., Reese, D. E. and Fahey Jr., G. C. (2001) In *Swine Nutrition*, Lewis, A. J. and Southern, L. L. (eds) CRC Press, Boca Raton, FL.

Hancock, D. D., Besser, T. E., Rice, D. H., Herriott, D. E., Tarr, P. I. (1997a) *Epidemiol. Infect.*, **118**, 193–195.

Hancock, D. D., Rice, D. H., Herriott, D. E., Besser, T. E., Ebel, E. D. and Carpenter, L. V. (1997b) *J. Food Prot.*, **60**, 363–366.

Hancock, D. D., Rice, D. H., Thomas, L. A., Dargatz, D. A. and Besser, T. E. (1997c) *J. Food Prot.*, **60**, 462–465.

Hancock, D., Besser, T., Lejeune, J., Davis, M. and Rice, D. (2001) *Int. J. Food Microbiol.*, **66**, 71–78.

Harmon, B. G., Brown, C. A., Tkalcic, S., Mueller, P. O. E., Parks, A., Jain, A. V., Zhao, T. and Doyle, M. P. (1999) *J. Food Prot.*, **62**, 574–579.

Herriott, D. E., Hancock, D. D., Ebel, E. D., Carpenter, L. V., Rice, D. H. and Besser, T. E. (1998) *J. Food Prot.*, **61**, 802–807.

Hollinger, K., Wray, C., Evans, S., Pascoe, S., Chappell, S. and Jones, Y. (1998) *JAVMA*, **213**, 1732–1733.

Hovde, C. J., Austin, P. R., Cloud, K. A., Williams, C. J. and Hunt, C. W. (1999) *Appl. Environ. Microbiol.*, **65**, 3233–3235.

Hungate, R. E. (1966) *The Rumen and its Microbes*, Academic Press, New York.

Huntington, G. B. (1997) *J. Anim. Sci.*, **75**, 852–867.
Huntington, G. B. and Reynolds, P. J. (1986) *J. Dairy Sci.*, **69**, 2428–2436.
Hurd, H. S., McKean, J. D., Griffith, R. W., Vesley, I. V. and Rostagno, M. H. (2002) *Appl. Environ. Microbiol.*, **68**, 2376–2381.
Hutchison, M. L., Walters, L. D., Avery, S. M., Munro, F. and Moore, A. (2005) *Appl. Environ. Microbiol.*, **71**, 1231–1236.
Jay, J. M. (2000) *Modern Food Microbiology*, Aspen Publishers, Gaithersburg, MD.
Jonsson, M. E., Aspan, A., Eriksson, E. and Vagsholm, I. (2001) *Int. J. Food Microbiol.*, **66**, 55–61.
Jordan, D. and McEwen, S. A. (1998) *J. Food Prot.*, **61**, 531–534.
Kabagambe, E. K., Wells, S. J., Garber, L. P., Salman, M. D., Wagner, B. and Fedorka-Cray, P. J. (2000) *Prev. Vet. Med.*, **43**, 177–194.
Keen, J. E., Uhlich, G. A. and Elder, R. O. (1999) *80th Conference Research Workers in Animal Diseases*, Nov. 7–9, Chicago, IL.
Khaitsa, M. L., Smith, D. R., Stoner, J. A., Parkhurst, A. M., Hinkley, S., Klopfenstein, T. J. and Moxley, R. A. (2003) *J. Food Prot.*, **66**, 1972–1977.
Kudva, I. T., Hatfield, P. G. and Hovde, C. J. (1995) *Appl. Environ. Microbiol*, **61**, 1363–1370.
Kuhnert, P., Dubosson, C. R., Roesch, M., Homfeld, E., Doherr, M. G. and Blum, J. W. (2005) *Vet. Microbiol.*, **109**, 37–45.
Laegreid, W. W., Elder, R. O. and Keen, J. E. (1999) *Epidemiol. Infect.*, **123**, 291–298.
LeJeune, J. T. and Christie, N. P. (2004) *J. Food Prot.*, **67**, 1433–1437.
Lema, M., Williams, L., Walker, L. and Rao, D. R. (2002) *Small. Rumin. Res.*, **43**, 249–255.
Losinger, W. C., Wells, S. J., Garber, L. P., Hurd, H. S. and Thomas, L. A. (1995) *J. Dairy Sci.*, **78**, 2464–2472.
Losinger, W. C., Garber, L. P., Smith, M. A., Hurd, H. S., Biehl, L. G., Fedorka-Cray, P. J., Thomas, L. A. and Ferris, K. (1997) *Prev. Vet. Med.*, **31**, 231–244.
Magnuson, B. A., Davis, M., Hubele, S., Austin, P. R., Kudva, I. T., Williams, C. J., Hunt, C. W. and Hovde, C. J. (2000) *Infect. Immun.*, **68**, 3808–3814.
McDermott, P. F., Zhao, S., Wagner, D. D., Simjee, S., Walker, R. D. and White, D. G. (2002) *Anim. Biotechnol.*, **13**, 71–84.
Mead, P. S., Slutsker, L., Dietz, V., McCaig, L. F., Bresee, J. S., Shapiro, C., Griffin, P. M. and Tauxe, R. V. (1999) *Emerg. Infect. Dis.*, **5**, 607–625.
Mechie, S. C., Chapman, P. A. and Siddons, C. A. (1997) *Epidemiol. Infect.*, **118**, 17–25.
Minihan, D., O'Mahony, M., Whyte, P. and Collins, J. D. (2003) *J. Vet. Med. B Infect. Dis. Vet. Pub. Health*, **50**, 378–382.
Miyao, Y., Kataoka, T., Nomoto, T., Kai, A., Itoh, T. and Itoh, K. (1998) *Vet. Microbiol.*, **61**, 137–143.
Nagaraja, T. G. and Chengappa, M. M. (1998) *J. Anim. Sci.*, **76**, 287–298.
Nielsen, E. M., Tegtmeier, C., Andersen, H. J., Gronbaek, C. and Andersen, J. S. (2002) *Vet. Microbiol.*, **88**, 245–257.
NOP (National Organic Program)/USDA (2000) *7 Code of Federal Regulations Part 205*. http://www.access.gpo.gov/nara/cfr/waisidx_06/7cfr205_06.htm
Owens, F. N., Secrist, D. S., Hill, W. J. and Gill, D. R. (1997) *J. Anim. Sci.*, **75**, 868–879.
Owens, F. N., Secrist, D. S., Hill, W. J. and Gill, D. R. (1998) *J. Anim. Sci.*, **76**, 275–286.
Padola, N. L., Sanz, M. E., Blanco, J. E., Blanco, M., Blanco, J., Etcheverria, A. I., Arroyo, G. H., Usera, M. A. and Parma, A. E. (2004) *Vet. Microbiol.*, **100**, 3–9.
Park, G. W. and Diez-Gonzalez, F. (2003) *J. Appl. Microbiol*, **94**, 675–685.
Rangel, J. M., Sparling, P. H., Crowe, C., Griffin, P. M. and Swerdlow, D. L. (2005) *Emerg. Infect. Dis.*, **11**, 603–609.
Ransom, J. R., Belk, K. E., Scanga, J. A., Sofos, J. N. and Smith, G. C. (2004) *J. Anim. Sci.*, **82 (Suppl. 1)**, 126.
Renter, D. G., Sargeant, J. M., Oberst, R. D. and Samadpour, M. (2003) *Appl. Environ. Microbiol.*, **69**, 542–547.

Riley, D. G., Gray, J. T., Loneragan, G. H., Barling, K. S. and Chase, C. C. J. (2003) *J. Food Prot.*, **66**, 1778–1782.
Rugbjerg, H., Nielsen, E. M. and Andersen, J. S. (2003) *Prev. Vet. Med.*, **58**, 101–113.
Sanz, M. E., Vinas, M. R. and Parma, A. E. (1998) *Europ. J. Epidem.*, **14**, 399–403.
Sargeant, J. M., Gillespie, J. R., Oberst, R. D., Phebus, R. K., Hyatt, D. R., Bohra, L. K. and Galland, J. C. (2000) *Am. J. Vet. Res.*, **61**.
Sargeant, J. M., Sanderson, M. W., Smith, R. A. and Griffin, D. D. (2003) *Prev. Vet. Med.*, **61**, 127–135.
Sato, K., Bartlett, P. C., Kaneene, J. B. and Downes, F. P. (2004a) *Appl. Environ. Microbiol.*, **70**, 1442–1447.
Sato, K., Bennedsgaard, T. W., Bartlett, P. C., Erskine, R. J. and Kaneene, J. B. (2004b) *J. Food Prot.*, **67**, 1104–1110.
Sofos, N. J. (2005) *Improving the Safety of Fresh Meat*, Woodhead, Cambridge, UK.
Shere, J. A., Bartlett, K. J. and Kaspar, C. W. (1998) *Appl. Environ. Microbiol.*, **64**, 1390–1399.
Sivapalasingam, S., Friedman, C. R., Cohen, L. and Tauxe, R. V. (2004) *J. Food Prot.*, **67**, 2342–2353.
Smith, D., Blackford, M., Younts, S., Moxley, R., Gray, J., Hungerford, L., Milton, T. and Klopfenstein, T. (2001) *J. Food Prot.*, **64**, 1899–1903.
Ternent, H. E., Innocent, G. T., Edge, V. L., Thomson-Carter, F. and Synge, B. (2001) *Conference on Epidemiology of VTEC*, Duffy, G. P. (ed.), Dublin, Ireland, pp. 5.
Van Baale, M. J., Sargeant, J. M., Gnad, D. P., DeBey, B. M., Lechtenberg, K. F. and Nagaraja, T. G. (2004) *Appl. Environ. Microbiol.*, **70**, 5336–5342.
Wallace, J., Falconer, M. L. and Bhargava (1989) *Curr. Microbiol.*, **19**, 277–281.
Warnick, L. D., Kaneene, J. B., Ruegg, P. L., Wells, S. J., Fossler, C., Halbert, L. and Campbell, A. (2003) *Prev. Vet. Med.*, **60**, 195–206.
Weigel, K. A., Kriegl, T. and Pohlman, A. L. (1999) *J. Dairy Sci.*, **82**, 191–195.
Wesley, I. V., Wells, S. J., Harmon, K. M., Green, A., Schroeder_Tucker, L., Glover, M. and Siddique, I. (2000) *Appl. Environ. Microbiol.*, **66**, 1994–2000.
Yilmaz, A., Gun, H. and Yilmaz, H. (2002) *J. Food Prot.*, **65**, 1637–1640.

11

Reducing antibiotic use for mastitis treatment in organic dairy production systems

Peter Klocke and Michael Walkenhorst, FiBL, Switzerland and Gillian Butler, Newcastle University, UK

11.1 Introduction

Effective management of mastitis is an essential precondition for successful dairy production, both from an animal welfare and economic point of view, since even minor udder inflammations lead to significant reductions in milk yield (Bartlett *et al.*, 1990; Hamann and Gyodi, 1994). In the last few decades, the main approach for the control mastitis in both conventional and organic dairy herds has been to develop and use veterinary products (primarily antibiotics) for both local and systemic administration. However, owing to (i) the often limited efficacy of antibiotic treatments (especially for environmental mastitis), (ii) the potential for development of antibiotic resistance and (iii) consumer concern about antibiotic residues in food, preventative management strategies and alternative treatments (which allow residue-free therapy) have received renewed interest (Walkenhorst *et al.*, 2001). Economic pressure on dairy farms caused by the inability to market milk during 'withdrawal periods' following antibiotic treatment and the ban on prophylactic antibiotic treatments (e.g. routine dry cow treatment) under organic farming standards (and in several European countries also in conventional farming) have also contributed to the search for alternatives. In organic farming this has resulted in a growing number of recommendations and guidance notes on the use of complementary/alternative products and therapies (e.g. teat sealers, homeopathic remedies and nosodes). However, there is very little sound scientific information on the efficacy and optimum use of these products within mastitis management plans. Although often sceptical, farmers as well as veterinarians are increasingly prepared to use these methods especially in organic farming systems, and homeopathy has become one of the preferred alternative methods (Hovi and Roderick, 1998). Homeopathy is defined within the EU organic farming regulations as one of the preferred alternative methods which can be considered for use before resorting to antibiotic therapy (Graf *et al.*, 1999; Hertzberg *et al.*, 2003).

The necessity to identify alternatives to antibiotics for the management of mastitis, together with the economic pressure on both organic and conventional milk producers, make it imperative that any alternatives introduced are demonstrated to have reproducible and verifiable levels of activity, in order to persuade a wider range of organic but also conventional farmers to adopt such innovative strategies. This chapter aims (i) to give a short overview of the causes and epidemiology of mastitis, (ii) to describe the symptoms of mastitis (iii) to give an overview of the currently available treatments for mastitis and their relative efficacy and (iv) to demonstrate the importance of integrating different approaches in farmer/farmer group specific management plans.

11.2 Causes and epidemiology of mastitis

Mastitis is an inflammation caused by microbial infections of the udder. Mastitis can be defined as an 'inflammation of the udder that affects at least one quarter of the mammary gland' (see Section 11.3 below for a detailed description of mastitis symptoms). Mastitis is most frequently caused by bacterial pathogens, but yeast, protozoal and algal infections may also cause the disease. Bacterial pathogens are usually classified into two groups: (i) animal associated bacteria, which are able to be transmitted from cow to cow during the time of milking and (ii) environment associated bacteria, which enter the mammary gland through exposure to a contaminated environment. The main species of cow associated pathogens are *Staphylococcus aureus*, *Streptococcus agalactiae* and *Streptococcus dysgalactiae* (Harmon, 1994; Smith and Hogan, 1993). The main species of environmental udder pathogens are *Streptococcus uberis*, Enterococci and other coliform bacteria like *E. coli* but also *Actinomyces pyogenes*, yeasts und fungi.

The infection process and disease development process can be related to a range of factors that affect various udder tissues (localisations). There are also a range of interactions between the udder tissues and mastitis pathogens. These were recently reviewed by Hamann 2001 (see also Table 11.1).

Hamann also reviewed a range of cow-specific and environmental factors that affect the risk of microbial infection and mastitis disease progression in dairy cows (see Table 11.2).

It is important to mention that during the last three decades the prevalence profile of different mastitis pathogens has changed significantly. For example, in Denmark incidence of infections with streptococci has been decreasing, while that of *S. aureus* has been increasing. Also, *S. agalactiae* infections, which were considered the greatest problem until the early 1980s, then

Table 11.1 Localisation and stages of germ–udder interaction depending on different factors (Hamann, 2001; modified)

Localisation	Interaction	Factors
Teat skin, teat canal	Contamination	Environment, udder anatomy, milking machine
Teat canal	Colonisation	Microbial species, cellular reaction, milking
Teat cistern	Invasion	Milking conditions, cellular and humoral defence
Teat cistern, glandular cistern, glandular lobuli	Infection, inflammation (infection disease)	Chemotaxis, cellular influx, immunoglobulins, complement, further mediators

Table 11.2 Interaction between cow-specific and environmental factors and mastitis risk (Hamann, 2001; modified)

Risk	Cow-specific and environmental factors affecting infection and disease progression
Contamination	Litter hygiene, air humidity, cleanliness, interim disinfection
Colonisation Invasion	Environmental pathogenes: hygiene, Cow associated pathogens: stereotypy
Infection	Immuno-deficiency, technical deficiency
Inflammation	Immuno-status, factors of microbial virulence

decreased owing to the introduction of new sanitation techniques, but were succeeded by *S. aureus* and later coagulase negative *Staphylococcus* (CNS), which have become the dominant problem in many herds (Madsen *et al.*, 1974; Myllys *et al.*, 1994; Schällibaum, 1999).

11.3 Symptoms of mastitis

11.3.1 Clinical mastitis

'Acute clinical' mastitis is characterised by a range of visible cardinal inflammation symptoms. These are used in the diagnosis of the disease and can be divided into *rubor* (redness), *tumor* (swelling), *dolor* (pain), *calor* (warmness) and *functio laesa* (dysfunction, represented by secretory alterations like flakes, clots and aqueous milk secret). However, not all five symptoms may be detected in cows suffering from acute mastitis, with *tumor* and *calor* being the most frequently observed symptoms. Historically, clinical mastitis is defined on the basis of the number of udder quarters affected and showing symptoms (Nickel *et al.*, 1984).

Besides the acute type of mastitis, which is generally accompanied by a more or less decrease in general health condition of the affected cow, less severe types of mastitis expressing only secretion symptoms are more common in dairy practice. These types are called chronic mastitis. An interim type is the so-called 'sub-acute mastitis' with slight inflammation symptoms. The severity levels and symptoms of mastitis are described in Table 11.3.

11.3.2 Sub-clinical mastitis

Sub-clinical mastitis, on the other hand, has no visible symptoms and can only be diagnosed with laboratory methods (Wendt *et al.*, 1994) (see below). The diagnosis of sub-clinical mastitis depends on two parameters, the microbiological profile of sampled milk and the somatic cell count (Hamann and Fehlings, 2002). The different severity levels of mastitis and symptoms used in diagnosis are described in Table 11.3.

Both clinical and sub-clinical mastitis are usually accompanied by an increase in the somatic cell count (SCC; a measure of white blood cells (95%) such as macrophages, segmented neutrophil granulocytes or lymphocytes) in milk. SCCs above a certain level are an indication of immune-system activation and are used as an indicator of 'sub-clinical', 'latent' or 'chronic' udder infections/mastitis (Concha, 1986; Östensson *et al.*, 1988), which were shown to affect milk quality, composition and shelf life. Dairy companies therefore impose SCC thresholds or reduce payments to farmers if certain thresholds are exceeded (Hamann, 2001; Urech *et al.*, 1999).

The SCC thresholds set by government regulation and/or dairy companies can differ significantly between countries. Swiss regulations on milk quality assurance define a SCC limit of 150,000 cells per ml milk, above which farmers have to do further diagnostic tests to ensure absence of mastitis. On the other hand, the German Veterinary Association (DVG) set a threshold of 100,000 cells per ml and the International Dairy Federation (IDF) proposed a historic threshold of 500,000 cells per ml (Hamann and Fehlings, 2002). Increasingly, a combination of microbiological tests is used in combination with the SCC levels for classification (see Table 11.4).

11.4 Mastitis management and treatment

If untreated, mastitis can cause prolonged acute or chronic illness and low productivity and even death of dairy cows. However, recovery after shorter periods (days to weeks) have been more frequently observed, but since animal welfare considerations require sick animals to be treated, very few

Table 11.3 Mastitis symptoms and severity levels

Mastitis					Mastitis?	No mastitis	
Clinical mastitis				Subclinical mastitis secretory disfunction	Unspecific mastitis/	Latent infection	Normal secretion
Severe mastitis		Mild mastitis					
Peracute mastitis	Acute mastitis	Subacute mastitis	Chronic mastitis				
All cardinal symptoms of a glandular inflammation; life-threatening; microbial; severe decrease of general condition	All cardinal symptoms of a glandular inflammation; generally not life-threatening; microbial	Regularly secretory alterations visible in milk; slight other clinical inflammation symptoms	Regularly secretory alterations visible in milk; covered or no other clinical inflammation signs, atrophy; microbial cause	No inflammation signs. Elevated somatic cell count >100,000 cells/ml up to 5000,000/ml; microbial cause	No inflammation signs. Elevated somatic cell count >100,000 cells/ml up to 5000,000/ml; no microbial culture possible	No inflammation signs. Normal SCC (<100,000 cells/ml?); microbial growth	No inflammation signs. Normal SCC (<100,000 cells/ml?); no microbial growth

Table 11.4 Definition of sub-clinical mastitis according to DVG guidelines[a] (Hamann and Fehlings, 2002)

Somatic cell count (SCC)	Diagnosis	
	Culture negative	Culture positive
< 100,000/ml	Normal secretion	Latent infection
> 100,000/ml	Unspecific mastitis/ secretory dysfunction	Sub-clinical mastitis

[a]Based on IDF basic definitions, but with revised SCC thresholds set by the DVG. The IDF suggests that two separate microbiological tests are carried out and if both samples lead to different results, a third sample should be taken after another week (IDF, 1987).

data are available. For example, for coagulase-positive *Staphylococcus* spp. (e.g. *S. aureus*), recovery rates in the absence of treatment have been estimated at between 15% and 70% (Rainard *et al.*, 1990; Rainard *et al.*, 1990; Timms and Schultz, 1987; Wilson *et al.*, 1998).

The objective of mastitis treatments is to cure the infected udders from the infection, but cure is defined in very different ways. For example, in economic terms, the farmer needs to achieve a level of udder health that allows expected milk yields and quality parameters specified by processors/ national regulations to be achieved. On the other hand, cure with respect to antibiotic treatments, is often defined in terms of absence of bacterial pathogens in milk (bacterial cure), with the proportion of cows without detectable pathogen presence following treatment being defined as the bacterial cure rate (BCR). The main problem with using BCR as the main indicator of cure is that it was frequently shown to include a proportion of cows with drastically elevated SCC values (indicative of sub-clinical mastitis) after treatment, but without clinical (sensory) symptoms.

Scientific studies therefore usually assess and aim for simultaneous bacterial cure and cytological normality; (i) absence of inflammatory symptoms within the udder, (ii) normal milk secretion (milky, no impurities, no abnormal smell or taste), (iii) absence of pathogenic microorganisms in milk and (iv) somatic cell count below a certain level (i.e. 100,000 cells/ml). If a cow matches all four criteria, the case is assessed as completely cured and the percentage of completely cured cows divided by all treated cows is called the total cure rate (TCR). The main mastitis management and treatment innovations currently in use or under investigation in organic farming systems are described in separate sections below.

11.4.1 Antibiotic treatment

Although the total replacement of antibiotic treatments is a major target in organic dairy production, it is generally accepted that this will not be possible in the short term. In fact, the dogmatic abandonment of highly effective veterinary medicines would be expected to have significant negative animal health and welfare implications. The most immediate emphasis in European organic farming systems is therefore the minimisation and a more strategic or targeted use of antibiotics, rather than a complete ban, as currently implemented under US organic farming standards. This should involve restricting antibiotic treatment to specific circumstances such as:

- cow specific dry-off therapy after microbial tests have confirmed infection
- an eradication measure for *S. agalactiae* epidemics
- treatment of defined primary infections in younger cows known to be susceptible to antibiotics (e.g. most infections caused by streptococci)
- second-line therapy where alternative treatments/therapies failed.

The use of more selective antibiotics should be considered where diagnostic tests have accurately identified the disease-causing organism. Appropriate antibiotic therapy is complex and the reader is referred to relevant veterinary textbooks such as Andrews (2000) for a more detailed description. However, several important principles and approaches to antibiotic therapies used for mastitis are described below.

The objective of antibiotic treatments is to reduce the density of microbial pathogen in infected udder tissues and thereby improve the capacity of the animal's immune system to deal with the infection. The effect of a successful antibiotic treatment is therefore 'self-cure' of mastitis (Hamann and Krömker, 1999). However, some antibiotics (e.g. tetracycline and gentamycine) may also have negative side effects on the animal's immune response to udder infection, as they have been shown to inhibit/reduce phagocytosis of the animal's own defence cells (Nickerson *et al.*, 1986).

Since a wide range of different Gram-positive and Gram-negative bacteria (and yeast, algal and protozoal pathogens which are not inhibited by antibacterial antibiotics) can cause mastitis, it is recommended that antibiotics are only used after the disease-causing organism(s) have been identified by microbiological laboratory tests. Laboratory tests should involve both the identification of the pathogen to an appropriate level and the determination of antibiotic sensitivity profiles, to ensure that the antibiotics and the dose level chosen are effective.

The efficacy of antibiotics may be limited, if drug-specific veterinary recommendations relating to dose level and the length of therapy are not complied with, and may lead to persisting infections, rapid re-infection and resistance development in the pathogen population. For many mastitis pathogens, veterinary guidelines stipulate a once a day treatment (after evening milking) and minimum therapy length of between 3 to 5 days, depending on pathogenity and/or tissue adherence ability of the microbial agent. Generally, longer treatment periods lead to better cure results.

There are two principal ways of antibiotic administration, which may be used: (i) the local, intracisternal application and/or (ii) systemic injection or infusion. For clinical mastitis, cure rates based on visual assessment of symptoms after either local or systemic administration of antibiotics were between 50% and 90%, while bacteriological cure rates (based on microbiological testing) were often lower and ranged between 20% and 70%. The limitations of currently available antibiotic treatments are demonstrated by the fact that complete cure of mastitis cases caused by *S. aureus* (one of the main cow-related bacterial pathogens) is achieved in only around 10% of the cases (Deluyker *et al.*, 1999; Guterbock *et al.*, 1993; Merck *et al.*, 1989; Reinhold *et al.*, 1986; Seymour *et al.*, 1989; Timms and Schultz, 1984; Winter *et al.*, 1997). Combined use of both local and systemic applications of antibiotics was described to only slightly increase the efficacy from 62% to 74% (Walkenhorst, 2006). For sub-clinical mastitis cure rates were found to be more variable (0% to 100%) depending on therapy regimen and the disease causing organisms (Friton *et al.*, 1998). For sub-clinical *S. aureus* infections, cure rates were between 0% (Oldham and Daley, 1991) and 77% (Friton *et al.*, 1998).

Given the diversity of disease-causing organisms, the interactions between farm-specific environmental factors and the types of mastitis causing pathogens found, the potential efficacy of antibiotic treatments can only be assessed accurately in the context of the specific on-farm conditions. Furthermore, many recent investigations concluded that, except for some specific infections caused by streptococci, a prophylactic and or longer-term use of antibiotics for chronic and sub-clinical mastitis should not recommended in the future (Deluyker *et al.*, 2005).

However, in case of acute clinical mastitis, it is widely accepted that animal welfare considerations should take prevalence. If both farmer and veterinarian are not familiar with non-antibiotic treatments, they should be advised to use broad-spectrum antibiotics immediately, because any delay (e.g. the 2–3 days it often takes between diagnosis and the return of microbiological test results) may seriously harm the animal. This approach should, however, only be taken after a sound clinical diagnosis, since antibiotic treatments themselves may lead to dramatic aggravation of the condition. For example, *E. coli* inflammations are able to develop into severe toxaemia, because increased levels of toxins are released into the animal tissues when *E. coli* cells are killed or stressed by antibiotic treatments. Also, if yeasts are the main cause or form part of the pathogen complex that causes mastitis, their growth and proliferation may be supported by the administration of anti-bacterial antibiotics (Crawshaw *et al.*, 2005).

11.4.2 Homeopathic therapy and prophylaxis

Homeopathic remedies are very low dose therapeutical preparations produced from different source materials/substances (e.g. extracts from plants or animal tissues, animal secretions, minerals or chemical substances). The source substances are processed and diluted with water under defined conditions. The diluted homeopathic remedy (also called preparation) is then applied to the animal orally either as a liquid or after application of the remedy onto

Table 11.5 Source materials used to prepare homeopathic remedies for different stages of mastitis development

Stage of mastitis	Plants genera and minerals used to prepare the initial extracts/ solutions for remedies
Beginning of inflammation, acute	*Aconitum napellus*
Acute inflammation	*Atropa belladonna, Bryonia alba, Phytolacca decandra Apis mellifica, Lachesis mutus, Pyrogenium, Laccaninum, Urtica urens*
Sub-acute, chronic mastitis	*Hepar sulfuris, Mercurius solubilis*, Silicea, *Calcarea carbonica, Carbo vegetabilis, Conium maculatum*
Traumatic mastitis	*Arnica montana, Bellis perennis*

small sugar pills (globuli). Since homeopathic preparations are produced by repeated serial dilution (see below) of different source substances (see Table 11.5) they should be 'by definition' free of residue levels that can be detected by the currently available analytical methodologies. In contrast to antibiotic treatments they are not considered to pose residue or any other food safety risks (Vaarst *et al.*, 2005). However, their efficacy has frequently been questioned by both human and veterinary scientists (Shang *et al.*, 2005).

In line with the expansion of organic farming, the use of homeopathy for the treatment of mastitis has increased rapidly and even some conventional farms now use homeopathy as part of their animal health management plans. As a consequence of its wide acceptance in the organic farming industry, the use of homeopathic remedies (but also other non-chemical treatment methods) is now recommended within organic farming regulations in both the EU and Switzerland (Graf *et al.*, 1999; Hertzberg *et al.*, 2003).

A significant number of investigations have attempted to quantify the efficacy of homeopathy. When the effect of acute mastitis was assessed, clinical cure rates between 48% and 92% were reported. However, bacterial cure rates were lower at between 29% and 42% and complete cure (return to low SCCs) was reported for only 19–34% of cows (Garbe, 2003; Hektoen *et al.*, 2004; Merck *et al.*, 1989; Otto, 1982). Overall, these studies suggested that the level of efficacy against *E. coli* or minor pathogens was similar to antibiotic treatment in contrast to those obtained for mastitis caused by staphylococci and streptococci (Garbe, 2003; Merck *et al.*, 1989; Otto, 1982). On the other hand, therapy of sub-clinical mastitis by standardized methods was found to be poor in several studies. There appeared to be no relevant effects of either systemically or intramammary administered homeopathy (Garbe, 2003; Leon *et al.*, 1999; Velke, 1988). Despite these results 'classical' homeopathic methods are still widely used for the therapy of sub-clinical mastitis. If such treatment were confirmed not to provide significant levels of activity, this would be of concern since chronic, sub-clinical mastitis is associated with massive aberrations in the udder tissue and culling rather than treatment of such cows should be the preferred course of action.

Despite the benefits on acute mastitis reported in the studies described above and the relatively widespread adoption of homeopathy in organic farming practice, many veterinary scientists still question whether there is sound scientific evidence for the efficacy of homeopathic remedies. The lack of studies which include 'un-treated' controls is a main reason why the currently available evidence is questioned. However, because the non-treatment of sick animals would violate ethical and animal welfare guidelines for animal experimentation, the inclusion of such controls will also not be possible in the future. Some additional indirect evidence may be gained by comparing farms using homeopathy or not with respect to the level of antibiotic use (e.g. in European organic farms) and/or udder health levels (e.g. US organic farms, which are not permitted to use antibiotics). However, given the wide range of potential confounding factors (see above), such survey-based studies may remain unconvincing. Nevertheless, based on the currently available scientific evidence, it can be concluded that the use of homeopathy for acute mastitis is promising, especially when combined with additional preventive methods (see below). In contrast, limited success should be expected for chronic and sub-clinical mastitis cases.

The homeopathy methods that are used differ significantly between (but often also within) countries and regions of Europe and the rest of the world. For example, in Germany the use of homeopathy against clinical mastitis is commonly based on low dilutions (also termed potencies) (i.e. the 6th to 12th dilution), while in Switzerland and UK more diluted (higher potencies, 30th to 1000th dilution) are often used. Often, no general recommendations on the most appropriate level of dilution are given and the use of differently potentised remedies depends only on the demand pattern in the 'alternative pharmaceutical' markets and the homeopathy traditions in the respective countries.

A range of attempts have been made to standardise the homeopathy methods used in different countries and regions of Europe. However, given the different homeopathy traditions, it has until now proved difficult to agree standardised recommendations for the processing and use of homeopathic remedies. Apart from the different dilution levels, there are also at least three different types of remedies used: (i) 'classical' individual homeopathy made from individual source substances (see above), (ii) 'clinical' homeopathy with remedies made from individual source materials (as described in most of the homeopathy textbooks, e.g. MacLeod, 1997) and (iii) 'complex' homeopathy based on mixtures of preparations made from different source materials. The latter are often used by inexperienced therapists, who appear to favour and 'feel safer with' homeopathic remedies based on a broader range of source materials, in analogy to the broad spectrum and combination antibiotic treatments often recommended in conventional farming practice after 'acute clinical mastitis' has been diagnosed. However, homeopathic remedies are based on the same source materials (see Table 11.4) in all countries.

A more detailed description of the different homeopathic remedies and their use for different types and stage of mastitis can be found in specialised textbooks (MacLeod, 1997).

11.4.3 Nosodes

Nosodes are homeopathically diluted (potentised) remedies made from pathogenic microorganisms and/or secretions from diseased animal tissues or diseased organs (including tumours). They are often described as 'homeopathic vaccines' in analogy with the more recently developed vaccine-based disease prevention strategies. Nosodes for mastitis treatment are often produced from milk for cows which have mastitis (Fidelak *et al.*, 2007) and commercial standardised formulations are available in some countries (May and Reinhart, 1993). The main target of these nosode treatments is sub-clinical mastitis and/or the prevention of mastitis in dry cows. While there are reports of 'positive experience' with respect to the use of nosodes in dairy herd health management (Day, 1995; May and Reinhart, 1993), more controlled trials could not detect significant effects of nosode treatments against mastitis (Fidelak *et al.*, 2007; Holmes *et al.*, 2005; Meany, 1995). However, many organic farmers continue to use nosode treatments as part of their mastitis management plans and report positive experiences, particularly in the UK.

11.4.4 Other complementary treatment methods

There are a range of traditional and novel approaches for the control of mastitis. For example, some immunomodulators proved to be effective in treatment of mastitis during the lactation period. Promising results were also described for the use of ginseng saponin, herbal gel, herbal extracts, propolis, lysosubtilin, antibacterial proteins and especially the lysozyme dimer (Malinowski, 2002). Chinese studies report a benefit of Ginseng injections on sub-clinical mastitis caused by *S. aureus* (Hu *et al.*, 2001). Also, acupuncture was discussed as a promising treatment method, but only if used on individual infected cow basis, owing to the high costs of this method (Kendall, 1988). Milking out often in conjunction with udder massage (which is thought to increase blood flow in the udder) has been reported to result in cure rates of >80% for clinical mastitis, even when used without other treatments (Opletal and Sladky, 1985). However, other investigations reported much lower cure rates for mastitis cases caused by environmental bacterial pathogens (Roberson, 1997) and udder inflammations caused by coliform microorganisms (Leininger *et al.*, 2003).

11.4.5 Teat care and disinfection

The disinfection of teats has become common practice in dairy production and provides a temporary barrier against both cow-related and environmental mastitis pathogens. The action of disinfection involves the immersion or spraying of the teat with an effective disinfectant solution. This is carried out during the milking process and can be applied before milking (pre-dipping) or after milking (post-dipping). In Europe pre-dipping is uncommon. Post-dipping is carried out immediately after removal of the milking aggregate and is aimed at contagious microorganisms. The dipping solutions consists of two components (i) a disinfectant with broad spectrum activity against classical mastitis pathogens and (ii) a skin care component (moisturiser) to ensure the integrity of the teat skin and thereby decrease bacterial adhesion.

The most widely used and effective disinfectant solutions are based on iodine (iodophor) with concentrations ranging between 0.05% and 0.1%, but sometimes higher concentrations are recommended. Other agents such as chlorhexidine or chlorine dioxide, peroxide, sodium chloride and lactic acid may also be effective (Wilson *et al.*, 1997) but are not common. Recent trials show positive effects of aloe vera-based dipping agents (Leon *et al.*, 2004). One problem of iodine containing products is their low pH value (<4.0), which is necessary for their antimicrobial activity (Hansen and Hamann, 2003).

Under current EU regulations it is not allowed to use teat disinfection prior to milking to minimise risk of disinfectant residues in milk. However, this pre-dipping is routinely used in other milk-producing countries, such as Australia, New Zealand and the USA (Falkenberg, 2002; Pankey, 1989) and is able to reduce new infection rates with Gram-negative microorganisms by more than 50% (Falkenberg, 2002).

Milk contamination was shown to be effectively minimised by wet cleaning of the teat after disinfection and before milking (Wildbrett, 1996). The efficacy of this simple measure is influenced by time, water quantity and cleaning technique. However, dry cleaning is the generally preferred method in order to reduce dung, dust, organic material or microbial pathogens (Fox and Cumming, 1996; Fox and Norell, 1994).

There are reports that the use of robotic milking systems will significantly reduce mastitis, linked at least partially to the machines' more efficient teat disinfection compared to human operators. However, the more frequent visit to be milked and more gentle application and removal of milking equipment from the teats may also contribute to reduced mastitis incidence. On the other hand, the absence of human operators in robotically milked herds may lead to late detection and treatment of mastitis. However, this may be overcome by the introduction of 'conductivity detection systems' which allow early signs of mastitis to be detected (Hogeveen and Meijering, 2000).

11.4.6 Teat sealers

Microbial contamination of the external aperture of the teat canal often leads to colonisation of the whole teat canal, which can persist for long periods during the lactation (Persson, 1990). The common characteristic of this colonisation is that the microorganisms do not penetrate the mucosa to the basal membrane and therefore do not trigger an immune system response. During lactation the keratine coating of the teat canal provides an important barrier against invasion by pathogens and it is now thought that a main function of the keratine membrane is to block the teat canal physically between the milkings (O'Brien, 1989; Schultze and Bright, 1983).

However, during the dry period the teat canal is of particular risk from environmental infection. Teat sealants represent preventive measures against new teat infections (especially from environmental pathogens) during the dry period. The mode of action of teat sealers is the introduction of a mechanical barrier that prevents bacterial penetration/colonisation of the teat canal. The first products investigated were latex-based external teat sealers (ETS), which were applied by dipping and subsequent hardening of the fluid. They initially showed satisfactory results with respect to skin tolerance and protection against mastitis (Farnsworth *et al.*, 1980). However, this could not be verified in subsequent studies (McArthur *et al.*, 1984). Since then, new products were released at the end of the 1990s and are now recommended as additional measures to antibiotic dry cow treatment in conventional farming and/or as stand-alone treatments in organic dairy systems (Hayton and Bradley, 2004).

The external teat sealers were more recently succeeded/replaced by novel internal teat sealants (ITS) developed in New Zealand (Woolford *et al.*, 1998) which are now widely used in Europe. However, teat sealers were initially thought to be an unsuitable alternative to dry cow treatment for organic systems, because they were thought to be dependent on simultaneous prophylactic antibiotic treatments. Furthermore, there were problems with the disposal and residues of sealant in milk at the beginning of the next lactation, because farmers reported that the residuals of the tough sealant substance can be mistaken for mastitis secretions. However, a recent study showed certain benefits of ITS against environmental infections such as coliform bacteria, *S. uberis* and Enterococci (Klocke *et al.*, 2006). The strategic use of ITS can therefore now be recommended, but only if accompanied by detailed instructions to the farmer and only in the case of problems with environmental mastitis pathogens. Under such circumstances teat sealer should be applied in healthy cows (cows with no positive cultural findings and normal somatic cell counts below 200,000 cells per ml) in herds exposed to increased risk by environment associated bacteria.

11.4.7 Vaccines

The development and feasibility of using vaccines to reduce mastitis in dairy cows was investigated extensively in the 1980s and 1990s. Some studies showed promising results for prevention of mastitis caused by *S. aureus* and *E. coli* (Hogan *et al.*, 1992; Watson, 1992). However, subsequent field studies showed low levels of efficacy (Krömker and Hamann, 1999) and there are currently no systemic vaccines that provide high enough activity to be considered as strategic tools in udder health management. Studies into the use of intramammary vaccines also showed insufficient efficacy (Finch *et al.*, 1994).

11.5 Husbandry and environmental improvement

The level of outdoor access, length of grazing periods and conditions during the indoor periods are known to affect mastitis levels. Fresh grass-based diets result in a cows receiving higher levels of α-tocopherol and other antioxidants/vitamins (e.g. carotenoids), which are known to improve the resistance of cows to infections and decrease mastitis incidence (Butler *et al.*, 2007a, b). It is also known that synthetic vitamin feed supplements, which are widely used in conventional farming systems, are not an efficient substitute for fresh forage-based vitamin/antioxidant intake (Butler *et al.*, 2007b). Cows kept inside are at a greater risk of udder injuries and show increased infection rates, owing to increased microbial pressure. In countries where cows are continuously grazed outdoors such as New Zealand and Australia, environmental mastitis caused by coliforms is rare (Daniel *et al.*, 1982). Clearly, organic regulations that prescribe/recommend maximum outdoor grazing periods contribute to reducing environmental mastitis.

The type of housing system is known to affect mastitis incidence (Faye *et al.*, 1997a; Wilson *et al.*, 1997). However, there is no evidence for the frequently described hypothesis that mastitis is more frequent in loose housing, because the microorganisms are more easily transmitted from one cow to another. However, herds kept at high cow density in loose housing systems show increased levels of mastitis (Faye *et al.*, 1997b) and another study identified straw-bedded loose housing systems as a risk factor for mastitis (Barnouin *et al.*, 2005). On the other hand, under loose housing conditions cows are less likely to incur udder injuries. This should reduce mastitis incidence, since udder injuries are recognised to be a main risk factor for environmental mastitis (Regula *et al.*, 2004). The greatest disadvantage in tied housing systems is that the animal's vertical movements are restricted, particularly when getting up or lying down. Partitioning between the stalls reduces the incidence of mastitis, as a result of abrupt movements of neighbouring animals and the likelihood of teat tramps (Matzke *et al.*, 1992).

11.6 Breeding strategies

It has been recognised that cross-breeds and traditional breeds often show lower incidences of mastitis than highly productive Holstein Friesian cows (Butler *et al.*, 2007a). However, since feeding regimes of traditional and cross-breeds are often based on higher levels of grazing, lower levels of concentrate and conserved forage intake and result in lower milk yield, it has been difficult to identify the breed effect in many studies (Butler *et al.*, 2007a). Breedings organisations still have to identify the breeding goals/traits to allow targeted selection for mastitis resistance, but further research is required to identify the potential for reducing mastitis via breeding for resistance (Morris, 2006).

11.7 Integration of management and treatment approaches: farm specific mastitis management plans

The development and implementation of 'farm-specific mastitis management plans' were recently shown to allow significant improvements in udder health and reductions in antibiotic use in organic farming systems in Switzerland (Heil *et al.*, 2006). The programme to develop such plans was initiated when it was realised that it is extremely difficult for farmers who have recently converted to organic farming practice (but also some long-standing organic farmers) to identify the underlying farm-specific reasons for their mastitis problems and access information on new approaches for the control of mastitis described in Sections 11.4.1 to 11.4.7. A particular problem was that the veterinarians (used by farmers) were also not trained or had limited knowledge of the alternative management and treatment approaches (available for use in organic farming) and therefore often used the safe 'antibiotic' option. The approaches taken in the development of farm-specific mastitis management plans are only outlined here and the reader is referred to the detailed report by Heil *et al.* (2006) for further details.

Important preconditions for the successful development and implementation of farm-specific mastitis management plans are that (i) the farmer realises that there is a problem and is committed to reducing both mastitis prevalence and antibiotic use and (ii) that the farm's veterinarian is involved and committed to the improvement plan. Once these preconditions are met, development and implementation of plans then involves four stages.

11.7.1 Stage 1: farm evaluation stage

This involves the recording and analysis of a wide range of farming systems, milk quality and environmental data on the farm. Of particular importance are records and analyses of (i) feeding regimes (levels and quality of fresh and conserved forage, mineral supplements and concentrate used) and history of micronutrient and other dietary deficiencies, (ii) milk and animal quality parameters (see Table 11.6) many of which are part of the standard assessment regime in dairy production, (iii) hygiene-related issues/data (type and management of litter/bedding, state of pathways, length and management of outdoor periods, air flow, humidity and lighting in the barn), (iv) points where opportunities for antagonistic behaviour between cows can arise (e.g. small alleyways, feeding areas, dead ends in barns and pathways, (v) points

Table 11.6 Individual cow parameters to evaluate feeding adequacy

Data	Source	Parameter	Indicator[a]	Problem area
Milk records	Routine data	Milk yield	High (> 40 litres)	General health and susceptibility to metabolic disorders
			Low (<15 litres)[b] in early or mid lactation	Deficiency of energy, protein or water
		Milk fat	Very high (>5.0%)[b] in early lactation	Body fat mobilisation, risk of ketosis
			Very low (<3.5%)	Lack of fibre or excess oil in ration
		Milk protein	Very low (<3.0)	Energy deficiency, especially from starch sources
		Milk urea	Very low (<15 mg/dl)	Protein deficiency
			Very high (>35 mg/dl)	Protein excess in relation to energy
Body condition score (BCS)	Farm visits, regular farmer's evaluation	BCS	Score >3.5[c]	Dystocia, ketosis and fatty liver in early lactation
			Score <2.0[c]	Energy deficiency
Faeces consistency score (FCS)	Farm visits, regular farmer's evaluation	FCS	Very fluid	High protein intake

[a] Threshold of individual values; values below or above the acceptable level indicate an individual may be at risk of problems listed.
[b] Milking record indicators are dependent on breed and age of cow.
[c] Body condition score will vary throughout lactation.

where traumatic incidents may occur (e.g. edges and steps on the way to the milking parlour), (vi) equipment and routines (e.g. teat disinfection, operator hygiene) used in the milking process, (vii) individual cow investigations (body condition, hygiene score, lameness score, injuries score, claw care state, general health state) and (viii) udder investigations during milking time (udder tissue considering nodules, fibrotic aberrations, traumatic injuries, acute mastitis symptoms, sensory milk findings, somatic cell count screening by Californa mastitis test) including quarter milk samples for cultural investigation.

11.7.2 Stage 2: design and implementation of improved mastitis management plans

Based on the analyses carried out, changes to current feeding regimes, management practices and facilities will be identified in line with the following underlying objectives:

- increase udder health
- reduce the need for mastitis-related treatments, especially antibiotics
- maintain the economic viability of the farm by increasing longevity and milk yield and quality.

Plans usually involved the identification of short- and long-term measures, especially where recommendations result in increased (management-related) labour costs and/or capital investments. Plans also involve the agreement and establishment of advisor and/or veterinarian visits and data recording systems to monitor the impact of new measures and help with the implementation and continuous revision of the programme of measures agreed with the farmer. Therefore, an adequate recording system based on the use of database programmes is highly recommended.

11.7.3 Stage 3: consolidation and benchmarking

Once all immediately applicable management, training and facility improvements have been introduced, a programme of continuous monitoring and benchmarking (against other similar enterprises) needs to be set up along with activities designed to introduce new proven emerging technologies (e.g. methods such as homeopathy, acupuncture, novel teat sealers, vaccines).

11.7.4 Stage 4: optimised mastitis management plans introduced

Once the desired improvements have been achieved, the level of support from extension advisors and veterinarians is reduced. A regular monitoring system including a defined milk sample schedule (for cell count elevated cows at drying off and after calving and those showing clinical mastitis) and regular farm visits should be introduced. A simple method of external monitoring is the regular analysis of monthly milk test data. A respective module is to be implemented in the database. The success of strategic programmes based on farm-specific mastitis management plans in improving udder health and reducing antibiotic use in organic dairy production in Switzerland, shows the importance of following up technology development with appropriate extension and advisory programmes.

11.8 Acknowledgement

The work concerning farm-specific health plans and on-farm measures was funded by the European Commission in the framework of the Integrated project QualityLowInputFood (QLIF) and the Naturaplan Fund of the COOP supermarket group (Switzerland).

11.9 References

Andrews, A.H., (2000), *The Health of Dairy Cattle* (Veterinary Health Series), 1st Edition. Blackwell Publishing Limited, 359 pp.

Barnouin, J., Bord, S., Bazin, S. and Chassagne, M., (2005), 'Dairy management practices associated with incidence rate of clinical mastitis in low somatic cell score herds in France'. *Journal of Dairy Science*, **88**, 3700–3709.

Bartlett, P.C., Miller, G.Y., Anderson, C.R. and Kirk, J.H. (1990), 'Milk production and somatic cell count in Michigan dairy herds'. *Journal of Dairy Science*, **Oct**, 2794–2800.

Butler, G., Nielsen, J.H., Slots, T., Sanderson, R.A., Eyre, M.D. and Leifert, C. (2007a), 'Effect of low input dairy management systems on milk composition – I. Fatty acid profiles'. *Journal of Dairy Science*, submitted.

Butler, G., Nielsen, J.H., Slots, T., Sanderson, R.A., Eyre, M.D. and Leifert, C. (2007b), 'Effect of low input dairy management systems on milk composition – II. Fat soluble antioxidants/vitamins'. *Journal of Dairy Science*, submitted.

Concha, C. (1986), 'Cell types and their immunological functions in bovine mammary tissues and secretions – a review of literature'. *Nordisk Veterinaermedicin*, **38**, 257–272.

Crawshaw, W.M., MacDonald, N.R. and Duncan, G. (2005), 'Outbreak of *Candida rugosa* mastitis in a dairy herd after intramammary antibiotic treatment'. *Veterinary Record*, **156**, 812–813.

Daniel, R.C.W., O'Boyle, D., Marek, M.S. and Frost, A.J. (1982), 'A survey of clinical mastitis in south-east Queensland dairy herds'. *Australian Veterinary Journal*, **58**(4), 143–147.

Day, C. (1995), *The Homoeopathic Treatment of Beef and Dairy Cattle*, Beaconsfield Publishers Ltd., Beaconsfield, UK.

Deluyker, H.A., Chester, S.T. and Van Oye, S.N. (1999), 'A multilocation clinical trial in lactating dairy cows affected with clinical mastitis to compare the efficacy of treatment with intramammary infusions of a lincomycin/neomycin combination with an ampicillin/cloxacillin combination'. *Journal of Veterinary Pharmacology and Therapy*, **22**, 274–282.

Deluyker H.A., Van Oye, S.N. and Boucher, J.F. (2005). 'Factors affecting cure and somatic cell count after pirlimycin treatment of subclinical mastitis in lactating cows'. *Journal of Dairy Science*, **88**, 604–614

Falkenberg, U. (2002), *Untersuchungen zum Einsatz verschiedener Zitzendippverfahren*. Doctoral Thesis, Freie Universität Berlin.

Farnsworth, R.J., Wyman, L. and Hawkinson, R. (1980), 'Use of a teat sealer for prevention of intramammary infections in lactating cows'. *Journal of the American Veterinary Medical Association*, **177**, 441–444.

Faye, B., Lescourret, F., Dorr, N., Tillard, E., MacDermott, B. and McDermott, J., (1997a), 'Interrelationships between herd management practices and udder health status using canonical correspondence analysis'. *Preventive Veterinary Medicine*, **32**, 171–192.

Faye, B., Lescourret, F., Dorr, N., Tillard, E., MacDermott, M. and McDermott, J., (1997b), 'Interrelationships between herd management practices and udder health status using canonical correspondence analysis'. *Preventive Veterinary Medicine*, **32**, 171–192.

Fidelak, C., Berke, M., Klocke, P., Spranger, J., Hamann, J. and Heuwieser, W. (2007). 'Nosoden zum Trockenstellen – eine placebokontrollierte Blindstudie', in *Wissenschaftstagung*, Hohenheim.

Finch, J.M., Hill, A.W., Field, T.R. and Leigh, J.A. (1994), 'Local vaccination with killed *Streptococcus uberis* protects the bovine mammary gland against experimental intramammary challenge with the homologous strain'. *Infection and Immunity*, **62**, 3599–3603.

Fox, L.K. and Norell, R.J. (1994), '*Staphylococcus aureus* colonization of teat skin as affected by postmilking teat treatment when exposed to cold and windy conditions'. *Journal of Dairy Science*, **77**, 2281–2288.

Fox, L.K. and Cumming, M.S. (1996), 'Relationship between thickness, chapping and *Staphylococcus aureus* colonization of bovine teat tissue'. *Journal of Dairy Research*, **63**, 369–375.

Friton, G.M., Sobiraj, A. and Richter, A. (1998), 'Über den Erfolg verschiedener antibiotischer Therapieformen bei laktierenden Kuhen mit subklinischer Mastitis'. *Tierärztliche Praxis, (G)*, **26**, 254–260.

Garbe, S. (2003), *Untersuchungen zur Verbesserung der Eutergesundheit bei Milchkühen unter besonderer Berücksichtigung des Einsatzes von Homöopathika*, Doctoral Thesis, Freie Universität Berlin.

Graf, S., Haccius, M. and Willer, H., (1999), Die EU-Verordnung zur ökologischen Tierhaltung – Hinweise und Umsetzung, SÖL-Sonderausgabe, Stiftung Ökologie & Landbau (SÖL), 72.

Guterbock, W.M., Van Eenennaam, A.L., Anderson, R.J., Gardner, I.A., Cullor, J.S. and Holmberg, C.A. (1993), 'Efficacy of intramammary antibiotic therapy for treatment of clinical mastitis caused by environmental pathogens'. *Journal of Dairy Science*, **76**, 3437–3444.

Hamann, J. (2001), Relationships between somatic cell count and milk composition, in *Proceedings of the IDF World Summit*, Auckland/New Zealand.

Hamann, J. and Fehlings, K. (2002), 'Leitlinien zur Bekämpfung der Mastitis des Rindes als Bestandsproblem'. *Sachverständigenausschuss 'Subklinische Mastitis' der Deutschen Veterinärmedizinischen Gesellschaft*, Giessen/DE.

Hamann, J. and Gyodi, P. (1994), 'Effects on milk yield, somatic cell count and milk conductivity of short-term, non-milking of lactating quarters of cows'. *Journal of Dairy Research*, **61**, 317–322.

Hamann, J. and Krömker, V. (1999), 'Mastitistherapie – Hilfe zur Selbsthilfe'. *Praktischer Tierarzt*, **80**, 38–42.

Hansen, S. and Hamann, J. (2003), 'Massnahmen zur Desinfektion der bovinen Zitze – Ziele Verfahren und Produkte'. *Praktischer Tierarzt*, **84**, 780–793.

Harmon, R.J. (1994), 'Physiology of mastitis and factors affecting somatic cell counts'. *Journal of Dairy Science*, **77**, 2103–2112.

Hayton, A.J. and Bradley, A.J. (2004), 'The control of mastitis on organic units'. in *British Mastitis Conference 2004*, Stoneleigh, Oct 13th.

Hell, F., Ivemeyer, S., Klocke, P., Notz, C., Maeschli, A., Schneider, C., Spranger, J. and Walkenhorst, M. (2006), *Pro-Q: Förderung der Qualitüt biologisch erzeugter Milch in der Schweiz durch Prävention und Antibiotikaminimierung*, Final Report, May 2003 to April 2006, FiBL, Frick, Switzerland. http://orgprints.org/9924/

Hektoen, L., Larsen, S., Odegaard, S.A. and Loken, T. (2004), 'Comparison of homeopathy, placebo and antibiotic treatment of clinical mastitis in dairy cows – methodological issues and results from a randomized-clinical trial. *Journal of Veterinary Medicine A Physiology, Pathology and Clinical Medicine*, **51**, 439–446.

Hertzberg, H., Walkenhorst, M. and Klocke, P. (2003), 'Tiergesundheit im biologischen Landbau: Neue Richtlinien und Perspektiven für die Nutztierpraxis'. *Schweizer Archiv für Tierheilkunde*, **145**, 519–525.

Hogan, J.S., Weiss, W.P., Todhunter, D.A., Smith, K.L. and Schoenberger, P.S. (1992), 'Efficacy of an *Escherichia coli* J5 mastitis vaccine in an experimental challenge trial'. *Journal of Dairy Science*, **75**(2) 415–422.

Hogeveen, H. and Meijering, A. (2000), *Robotic Milking*, Wageningen Academic, Wageningen, The Netherlands, 320 pp.

Holmes, M.A., Cockcroft, P.D., Booth, C.E. and Heath, M.F. (2005), 'Controlled clinical trial of the effect of a homoeopathic nosode on the somatic cell counts in the milk of clinically normal dairy cows'. *Veterinary Record*, **156**, 565–567.

Hovi, M. and Roderick, S. (1998), Mastitis therapy in organic dairy herds, in *British Mastitis Conference 1998*, Crewe (United Kingdom).

Hu, S., Concha, C., Johannisson, A., Meglia, G. and Waller, K.P. (2001), Effect of

subcutaneous injection of ginseng on cows with subclinical *Staphylococcus aureus* mastitis. *Journal of Veterinary Medicine B Infection Diseases, Veterinary Public Health*, **48**, 519–528.
Kendall, D. (1988), 'Acupuncture beats antibiotics.' *The New Farm*, **7/8–88**, 1418.
Klocke, P., Ivemeyer, S., Walkenhorst, M., Maeschli, A. and Heil, F. (2006), 'Handling the dry-off problem in organic dairy herds by teat sealing or homeopathy compared to therapy omission' in *Organic Joint Congress*, Odense (DK), May 30–31, FiBL, Frick, Switzerland.
Krömker, V. and Hamann, J. (1999), 'Nichtantibiotische Mastitistherapie – Einordnung und Beurteilung'. *Praktischer Tierarzt*, **80**, 48–51.
Leininger, D.J., Roberson, J.R., Elvinger, F., Ward, D. and Akers, R.M. (2003), 'Evaluation of frequent milkout for treatment of cows with experimentally induced *Escherichia coli* mastitis'. *Journal of the American Veterinary Medical Association*, **222**, 63–66.
Leon, L., Sommer, H. and Andersson, R. (1999), 'Intrazisternale Behandlung boviner subklinischer Mastitiden mit dem Homoepathikum Lachesis D8', in *Beitrage zur 5. Wissenschaftstagungen zum Okologischen Landbau*, Berlin/DE, 372–375.
Leon, L., Beer, C., Waecken, H., Nuernberg, M. and Andersson, R. (2004), 'Mastitisprophylaxe mit einem Dippmittel auf der Basis von Aloe vera'. *Tieraerztliche Umschau*, **59**, 237–244.
MacLeod, G. (1997), *The Treatment of Cattle by Homeopathy*, Beekman Books, Wappingers Falls, NY.
Madsen, P.S., Klastrup, O., Olsen, S.J. and Pedersen, P.S. (1974), 'Herd incidence of bovine mastitis in four Danish dairy districts. i. The prevalence and mastitogenic effect of microorganisms in the mammary glands of cows'. *Nordisk Veterinaermedicin*, **26**, 473–482.
Malinowski, E. (2002), 'The use of some immunomodulators and non-antibiotic drugs in a prophylaxis and treatment of mastitis'. *Polish Journal of Veterinary Sciences*, **5**, 197–202.
Matzke, P., Holzer, A. and Deneke, J. (1992), (The effect of environmental factors on the occurrence of udder diseases). *Tierarztliche Praxis*, **20**, 21–32.
May, T. and Reinhart, E. (1993), 'Feldversuch zur Bestandsbehandlung bei erhöhten Milchzellzahlen mit Nosoden'. *Biologische Tiermedizin*, **10**, 4–12.
McArthur, B.J., Fairchild, T.P. and Moore, J.J. (1984), 'Efficacy of a latex teat sealer'. *Journal of Dairy Science*, **67**, 1331–1335.
Meany, W.J. (1995), 'Treatment of mastitis with homeopathic remedies'. *IDF Mastitis Newsletter*, 5–6.
Merck, C. and Sonnenwald, B., H., R. (1989), 'Untersuchung über den Einsatz homöopathischer Arzneimittel zur Behandlung akuter Mastitiden beim Rind'. *Berlin-Münchener Tierarztliche Wochenschrift*, **102**, 266–272.
Morris, C.A. (2006), 'A review of genetic resistance to disease in *Bos taurus* cattle'. *Vet J.*, November (Epub ahead of print).
Myllys, V., Honkanen-Buzalski, T., Huovinen, P., Sandholm, M. and Nurmi, E. (1994), 'Association of changes in the bacterial ecology of bovine mastitis with changes in the use of milking machines and antibacterial drugs'. *Acta Veterinaria Scandinavica*, 363–369.
Nickel, R., Schummer, A. and Seiferle, E. (1984), 'Lehrbuch der Anatomie der Haustiere, Band III, Kreislaufsystem', *Haut und Hautorgane*, Verlag Paul Parey, Berlin and Hamburg/DE.
Nickerson, S.C., Paape, M.J., Harmon, R.J. and Ziv, G. (1986), 'Mammary leukocyte response to drug therapy'. *Journal of Dairy Science*, **69**, 1733–1742.
O'Brien, B. 1989, 'Teat canal penetrability and mastitis'. *Farm and Food Research*, **20**, 6–7.
Oldham, E.R. and Daley, M.J. (1991), 'Lysostaphin: use of a recombinant bactericidal enzyme as a mastitis therapeutic'. *Journal of Dairy Science*, **74**, 4175–4182.

Östensson, K., Hageltorn, M. and Aström, G. (1988), 'Differential cell counting in fraction-collected milk from dairy cows'. *Acta Veterinaria Scandinavica*, **29**, 493–500.
Otto, H. (1982), 'Erfahrungen mit der homöopathischen Therapie akuter parenchymatöser Mastitiden des Rindes'. *Tierärztliche Umschau*, **37**, 732–734.
Pankey, J.W. (1989), 'Teat dips and the practitioner'. *Proceedings of the Annual Convention of the American Association of Bovine Practitioners*, Stillwater, Okla, The American Association of Bovine Practitioners, April, 115–118.
Persson, K. (1990), 'Inflammatory reactions in the teat and udder of the dry cow'. *Zentralblatt Veterinarmedizin* B, **37**, 599–610.
Rainard, P., Ducelliez, M. and Poutrel, B. (1990), 'The contribution of mammary infections by coagulase-negative staphylococci to the herd bulk milk somatic cell count'. *Veterinary Research Communications*, **14**, 193–198.
Regula, G., Danuser, J., Spycher, B. and Wechsler, B. (2004), 'Health and welfare of dairy cows in different husbandry systems in Switzerland'. *Preventive Veterinary Medicine*, **66**, 247–264.
Reinhold, P., Schulz, J., Beuche, W. and Jakel, L. (1986), 'Zur Behandlung akuter Mastitiden des Rindes. 1. Therapeutischer Einsatz von Glukose-Lösungen'. *Archiv für Experimentelle Veterinärmedizin*, **40**, 627–638.
Roberson, J.A. (1997), 'Frequent milk-out as a treatment for subacute clinical mastitis'. in *Proceedings of the National Mastitis Council Annual Meeting*, Arlington/USA, 152–157.
Schällibaum, M. (1999), 'Mastitis pathogens isolated in Switzerland, 1987–1996. *IDF Mastitis Newsletter*, **23**, 14–15.
Schultze, W.D. and Bright, S.C. (1983), 'Changes in penetrability of bovine papillary duct to endotoxin after milking'. *American Journal of Veterinary Research*, **44**, 2373–2375.
Seymour, E.H., Jones, G.M. and McGilliard, M.L. (1989), 'Effectiveness of intramammary antibiotic therapy based on somatic cell count'. *Journal of Dairy Science*, **72**, 1057–1062.
Shang, A., Huwiler-Muntener, K., Nartey, L., Juni, P., Dorig, S., Sterne, J.A., Pewsner, D. and Egger, M. (2005), 'Are the clinical effects of homoeopathy placebo effects? Comparative study of placebo–controlled trials of homoeopathy and allopathy'. *Lancet*, **366**, 726–732.
Smith, K.L. and Hogan, J.S. (1993), 'Environmental mastitis. The Veterinary Clinics of North America'. *Food Animal Practice*, **9**, 489–498.
Timms, L.L. and Schultz, L.H. (1984), 'Mastitis therapy for cows with elevated somatic cell counts or clinical mastitis'. *Journal of Dairy Science*, **67**, 367–371.
Timms, L.L. and Schultz, L.H. (1987), 'Dynamics and significance of coagulase-negative staphylococcal intramammary infections'. *Journal of Dairy Science*, **70**, 2648–2657.
Urech, E., Puhan, Z. and Schallibaum, M. (1999), 'Changes in milk protein fraction as affected by subclinical mastitis'. *Journal of Dairy Science*, **82**, 2402–2411.
Vaarst, M., Padel, S., Hovi, M., Younie, D. and Sundrum, A. (2005), 'Sustaining animal health and food safety in European organic livestock farming'. *Livestock Production Science*, **94**, 61–69.
Velke, H. 1988, 'Einsatz verschiedener Homöopathika zur Prophylaxe von Erkrankungen des Partussyndroms als Instrument zur Gesunderhaltung von Milchviehherden. IV. Mitteilung: Der Einsatz von Traumeel bei Kühen mit erhöhten Zellzahlen – Erste Erfahrungen'. *Biologische Tiermedizin*, **4**, 113–116.
Walkenhorst, M., Garbe, S., Klocke, P., Merck, C.C., Notz, C., Rüsch, P. and Spranger, J. (2001), 'Strategies for prophylaxis and therapy of bovine mastitis'. in *Positive Health: Preventive Measures and Alternative Strategies*, Proceedings of the Fifth NAHWOA Workshop, Rodding/DK, 27–32.
Watson, D.L. (1992), 'Vaccination against experimental staphylococcal mastitis in dairy heifers'. *Res Vet Sci. London : British Veterinary Association*, **Nov**, 346–353.

Wendt, K., Bostedt, H., Mielke, H. and Fuchs, H.W. (1994), *Euter- und Gesäugekrankheiten*, Gustaf-Fischer-Verlag, Stuttgart/DE.

Wildbrett, G. (1996), *Reinigung und Desinfektion in der Lebensmittelindustrie*, Behr's Verlag, Hamburg.

Wilson, D.J., Das, H.H., Gonzalez, R.N. and Sears, P.M. (1997), 'Association between management practices, dairy herd characteristics, and somatic cell count of bulk tank milk'. *Journal of the American Veterinary Medical Association*, **210**, 1499–1502.

Wilson, D.J., Case, K.L., Gonzalez, R.N. and Han, H.R. (1998), 'Bacteriologic cure rates for bovine mastitis cases with no treatment or with eight different antibiotics'. in *Proceedings of the National Mastitis Council Annual Meeting*, Arlington/USA, 273–274.

Winter, P., Spielleutner, Kussberger, Rockenschaub, Petracek, (1997), 'Zum Einsatz von Cefoperazon (Percef) bei subklinischen und klinischen Mastitiden bei Kühen'. *Tierärztliche Umschau*, **52**, 577–583.

Woolford, M.W., Williamson, J.H., Day, A.M. and Copeman, P.J. (1998), 'The prophylactic effect of a teat sealer on bovine mastitis during the dry period and the following lactation'. *New Zealand Veterinary Journal*, **46**, 12–19.

12

Reducing anthelmintic use for the control of internal parasites in organic livestock systems

Veronika Maurer, Philipp Hördegen and Hubertus Hertzberg, FiBL, Switzerland

12.1 Introduction

This chapter covers the most important livestock species (cattle, sheep, goats, poultry and pigs). A focus is given to gastrointestinal nematodes (GIN) of cattle and small ruminants, mainly covering the situation in Western Europe. Until recently, organic animal husbandry has relied largely on conventional veterinary approaches for the control of diseases and parasites. Today, outlines for organic approaches to these problems are emerging, but much more work needs to be done in this area. The basic approach is illustrated in Fig. 12.1.

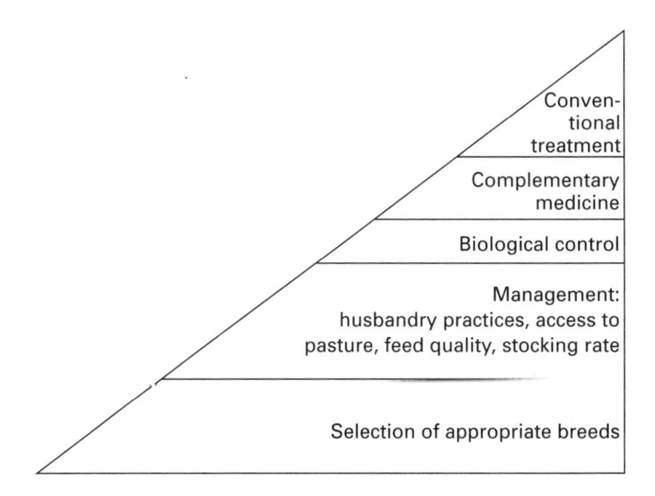

Fig. 12.1 Approach to secure animal health on organic farms.

As a first step, preventive measures should be taken. This basically includes selection of animals adapted to the farm conditions. In addition, attendance should be paid to general aspects such as, appropriate herd size, housing, feed and proper use of technical installations, as well as measures, such as grazing management, aimed directly at the prevention of parasite infections. As a second step, biological control can be used. As a third step, complementary medicine and direct control measures by means of natural compounds should be applied. A number of natural compounds are commercially available, mainly for the control of external parasites. These include pyrethrins, plant extracts, silica and several acids for use against varroa mites. As the last step only, recourse may be taken to chemically synthesised veterinary products or antibiotics, if other methods are not appropriate or successful. In contrast to other veterinary treatments, more than three courses of antiparasitic treatments within one year are permitted in organic agriculture, according to the council regulation No 1804/1999. Although these products are not of natural origin, their use as a last option is sometimes necessary for the sake of animal welfare.

Among the numerous species of arthropods, helminths and protozoa living as parasites in or on farm animals, helminths present the most important group from an animal welfare as well as an economical standpoint. The quantities of antiparasitic drugs sold worldwide in the farm animal sector are almost equal to the quantities of antibiotics. Also comparable to antibiotics, the widespread use of anthelmintics has led to a severe problem with resistance to these compounds. Development of anthelmintic resistance is a major problem particularly in the GIN of small ruminants. Currently, all marketed chemical groups are affected, including the macrocyclic lactones, which are the most recent group, introduced into the market in the 1980s. Since then, no new classes of anthelmintics have been introduced into practice. Therefore, the problem of resistance is expected to increase in the future. The use of anthelmintics bears a risk of residues in animal products, mainly derived from living animals (milk and eggs).

12.2 Ruminants

12.2.1 Important helminth infections of cattle, sheep and goats

This section only gives a brief description of the parasite epidemiology and control. The interested reader is asked to refer to parasitological textbooks for more details.

Gastrointestinal nematodes (GIN) are the most significant parasites in grazing ruminants on organic as well as on low input or conventional farms. Several genera (e.g. *Haemonchus*, *Ostertagia*, *Cooperia*, *Trichostrongylus*, *Oesophagostomum*) comprising various species with similar direct life cycles, parasitise the abomasum or the intestine of their hosts. In general, GIN species are specific for large or small ruminants. GIN have achieved a wide dissemination, being present in virtually all farms where grazing is practised. Whereas in cattle, severe infections are restricted to animals in their first grazing season, sheep and goats may also be highly susceptible as older animals, owing to an unsatisfactory immune response. While it is normal for grazing ruminants to host these parasites, higher levels of infestation can lead to the development of clinical symptoms, apparent as diarrhoea, inappetence and weight loss.

Lungworm infection in cattle is caused by *Dictyocaulus viviparus* and is much less prevalent compared with GIN, but its occurrence may vary considerably from region to region. The bovine lungworm has a direct life cycle and several generations may develop during the grazing season. Parasitic bronchitis typically occurs in late summer and autumn and is characterised by severe respiratory disorder, which can have fatal consequences if treatments are lacking. In small ruminants, lungworms are much more widespread compared with cattle, but the prevailing species are of less clinical importance.

Infections with liver flukes are caused by two species, the large liver fluke, *Fasciola hepatica*, and the small liver fluke, *Dicrocoelium dendriticum*. *F. hepatica* is infective for a wide range of grazing animals and has a widespread dissemination in the temperate climatic zones. The life cycle is dependent on aquatic snails and therefore the occurrence of infections is focussed regionally. In cattle, chronic infections are dominant, whereas subacute and acute disease may occur in sheep and goats. Depending on the level of infection, migration of immature flukes through the liver parenchyma is primarily responsible for the pathological consequences of the infection. Fasciolosis can affect all types of grazing animals from cattle to small ruminants, but numerically the economic consequences in dairy cows are probably the most important ones. The development of the small liver fluke, *D. dendriticum*, depends on certain species of terrestrial snails and ants. Therefore, its occurrence in Europe is restricted to isolated regions. *Dicrocoelium* can affect a wide range of grazing animals but, compared with *Fasciola*, its pathogenicity is much lower, so that infections mainly remain undetected until slaughter, when the affected livers will be condemned. Separate control measures against this parasite are normally not performed.

Infections with tapeworms (mainly *Moniezia* spp.) do not generally represent an isolated problem in grazing ruminants, with the exception of young lambs, which may occasionally suffer from severe infections.

12.2.2 Gastrointestinal nematodes

Breeding for parasite resistance

Within cattle breeds, there is genetic variation in terms of resistance to GIN infection and the heritability is great enough to enable successful selection for GIN resistance (Gasbarre and Miller, 2000). Since anthelmintic resistance is not a major issue in bovine GIN and grazing management measures are effective in reducing GIN infections in cattle, the establishment of a selection programme for cattle is currently less attractive from an economic standpoint; however, breeding for GIN resistance can be an option in small ruminants. Evaluation of sheep breeds in different breeding situations and environments has demonstrated that breeding for nematode resistance is possible. Selective breeding of lambs reduces faecal nematode egg count and can lead to a substantial reduction in the population density of infective larvae on pasture (Gauly and Erhardt, 2001). Flocks of sheep with superior immunity to worm infection have been bred in several countries by selecting rams that have low faecal egg counts or high anti-parasite antibodies. Sheep with improved immunity to worms carry fewer worms, pass fewer eggs and need to be treated less often, but may carry a production penalty (Sangster, 2001). Reduced faecal egg counts and increased body weight gain were revealed in Merino (Eady *et al.*, 2003; Kahn *et al.*, 2003) and Rhön sheep (Gauly and Erhardt, 2001, 2002) and additionally there was a lowered adult worm burden in Menz (Haile *et al.*, 2002; Rege *et al.*, 2002), Black Belly (Gruner *et al.*, 2003) and in INRA 401 sheep (Gruner *et al.*, 2002). Some resistance to gastrointestinal nematode infection was also shown for Dorper crossbred, Katahdin and St Croix ewes (Burke and Miller, 2002).

Grazing management

Grazing management strategies aim to reduce the contact of hosts to infectious GIN larvae on pasture. They can be divided into three categories (Barger, 1997): the preventive procedure, evasive grazing and dilusive grazing.

The *preventive procedure* relies on putting worm-free animals on a parasite-free or a clean pasture. Clean areas may be new leys or can be provided on permanent pasture by alternating with other livestock species not sharing the same spectrum of parasites, or by use of aftermath after harvesting a hay or silage crop (this procedure is generally difficult to apply under practical farming conditions).

The *evasive grazing* approach means that susceptible animals will be moved to clean pastures before they can reinfect themselves. Evasive grazing does not restrict contamination of the pasture by worm eggs, but relies on movement of livestock to another pasture just before the larvae resulting from these eggs are likely to appear in significant numbers on the original pasture.

Dilusive grazing is practised by mixed grazing of susceptible animals together with immune animals of the same species or with other host species. Both groups can be regarded as helminthologically 'inert' animals which do not induce harmful herbage contamination for the susceptible stock (Jacquiet *et al.*, 1998; Giudici *et al.*, 1999). The efficacy of the approach is mainly dependent on the ratio of cattle equivalents of both groups. Dilusive grazing can also be attained by reducing the animal density or stocking rate, that is the number of animals per unit area (Barger, 1997; Thamsborg *et al.*, 1999; Eysker, 2001). Prealpine and alpine grazing conditions are an example where the stocking rate is lowered continuously with increasing altitude, which is mainly due to the decreased growth rate of herbage. Infections with GIN under these conditions are additionally reduced by the retarded development of infective larvae as a result of the lower mean temperatures at higher altitudes.

Nutritional approaches/bioactive forages

Fodder plants with a high content of condensed tannins have shown an effect on resistance and resilience against GIN of ruminant livestock in either indirect or direct ways. They can increase the supply and absorption of digestible protein, which will indirectly improve host resistance and resilience to GIN. There may also be direct effects on biological processes of GIN (Niezen *et al.*, 1998; Kahn and Diaz-Hernandez, 1999; Kahn *et al.*, 2000; Molan *et al.*, 2000). It has been postulated that the beneficial effects of tanniferous plants against parasites could be due to either one, or both, of these factors (Hoste *et al.*, 2006). Tanniferous plants increase the supply and absorption of digestible proteins by animals. This is achieved by tannins forming non-biodegradable complexes with protein in the rumen, which dissociate at low pH in the abomasum to release more protein for metabolism in the small intestine of ruminants. This indirectly improves host resistance and resilience to nematode parasite infections by enhancing the animal's nutritive status.

Tannins may have a direct anthelmintic effect on resident worm populations in animals and tannins and/or metabolites in faeces containing condensed tannins have a direct effect on the viability of the free-living stages (development of eggs to infective larval stages) (Min *et al.*, 2004, Waller *et al.*, 2001); however, the exact mechanism of this anthelmintic effect is not known. Condensed tannins are not absorbed from the digestive tract of ruminants in large quantities and become concentrated in the faeces. Thus, they may directly impair gastrointestinal parasite larval development within faeces. This was shown with *Dorycnium rectum* (Fabaceae), which reduced egg hatching and larval development of *Trichostrongylus colubriformis* in faeces. The composition of faeces could also alter the ability of infectious larvae to migrate from it (Niezen *et al.*, 2002b). The condensed tannins and crude sesquiterpene lactones of chicory (*Cichorium intybus*) have demonstrated inhibitory effects on the motility of L1 and L3 larvae of deer lungworm (*D. viviparus*) and L3 larvae of deer-origin GIN as evidenced by their ability to immobilise larvae and prevent larval migration (Molan *et al.*, 2003).

Resistance and resilience against nematode infections can be influenced by other nutritional approaches. Nutrient supplementation can be used both to ameliorate the effects of infection and to improve host immunity and lower parasite burdens (Coop and Kyriazakis, 1999). Weight gain responses to supplementation could be due to both improved nutrition and reduced parasite loads (Bransby, 1993). Another aspect is that supplementary-fed animals will consume less herbage, resulting in a lower infection pressure (Eysker, 2001). Protein supplementation improves the resilience and resistance of lambs to single and mixed gastrointestinal nematode species infections (Haile *et al.*, 2002). Animals at high nutritional levels are far better able to cope with nematode infections than similar animals at low nutritional levels. In particular, the protein content of the diet is important (Eysker, 2001). Many components of the immune system are proteinaceous and are expected to draw on metabolisable protein resources (MacRae, 1993). Nutritional supplementation with concentrates may explain the low to moderate gastrointestinal nematode infections in cattle on organic farms in Sweden (Höglund *et al.*, 2001). A higher proportion of white clover (*Trifolium repens*) on pasture can help alleviate the production losses of lambs caused by gastrointestinal parasitism (Niezen *et al.*, 2002a).

Copper supplements can also have an anthelmintic effect. Small amounts of copper oxide wire particles (COWP) were effective against both incoming and established *Haemonchus contortus* worms (Bang *et al.*, 1990) and a mineral supplement containing elemental copper reduced faecal egg counts in lambs (Lindqvist *et al.*, 2001). Also, Burke *et al.* (2004) showed a great reduction of faecal egg counts and adult worm burden of *H. contortus* in COWP-treated lambs. However, Dimander *et al.* (2003) could not show an effect against GIN in grazing cattle.

Changes in dietary metabolisable energy supply do not seem to affect the expression of immunity to GIN (Donaldson *et al.*, 1997; Houdijk *et al.*, 2001). However, the improvement in energy supply clearly enhanced the development of resistance to GIN infection in lambs (Valderrábano *et al.*, 2002).

Biological control

Biological control with macro- and microorganisms is compatible with organic farming. Whenever biological control has fewer side effects on the environment than plant- or mineral-based products, it should be favoured. However, case-by-case evaluations are necessary to eliminate unwanted effects, for example, agents attacking a large number of non-target species (Speiser *et al.*, 2006). At the moment, one microbial biocontrol agent and three predators/parasites against house and stable flies are commercially available. Biocontrol agents against other parasite species are currently under development.

Biological control by means of nematophagous fungi is a promising element to be incorporated as a first step into a future control strategy against GIN in ruminants. Attempts to control parasitic nematodes of livestock by predacious or parasitic fungi have been made since the 1930s. While endoparasitic fungi are less promising for use in grazing animals (Larsen, 1999), *Duddingtonia flagrans*, a predacious nematophagous fungus, has shown good efficacy against parasitic nematodes in cattle and horses, whereas the effect was less marked in field studies with sheep (Larsen, 2000) and goats (Wright *et al.*, 2003). The poorer effect in the small ruminants may be due to the structure of their faeces and the extreme reproductive potential of their major nematode species, *H. contortus*. Spores of the fungus (chlamydospores) are fed to the grazing animal daily for approximately three months. The spores are able to survive the passage through the gastrointestinal tract and to reduce free living larval stages of parasitic nematodes in the dung. Thus *D. flagrans*, if deposited at the same time as parasite eggs, largely prevents transmission of third-stage larvae from the faecal deposit onto pasture, but does not have an effect on third-stage larvae in the surrounding soil (Faedo *et al.*, 2000). No negative effects of *D. flagrans* were found up to now on free-living soil nematodes in and around treated dung pats (Yeates *et al.*, 1997; Baudena *et al.*, 2000), earthworms (Gronvold *et al.*, 2000) or dung degradation on pasture (Fernandez *et al.*, 1999a, b, c); however, clarifications of long-term environmental effects are still under study (Yeates *et al.*, 2003).

The lack of simple and reliable application systems is a major problem to be solved before the introduction of this biocontrol agent into practical control strategies. Mixing fungal chlamydospores into a feed supplement was used in most plot and field studies as an application system (Larsen, 2000). Incorporation into various types of feed blocks or mineral licks, as well as slow-release devices, may also become feasible (Thamsborg *et al.*, 1999; Chandrawathani *et al.*, 2003).

Other potential biocontrol agents are of less practical relevance, but may become more important in the future. Studies done with the toxin of the bacteria *Bacillus thuringiensis* (*Bt*) *israelensis* showed that the toxin could kill eggs and larvae of *T. colubriformis* (Bone *et al.*, 1986; Bottjer and Bone, 1987). Ciordia and Bizzell (1961) also revealed an effect of *Bt* toxin on free living stages of some parasitic cattle nematodes. But further work with these bacteria against parasites of livestock is lacking. Although *Bt* toxins are very specific to certain harmful insects (Van Frankenhuyzen, 1993), they can affect a variety of non-target species (Gould and Keeton, 1996; Johnson *et al.*, 1995; Hilbeck *et al.*, 1998; Boulton, 2004; Wei *et al.*, 2003; Zwahlen *et al.*, 2003). Thus, *Bt* toxins cannot be considered as a promising biological control agent for use against GIN.

Earthworms, dung beetles, predatory mites and other organisms can play an important role in controlling GIN by predation or by rapidly disintegrating the dung and thus affecting the development of infective GIN larvae. However, conservation biocontrol (defined as deliberate modification of the environment or management practices to enhance specific, natural enemies by Eilenberg *et al.*, 2001) is not a practice applied explicitly for the control of GIN.

Earthworms (Lumbricidae) play an important role in the degradation of cattle dung on pasture. In temperate regions they account for about 50% of the dung disappearance (Holter, 1979). An investigation showed that rapid disintegration of cow pats, facilitated by earthworms in particular, resulted in a significantly lower migration of infective larvae of the species *Cooperia oncophora* (Grønvold, 1987) and *Ostertagia circumcincta* (Waghorn *et al.*, 2002). Also dung beetles, which can be responsible for 14–20% of the dung degradation, can indirectly reduce the transmission of infective GIN larvae on to herbage (Holter, 1979). Earthworms, together with dung beetles, may be significant by breaking up and distributing faecal material around the pasture. This allows drying of the faecal material, which adversely affects both development and survival of trichostrongylid larvae (Stromberg, 1997).

Predatory mites of the family Macrochelidae could also be an option to control parasitic nematodes. They reach the dung pat by phoresis (flies, beetles) and include nematodes in their nutrition. One study with *Macrocheles glaber*, added to cattle dung, showed a significant reduction of infective nematode larvae (Waller and Faedo, 1996). But little research has been performed in this field.

Phytotherapy

There is a renaissance in the use of herbal medicine against helminths in livestock in the western world (Danø and Bøgh, 1999). A lot of research was done in developing countries (Hammond *et al.*, 1997; Akhtar *et al.*, 2000), because chemical anthelmintics were often not available and very expensive in rural areas. Interest in anthelmintic plants is also growing in developed countries (Waller *et al.*, 2001; Waller and Thamsborg, 2004), mainly owing to the rising anthelmintic resistance of helminths and because of concerns about synthetic drug residues in the environment and the food chain. Many plants with anthelmintic properties have been listed, but scientific validation is mostly lacking. Scientific evaluations comparing the anthelmintic efficacy of plants to commercial anthelmintics are limited (Akhtar *et al.*, 2000). Thus, a standard procedure, based on evidence-based medicine, is necessary for evaluating efficacy, toxicity and environmental sustainability of herbal remedies (Cabaret *et al.*, 2002). Chemical constituents can vary considerably between individual plants owing to genetic or environmental differences, the development stages of the plant at harvesting, the drying process, storage techniques and the different origins of the plant material (Croom, 1983). The type of solvent can also influence the amount and spectrum of the active components in the final extract. Recently, reliable studies demonstrated the anthelmintic effect of traditional herbal remedies against *H. contortus* and *T. colubriformis* (Pessoa *et al.*, 2002; Assis *et al.*, 2003; Hördegen *et al.*, 2003), but in some studies no effects were revealed (Ketzis *et al.*, 2002; Githiori *et al.*, 2002; Githiori *et al.*, 2003).

Vaccines

There are no commercially available vaccines for the control of helminth infections in ruminants with the exception of a vaccine against the bovine lungworm, *D. viviparus*. However, the development of vaccines against GIN in sheep and cattle is in progress (Smith, 1999; Dalton and Mulcahy, 2001). Currently, the most promising approach is based on the use of hidden antigens in the nematode's gut, which are normally not presented to the host's immune system. This strategy is especially effective in blood-feeding nematodes like *H. contortus*, which ingest antibodies developed by the immune system against these antigens. Binding of the antibodies to the specific epitopes in the intestine will finally damage the epithelium and kill the parasite. Besides defending the host from the adult worm population, the benefits of vaccination also include reduction of herbage contamination and thus reinfection (Sangster, 2001). Vaccines have the advantages that they are generally safe, leave no chemical residues and usually have no withdrawal periods for animal products, and are environmentally friendly.

Homeopathy

The few homeopathic products which were tested against GIN in sheep and cattle were based on plants and were not associated with a decrease in the worm burdens or faecal egg counts. However, considering the basic principles of homeopathy, it is likely that this approach reacts via modulation of the host's immune system (Cabaret *et al.*, 2002). The use of non-classical parameters for assessing an antiparasitic effect should therefore be considered in future studies of homeopathy.

Monitoring and targeted treatment

Organic and low input farmers who are dependent on anthelmintics should optimise the benefit of these treatments by monitoring levels of infestation on their farms. By minimising treatment frequency, optimal timing of the treatments should result in a parasitological status that is both tolerable for herd health and also economically justified. With the exception of the large liver fluke, *F. hepatica*, the parasites treated in this chapter should therefore not be controlled on the basis of a standardised schedule, which most likely leads to excessive use of anthelmintics.

In organic farms regulations require an exact diagnosis, usually performed by a faecal examination, before treatment may be given. The usual situation is that farmers organise the timing of coprological examination in collaboration with their veterinarian or health adviser. Owing to a lack of epidemiological knowledge and/or interpretation of the data, the organisation of a farm-specific monitoring programme is often complicated. In some countries, monitoring programmes for sheep and goat farmers have been organised on a small or medium scale basis, to support owners in managing their parasite problem. This is primarily important for sheep and goat farms, as the development of the seasonal dynamic of the disease is difficult to predict and the situation may change rapidly. In Switzerland, an endoparasite-monitoring programme was established in 1994 by the Extension and Health Service for Small Ruminants (BGK/SSPR, Herzogenbuchsee), an organisation funded by Swiss livestock owners. Faecal examinations can be performed three to five times per season depending on the animal species and age category. On the basis of the results and the owner's assessment of the clinical status of his/her flock, recommendations are given on treatment and subsequent examinations. The programme has achieved a high degree of acceptance among the participating farmers and the vast majority of them regard their helminth problem as under good or adequate control. Participation resulted in a significant reduction in anthelmintic treatments per year compared with non-participants. Organic farmers were able to manage their flock with two treatments per year, which is low compared to the other production systems. Thus, the programme fulfils the requirements with respect to optimising the treatment schedule and is furthermore cost-effective for most farmers.

In those countries where *H. contortus* represents the major parasite pathogen, selective treatments can be performed according to the FAMACHA©-system, which is based on the degree of anaemia in the ocular conjunctiva (van Wyk and Bath, 2002). In young cattle the endoparasite problem is more predictable than in sheep, especially in extensive grazing systems; therefore monitoring programmes covering GIN in cattle have so far not been very attractive.

Monitoring programmes play an important role in the prevention of anthelmintic resistance. 'Dose and move' strategies have been associated with development of anthelmintic resistance; it is thought that any nematodes capable of surviving the anthelmintic treatment may set up new resistant populations on the new pasture (Cawthorne and Cheong, 1984). Selection of resistant populations can be retarded by enlarging so-called 'refugia' (Van Wyk, 2001). 'Refugia' are fractions of a population of parasites not exposed to selection for resistance by anthelmintic treatments. The sources for 'refugia' are worms from untreated animals, inhibited stages surviving treatment in the host and continuously grazed pastures, which are less likely to permit the development of resistance, providing that there is no heavy reliance on anthelmintic treatments (Coles, 2002).

12.2.3 Lungworms

Lungworm control is a significant task in young cattle, whereas it is of minor importance in sheep and goats. Compared with the GIN, the spectrum of complementary control strategies against these parasites is very limited. The fact that pasture infectivity may build rapidly and that relatively few larvae may trigger severe clinical disease, makes control measures on the basis of grazing management difficult, and thus widely impracticable. Also biological control with *D. flagrans* has so far not proved to be a successful approach against the bovine lungworm, possibly because the sluggish movement of the infective larvae does not trigger fungal growth sufficiently (Fernandez *et al.*, 1999b). In those countries where it is available, the oral lungworm vaccine is an important factor for the control of *D. viviparus*. Currently, this is the only commercially available vaccine for the control of helminth infections. The vaccine is safe and free of chemically synthesised components and therefore fully compatible with organic guidelines. It should be the strategy of choice for low input and organic husbandry in those countries where it is marketed.

12.2.4 Liver flukes

Control of liver flukes is hampered by their complex biology and the presence of intermediate hosts. Significant efforts for control have only been made for the large liver fluke, *F. hepatica*, as it is much more pathogenic and can be the cause of severe economic damage. Even on conventional farms, strategic control, based on anthelmintics, is a difficult multi-year-task needing considerable effort by the farmer to lower the infection risk. Generally, eradication of the parasite from single farms is not a realistic goal. Control of fasciolosis on low input or organic farms is likely to be even more complex than on conventional farms and is generally not possible without the use of anthelmintics. As an important part of the control strategy, pasture hygiene measures and grazing management have to be incorporated. If the source of infection is located in permanent water sources, like ponds or ditches, animals should be kept at distances of about 2 m to avoid contact with the snail habitat. If the source of infection is more diffuse and whole pastures represent habitats for the intermediate hosts, grazing has to be avoided on these areas and instead the herbage should be conserved for hay or silage. If a farmer has both safe and *Fasciola*-affected pastures, a rotation can be performed between both areas, combined with strategic anthelmintic treatment (Boray, 1971). In this case, treated animals may graze on infested pastures for no longer than 8 weeks, corresponding to the period during which the parasite reaches maturity. Afterwards, animals are transferred to safe pastures where the deposited eggs will not succeed in continuing the life cycle. Another treatment has to be given before animals may be transferred back to the 'dangerous' pastures. Depending on how many changes from safe to unsafe pastures are necessary in the grazing season, the frequency of treatments can be reduced substantially. Farmers dealing with uniformly wet grassland, currently have no choice but to base the control of *F. hepatica* entirely on anthelmintic treatments. Many efforts have been undertaken recently to develop a vaccine against *F. hepatica* (Dalton and Mulcahy, 2001), but a marketed product is not yet available.

12.3 Non-ruminants

12.3.1 Important helminth infections of pigs and poultry

The risk of parasitic infestations is elevated in hens and pigs in free-range systems compared to systems without outdoor runs (Permin *et al.*, 1999; Thamsborg *et al.*, 1999). However, effective alternative methods for parasite control in monogastrics are almost completely lacking and the use of conventional antiparasitic drugs is the rule on organic as well as on conventional farms, although the extent of use may vary. The situation is particularly delicate in the control of endoparasites of laying hens, because only one anthelmintic (Flubendazole) is registered for this indication. This extensively used anthelmintic causes residues in eggs; managing this issue presents a major problem for organic egg production. The roundworms *Ascaridia galli* and *Heterakis gallinarum* are the most important parasitic species in the intestines of poultry. In addition, free range hens are at an increased risk of being infected by the protozoan *Histomonas meleagridis* (blackhead disease), which is transmitted by *H. gallinarum*. At present, no feed additives against blackhead disease are registered (Council Regulation (EC) No 1756/2002 of 23 September 2002) and helminth control is probably the most effective control measure.

In pigs, the large roundworm, *Ascaris suum*, is common throughout the world and often present in high prevalence (Roepstorff, 2003). Other widespread nematode species are *Oesophagostomum* spp. and *Trichuris suis*. Both *A. suum* and *T. suis* have egg dwelling infective stages, whereas free living infective *Oesophagostomum* spp. larvae are present in the faeces. In contrast to poultry, several active substances are available for the control of internal parasites in pigs. Routine anthelmintic treatments are usually applied in both monogastric species in order to control roundworms, whose eggs generally have a long survival rate in the environment and thus a high infection potential.

12.3.2 Approaches for reducing anthelmintic use in non-ruminants

Breeding for parasite resistance

Results of a Danish study (Schou *et al.*, 2003) suggest that the epidemiology of *A. galli* infections in chicken may be influenced by a genetic component. Gauly *et al.* (2002) observed significantly higher faecal egg outputs in white laying hens than in brown hens; they estimated sufficiently high heritabilities for faecal egg counts to allow for selection for *A. galli* resistance in chicken. However, parasite resistance is not a criterion in poultry breeding at present. Studies on genetic resistance to parasites in pigs are scarce; Nejsum *et al.* (2005) observed differences in parasite load between litters, which are considered to be related to host genotype.

Preventive management

Preventive strategies are less effective in monogastrics than in ruminants, because of the different epidemiology of the parasite species involved. Hygiene and proper management of the runs and pastures are the basis for the prevention of helminth infections. Good hygiene of houses is essential for the prevention of an accumulation of the long-lasting infective parasite eggs over time. In pigs, cleaning indoor pens and runs with solid floors regularly (every three weeks), is probably the only method to achieve an *Ascaris* free status without medication (Roepstorff and Nansen, 1994); however, maintaining this schedule is difficult and time consuming. In poultry, thorough cleaning of the house and 'wintergarden' is only possible between flocks. The outdoor run cannot be disinfected in order to destroy infectious stages and, therefore, different strategies have to be elaborated. In some production systems (e.g. free ranging broilers), an all-in-all-out system where new areas are provided for each batch of animals, is feasible (Thamsborg *et al.*, 1999). Where hens are kept in solid henhouses with surrounding runs, a rotation scheme with sufficient resting time between flocks is nearly impossible to achieve. A rotation scheme where the hen flock returns to the same surface during one season, helps maintain the sward, but is not effective as a preventive measure against helminths.

It is known that the survival of eggs of the pig parasites *A. suum* and *T. suis* is better in soil than in short swards (Larsen and Roepstorff, 1999). Using nose-rings in pigs is sometimes recommended as a preventive measure; however, this practice interferes with the normal rooting behaviour of the pig and it is therefore banned by national organic farming regulations in some countries (e.g. UK, NL, CH) for animal welfare reasons. Carstensen *et al.* (2002) recommend starting pig production on a farm with helminth-free sows and then keeping the accumulation and transmission of helminth eggs/larvae low by strict pasture rotation.

Quintern (2005) shows several possibilities for the integration of fattening pigs into the crop rotation of an organic farm, with the goal of minimising nutrient accumulation and nitrate leaching. Mejer and Roepstorff (2005), however, found that a pasture rotational scheme of three years was not sufficient to control *A. suum* and *T. suis* in a Danish experiment and this factor should be considered when including pigs in the crop rotation.

Roepstorff *et al.* (2002) showed that sows and particularly heifers had reduced parasite loads when co-grazing, possibly due to both a lower effective stocking rate and a mutual elimination of foreign parasites. The latter may be of particular importance for the heifers, as the sows grazed closely around the cow pats, thereby either exposing the infectious larvae in faeces to adverse environmental conditions or ingesting cattle parasite larvae, or both (Fernandez *et al.*, 2001). Moreover, an extremely poor rate of development and survival of both bovine nematode eggs and infective larvae was demonstrated after passage in pigs (Steenhard *et al.*, 2001).

Nutritional approaches/bioactive forages

Several components in the diet have been observed to affect helminth infections in ruminants (see Section 12.2.2), but only a few studies have been carried out in pigs and none in poultry. In pigs, high levels of insoluble dietary fibres have resulted in higher establishment rates and better fecundity of *Oesophagostomum dentatum* compared to diets of similar protein and energy levels that are rich in soluble carbohydrates. Feeding infected pigs with easily degradable and rapidly fermentable carbohydrates has been shown to reduce the worm burdens and female fecundity of *O. dentatum* significantly; *A. suum* infections were not similarly affected (Petkevicius *et al.*, 1997). This principle has been further investigated using diets for pigs containing inulin, chicory roots or sugar beet fibre (Petkevicius *et al.*, 2003; Mejer *et al.* 2005). In the future, traditional organic pig diets with high contents of fibre and partly indigestible carbohydrates, may have to be replaced by novel feeding strategies including easily fermentable carbohydrates.

Biological control
In contrast to the GIN in ruminants, the main parasitic nematodes of monogastric animals have egg dwelling infective stages. Until now, no promising biocontrol agents have been identified for use against these species, although the nematophagous fungus, *D. flagrans*, was shown to significantly reduce *O. dentatum* infections in grazing pigs (Nansen *et al.*, 1996).

Phytotherapy
Worldwide, a variety of plants have been shown to affect survival and/or reproduction of helminths of chicken and pigs *in vitro* or *in vivo*; in some cases, severe side effects on the host have been observed after use of some plant products (e.g. Akhtar and Riffat, 1985; Javed *et al.*, 1994; Satrija *et al.*, 1994).

12.4 Future trends

A number of promising approaches for reducing the intensity of antiparasitic drug treatments in low input and organic farms using complementary helminth control exist, but most of them are still at an experimental stage. Even those strategies where good progress has been made and which have proven to be beneficial (i.e. grazing management in cattle), still have not been adopted in practice to a wide extent. Thus, besides ongoing research activities, the education of farmers and transfer of the techniques into the field is a critical issue, which has not been successfully addressed in the past. It is primarily the responsibility of the extension services to explain the benefits of these strategies and thus serve as an interface between research and practice. In addition, disadvantages of the proposed strategies, which diminish their appeal and acceptance, have to be recorded and processed. There is evidence that the obligation to perform faecal examinations prior to antiparasitic treatment in organic farms, in order to fulfil the demand for making an exact diagnosis, is disobeyed knowingly or unknowingly in many cases. In this area further education is essential, as it is evident that a large number of treatments can be omitted when faecal examinations are adequately used. Experiences from Switzerland have clearly shown that regular faecal monitoring may reduce the drenching frequency substantially. With the same priority as developing new strategies, the integration of the existing knowledge into herd health management is of major importance.

12.5 References

Akhtar, M. S., Iqbal, Z., Khan, M. N. and Lateef, M. (2000). 'Anthelmintic activity of medicinal plants with particular reference to their use in animals in the Indo-Pakistan subcontinent'. *Small Ruminant Research*, **38**, 99–107.

Akhtar, M. S. and Riffat, S. (1985). 'Evaluation of *Melia azedarach* Linn. Fruit (Bakain) against *Ascaridia galli* infection in chickens'. *Pakistan Veterinary Journal*, **5**(1), 34–37.

Assis, L. M., Bevilaqua, C. M., Morais, S. M., Vieira, L. S., Costa, C. T. and Souza, J. A. (2003). 'Ovicidal and larvicidal activity in vitro of *Spigelia anthelmia* Linn. extracts on Haemonchus contortus'. *Veterinary Parasitology*, **117**(1–2), 43–49.

Bang, K. S., Familton, A. S. and Sykes A. R. (1990). 'Effect of copper oxide wire particle treatment on establishment of major gastrointestinal nematodes in lambs'. *Research in Veterinary Science*, **39**(2), 132–137.

Barger, I. (1997). 'Control by management'. *Veterinary Parasitology* **72**(3–4), 493–500.

Baudena, M. A., Chapman, M. R., Larsen, M. and Klei T. R. (2000). 'Efficacy of the nematophagous fungus *Duddingtonia flagrans* in reducing equine cyathostome larvae on pasture in south Louisiana'. *Veterinary Parasitology*, **89**, 219–230.

Bone, L. W., Bottjier, W. K. P. and Gill, S. S. (1986). '*Trichostrongylus colubriformis*: isolation and characterization of ovizidal activity from *Bacillus thuringiensis israelensis* toxin'. *Exp. Parasitology*, **62**, 247–253.

Boray J. (1971). 'Fortschritte in der Bekämpfung der Fasciolose'. *Schweizer Archiv für Tierheilkunde*, **113**, 361–386

Bottjer, K. P. and Bone, L. W. (1987). 'Changes in morphology of *Trichostrongylus colubriformis* eggs and juveniles caused by *Bacillus thuringiensis israelensis*'. *Journal of Nematology*, **19**, 282–286.

Boulton, T. J. (2004). 'Responses of nontarget Lepidoptera to Foray 48B *Bacillus thuringiensis* var. *kurstaki* on Vancouver Island, British Columbia, Canada'. *Environmental Toxicology and Chemistry*, **23**(5), 1297–304.

Bransby, D. I. (1993). 'Effects of grazing management practices on parasite load and weight gain of beef cattle'. *Veterinary Parasitology*, **46**(1–4), 215–221.

Burke, J. M. and Miller, J. E. (2002). 'Relative resistance of Dorper crossbred ewes to gastrointestinal nematode infection compared with St. Croix and Katahdin ewes in the southeastern United States'. *Veterinary Parasitology*, **109**, 265–275.

Burke, J. M., Miller, J. E., Olcott, D. D., Olcott, B. M. and Terrill, T. H. (2004). 'Effect of copper oxide wire particles dosage and feed supplement level on *Haemonchus contortus* infection in lambs'. *Veterinary Parasitology*, **123**, 235–243.

Cabaret, J., Bouilhol, M. and Mage, C. (2002). 'Managing helminths of ruminants in organic farming'. *Veterinary Research*, **33**, 625–640.

Carstensen, L., Vaarst, M. and Poepstorff, A. (2002). 'Helminth infections in Danish organic swine herds'. *Veterinary Parasitology*, **106**, 253–264.

Cawthorne, R. J. and Cheong, F. H. (1984). 'Prevalence of anthelmintic resistant nematodes in sheep in south-east England'. *Veterinary Record*, **114**(23), 562–4.

Chandrawathani, P., Jamnah, O., Waller, P. J., Larsen, M., Gillespie, A. T. and Zahari, W. M. (2003). 'Biological control of nematode parasites of small ruminants in Malaysia using the nematophagous fungus *Duddingtonia flagrans*'. *Veterinary Parasitology*, **117**, 173–183.

Ciordia, H. and Bizzell, W. E. (1961). 'A preliminary report on the effects of *Bacillus thuringiensis* Berliner on the development of free-living stages of some cattle nematodes'. *Journal Parasitology*, **47**, 41.

Coles, G. C. (2002). 'Cattle nematodes resistant to anthelmintics: why so few cases?' *Veterinary Research*, **33**, 481–489.

Coop, R. L. and Kyriazakis, I. (1999). 'Nutrition–parasite interaction'. *Veterinary Parasitology*, **84**, 187–204.

Croom, E. M. (1983). 'Documenting and evaluating herbal remedies'. *Economic Botany*, **37**(1), 13–27.

Dalton, J. P. and Mulcahy, G. (2001). 'Parasite vaccines – a reality?' *Veterinary Parasitology*, **98**, 149–167.

Danø, A. R. and Bøgh, H. O. (1999). 'Use of herbal medicine against helminths in livestock-renaissance of an old tradition'. *World Animal Review*, **93**, 60–67.

Dimander, S., Höglund, J., Uggla, A., Spörndly, E. and Waller, P. J. (2003). 'Evaluation of gastro-intestinal nematode parasite control strategies for first-season grazing cattle in Sweden'. *Veterinary Parasitology*, **111**, 193–209.

Donaldson, J., Van Houtert, M. F. J. and Sykes, A. R. (1997). 'The effect of protein supply on the periparturient parasite status of the mature ewe'. *Proceedings of the New Zealand Society of Animal Production*, **57**, 186–189.

Eady, S. J., Woolaston, R. R. and Barger, I. A. (2003). 'Comparison of genetic and nongenetic strategies for control of gastrointestinal nematodes of sheep'. *Livestock Production Science*, **81**, 11–23.

Eilenberg, J., Hajek, A. and Lomer, C. (2001). 'Suggestions for unifying th terminology in biological control'. *Biocontrol*, **46**, 387–400.

Eysker, M. (2001). 'Strategies for internal parasite control in organic cattle', in Hovi, M. and Vaarst, M., *Positive Health: Preventive Measures and Alternative Strategies*. Proceedings of the Fifth NAHWOA Workshop Rødding, Denmark–November, 59–71.

Faedo, M., Larsen, M. and Thamsborg, S. (2000). 'Effect of different times of administration of the nematophagous fungus *Duddingtonia flagrans* on the transmission of ovine parasitic nematodes on pasture – a plot study'. *Veterinary Parasitology*, **94**, 55–65.

Fernandez, A. S., Henningsen, E., Larsen, M., Nansen, P., Grønvold, J. and Sondergaard, J. (1999a). 'A new isolate of the nematophagous fungus *Duddingtonia flagrans* as biological control agent against free-living larvae of horse strongyles'. *Equine Veterinary Journal*, **31**, 488–491.

Fernandez, A. S., Larsen, M., Nansen, P., Grønvold, J., Henriksen, S. A., Bjørn, H. and Wolstrup, J. (1999b). 'The efficacy of two isolates of the nematode-trapping fungus *Duddingtonia flagrans* against *Dictyocaulus viviparus* larvae in faeces'. *Veterinary Parasitology*, **85**, 289–304.

Fernandez, A. S., Larsen, M., Nansen, P., Henningsen, E., Grønvold, J., Wolstrup, J., Henriksen, S. A. and Bjørn, H. (1999c). 'The ability of the nematode-trapping fungus *Duddingtonia flagrans* to reduce the transmission of infective *Ostertagia ostertagi* larvae from faeces to herbage'. *Journal of Helminthology*, **73**, 115–122.

Fernandez, S., Sarkunas, M. and Roepstorff, A. (2001). 'Survival of infective *Ostertagia ostertagi* larvae on pasture plots under different simulated grazing conditions'. *Veterinary Parasitology*, **96**, 291–299.

Gasbarre, L. C. and Miller, J. E. (2000). 'Genetics of helminth resistance'. in Axford, R. F. E., Bishop, S. C., Nicholas, F. W. and Owen, J. B. *Breeding for Disease Resistance in Farm Animals*, CABI International Oxon, New York, 129–152.

Gauly, M. and Erhardt, G. (2001). 'Genetic resistance to gastrointestinal nematode parasites in Rhön sheep following natural infection'. *Veterinary Parasitology*, **102**, 253–259.

Gauly, M. and Erhardt, G. (2002). 'Changes in faecal trichostrongyle egg count and haematocrit in naturally infected Rhön sheep over two grazing periods and associations with biochemical polymorphisms'. *Small Ruminant Research*, **44**, 103–108.

Gauly, M., Bauer, C. Preisinger, R. and Erhardt, G. (2002). 'Genetic differences of *Ascaridia galli* egg output in laying hens following a single dose infection'. *Veterinary Parasitology*, **103**, 99–107.

Githiori, J. B., Hoglund, J. Waller P. J. and Baker, R. L. (2002). 'Anthelmintic activity of preparations derived from *Myrsine africana* and *Rapanea melanophloeos* against the nematode parasite', *Haemonchus contortus*, of sheep. *Journal of Ethnopharmacology*, **80**(2–3), 187–191.

Githiori, J. B., Hoglund, J. Waller, P. J. and Baker, R. L. (2003). 'The anthelmintic efficacy of the plant, *Albizia anthelmintica*, against the nematode parasites *Haemonchus contortus* of sheep and *Heligmosomoides polygyrus* of mice'. *Veterinary Parasitology*, **116**(1), 23–34.

Giudici, C., Aumont, G., Mahieu, M., Saulai, and Cabaret, J. (1999). 'Changes in gastro-intestinal helminth species diversity in lambs under mixed grazing on irrigated pastures in the tropics (French West Indies)'. *Veterinary Research*, **30**, 573–581.

Gould, J. L. and Keeton, W. T. (1996). *Biological Science*, 6th edition, W. W. Norton and Company, New York.

Grønvold, J. (1987). 'Field experiment on the ability of earthworms (Lumbricidae) to reduce the transmission of infective larvae of *Cooperia oncophora* (Trichostrongylidae) from cow pats to grass'. *Journal of Parasitology*, **73**(6), 1133–1137.

Grønvold, J., Wolstrup, J., Larsen, M., Nansen, P. and Bjorn, H. (2000). 'Absence of obvious short term impact of the nematode-trapping fungus *Duddingtonia flagrans* on survival and growth of the earthworm *Aporrectodea longa*'. *Acta Veterinaria Scandinavica*, **41**, 147–151.

Gruner, L., Cortet, J., Sauvé, C., Limouzin, C. and Brunel, J. C. (2002). 'Evolution of nematode immunity in grazing sheep selected for resistance and susceptibility to *Teladorsagia circumcincta* and *Trichostrongylus colubriformis*: a 4-year experiment'. *Veterinary Parasitology*, **109**, 277–291.

Gruner, L., Aumont, G., Getachew, T., Brunel, J. C., Pery, C., Cognié, Y. and Guérin, Y. (2003). 'Experimental infection of Black Belly and INRA 401 straight and crossbred sheep with trichostrongyle nematode parasites'. *Veterinary Parasitology*, **116**, 239–249.

Haile, A., Tembely, S., Anindo, D. O., Mukasa-Mugerwa, E., Rege, J. E. O., Yami, A. and Baker, R. L. (2002). 'Effects of breed and dietary protein supplementation on the responses to gastrointestinal nematode infections in Ethiopian sheep'. *Small Ruminant Research*, **44**, 247–261.

Hammond, J. A., Fielding, D. and Bishop, S. C. (1997). 'Prospects for plant anthelmintics in tropical veterinary medicine'. *Veterinary Research Communications*, **21**(3), 213–228.

Hilbeck, A., Moar, W. J., Pusztai-Carey, M., Filippini, A. and Bigler, F. (1998). 'Toxicity of *Bacillus thuringiensis* Cry 1Ab to the predator *Chrysoperla carnea* (Neuroptera: Chrysopidae)'. *Environmental Entomology*, **27**, 1–9.

Höglund, J., Svensson, C. and Hessle, A. (2001). 'A field survey on the status of internal parasites in calves on organic dairy farms in southwestern Sweden'. *Veterinary Parasitology*, **99**, 113–128.

Holter, P. (1979). 'Effect of dung beetles (*Aphodius* spp.) and earthworms on the disappearance of cattle dung'. *Oikos*, **32**, 393–402.

Hördegen, P., Hertzberg, H., Heilmann, J., Langhans, W. and Maurer, V. (2003). 'The anthelmintic efficacy of five plant products against gastrointestinal trichostrongylids in artificially infected lambs'. *Veterinary Parasitology*, **117**(1–2), 51–60.

Hoste, H., Jackson, F., Athanasiadou, S., Thamsborg, S. M. and Hoskin, S. O. (2006). 'The effects of tannin-rich plants on parasitic nematodes in ruminants'. *Trends in Parasitology*, **22**, 253–261

Houdijk, J. G. M., Kyriazakis, I., Coop, R. L. and Jackson, F. (2001). 'The expression of immunity to *Teladorsagia circumcincta* in ewes and its relationship to protein nutrition depend on body protein reserves'. *Parasitology*, **122**, 661–672.

Jacquiet, P., Cabaret, J., Thiam, E. and Cheikh, D. (1998). 'Host range and the maintenance of *Haemonchus* spp. in an adverse arid climate'. *International Journal for Parasitology*, **28**(2), 253–261.

Javed, I., Akhtar, M. S., Rahman, Z. U., Khaliq, T. and Ahmad, M. (1994). 'Comparative anthelmintic efficacy and safety of *Caesalpinia crista* seed and piperazine adipate in chickens with artificially induced *Ascaridia galli* infection'. *Acta Veterinaria Hungarica*, **42**(1), 103–109.

Johnson, S. K., Scriber, M., Nitao, K. J. and Smitley, R. D. (1995). 'Toxicity of *Bacillus thuringiensis* var. *kurstaki* to three nontarget Lepidoptera in field studies'. *Annals of the Entomological Society of America*, **24**, 288–297.
Kahn, L. and Diaz-Hernandez, A. (1999). 'Tannins with anthelmintic properties'. *ACIAR Proceedings*, **92**, 130–139.
Kahn, L. P., Kyriazakis, I., Jackson, F. and Coop, R. L. (2000). 'Temporal effects of protein nutrition on the growth and immunity of lambs infected with *Trichostrongylus colubriformis*'. *International Journal for Parasitology*, **30**, 193–205.
Kahn, L. P., Knox, M. R., Gray, G. D., Lea, J. M. and Walkden-Brown, S. W. (2003). 'Enhancing immunity to nematode parasites in single-bearing Merino ewes through nutrition and genetic selection'. *Veterinary Parasitology*, **112**, 211–225.
Ketzis, J. K., Taylor, A., Bowman, D. D., Brown, D. L., Warnick, L. D. and Erb, H. N. (2002). '*Chenopodium ambrosioides* and its essential oil as treatments for *Haemonchus contortus* and mixed adult-nematode infections in goats'. *Small Ruminant Research*, **44**, 193–200.
Larsen, M. (1999). 'Biological control of helminths'. *International Journal for Parasitology*, **29**(1), 139–146.
Larsen, M. (2000). 'Prospects for controlling animal parasitic nematodes by predacious micro fungi'. *Parasitology*, **120**, S121–S131.
Larsen, M. N. and Roepstorff, A. (1999). *Seasonal Variation in Development and Survival of* Ascaris suum *and* Trichuris suis *Eggs on Pastures.* 17th International Conference of the World Association for the Advancement of Veterinary Parasitology, Copenhagen.
Lindqvist, Å, Ljungström, B.-L., Nilsson, O. and Waller, P. J. (2001). 'The dynamics, prevalence and impact of nematode infections in organically raised sheep in Sweden'. *Acta Veterinaria Scandinavica*, **42**(3), 377–389.
Mejer, H. and Roepstorff, A. (2005). 'Survival of the free-living stages of porcine helminths in relation to pasture management'. *Proceedings of the 20th International Conference of the WAAVP*, 16–20 October 2005. Christchurch, NZ, p 235.
Mejer, H., Roepstorff, A., Thamsborg, S. M., Hansen, L. L. and Knudsen, K. E. B. (2005). 'Effect of feeding with chicory roots on *Oesophagostomum dentatum* and *Ascaris suum* infections in pigs'. *Proceedings of the 20th International Conference of the WAAVP*, 16–20 October 2005. Christchurch, NZ, p 222.
MacRae, J. C. (1993). 'Metabolic consequences of intestinal parasitism'. *Proceedings of the Nutrition Society*, **52**, 121–130.
Min, B. R., Pomroy, W. E., Hart, S. P. and Sahlu, T. (2004). 'The effect of short-term consumption of a forage containing condensed tannins on gastro-intestinal nematode parasite infections in grazing wether goats'. *Small Ruminant Research*, **44**, 81–87.
Molan, A. L., Hoskin, S. O., Barry, T. N. and McNabb, W. C. (2000). 'Effect of condensed tannins extracted from four forages on the viability of the larvae of deer lungworms and gastrointestinal nematodes'. *Veterinary Record*, **147**(2), 44–8.
Molan, A. L., Duncan, A. J., Barry, T. N. and McNabb, W. C. (2003). 'Effects of condensed tannins and crude sesquiterpene lactones extracted from chicory on the motility of larvae of deer lungworm and gastrointestinal nematodes'. *Parasitology International*, **52**, 209–218.
Nansen, P., Larsen, M., Roepstorff, A., Gronvold, J., Wolstrup, J. and Henriksen, S. A. (1996). 'Control of *Oesophagostomum dentatum* and *Hyostrongylus rubidus* in outdoor–reared pigs through daily feeding with the microfungus *Duddingtonia flagrans*'. *Parasitology Research*, **82**, 580–584.
Nejsum, P., Jørgensen, C., Fredholm, M., Roepstorff, A. and Thamsborg, S. M. (2005). 'Helminth infections in pigs: possible effects of host genotype'. *Proceedings of the 20th International Conference of the WAAVP*, 16–20 October 2005. Christchurch, NZ, p 239.
Niezen, J. H., Robertson, H. A., Waghorn, G. C. and Charleston, W. A. G. (1998). 'Production, feacal egg counts and worm burdens of ewe lambs which grazed six contrasting forages'. *Veterinary Parasitology*, **80**, 15–27.

Niezen, J., Ha., R., Sidey, A. and Wilson, S. (2002a). 'The effect of pasture species on parasitism and performance of lambs grazing one of three grass – white clover pasture swards'. *Veterinary Parasitology*, **105**, 303–215.

Niezen, J., Waghorn, G., Graham, T., Carter, J. and Leathwick, D. (2002b). 'The effect of diet fed to lambs on subsequent development of *Trichostrongylus colubriformis* larvae in vitro and on pasture'. *Veterinary Parasitology*, **105**, 269–283.

Permin, A., Bisgaard, M., Frandsen, F., Pearman, M., Kold, J. and Nansen, P. (1999). 'Prevalence of gastrointestinal helminths in different poultry production systems'. *British Poultry Science*, **40**(4): 439–43.

Pessoa, L. M., Morais, S. M., Bevilaqua, C. M. L. and Luciano, J. H. S. (2002). 'Anthelmintic activity of essential oil of *Ocimum gratissimum* Linn. and eugenol against *Haemonchus contortus*'. *Veterinary Parasitology*, **109**, 59–63.

Petkevicius, S., Bach Knudsen, K. E., Nansen, P., Roepstorff, A., Skjoth, F., Jensen, K. and Knudsen, K. E. B. (1997). 'The impact of diets varying in carbohydrates resistant to endogenous enzymes and lignin on populations of *Ascaris suum* and *Oesophagostomum dentatum* in pigs'. *Parasitology*, **114**(6), 555–568.

Petkevicius, S., Bach Knudsen, K. E., Murrell, K. D. and Wachmann, H. (2003). 'The effect of inulin and sugar beet fibre on *Oesophagostomum dentatum* infection in pigs'. *Parasitology*, **127**, 61–68.

Quintern, N. (2005). 'Integration of organic pig production within crop rotation: Implications on nutrient losses'. *Landbauforschung Völkenrode*, Special Issue, **281**, 31–40.

Rege, J. E. O., Tembely, S., Mukasa-Mugerwa, E., Sovani, S., Anindo, D., Lahlou-Kassi, A., Nagda, S. and Baker, R. L. (2002). 'Effect of breed and season on production and response to infections with gastro-intestinal nematode parasites in sheep in the highlands of Ethiopia'. *Livestock Production Science*, **78**, 159–174.

Roepstorff, A. (2003). Ascaris suum *in Pigs: Population Biology and Epidemiology*. Danish Center for Experimental Parasitology, Copenhagen, 210pp.

Roepstorff, A. and Nansen, P. (1994). 'Epidemiology and control of helminth infections in pigs under intensive and non-intensive production systems'. *Veterinary Parasitology*, **54**(1–3), 69–85.

Roepstorff, A., Monrad, J., Sehested, J., Søegaard, K. and Danielsen, V. (2002). 'Mixed grazing with sows and heifers – as a mean of parasite control', in *Novel Approaches Meeting III – A Workshop Meeting on Helminth Control in Livestock in the New Millenium*. Third International conference, 1–5 July 2002, 41.

Sangster, N. C. (2001). 'Managing parasiticide resistance'. *Veterinary Parasitology*, **98**, 89–109.

Satrija, F., Nansen, P., Bjorn, H., Martini, S. and He, S. (1994). 'Effect of *Papaya latex* against *Ascaris suum* in naturally infected pigs'. *Journal of Helminthology*, **68**(4), 343–346.

Schou, T., Permin, A., Roepstorff, A., Sorensen, P. and Kjaer, J. (2003). 'Comparative genetic resistance to Ascaridia galli infections of 4 different commercial layer-lines'. *British Poultry Science*, **44**(2), 182–185.

Smith, W. D. (1999). 'Prospects for vaccines of helminth parasites of grazing ruminants'. *International Journal of Parasitology*, **29**, 17–24.

Speiser, B., Wyss, E. and Maurer, V. (2006). 'Biological control in organic production: First choice or last option?' in Eilenberg, J. and Hokkanen, H. T. M. *An Ecological and Societal Approach to Biological Control*, Springer, Dordrecht, pp 27–46.

Steenhard, N. R., Roepstorff, A., and Thamsborg, S. M. (2001). 'Inactivation of eggs and larvae of the cattle nematodes *Ostertagia ostertagi* and *Cooperia oncophora* after passage in pigs'. *Veterinary Parasitology*, **101**, 137–142.

Stromberg, B. E. (1997). 'Environmental factors influencing transmission'. *Veterinary Parasitology*, **72**(3–4), 247–264.

Thamsborg, S. M., Roepstorff, A. and Larsen, M. (1999). 'Integrated and biological control of parasites in organic and conventional production systems'. *Veterinary Parasitology*, **84**, 169–186.

Valderrábano, J., Delfa, R. and Uriarte, J. (2002). 'Effect of level of feed intake on the development of gastrointestinal parasitism in growing lambs'. *Veterinary Parasitology*, **104**, 327–338.

Van Frankenhuyzen, K. (1993). 'The challenge of *Bacillus thuringiensis*' in Entwistle, P. E., Cory, J. S., Bailey, M. J., and Higgs, S. Bacillus thuringiensis, *An Environmental Biopesticide: Theory and Practice*, John Wiley & Sons, Chichester, UK, 1–35.

Van Wyk, J. A. (2001). 'Refugia – overlooked as perhaps the most potent factor concerning the development of anthelmintic resistance'. *Onderstepoort Journal of Veterinary Research*, **68**, 55–67.

Van Wyk, J. A. and Bath, G. F. (2002). 'The FAMACHA© system for managing haemonchosis in sheep and goats by clinically identifying individual animals for treatment'. *Veterinary Research*, **33**, 509–529.

Waghorn, T. S., Leathwick, D. M., Chen, L.-Y., Gray, R. A. J. and Skipp, R. A. (2002). 'Influence of nematophagous fungi, earthworms and dung burial on development of the free-living stages of *Ostertagia* (*Teladorsagia*) *circumcincta* in New Zealand'. *Veterinary Parasitology*, **104**, 119–129.

Waller, P. J. and Faedo, M. (1996). 'The prospects for biological control of the free-living stages of nematode parasites of livestock'. *International Journal for Parasitology*, **26**(8–9), 915–925.

Waller, P. J. and Thamsborg, S. M. (2004). 'Nematode control in 'green' ruminant production systems'. *Trends in Parasitology*, **20**(10), 493–497.

Waller, P. J., Bernes, G., Thamsborg, S. M., Sukura, A., Richter, S. H., Ingebrigtsen, K. and Höglund, J. (2001). 'Plants as de-worming agents of livestock in the Nordic countries: historical perspective, popular beliefs and prospects for the future'. *Acta Veterinary Scandinavia*, **42**(1), 31–44.

Wei, J. Z., Hale, K., Carta, L., Platzer, E., Wong, C., Fang, S. C. and Aroian, R. V. (2003). '*Bacillus thuringiensis* crystal proteins that target nematodes'. *Proceedings National Academy Science USA*, **100**(5), 2760–2765

Wright, D. A., McAnulty, R. W., Noonan, M. J. and Stankiewicz, M. (2003). 'The effect of *Duddingtonia flagrans* on trichostrongyle infections of Saanen goats on pasture'. *Veterinary Parasitology*, **118**, 61–69.

Yeates, G. W., Waller, P. J. and King, K. L. (1997). 'Soil nematodes as indicators of the effect of management on grasslands in the New England Tablelands (NSW): effect of measures for control of parasites of sheep'. *Pedobiologia*, **41**(6), 537–548.

Yeates, G., Dimander, S.-O., Waller Peter, J. and Höglund, J. (2003). 'Soil nematode populations beneath faecal pats from grazing cattle treated with the ivermectin sustained-release bolus or fed the nematophagous fungus *Duddingtonia flagrans* to control nematode parasites'. *Acta Agricultural Scandinavica, Section A, Animal Science*, **53**, 197–206.

Zwahlen, C., Hilbeck, A., Howald, R. and Nentwig, W. (2003). 'Effects of transgenic *Bt* corn litter on the earthworm *Lumbricus terrestris*'. *Molecular Ecology*, **12**(4), 1077–1086.

13

Alternative therapies to reduce enteric bacterial infections and improve the microbiological safety of pig and poultry production systems

Bruno Biavati and Cecilia Santini, Bologna University, Italy and Carlo Leifert, Newcastle University, UK

13.1 Introduction

While the acidic environment of the stomach is a hostile environment for most microorganisms and usually considered 'the line' of defence against foodborne pathogens, the intestine has a higher pH and is colonised by dense and diverse microflora, which may be commensal or beneficial, but may also include pathogens and parasites. While the disinfection function of the stomach acid has long been recognised, no efforts have until recently been made to develop strategies to increase the efficacy. However, more recently a range of compounds naturally found in foods (e.g. nitrate, nitrite and isothiocyanate) were shown significantly to increase the antimicrobial activity of the stomach acid (Dykhuizen *et al.*, 1996, 1998). This may provide a novel strategy to increase the disinfection capacity of the stomach and thereby prevent or reduce the transfer of food or waterborne enteric pathogens into the intestine.

Many microorganisms in the natural intestinal flora are non-pathogenic bacteria and have been shown to have preventive or even therapeutic effects on pathogens. The most commonly studied groups are the lactobacilli and bifidobacteria, which are also increasingly used and/or developed as probiotic treatments. Probiotics are defined as selected, viable microbial dietary supplements which, when introduced in sufficient quantities, beneficially affect animals or human through their effects in the intestinal tract, they may improve both the health and the growth rate of the animal and its immune status (Choct, 2001). The protection afforded by both indigenous microbiota and probiotics is thought to be related to competitive exclusion and interference with attachment to mucosal surfaces (Tannock, 2005).

Prebiotics, on the other hand, are non-digestible for the host but benefit the host by selectively stimulating the growth and/or activity of one or a limited number of bacteria in the intestine and colon (Tannock, 2005; Newman, 1994; Pluske *et al.*, 1996). They have been shown to improve intestinal and colonic microbial ecology and enhance stool quality in livestock and poultry, control pathogenic bacteria, reduce faecal odour and enhance growth performance (Newman, 1994; Pluske *et al.*, 1996; Sinlac and Choct, 2000; Tannock, 2005). Another option for intestinal microbiota management is the use of synbiotics, in which probiotics and prebiotics are used in combination (Tannock, 2005).

Enteric bacterial infections cause extensive morbidity and loss of production in the pig and poultry industry (Inborr, 2000). As a result, both broad spectrum and selective antibiotic treatments are widely (and often prophylactically) used in monogastric livestock production systems (Best, 1997; Inborr, 2000). As a direct consequence of over-use, transmissible antibiotic resistance genes are often present in both the pathogenic and commensal bacterial flora of the gastrointestinal tract of farm animals (Cohen, 1992; Gill and Best, 1998; Collignon, 1999; Newman, 2002). This has resulted in public health concern about the risks associated with excessive use of antibiotics and in new national and EU legislation that will curtail their future use in veterinary and agricultural practices (Inborr, 2000; Radcliffe, 2000). Research directed towards establishing alternative strategies to combat widespread endemic diseases normally controlled by antibiotics and other chemotherapeutics is now recognised as a high priority.

This chapter reports the progress made in the development of probiotic, prebiotic and other preventative treatments for gastrointestinal diseases, which have potential for use in organic and low input monogastric livestock production systems. Most of the results discussed focus on research with pigs, but most results should be transferable to other monogastric animals including poultry.

13.2 Anatomy and physiology of digestive tracts of monogastric livestock

13.2.1 The pig (*Sus scrofa domestica*)

The pig is widely used in research and testing since it shares many anatomical and physiological characteristics with humans. This makes it a unique and viable model for biomedical research. In fact, most relevant models have been reviewed in a series of technical proceedings and books in the last two decades (Swindle, 1998).

The pig's digestive tract may be thought of as a long tube through which food passes. As food passes through the digestive tract, it is broken down into smaller and smaller units. These small units of food are absorbed as nutrients or pass out of the body as urine and faeces. The digestive tract of the pig has five main parts: the mouth, oesophagus, stomach, and small and large intestines.

The oesophagus is a tube that carries the food from the mouth to the stomach. It forms one morphological unit, which is compartmentalised for its different functions by different types of mucosa. The functions of the stomach are storage and successive release of small portions of ingesta to the intestine, carbohydrate digestion and fermentation, and enzymatic and hydrolytic degradation of proteins as preparation for intestinal digestion and absorption. Because of its low pH (between 1.5 and 4), the stomach is also known as the first barrier to microbial passage into the intestine. The intestine is divided into the small intestine and large intestine. Each consists of three different sections: the duodenum, jejunum and ileum of the small intestine and the caecum, colon and rectum of the large intestine. Gastrointestinal diseases usually initially establish and manifest themselves in the upper small intestine, but may also spread to other areas of the intestine (Tannock, 2005). It is therefore thought that the successful control of enteric pathogen infections requires the prevention of infections in the small intestine. This is particularly true for therapies based on lactic acid bacteria, for which the still slightly lower pH (owing to the influence of the stomach) provides a selective advantage. For additional details on the digestive system of pigs, see Schantz *et al.* (1996).

13.2.2 Digestive system of poultry

The digestive system of poultry is also basically a tube, beginning at the mouth and ending at the vent (Fig. 13.1). Food is taken in via the mouth and mixed with saliva to lubricate it. The bolus moves down the oesophagus by gravity and a wave-like contraction of the muscles (peristalsis). Then it enters the crop where it is stored if the stomach is full. A certain amount of softening and fermentation may occur here. The crop is situated just outside the entrance of the chest cavity (thoracic inlet) and is an expansion of the oesophagus.

The food enters the proventriculus, the ‘true stomach’, where the process of breaking down the protein constituent of food into simple components begins. As in pigs, the poultry stomach has a very low pH (1.5–4) and is thought to be the first line of defence against foodborne pathogens. The food, mixed with these gastric juices passes to the gizzard, or ventriculus, then arrives at the intestine. The duodenal loop is a gland that secretes pancreatic juice into the intestines, a solution comprising digestive enzymes and sodium bicarbonate, which neutralises the acidity of the food leaving the stomach. As in pigs, the intestine in poultry has a more neutral pH and therefore allows the establishment of enteric pathogens. For additional details on the digestive system of poultry, see Grist (2004).

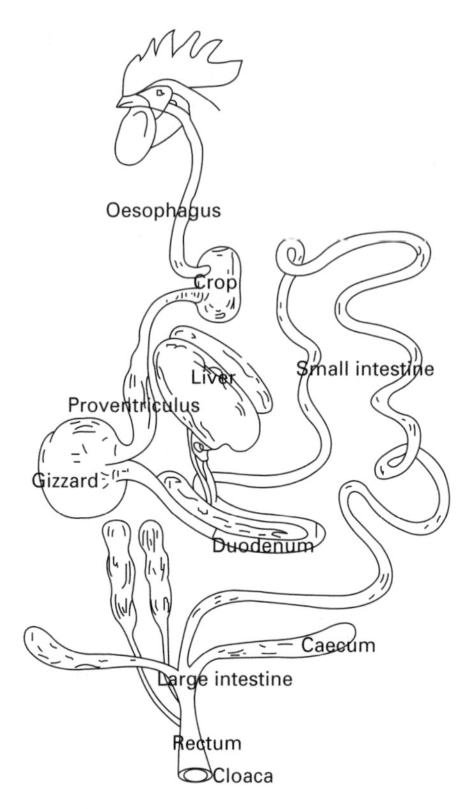

Fig. 13.1 The digestive system of poultry.

13.2.3 Role of the intestinal microbiota

Each section of the gastrointestinal tract harbours a mixed microbial population that lives in equilibrium with the host animal. The rich and diverse population of anaerobic bacteria present in the large intestine of pigs carries out the hydrolytic digestive function. Characterisation of the intestinal microbiota of pigs has been done by anaerobic culturing techniques and several studies have shown that the predominant bacterial genera in the intestine are *Streptococcus*, *Lactobacillus*, *Prevotella*, *Selenomonas*, *Mitsuokella*, *Megasphaera*, *Clostridium*, *Eubacterium*, *Fusobacterium*, *Acidaminococcus* and *Enterobacter* (Mitsuoka, 1992; Tannock, 1992).

The chicken intestinal microbiota is composed principally of Gram-positive bacteria (Gong *et al.*, 2002). The lactobacilli are predominant in the small intestine (with smaller numbers of streptococci and enterobacteria), whereas the caecal flora is composed mainly of anaerobes and fewer numbers of facultative bacteria. Predominant cultured flora of the ileum of chicken include *Lactobacillus*, *Streptococcus*, *E. coli* and *Eubacterium*, while *Eubacterium* and *Bacteroides* dominated the caecum flora.

Although the microbiota are not as essential for nutrient supply to the pig as they are in ruminants, these organisms do provide some benefit to their host. This includes (i) synthesis of vitamins and growth factors (biotin, riboflavin, vitamin K and others), (ii) degradation of less digestible food and the production of volatile fatty acids), (iii) control of epithelial cell proliferation and differentiation, (iv) metabolism and enterohepatic circulation of xenobiotics, (v) immunostimulation, (vi) pro- and anticarcinogenesis and mutagenesis activities, (vii) control of ion concentrations and absorption and pH and (viii) competition or direct antagonism of enteric pathogens contributing to an overall increased resistance to gastrointestinal infection (Dunne *et al.*, 1999; Naidu *et al.*, 1999; Kasper 1998). Rapid changes in diet, or antibiotic treatments may temporarily reduce the numbers or activities of the normal microbiota, which may lead to a greater risk of colonisation by enteric pathogens (MacClane *et al.*, 2005; Guarner and Malagelada, 2003).

13.3 Intestinal bacteria and their potential as probiotics

The ability of bacteria to colonise the intestine is an essential prerequisite for their development as probiotics. Colonisation capacity is a composite process of natural selection and ecological succession and is influenced by numerous regulatory factors of both bacterial and host origin including animal genotype, bacterial antagonisms, animal physiology and nutrition (Tannock, 2005). It is also important to point out that several microhabitats exist within the intestine, which favour different microbial populations (Tannock, 2005; MacClane *et al.*, 2005; Falk *et al.*, 1998). These microniches, which exert a selective influence on the local composition and metabolic action of the microbiota, are found in the intestine associated with the villus surface, epithelial associated mucins, crypts and luminal mucus. Variables that contribute to the regional compositional diversity comprise available molecular oxygen, nutrient availability and composition, the flow of digesta, pH and Eh (oxidation/reduction potential), the presence of gut receptors and immune reactivity (Blum *et al.*, 1999; Falk *et al.*, 1998).

Several bacteria in the natural gut flora or non-pathogenic bacteria which can colonise the gut have been shown to have preventive or even therapeutic effects on pathogens. Most commonly used and studied are bifidobacteria and lactobacilli (see Fig. 13.2); they have been shown to stimulate the innate immune system to produce cytokines, antimicrobial compounds and other metabolites affecting either the host and/or enteric bacteria (Aattouri *et al.*, 2001; Xuan *et al.*, 2001; Kralik *et al.*, 2004; Scharek *et al.*, 2005; Tannock, 2005; Davis *et al.*, 2006).

13.3.1 Bifidobacteria

In the last on-line edition of *The Prokaryotes* (Biavati and Mattarelli, 2001) the genus *Bifidobacterium* is associated with the *Bifidobacteriaceae* family. Within the genus *Bifidobacterium* and the relevant *Bifidobacteriaceae* family,

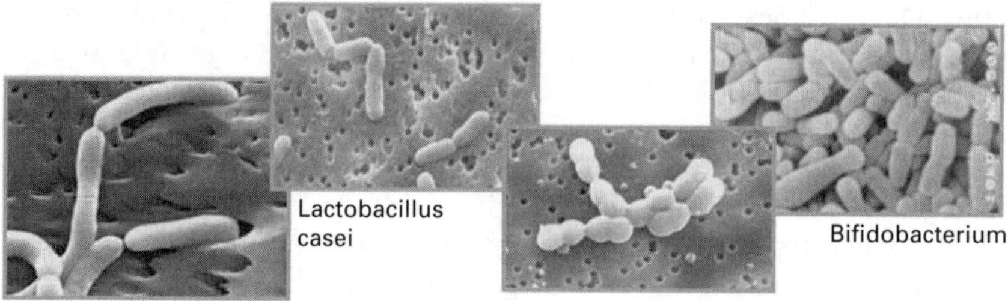

Fig. 13.2 Cellular morphology of *Lactobacillus* (*L. delbrueckii* subsp. *bulgaricus* and *L. casei*), *Streptococcus thermophilus* and *Bifidobacterium* sp.

some changes in their taxonomy are likely to be made in the near future based on results from new molecular techniques, which have been able to detect differences that were not recognised until now.

Bifidobacteria are non-motile and non-spore-forming rods of variable appearance. Freshly isolated strains may have forms ranging from uniform to branched, bifurcated Y and V forms, spatulate or club shapes. Bifidobacteria are Gram-positive, they are strictly anaerobic bacteria, but may sometimes tolerate oxygen in the presence of CO_2; there are also great differences in sensitivity to oxygen between different strains. Bifidobacteria are nutritionally a fairly heterogeneous group and some strains use ammonium salts as a source of nitrogen whereas others require organic nitrogen. Bifidobacteria are saccharolytic organisms and all strains ferment glucose, galactose and fructose. There are differences between various species and biotypes in the fermentation of other carbohydrates and alcohols. The main species currently considered or used for the development of probiotic products are: *B. animalis*, *B.suis*, *B. choerinum*, *B. pseudolongum* subsp. *globosum*, *B. pseudolongum* subsp. *pseudolongum*, *B. thermophilum* and *B. boum* (Babinska *et al.*, 2005; Lepercq *et al.*, 2005). The main species used in poultry are: *B. gallinarum*, *B. pullorum*, *B. pseudolongum* subsp. *globosum*, *B. pseudolongum* subsp. *pseudolongum* and *B. thermophylum* (Bilgili and Moran, 1990).

13.3.2 Lactobacilli

The genus *Lactobacillus* constitutes, together with the genus *Pediococcus*, the family *Lactobacillaceae* and presently comprises 80 recognised species and 15 subspecies which are distinct on the basis of the results of genotypic studies. Modern classification, mainly based upon comparative sequence analysis of 16S ribosomal ribonucleic acid (16S rRNA) and 23S rRNA sequences (partial or complete), has determined phylogenetic relationships of lactic acid bacteria. This analysis permitted allotment of the lactobacilli to the following groups: *L. casei* group, *L. plantarum* group. *L. sakei* group, *L. buchneri* group, *L. delbrueckii* group, *L. reuteri* group and *L. salivarius* group; *L. brevis*, *L. perolens*, *L. bifermentans* and *L. coryneformis* are uniquely positioned among the lactobacilli.

Traditionally, the lactic acid bacteria are defined by formation of lactic acid as a sole or main end product from carbohydrate metabolism. Lactic acid bacteria comprise a diverse group of Gram-positive, non-spore forming bacteria. They occur as cocci or rods and are generally lacking catalase, although pseudocatalase can be found in rare cases. They are chemo-organotrophic and grow only in complex media. Fermentable carbohydrates are used as an energy source. Hexoses are degraded mainly to lactate (homofermentatives) or to lactate and additional products such as acetate, ethanol, CO_2, formate or succinate (heterofermentatives).

The main species currently considered or used for the development of probiotic products in pig and poultry are: *L. reuteri*, *L. acidophilus*, *L. rhamnosus*, *L. plantarum*, *L. delbrueckii* subsp. *bulgaricus*, *L. lactis* and *L. brevis* (Massi *et al.*, 2006; Bilgili and Moran, 1990).

13.4 Probiotics for farm animals

Probiotics are defined by many authors as selected, viable, microbial dietary supplements which, when introduced in sufficient quantities, beneficially affect human organisms through their effects in the intestinal tract (Zubillaga *et al.*, 2001; Holzapfel and Schillinger, 2002). Also the United Nations Food and Agriculture Organisation and the World Health Organisation (FAO/WHO) have adopted the definition of probiotics as 'Live microorganisms which when administered in adequate amounts confer a health benefit on the host' (FAO/WHO, 2002). A large number of probiotics are currently used. Some selected strains of *Lactobacillus*, *Bifidobacterium*, *Streptococcus*, *Lactococcus* and *Saccharomyces* have been promoted in food products because of their reputed health benefits (Puupponen-Pimia *et al.*, 2002).

Many claims about the potential protective activities of probiotics in animals (including humans) have been made, but it is not always possible to provide good scientific evidence to support them. Figure 13.3 shows the benefits that can arise from applications of the probiotic concept. Probiotics were shown to be involved in protection against a variety of pathogens including *Escherichia coli*, *Salmonella*, *Campylobacter* and *Clostridium* (Choct, 2001; Tannock, 2005) and the probiotic approach may therefore provide an effective alternative to antibiotics in the prevention (and/or therapy) of gastrointestinal infections in organic monogastric and other animal production systems. In humans, lactic acid bacteria were also shown to have other health benefits such as reduction in large bowel (colon) carcinogens and mutagens, increased lactose digestion, antitumour properties, relief from constipation, stimulation of immunocompetent cells and enhancement of phagocytosis.

While the efficacy of probiotics was tested in a wide variety of animals including pets, horses and farm animals, the main potential application will

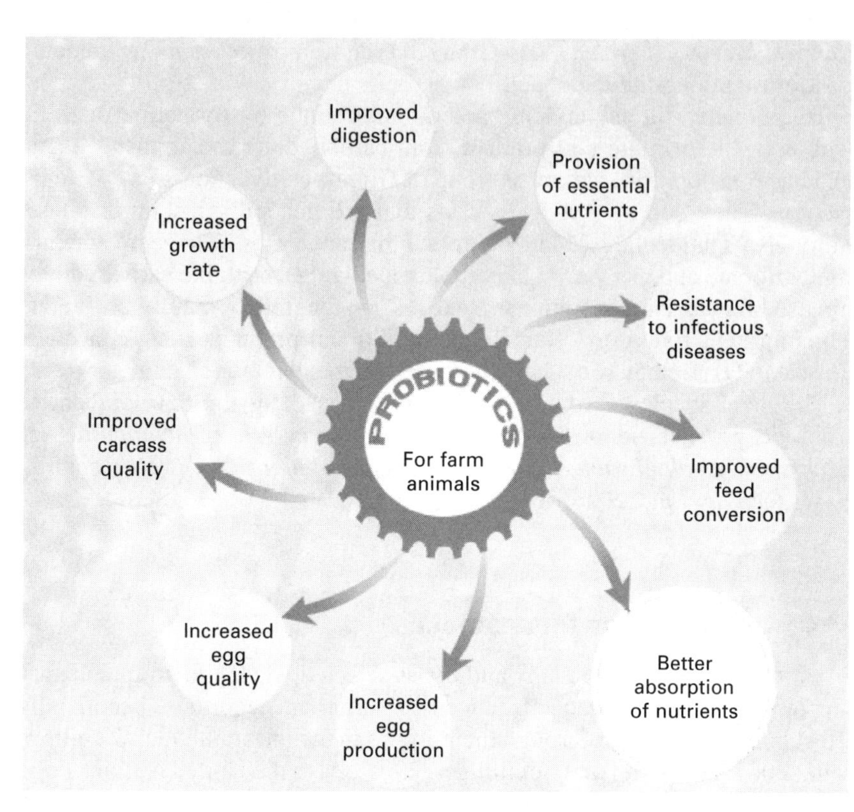

Fig. 13.3 Definition of the term 'probiotic'.

be for the prevention of diarrhoea in chickens, pigs and calves. It is important to note that the composition of naturally occurring bifidobacteria and lactobacilli were shown to differ between animals, diet and the age of animals. The microbial groups of species used for probiotics may therefore need to be adapted to individual animal species, diets and growth stages (Garcia-Rodenas *et al.*, 2006). Also, young pre- and post-weaning pigs may have different requirements and the newly hatched chick may respond more readily than the older chicken where the microbiota has become more stable and is therefore more difficult to influence.

The modes of action related to probiotic bacteria include (i) production of antibacterial substances (e.g. organic acids, bacteriocins, hydrogen peroxide, diacetyl, acetaldehyde, lactoperoxidase system, lactones and other unidentified substances), (ii) competition for nutrients, (iii) competition for adhesion receptors on the gut epithelium, (iv) reconstruction of normal intestinal microbiota after antibiotic therapy and (v) stimulation of immune function (Zubillaga *et al.*, 2001; Holzapfel and Schillinger 2002). Orally administered lactobacilli can improve immune status by increasing circulation and local antibody levels, gamma interferon concentration, macrophage activity and the number of natural killer cells (Fuller and Perdigon, 2000). Three important modes of action underlying the prevention of gastrointestinal diseases by probiotics are described in detail in separate sections below.

13.4.1 Competitive exclusion

Following administration, the probiotic, together with the resident microbiota, is thought to exist in a symbiotic relationship with the host and receives a rich and continuous nutrient supply, complementing this by improving nutrient bioavailability and augmenting disease resistance mechanisms of the host. The protection afforded by the probiotic/indigenous flora is also thought to be related to competitive exclusion and interference with attachment to mucosal surfaces (Tannock, 2005; Jin *et al.*,1996; Nisbet, 2002). Many pathogenic bacteria specifically adhere to complex oligosaccharides associated with proteins and/or lipids of intestinal membranes and secreted mucin glycoconjugates. The inter- and intra-species diversity of intestinal glycosylation patterns/carbohydrate epitopes is well established (Kelly and King, 1991; King, 1995). Furthermore, it is possible that the ability of lactic acid bacteria to compete with pathogens for adhesion to the intestinal wall is influenced by their membrane fluidity: in fact, the type and quantities of polyunsaturated fatty acids in the extracellular milieu influence the adhesive properties of lactic acid bacteria to the epithelium (Kankaanpaa *et al.*, 2001).

A *Lactobacillus* strain was recently shown to inhibit competitively adhesion of enteropathogenic *E. coli* to pig ileum and interfered with bacterial attachment to the mucosal layer of ileal conducts (Blomberg *et al.*, 1993). Although *L. acidophilus* inhibits the adhesion of several enteric pathogens to human intestinal cells in culture, when pathogen attachment preceded *L. acidophilus* treatment, no inhibitory interference occurred indicating that steric hindrance of site occupation is important in the inhibition of adhesion. Thus, therapeutic use is likely to be limited to preventive application of probiotics.

13.4.2 Stimulation of host immune responses

The main site of the mucosal immune system in the gut is referred to as gut-associated lymphoid tissue (GALT), which can be divided into inductive and effector sites. In the small intestine, the inductive sites are in the Peyer's patches, which consist of large lymphoid follicles in the terminal small intestine. The contact with external antibodies triggers a series of cascade events in the body based on immune response (Brandtzaeg *et al.*, 1999).

Since the gastrointestinal tract is the largest interface between our body and the outer world, the GALT is involved in a large number of events related to the identification of incoming antigens and the elaboration of immune response (Ouwehand *et al.*, 2004). Mucous membranes act as a defensive barrier aimed at preventing pathogenic bacteria and other agents from entering the tissues. Within such a complex system, the Peyer's patches play a primary role, being the site where the immune system and bacterial antigens come into direct contact, with lymphocytes spreading throughout the intestinal walls in order to make up the immune barrier (Tlaskalova-Hogenova *et al.*, 1995; Nagura and Sumi, 1988). From the intestinal walls the antibodies reach the systemic immune system and spread all over the body and into the other exocrine glands (Ouwehand *et al.*, 2004).

The best defined effector component of the mucosal adaptive immune system is secretory immunoglobulin A (sIgA). sIgA is the main immunoglobulin of the humoral immune response, which together with the innate mucosal defences provides protection against microbial antigens at the intestinal mucosal surface (Brandtzaeg *et al.*, 1999). It inhibits the colonisation of pathogenic bacteria in the gut, as well as the mucosal penetration of pathogenic antigens. At least 80% of all the body's plasma cells, the source of sIgA, are located in the intestinal lamina propria throughout the length of the small intestine. The production of intestinal sIgA requires the presence of commensal microbiota (Macpherson *et al.*, 2005), which indicates that the production of intestinal sIgA is induced in response to antigenic stimulation.

13.4.3 Production of antimicrobials

Many lactobacilli and bifidobacteria species are able to excrete natural antibiotics, which can have a broad spectrum of activity (e.g. lactocins, helveticins, lactacins, curvacins, nisin or bifidocin). In particular, for the bifidobacteria, some species are able to exert antimicrobial effects on various gram-positive and gram-negative intestinal pathogens including genera *Salmonella*, *Campylobacter* and *Escherichia* (Gibson and Wang, 1994).

13.4.4 Probiotics: experimental evidence in pigs and poultry

Several studies demonstrated that administering lactic acid bacteria may increase the population density of both bifidobacteria and lactobacilli and this is thought to provide protection against enteric pathogens in young pigs (Xuan *et al.*, 2001). Also, Deprez *et al.* (1986) demonstrated that administration of streptococci and *Enterococcus faecium* strain 68 to piglets reduced faecal *E. coli* and haemolytic *E. coli* levels. Several other reports demonstrated that administration of probiotics to newly weaned pigs increases growth rates (Kyriakis *et al.*, 1999, Collinder *et al.*, 2000, Neupert, 1988), feed use efficiency (Fialho *et al.*, 1998; Kyriakis *et al.*, 1999), feed intake (Lessard and Brisson, 1987), feed conversion ratio (FCR) (Neupert, 1988), nitrogen retention and biological value (Fialho *et al.*, 1998) and nitrogen and fibre digestibility (Hale and Newton, 1979). Kyriakis *et al.* (1999) noted a reduction in diarrhoea and mortality in weanlings following administration of *Bacillus toyoi* treatments.

More recently, work of Scharek *et al.* (2005) has shown that supplementing the feed of pregnant sows and piglets with a probiotic strain of *E. faecium* produces an immune-stimulatory effect. Also, Davis *et al.* (2006) described that application of *L. brevis* to the diet of newly weaned piglets decreases the expression of several genes within the jejunum involved in the toll-like receptor pathway. This clearly indicated an influence of probiotic application on the developing innate immune system of the porcine gastrointestinal tract. They also described how the use of a probiotic based on both *L. brevis* and *Bacillus,* direct-fed after weaning, resulted in a decrease in the proportion of phagocytic monocytes in the gastrointestinal tract, similar to that observed following antibiotic (Carbadox) supplementation. The change in gastrointestinal microbial populations following dietary probiotic supplementation also appears to elicit anti-inflammatory effects and immune characteristics similar to those observed with antibiotic administration.

The effect of probiotic use in chicken production has also been studied by several groups. For example, Kralik *et al.* (2004) have studied the effect of adding probiotics to the drinking water of avian broilers. They reported an increase of live weights of broilers, an improvement in feed conversion ratios and lower population densities of the pathogenic bacteria *E. coli* and *Staphylococcus aureus* in the intestinal contents.

Kalavathy *et al.* (2003) studied the effect of a mixture of 12 *Lactobacillus* strains on the growth rate, abdominal fat deposition, serum lipids and weight of organs of broiler chickens between the 1st and 42nd day of age. The *Lactobacillus* supplementation of broiler diets increased the body weight gain, improved the feed conversion ratio and was effective in reducing abdominal fat deposition. Results also showed a reduction in total cholesterol, low density lipoprotein (LDL), cholesterol and triglycerides levels in the serum of broilers between days 21 and 42.

Jin *et al.* (1996), studied the effect of different *Lactobacillus* preparations on the growth rate of chickens. They isolated *Lactobacillus* strains from jejunum, ileum and caecum tissue fragments and tested the ability of strains to adhere to chicken ileal epithelial cells in an *in vitro* assay. A single strain of *L. acidophilus* (I 26), which showed the greatest capacity to adhere to cells, and a mixture of 12 strains showing higher adhesion ability were tested for their probiotic activity in two separate experiments. Both probiotic preparations were processed into freeze-dried cultures and subsequently mixed into the diet. In the first study, Jin *et al.* (1998) investigated the effects of the two probiotic preparations on growth, organ weight, intestinal microflora and intestinal volatile fatty acids (VFA) in broilers. In the second study, Jin *et al.* (2000) looked at the impact of the two probiotic treatments on growth characteristics and the levels of digestive and bacterial enzyme activities in broilers. The probiotic treatments in both studies significantly increased body weights. The multistrain preparation, but not the monostrain probiotic, also tended to reduce mortality. The first study showed that probiotic treatments resulted in a significantly lower pH in the caecum associated with an increased concentration of total VFA in ileal and caecal contents. They also observed a decrease in coliform numbers in the caecum after 10 and 20 days and suggested that this may have been a consequence of the higher VFA level. In the second study it was found that supplementation with the probiotics significantly increased amylolytic activity in the small intestine. Overall, the two experiments indicate that growth performance of the chickens can be improved by both multi- and monostrain probiotics, but that multistrain probiotics may be more effective than the monostrain probiotics.

13.5 Prebiotics for farm animals

The term 'prebiotic' was introduced by Gibson and Roberfroid (1995) who defined prebiotics as 'a non-digestible food ingredient that beneficially affects the host by selectively stimulating the growth and/or activity of one or a limited number of bacteria in the colon'.

Prebiotics may be considered as functional food ingredients. They are attracting considerable interest from pet food manufacturers, livestock producers and feed manufacturers. The most common forms of prebiotics are non-digestible oligosaccharides (NDO), including inulin, oligofructose mannanoligosaccharides, gluco-oligosaccharides and galacto-oligosaccharides. These NDO are non-digestible by enzymes present in the mammalian small intestine, but are fermented by bacteria present in the intestines of monogastric animals (Choct, 2001; Tannock, 2005). Consumption of prebiotic oligosaccharides elicits several purported health benefits. In the pig and poultry, prebiotics were shown to control gastrointestinal diseases, reduce faecal odour and enhance growth performance, but their efficacy was usually lower than that of antibiotic treatments (Choct, 2001).

The use of prebiotics in pig and poultry production is not described in detail here, because an excellent recent textbook on prebiotics is available (Tannock, 2005). See also Section 13.6 on synbiotics below.

13.6 Synbiotics

Another possibility in microbiota management procedures is the use of synbiotics, in which probiotics and prebiotics are used in combination (Gibson and Roberfroid, 1995). The live microbial additions (probiotics) may be used in conjunction with specific substrates (prebiotics) for growth. This combination could improve the survival of the probiotic organism, because a specific nutrient substrate (the prebiotic) is readily available and results in selective advantages to the probiotic strain. Because the word alludes to synergism, this term should be reserved for products in which the prebiotic compound selectively favours the probiotic strain(s) added. In this strict sense, for example, the combination of a fructose-containing oligosaccharide with *Bifidobacterium* strain is a potentially effective synbiotic, as is the use of lactilol or lactulose in conjuction with lactobacilli (Fuller and Perdigon, 2000).

The results of many studies (Perin *et al.*, 2000; Schrezenmeir and de Vrese, 2001; Shin *et al.*, 2000; Gmeiner *et al.*, 2000) point to a synergistic effect of probiotic and prebiotic combinations on the faecal microbiota of experimental animals. This effect was demonstrated by increased total anaerobes, aerobes, lactobacilli and bifidobacteria counts as well as by decreased clostridia, *Enterobacteriaceae* and *E. coli* counts. The combination of probiotics and non-digestible carbohydrates may be a way of stabilisation and/or improvement of the probiotic effect.

The idea that inulin-type fructans are fermented by bacteria colonising the large bowel is supported by many *in vitro* (both analytic and microbiological) and *in vivo* studies, which, in addition, confirm the production of lactic and short-chain carboxylic acids as end products of the fermentation (Tanner, 2005). Furthermore, it was shown in human *in vivo* studies that this fermentation leads to the selective stimulation of growth of the bifidobacteria population, making inulin-type fructans the prototypes of prebiotics (Roberfroid, 1997; Roberfroid, 2001).

An indirect consequence of the stimulation of bifidobacterial growth is faecal bulking, for which strong evidence was published previously. A second effect worth considering when discussing potential functional effects of prebiotics such as inulin-type fructans, is the increased bioavailability of minerals (Scholz-Ahrens *et al.*, 2001). Experimental data support the hypothesis that one prebiotic, oligofructose, inhibits hepatic lipogenesis in rats and consequently induces a significant hypotriglyceridemic effect (Delzenne and Kok, 2001). The potential mechanisms of this effect include metabolic or genetic effects of short-chain carboxylic acids, low glycaemia and insulinaemia, or both. Thus, there is preliminary evidence of a hypotriglyceridemic effect of inulin-type fructans.

Among the large number of bacterial species present in the colon, three groups can be distinguished. The first one is the beneficial group, consisting of bifidobacteria, lactobacilli and other genera of the lactic acid bacteria. These bacteria are thought to be health improving. The second group comprises Enterobacteriaceae and species from the genus *Clostridium*, which are both considered negative for general health. The third group comprises all the other bacteria and are considered neutral.

Although any food ingredient that enters the large intestine is a candidate prebiotic, it is the selectivity of the fermentation in the mixed culture environment that is critical. At present, most searches for prebiotics are directed toward the growth of lactic acid-producing microorganisms. This is due to their purported health-promoting properties. As important as the impact of prebiotics on selective proliferation of beneficial microbial populations is their influence on the metabolic activity of the microbiota. Prebiotics may stimulate autochthonous bacteria not only to grow, but also to produce compounds beneficial to the host.

13.7 Acid activated antimicrobials (AAA)

Apart from probiotics, a range of acid-activated chemicals (in particular nitrite, nitrate and isothiocyanate), which naturally occur in foods and feeds, were recently shown to increase the antimicrobial activity of the stomach acid and thereby the resistance of monogastric animals to bacterial pathogens (Duncan *et al.*, 1995, 1997; Dykhuizen *et al.*, 1996, 1998; McKnight *et al.*, 1999). Acidification of ingested food in the stomach was until recently believed to be an effective primary resistance mechanism preventing foodborne pathogens from entering the more distal intestine. However, recent studies revealed that two-hour exposure (the period of time food remains in the stomach) to pH values of between 2 and 4 (the pH values commonly found in the stomach) did not have a bacteriocidal effect on pathogens such as *Campylobacter*, *Salmonella*, *Shigella* and *E. coli* (Dykhuizen *et al.*, 1996, 1998). Sufficient numbers of the pathogens would therefore be expected to survive the acid treatment in the stomach and regrow and cause disease when reaching the neutral environment of the more distal intestine.

Addition of nitrite (at concentrations commonly found in the saliva) to hydrochloric acid solutions with pH values of between 2 and 4, increased its antimicrobial activity by up to 100-fold against a range of gastrointestinal pathogens including *Salmonella enteriditis*, *Salmonella typhimurium*, *Yersinia enterocolitica*, *Shigella sonnei*, and *E. coli* 0157 (Dykhuizen *et al.*, 1996, 1998). However, food pathogens differed in sensitivity to acidified nitrite (*Y. enterocolitica* > *S. enteriditis* > *S. typhimurium* = *S. sonnei* > *E. coli* 0157). Most recently it was demonstrated that *E. coli* O157 strains which express the *Shigella*-like toxin 1 ($ST1^+$) are more resistant to acidified nitrite than $ST1^-$ *E. coli* O157 strains (Fraser, unpublished data). The relative resistance of *E. coli* O157 to acidified nitrite may explain the increase in O157 epidemics (many of which were associated with contaminated meat/meat products) since the early 1970s when concern about dietary nitrate intake resulted in a reduction or removal of nitrite and nitrate preservatives in fresh and processed meat products.

Preliminary *in vivo* experiments have confirmed *in vitro* results. In these studies, bacterial suspensions of *Salmonella typhi* live attenuated oral vaccine strain Ty21 were enclosed in a short segment of dialysis tubing, impermeable to high molecular weight molecules such as immunoglobulins and interferon, and capsules then swallowed into the stomach on a length of soft dental floss by healthy volunteers. Survival rates of bacteria exposed to the stomach environment of volunteers who had been administered a drink containing the amount of nitrate commonly found in half a lettuce were significantly lower than those of bacteria which were exposed to the stomach of volunteers which had a nitrate-free drink (Ndebele, 1996). These results were recently confirmed when the time to achieve complete kill of *S. typhi* was shown to be significantly ($P = 0.017$) shorter with the nitrate treatment (11.8 minutes following a drink of 2 mmol potassium nitrate compared to 17 minutes in the control (2 mmol potassium chloride)).

A range of other chemicals known to be present in saliva, food or stomach secretions (e.g. ascorbic acid, glutathione, thiocyanate and iodide) were recently tested in *in vitro* assays for their impact on the antimicrobial activity of acidified nitrite solutions (Leifert *et al.*, 1998). Reducing agents (ascorbic acid, glutathione) significantly reduced the antimicrobial activity of acidified nitrite solutions. In contrast thiocyanate and iodide (two compounds which are known to be concentrated in the saliva by the same mechanism as nitrate (McKnight *et al.*, 1999)) significantly increased the antimicrobial activity.

This clearly indicates that the antimicrobial activity of the stomach acid is effected by a range of dietary compounds and may be regulated by both salivary (NO_3, thiothyanate, iodide) and stomach (e.g. ascorbic acid) secretions.

Potential beneficial effects of nitrite were first hypothesised when the metabolism and fate of dietary nitrate was studied. It was found that, when dietary nitrate is ingested, the nitrate is rapidly absorbed into the bloodstream through the intestinal mucosa, primarily in the upper small intestine and to a lesser extent in the stomach (McKnight *et al.,* 1999). Nitrate is then circulated in the blood and actively assimilated into the salivary glands by an active transport mechanism, with salivary concentrations increasing to approximately ten times those found in the plasma. Up to 25% of circulating nitrate from exogenous sources is taken up by the salivary glands (see Duncan *et al.*, 1997 for a recent review).

Nitrate is converted to nitrite in the oral cavity and it is now firmly established that this is due to the action of nitrate reductase enzymes of the oral microflora (Duncan *et al.*, 1997; Li *et al.*, 1997, 1999, 2006). The bacteria species responsible for nitrate reduction in the oral cavity of rats and pigs have been characterised (*Li et al.*, 1997). The majority of isolates from both animal species were Gram-positive coccci (>60% of isolates). In the pig *Staphylococcus sciuri* and *S. intermedius* were the main species identified, while in rats *Micrococcus* spp. dominated. Gram-negative bacteria with nitrate reduction activity (mainly *Pasteurella* spp.) were also also isolated, but these are thought not to contribute significantly to nitrite production in the oral cavity. Exposure of oral tissues to antibiotics specific for Gram-positive bacteria resulted in a >90% decrease in nitrite production in oral tissues, while the application of Gram-negative specific antibiotics did not significantly reduce oral nitrite generation (Duncan, 2000). The nitrate reductase enzymes responsible for nitrite generation in the oral cavity were shown to be respiratory nitrate reductase enzymes, which enable bacteria to switch to nitrate as a terminal respiratory electron acceptor under low oxygen conditions. The highest level of nitrite generation and density of nitrate-reducing (denitrifying) bacteria can be found on the posterior tongue of rats in deep clefts in the tongue surface where oxygen levels are thought to be low (Li *et al.*, 1997).

It is thought that bicarbonate excretions by oral tissues and the neutral pH of the saliva maintain a neutral pH in the areas inhabited by nitrite-producing bacteria, thus preventing nitrite becoming toxic to the resident bacterial flora in the oral cavity. As with denitrifying bacteria isolated from other environments (e.g. soil), nitrate reductase activity is induced by nitrate (e.g. removal of nitrate from the diet resulted in the loss of nitrate reduction capacity in tongue tissues in the rat; Duncan, 2000). However, in contrast to most denitrifying bacteria isolated from soil, most oral denitrifyers do not reduce nitrite further (to NH_4, N_2O or N_2). This indicates a very close (symbiotic?) relationship in which the host animal maintains the availability of nitrate (via the enterosalivary circulation of nitrate) and protects the bacteria against a lowering of the pH, allowing nitrite to accumulate without a negative impact on the oral microflora. The potential beneficial impact of nitrite produced by the oral microflora on the host animal is described below.

The use of AAA feed supplements has been proposed as a method to increase microbial kill in the stomach (the first barrier encountered by foodborne bacteria in the gatrointestinal system). If significant reductions in pathogen numbers could be obtained, this would provide an opportunity to develop a two-barrier strategy for the prevention of gastrointestinal infections, where AAA decrease the levels of enteric pathogens at the stomach level (barrier 1) and pre-, pro- or synbiotics provide a second barrier in the upper intestine (Duncan *et al.*, 1997). Since pro-, pre- and symbiotic treatments were often found to fail when too high pathogen loads reach the intestine (Tannock, 2005) this two-barrier approach would be expected to have a synergistic effect.

An alternative approach to AAA feed supplementation is the use of feeds which are naturally high in nitrate and/or isothiocyanate (e.g. green plant materials, *Brassica* and/or cassava). The mechanism may be exploited in organic and many 'low input' pig production systems by increasing the access to forage (which contains substantial amounts of nitrate and is already encouraged in organic farming systems is some EU countries) and by utilising oil seed rape protein or *Brassica* processing waste.

13.8 Conclusion

The US Department of Agriculture (USDA) estimates that bacteria account for 3.6–7.1 million foodborne illnesses each year. Among the leading causes are *Salmonella*, *Campylobacter* and *E. coli* O157:H7, which colonise the intestines of farm animals and subsequently contaminate meat products during processing. Clearly, to reverse this trend, control at several critical control points in the food chain needs to be improved. This includes: (i) improved prevention of disease in animals at the farm level, (ii) improved methods/systems to reduce faecal contamination of meat during processing at the slaughterhouse and (iii) improved storage, transport and food preparation standards at the processer, retailer, caterers, and also the end consumer level (see Chapter 23). The continuous use of antibiotics may contribute to a reservoir of drug-resistant bacteria and is therefore increasingly difficult to justify (Cohen, 1992; Radcliffe, 2000; Newman, 2002).

The use of 16S rRNA technology combined with the polymerase chain reaction will increase the monitoring accuracy of probiotic and prebiotic efficacy. The use of both probiotics and prebiotics or their combination (synbiotics) will probably increase dramatically worldwide because of strong commercial interests in providing these supplements to both humans and animals. In contrast to antibiotics, pre- and probiotics stimulate rather than reduce microbial activity in the gastrointestinal systems and do not carry the risk of introducing antibiotic resistance and cross resistance into the gut microflora. Pro-, pre- and synbiotics therefore currently provide the most promising treatment-based alternative to antibiotics.

The use of AAAs as feed supplements appears an attractive approach either in itself or in combination with pro-, pre- and synbiotics, but the lack of animal feeding studies in which the efficacy of this approach could be determined makes it difficult to assess to what extent a two-barrier (targeting control at both the stomach and intestinal level) approach is commercially feasible. Also, since the viability of probiotics may also be affected by the use of AAA to increase the disinfection activity of the stomach, probiotics may need to be formulated in a way that protects them during stomach transfer. However, its potential should be determined in future research.

13.9 References

Aattouri N., Bouras M., Tome D., Marcos A. and Lemonnier D. (2001). 'Oral ingestion of lactic acid bacteria by rats increases lymphocyte proliferation and interferon – production'. *Br J Nutr*, **87**, 367–373.

Babinska I., Rotkiewicz T. and Otrocka-Domagala I. (2005). 'The effect of *Lactobacillus acidophilus* and *Bifidobacterium* spp. administration on the morphology of the gastrointestinal tract, liver and pancreas in piglets'. *Pol J Vet Sci*, **8**(1), 29–35.

Best P. (1997). 'Growth promoters: European battle rages'. *Feed Int*, **11**, 8–11.

Biavati B. and Mattarelli P. (2001). 'The family *Bifidobacteriaceae*'. in (M. Dworkin, S. Falkow, E. Rosenberg, K-H. Schleifer, E. Stackebrandt, *The Prokaryotes. An Evolving Electronic Resource for Microbiological Community*, third edition (latest update release 3.7, September 2001). New York, Springer-Verlag, (2001) [on line] http:// 141.150.157.117:8080//prokPuB/index.htm.

Bilgili S.F. and Moran E.T.J. (1990). 'Influence of whey and probiotic-supplemented withdrawal feed on the retention of *Salmonella* intubated into market age broilers'. *Poult Sci*, **69**(10), 1670–1674.

Blomberg L., Henmiksson A. and Conway P.L. (1993). 'Inhibition of adhesion of *Escherichia coli* K88 to piglet ileal mucus by *Lactobacillus* spp'. *Appl Environ Microb*, **59**, 34–39.

Blum S., Alvarez S., Haller D., Perez P. and Schiffrin E.J. (1999). Intestinal micro-flora and the interaction with immunocompetent cells. *Antoine van Leeuwenhoek*, **76**, 199–205.

Brandtzaeg P., Baekkevold E.S. and Farstad I.N. (1999). 'Regional specialization in the mucosal immune system: what happens in the microcompartments?' *Immunol Today*, **20**, 141–51.

Choct M. (2001). 'Alternatives to in-feed antibiotics in monogastric animal industry'. *ATS Technical Bulletin*, **30**, 1–6.

Cohen M.L. (1992). 'Epidemiology of drug resistance: implications for the post-antimicrobial era'. *Science*, **257**, 1050.

Collignon P.J. (1999). 'Vancomycin-resistant enterococci and use of avoparcin in animal feed; is there a link?' *Med J Austral*, **21**, 144–146.

Collinder E., Berge G.N., Cardona M.E., Norin E., Stern S. and Midtvedt, T. (2000). 'Feed additives to piglets, probiotics or antibiotics'. in *Proceedings of the 16th International Pig Veterinary Society Congress*, Melbourne, Australia. 17–20 Sept. 2000, pp 257.

Davis M.E., Brownb D.C., Dirainb M.S., Dawsonc H.D., Maxwellb C. and Rehbergera T. (2006). 'Comparison of direct-fed microbial and antibiotic supplementation on innate and adaptive immune characteristics of weanling pigs'. *Reprod Nutr Dev*, **46**, S63–S77.

Delzenne N.M. and Kok N. (2001). 'Effects of fructans-type prebiotics on lipid metabolism'. *Am J Clin Nutr*, **73**(2 Suppl), 456S–458S.

Deprez P., Van D.H.C. Muylle E. and Oyaert W. (1986). 'The influence of the administration of sow's milk on the postweaning excretion of hemolytic *E. coli* in pigs'. *Vet Res Comm*, **10**, 468–478.

Duncan C. (2000). *Effect of Nitrate Reduction in the Oral Cavity on the Antimicrobial Activity in the Stomach.* PhD Thesis, University of Aberdeen.

Duncan C., Dougal H., Johnston P., Green S., Brogan R., Leifert C., Smith L., Golden M. and Benjamin, N. (1995). 'Chemical generation of nitric oxide in the mouth from the enterosalivary circulation of dietary nitrate'. *Nature Med,* **1**, 546–551.

Duncan C., Li H., Dykhuizen R., Frazer R., Johnston P., MacKnight G., Smith L., Lamza K., McKenzie H., Batt L., Kelly D., Golden M., Benjamin N. and Leifert C. (1997). 'Protection against oral and gastrointestinal diseases: importance of dietary nitrate intake, oral nitrate reduction and enterosalivary nitrate circulation'. *Comparative Biochem Physiol*, **118**, 939–948.

Dunne C., Murphy L., Flynn S., O' Mahony L., O' Halloran S., Feeney M., Morrissey D., Thornton G., Fitzgerald G., Daly C., Kiely B., Quigley E.M.M., O'Sullivan G.C., Shanahan F. and Collins J.K. *et al.*, (1999). 'Probiotics, from myth to reality. Demonstration of functionality in animal models of disease and in human clinical trials'. *Antoine Van Leeuwenhoek*, **76**, 279–92.

Dykhuizen R., Frazer R., Benjamin N., Duncan C., Smith C.C., Golden M. and Leifert C. (1996). 'Antimicrobial effect of acidified nitrite on gut pathogens: the importance of dietary nitrate in host defence'. *J Antimicrob Agents Chemothe*, **40**, 1422–1425.

Dykhuizen R., Frazer A., MacKenzie H., Golden M., Leifert C. and Benjamin N. (1998). '*Helicobacter pylori* is killed by nitrite under acidic conditions'. *Gut*, **42**, 334–337.

Falk G., Hooper L.V., Midtvedt T. and Gordon J. (1998). 'Creating and maintaining the gastrointestinal ecosystem: what we know and need to know from gnotobiology'. *Microbiol Mol Biol Rev*, **62**(4), 1157–1170.

FAO/WHO (2002). *Guidelines for the evaluation of probiotics in food.* London, Ontario, Canada, April 30 and May 1, 2002. ftp:/ftp.fao.org/es/esn/food/wgreport2.pdf).

Fialho E.T., Vassalo M., Lima J.A.F. and Bertechine A.G. (1998). 'Probiotics utilization for piglets from 10 to 30 kg (performance and metabolism assay)'. in *Proceedings Contributed Papers – Vol. 1.* The 8th World Conference on Animal Production. June 28–July 4, 1998, Seoul National University, Seoul, Korea, 622–623.

Fuller R. and Perdigon G. (eds.) (2000). *Probiotics 3. Immunomodulation by the Gut Microbiota and Probiotics*, Kluwer Academic Publishers, Dordrecht.

Garcia-Rodenas C.L., Bergonzelli G.E., Nutten S., Schumsnn A., Cherbut C., Turini M., Ornstein K., Rochat F. and Corthesy-Theulaz I. (2006). 'Nutritional approach to restore impaired intestinal barrier function and growth after neonatal stress in rats'. *J Pediatr Gastroenterol Nutr*, **43**(1), 16–24.

Gibson G.R. and Roberfroid M.B. (1995). 'Dietary modulation of the human colonic microbiota. Introducing the concept of prebiotics'. *J. Nutr*, **125**, 1401–12.

Gibson G.R. and Wang X. (1994). 'Inhibitory effects of bifidobacteria on other colonic bacteria'. *J Appl Bacteriol*, **77**, 412–420.

Gill C. and Best P. (1998). 'Antibiotic resistance in USA: scientists to look more closely'. *Feed Inte*, **8**, 16–18.

Gmeiner M., Kniefel W., Kulbe K.D., Wouters R., De Boever P., Nollet L. and Verstreate V. (2000). 'Influence of a synbiotic mixture consisting of *Lactobacillus acidophilus* 74–2 and a fructooligosaccharide preparation on the microbial ecology sustained in a simulation of the human intestinal microbial ecosystem (SHIME reactor)'. *Appl Microbiol Biotechnol*, **53**, 219–223.

Grist A. (2004). *Poultry Inspection. Anatomy, Physiology and Disease Conditions*. Nottingham University Press, Nottingham.

Gong J., Forster R.J., Yu H., Chambers J.R., Sabour P.M., Wheatcroft R. and Chen S. (2002). 'Diversity and phylogenetic analysis of bacteria in the mucosa of chicken caeca and comparison with bacteria in the caecal lumen'. *FEMS Microbiol Lett*, **208**, 1–7.

Guarner F. and Malagelada J.R. (2003). 'Gut flora in health and disease'. *The Lancet*, **360**, 8–10.

Hale O.M. and Newton G.L. (1979). 'Effects of a nonviable *Lactobacillus* species fermentation product on performance of pigs'. *J Anim Sci*, **48**(4), 770–775.

Holzapfel W.H. and Schillinger U. (2002). 'Introduction to pre- and probiotics'. *Food Rese Int*, **35**, 109–116.

Hume M.E., Byrd III J.A., Stanker L.H. and Ziprin R.L. (1998). 'Reduction of caecal *Listeria monocytogenes* in leghorn chicks following treatment with a competitive exclusion culture'. *Lett Appl Microbiol*, **26**, 432–436.

Inborr J. (2000). 'Swedish poultry production without in-feed antibiotics – a testing ground or a model for the future'. *Austral Poult Scie Sympm*, **12**, 1–9.

Jin L.Z., Ho Y.W., Ali M.A., Abdullah N., Ong K.B. and Jalaludin S. (1996). 'Adhesion of *Lactobacillus* isolates to intestinal epithelial cells of chicken'. *Lett Appl Microbiol*, **22**, 229–232.

Jin L.Z., Ho Y.W., Abdullah N., Ali M.A. and Jalaludin S. (1998). 'Effects of adherent *Lactobacillus* cultures on growth, weight of organs and intestinal microflora and volatile fatty acids in broilers'. *Anim Feed Sci Technol*, **70**, 197–209.

Jin L.Z., Ho Y.W., Abdullah N. and Jalaludin S. (2000). 'Digestive and bacterial enzyme activities in broilers fed diets supplemented with *Lactobacillus* cultures'. *Poult Sci*, **79**, 886–891.

Kalavathy R., Abdullah N., Jalaludin S. and Ho Y.W. (2003). 'Effects of *Lactobacillus* cultures on growth performance, abdominal fat deposition, serum lipids and weight of organs of broiler chickens'. *Br Poult Sci*, **44**, 139–144.

Kankaanpaa P., Salminen S.J., Isolauri E. and Lee Y.K. (2001). 'The influence of polyunsaturated fatty acids on probiotic growth and adhesion'. *FEMS Microbiol Lett*, **194**, 149–53.

Kasper H. (1998) 'Protection against gastrointestinal diseases – present facts and future developments'. *Int J Food Microbiol*, **41**, 127–31.

Kelly D. and King T.P. (1991). 'The influence of lactation products on the temporal expression of histo-blood group antigens in the intestines of suckling pigs: lectin histochemical and immunohistochemical analysis'. *Histochem J*, **23**, 55–60.

King T.P. (1995). 'Lectin cytochemistry and intestinal epithelial cell biology' in Pusztai A. and Bardocz S., *Lectins: Biomedical Perspective*, Taylor and Francis, London and Bristol, 183–210.

Kralik G., Milakovié Z. and Ivankoviæ S. (2004). 'Effect of probiotic supplementation on the performance and the composition of the intestinal microflora in broilers chickens'. *Acta Agr Kapos*, **8**(2), 23–31.

Kyriakis S.C., Tsiloyiannis V.K., Vlemmas J., Sarris K., Tsinas A.C., Alexopoulos C. and Jansegers, L. (1999). 'The effect of probiotics LSP 122 on the control of post-weaning diarrhea syndrome of piglets'. *Res Vet Sci*, **67**, 223–228.

Leifert C., Li H., Duncan C. and Golden M. (1998). *Novel Probiotic, Animal Feed*

Supplement and Crop Protection Strategies. International Patent application 97044713, Aberdeen University.
Lepercq P., Hermier D., David O., Michelin R., Gibard C., Beguet F., Relano, P., Cavuela C. and Juste C. (2005). 'Increasing ursodeoxycholic acid in the enterohepatic circulation of pigs through the administration of living bacteria'. *Br J Nutr*, **93**(4), 457–469.
Lessard M. and Brisson G.J. (1987). 'Effect of a *Lactobacillus* fermentation product on growt immune response and fecal enzyme activity in weaned pigs'. *Can J Anim Sci*, **67**, 509–516.
Li H., Duncan C., Townend J., Smith L., Kilham K., Johnston P., Dykhuizen R., Kelly D., Golden M., Benjamin N. and Leifert C. (1997). 'Nitrate reducing bacteria on rat tongues'. *Appl Environ Microbiol*, **63**, 924–930.
Li H., Duncan C., Golden M. and Leifert C. (1999). 'Identification of nitrate reducing bacteria from the oral cavity of rats and pigs', in Wilson W.S., Ball A.S. and Hinton R.H., *Managing Risks of Nitrates to Humans and the Environment*, Royal Society of Chemistry, Cambridge, 259–268.
Li H., Thompson I., Carter P., Whiteley A., Bailey M., Leifert C. and Killham K. (2007). 'Salivary nitrate; an ecological factor in reducing oral acidity'. *Oral Microbiol Immunol*, **22**, 67–71.
MacClane B.A., Uzal F.A., Fernandez Miyakawa M.E., Lyerly D., Wilkins T. (2005). 'The Enterotoxic clostridia', in *The Prokaryotes* 3rd Edition. On-line.http://141.150.157.117.8080//prokPUB/index.htm.
Macpherson A.J., Geuking M.B. and McCoy K.D. (2005). 'Immune responses that adapt the intestinal mucosa to commensal intestinal bacteria'. *Immunology*, **115**, 153–163.
Marcato P.S., Zaghini L., Bettini G. and Della Salda L. (1998). *Patologia suina: testo e atlante*. Ed agricole, Bologna.
Massi M., Ioan P., Budriesi R., Chiarini A., Vitali B., Lammers K.M., Gionchetti P., Campieri M., Lembo A. and Brigidi P. (2006). 'Effects of probiotic bacteria on gastrointestinal motility in guinea-pig isolated tissue'. *World J Gastroenterol*, **7**, 12(37), 5987–5994.
McKnight G.M., Duncan C.W., Leifert C. and Golden M.H.N. (1999). 'Dietary nitrate in man – friend or foe?' *Br J Nutr*, **81**, 349–358.
Mitsuoka T. (1992). *The Lactic Acid Bacteria in Health and Disease*, B.J.B. Wood (ed.). Elsevier Applied Sciences, London, 69–114.
Nagura H. and Sumi Y. (1988). 'Immunological functions of the gut-role of the mucosal immune system'. *Toxicol Pathol*, **16**(2), 154–164.
Naidu A.S., Bidlack W.R. and Clemens R.A. (1999). 'Probiotic spectra of lactic acid bacteria (LAB)', *Criti Rev Food Sci Nutr*, **39**(1), 13–126.
Ndebele N. (1996). *In vivo Activity of Acidified Nitrite Against Enteric Pathogens*. MSc Thesis, University of Aberdeen.
Neupert B. (1988). 'Milchsäurebakterien CYCLACTIN in Verbinddung mit AVOTAN erhöhen den Deckungsbeitrag in der Schweinemast'. *Kraftfutter*, **4**, 126.
Newman K. (1994). 'Mannan-oligosaccharides: Natural polymers with significant impact on the gastrointestinal microflora and immune system', in Lyon, T.P. *Biotechnology in the Feed Industry*, Nottingham University Press, Nicolasville, Kentucky, 167–180.
Newman M.G. (2002). 'Antibiotic resistance is a reality: novel techniques for overcoming antibiotic resistance when using new growth promoters'. *Nutritional Biotechnology in the Feed and Food Industries*. Nottingham University Press, Nottingham.
Nisbet D. (2002). 'Defined competitive exclusion cultures in the prevention of enteropathogen colonisation in poultry and swine'. *Antonie van Leeuwenhoek*, **81**, 481–486.
Ouwehand A., Isolauri E. and Salminen S. (2004). 'The role of the intestinal microflora for the development of the immune system in early childhood'. *Europ J Nutr*, **41**, 3456–3464.
Perin S., Grill J.P. and Schneider F. (2000). 'Effects of fructooligosaccharides and their

monomeric components on bile salt resistance in three species of bifidobacteria', *J Appl Microbiol*, **88**, 968–974.

Pluske J.R., Siba P.M., Pethick D.W., Durmic Z., Mullan B.P. and Hampson D.J. (1996). 'The incidence of swine dysentery in pigs can be reduced by feeding diets that limit the amount of fermentable substrate entering the large intestine'. *J Nutr*, **126**, 2920–2933.

Puupponen-Pimia R., Aura A-M., Oksman-Caldentey K-M., Myllaerinen P., Saarela M., Mattila-Sandholm T. and Poutanen K. (2002). 'Development of functional ingredients for gut heath'. *Trends Food Sci Technol*, **13**, 3–11.

Radcliffe, J. (2000). 'British supermarkets forging changes in poultry nutrition'. *Austral Poult Sci Symp*, **12**, 25–31.

Roberfroid M.B. (1997). 'Health benefits of non-digestible oligosaccharides'. *Adv Exp Med Biol*, **427**, 211–9.

Roberfroid M.B. (2001). 'Prebiotics: preferential substrates for specific germs?' *Am J Clin Nutr*, **73**, 406–409.

Schantz L.D., Laber-Laird K., Bingel S. and Swindle M. (1996). 'Pigs: applied anatomy of the gastrointestinal tract', in Jensen S.L., Gregersen H., Moody F. and Shokouh-Amiri M.H. *Essentials of Experimental Surgery: Gastroenterology*, Harwood Academic Publishers NY, 2611–2619.

Scharek L., Guth J., Reiter K., Weyrauch K.D., Taras D., Schwerk P., Schierack P., Schmidt M.F.G., Wieler L.H. and Tedin K. (2005). 'Influence of a probiotic *Enterococcus faecium* strain on development of the immune system of sows and piglets'. *Vet Immunol Immunopathol*, **105**, 151–161.

Scholz-Ahrens K.E., Schaafsma G., van den Heuvel EGHM, Schrezenmeir J., (2001). 'Effects of prebiotics on mineral metabolism'. *Am J Clin Nut*, **73**, 459–464.

Schrezenmeir J. and de Vrese M. (2001). 'Probiotics, prebiotics, and synbiotics – approaching a definition'. *Am J Clini Nutr*, **73**, 361–364.

Shin H.S., Lee J.H., Pestka J.J., Ustunol Z. (2000). 'Growth and viability of commercial *Bifidobacterium* spp. in skim milk containing oligosaccharides and inulin'. *J Food Sci*, **65**, 884–887.

Sinlac M. and Choct M. (2000). 'Xylanase supplementation affects the caecal microflora of broilers'. *Austral Poult Sci Symp*, **12**, 209.

Swindle M.M. (1998). *Surgery, Anesthesia and Experimental Techniques*, Iowa State University Press, Swine, Ames, IA.

Tannock G.W. (1992). *The Lactic Acid Bacteria in Health and Disease, Vol 1*, Wood. B.J.B. (ed.), Elsevier Applied Sciences, London, 21–48.

Tannock G.W. (2005). *Probiotics and Prebiotics: Scientific Aspects.* Horizon Scientific Press, Caister.

Tlaskalova-Hogenova H., Farre-Castany M.A., Stepankova R., Kozakova H., Tuckova L., Funda D.P., Barot R., Cukrowska B., Sinkora J. and Mandel L. (1995). 'The gut as a lymphoepithelial organ: the role of the intestinal epithelial cells in mucosal immunity'. *Folia Microbiol (Praha)*, **40**(4), 385–91.

Xuan Z.N., Kim J.D., Heo K.N., Jung H.J., Lee J.H., Han Y.K., Kim Y.Y. and Han I.K. (2001). 'Study on the development of a probiotics complex for weaned pigs'. *Asian-Aust J Anim Sci*, **14**, 1425–1428.

Zubillaga M., Weil R., Postaire E., Goldman C., Caro R. and Boccio J. (2001). 'Effect of probiotics and functional foods and their use in different diseases'. *Nutr Res*, **21**, 569–579.

Part III

Organic crop foods

14

Dietary exposure to pesticides from organic and conventional food production systems

C. Benbrook, The Organic Center, USA

14.1 Introduction

Current interest in the United States in reducing dietary pesticide exposure through consumption of organic food arises from two milestones in the 1990s. In 1993, the National Academy of Sciences released the report *Pesticides in the Diets of Infants and Children* (National Research Council, 1993). The NRC report gave voice to a new consensus in the biomedical and regulatory communities regarding the primary focus of pesticide risk assessments and regulation. After assessing the science base supporting pesticide regulation, the committee that authored the NRC report reached several important conclusions:

- Pregnant women, infants and children face unique and possibly significant developmental and endocrine-system mediated risks from low-level pesticide exposure during critical windows of development, some with serious life-long consequences.
- Infants and children consume more food per kilogram of body weight than adults, and a less varied diet, increasing risks when pesticides are present in a food consumed by children.
- Children are less able to detoxify many chemicals and rapidly developing organ systems are highly vulnerable during critical stages of development.
- Government risk assessment methods used to determine acceptable residues in foods (governed by tolerances) were not designed to detect or quantify the majority of these unique risks.
- The then-current pesticide exposure data and risk assessment models failed to reflect real world exposures and risks. Deficiencies can only be rectified by carrying out cumulative risk assessments (CRA) across all

routes of exposure, encompassing residues of all pesticides that work through a common mechanism of action (e.g. the organophosphate (OP) insecticides, all of which are cholinesterase inhibitors).

The second milestone event in the USA in the 1990s was passage of the Food Quality Protection Act (FQPA) in 1996. The goal of the FQPA was to assure a 'reasonable certainty of no harm' as a result of exposure to pesticides for all US population groups. The FQPA incorporated into federal law the major recommendations of the 1993 National Academy of Sciences (NAS)/NRC report, as well as the recommendations of a 1987 NAS/NRC report entitled *Regulating Pesticides in Food: The Delaney Paradox* (National Research Council, 1987).

There were four major changes made by the FQPA in how the Environmental Protection Agency (EPA) evaluates pesticide dietary risks and makes tolerance decisions:

- Assure that pesticide tolerances are safe for vulnerable populations, in particular, infants, children and the elderly, based on a 'reasonable certainty of no harm' health-based standard, instead of the previous cost–benefit balancing standard;
- Aggregate exposure to a pesticide from all dietary sources: drinking water, residential and other routes must be taken into account;
- An added ten-fold safety factor shall be added in setting pesticide reference doses (RfDs) (i.e. acceptable daily intakes) to account for the unique risks faced by infants and children, unless the EPA administrator has solid data supporting a determination that existing RfDs are fully health protective, even for infants and that exposures are fully and accurately characterized; and
- For pesticides that pose human risks through a common biological mode of action (like the OP insecticides), aggregate exposures to all such pesticides must be evaluated together in determining whether a given tolerance is safe.

The EPA was also directed to review the 9721 tolerances on the books in 1996 to assure that they were in compliance with the FQPA's new safety standard. The agency was responsible for reviewing the riskiest one-third of pesticide tolerances within three years of passage (i.e. by summer 1999). Two-thirds of existing tolerances were to be reviewed and brought into compliance with the new statute six years after passage (summer 2002). Within 10 years, all tolerances were to be reviewed and adjusted as needed, or by August 2006.

As a result of these two milestone events in the 1990s, significant progress has been made in refining the accuracy of pesticide risk assessments (Consumers Union, 2001). Bigger and better pesticide exposure databases are now available. Government-sponsored research on the developmental impact of pesticides has deepened understanding of both the nature of risks stemming from pesticide exposures and the levels and distribution of those risks across exposure pathways, foods and types of pesticides. Progress in these areas has heightened consumer concern about pesticide exposure and motivated many people to seek out organic food as one step to lessen potential adverse health effects triggered by pesticide exposure.

14.2 Dietary exposure data sources

The US Congress funded the Pesticide Data Program (PDP) in order to improve the accuracy of pesticide dietary risk assessments carried out by the US EPA. The USDA's Agricultural Marketing Service carries out the program. As recommended in the 1993 NAS/NRC report, the PDP focuses on the foods consumed most heavily by children and food is tested, to the extent possible, 'as eaten' (Agricultural Market Service, 2004; National Research Council, 1987). For example, banana and orange samples are tested without the peel; processed foods are tested as they come out of a can, jar or freezer bag.

Since its inception, the PDP has tested nearly 200,000 samples of the 20-odd foods consumed most frequently by children: milk, apples, apple juice, pears, peaches, grapes, oranges, bananas, peas, green beans, carrots, tomatoes and strawberries have been in and out of the program two or three times. Less commonly consumed foods like nectarines and spinach have also been tested. In general, the more residues found in one round of PDP testing for a given food, the more likely that food will be added again to the program. About one-quarter of the samples in a given year are processed foods and juices.

Some 300 to 800 samples are tested of each fresh or processed food, although as few as 120 samples of some foods have been run. The sample design strives to reflect the composition of the food supply in terms of the geographic origin of food; the number of domestic versus imported samples is roughly proportional to their respective share of annual consumption. The USDA also records information on any market claims made on a given sample of food. Possible claims include 'organic', 'IPM grown', 'no detectable residues' or 'pesticide-free'. Foods representing each market claim are supposed to be sampled roughly in proportion to their occurrence in retail market channels. As a result, PDP results allow comparisons to be made of the frequency and levels of pesticide residues in domestic versus imported foods, across food groups, as well as by market claim.

14.2.1 Overview of differences in residues by market claim in the 1990s

A detailed overview of pesticide residue patterns in conventional, IPM grown and organic food has been published that draws on three datasets collected in the 1990s (Baker *et al.*, 2002). Six years of data from the USDA's PDP were analyzed (1993–1999), along with 10 years of California Department of Pesticide Regulation (DPR) data (1989–1998) and the results of a 1998 Consumers Union testing project that focused on four crops (apples, peaches, tomatoes, and peppers).

Some major food groups – most oils, dairy, meat and poultry products – contain few detectable pesticides and contribute modestly at the national level to dietary exposure and risk from contemporary pesticides. Grain products contain few pesticides other than insecticides used during storage. In a special survey of wheat flour in 2004, the PDP tested 725 samples and found two post-harvest storage insecticides in a significant share of samples: malathion (49.4% positive) and chlorpyrifos-methyl (20.8% positive) (Agricultural Market Service, 2006). Seven other pesticides were found in just one sample each, three were detected in 2–5 samples and four were detected in 10–21 samples. A special sampling of rice in 2000 also detected two post-harvest storage insecticides in 17% and 24% of samples, and just a few other samples had residues of different insecticides and herbicides (Agricultural Marketing Service, 2002).

Residues of long-banned organochlorine insecticides like DDT (dichlorodiphenyl trichloroethane) and dieldrin remain common in animal products, but are generally regarded by regulators to be below safety thresholds. Contemporary use pesticides are rarely detected in animal products. A special survey tested 480 samples of poultry adipose, liver, and muscle tissues. Other than low-levels of organochlorine residues (p,p′ DDE (dichlorodiphenyldichloroethylene), dieldrin), 11 samples were found to contain one of six pesticides (Agricultural Market Service, 2002). A 2001 special survey of beef detected only two pesticides (diazinon, endosulfan sulfate) in a handful of samples, other than organochlorine residues (Agricultural Marketing Service, 2003).

Baker *et al.* (2002) reported that nearly three-quarters of the fresh fruits and vegetables (F&V) consumed most frequently by children in the USA contain residues. In general, soft-skinned fruit and vegetables tend to contain residues more frequently than foods with thicker skins, shells or peels. Baker *et al.* present consistent and highly significant data from three sources that show that the pattern of residues found in organic foods differs markedly from the pattern in conventional samples.

In the case of foods tested by USDA's PDP, conventional fruits were 3.6 times more likely to contain residues than organic fruit samples. Conventional vegetables are 6.8 times more likely to have one or more detectable residue. Data from California's DPR show that conventional food is more than five times more likely to contain residues than organic samples and Consumers Union (CU) testing of four foods found residues in conventional foods three times more often (Baker *et al.*, 2002). Great Britain's pesticide sampling program found residues in conventional food 7.5 times more frequently than in organic samples of the same foods in 2001 testing (Pesticide Residue Committee, 2001).

14.2.2 Multiple residues

The frequency of multiple residues in the three datasets analyzed by Baker *et al.* is shown in Table 14.1. The PDP found that about 45% of conventional fruit and vegetable samples tested from 1994–1999 contained residues of two or more pesticides, while 7.1% of organic samples had multiple residues. The average conventional apple tested in this period by PDP contained residues of three different pesticides. In CU testing 62% of conventional samples contained multiple residues, compared to 6% of organic samples.

Remarkably, the PDP tested 530 apple samples in 1996 and found that the odds of buying a bag of apples with nine or more pesticide residues was as great as selecting a bag with no residues. In 2003 PDP testing, 744 samples of apples were analyzed. The odds were slightly higher (2.5%) that a consumer would be exposed to seven or more residues in an apple than no residues (2.3%).

Detailed information on multiple residues in different foods is reported each year in an appendix table in the PDP summary report entitled 'Number of pesticides detected per sample'. For example, Appendix K in the 2004 PDP report reports that almost 11% of the 12,446 samples tested contained four or more residues, while over 12% of the sweet bell peppers tested contained seven or more residues.

14.2.3 Multiple exposures occur daily

Few people are aware of how frequently infants and children are exposed to pesticides via their diet and drinking water. According to USDA food consumption surveys, the average American consumes about 3.6 servings of fresh and processed fruits and vegetables per day, of which about two are fresh fruits and vegetables. There are about 75 million Americans under the age of 20. About 70% of the samples of fresh fruits and vegetables most commonly consumed in America contain one or more pesticide residues (Agricultural Market Service, 2002; Baker *et al.*, 2002). Since the average piece of fruit or vegetable contains about two different pesticides, most children are consuming three to four residues daily just through fresh fruits and vegetables.

Drinking water is another major source of pesticide exposure, particularly for children living in the midwest and other farming regions. In recent years the PDP has also tested drinking water as it comes out of the tap. About 54% of drinking water samples tested positive for one or more pesticides and pesticide metabolites in 2004 (see Appendix M for detailed findings (Agricultural Market Servicc, 2006)). Individuals under 20 years of age in the USA consume about six servings of drinking water per day, about half of which contain pesticides. Accordingly, the average young American is exposed to more than six servings of food and water daily that contains pesticide residues. Fortunately, the levels are very low in most cases and the residues pose modest if any risks to healthy young people.

Table 14.1 Samples containing multiple residues by market claim in three datasets

Data set	Organic			IPM/NDR			No market claim		
	Number of samples	Samples with multiple residues	Samples with multiple residues (%)	Number of samples	Samples with multiple residues	Samples with multiple residues (%)	Number of samples	Samples with multiple residues	Samples with multiple residues (%)
PDP 20 crops	128	9	7.1	195	46	23.6	26,571	12,1	45.5
DPR 19 crops	609	8	1.3	0			34,003	4,055	11.9
CU 4 crops	67	4	6.0	45	20	44.4	68	42	62.0

Source: Baker *et al*., 2002.

Unfortunately, this is not always the case. Some of the residues are in the range where the weight of the evidence points to potential for adverse biological impact, particularly when exposures occur at vulnerable periods of development, or during an illness (National Research Council, 1993; Consumers Union, 2001; Curl *et al.*, 2003; Lu *et al.*, 2006; Landrigan and Benbrook, 2006). This conclusion is based on comparisons of the high-end residue levels sometimes found in the PDP and the maximum amount of pesticide that can be present in a serving of a given food, for a child of a known weight, a concept often called the 'reference concentration' or RfC (Groth *et al.*, 2000).

When a pesticide is present in food at the RfC value, the residues in a single serving of the food (e.g. 100 g of fresh apple) would deliver to a child of known weight a level of exposure equal to the child's reference dose. When residues are present in a typical serving of food above the RfC, children would be exposed at a level above what EPA regards as consistent with the FQPA's 'reasonable certainty of no harm' standard. Each year, the PDP finds several dozen residues at levels well above the applicable RfC. These residues fall in a grey area – they are higher than what EPA regards as safe, yet are generally below the levels known to cause adverse impacts in experimental animals.

14.3 Organic food and pesticide residues

Consistently over the last decade, between two-thirds and three-quarters of US consumers have voiced strong or very strong concerns over pesticide residues in food. Seeking out organic food is one step that a growing percentage of consumers are taking in the hope of reducing pesticide exposures.

The US Department of Agriculture (USDA) regulates organic foods under standards set forth by the National Organic Program (NOP). NOP standards describe the core features of farming systems eligible for organic certification and also establish the basis for a list of materials that are allowed for use in organic farming and food processing. In general on organic farms, synthetic chemical substances, including most pesticides, are prohibited and natural substances including botanicals, copper fungicides and sulfur, are allowed. A small number of exceptions to this 'synthetic versus non-synthetic' rule are included on the national list of approved substances.

On the farm, a crop can be labeled as organic if it is produced on land where no prohibited substances have been applied for a minimum of three years prior to harvest. Animal products labeled as organic must come from livestock that have been fed organic feed, raised using humane animal husbandry systems, and that have not be treated with synthetic medications, except under very limited conditions set forth in the national list.

Processed food products that are labeled as '100% organic' must be made from agricultural commodities and ingredients that are all certified organic.

Products may be labeled 'organic' if they contain at least 95% certified organic ingredients; up to 5% of an 'organic' product can be composed of ingredients like pH buffers, enzymes, flavorings and filtering aids that are on the NOP's national list of allowed synthetic substances. The label on products with a minimum organic content of 70%, not including water and salt, can only state that the product is 'made with' specific organic ingredients. No ingredients in products that bear an organic label, including the non-organic ingredients in a 70% 'made with organic' product, may be produced or handled using 'excluded methods'. These include genetically modified organisms, irradiation and sewage sludge. The food chain must prevent the commingling of organic with non-organic products and protect organic products from contact with prohibited substances.

Consuming organic food will not eliminate pesticides from the diet, as shown in the following tables. Table 14.2 provides an overview of residues in organic, integrated pest management (IPM) 'IPM grown' or 'no detectable residue' (NDR), and conventional (no market claim) samples of fresh fruits and vegetables tested by PDP from 1993–2004. Residues of long-banned organochlorines are excluded from this table.

Over this 12-year period, 66% of the conventional samples contained one or more residues, while 17% of the organic samples tested positive for one or more pesticide. Accordingly, residues are 3.9 times more likely in conventional samples compared to organic samples. Residues were present in 45% of the IPM/NDR samples. This finding suggests that the patterns of pesticide use and resulting residues on food harvested from land farmed under IPM or NDR programs are closer to conventional production systems than organic farming.

Table 14.3 presents the most recent findings of the PDP, covering testing done in 2004. Residues found in organic and conventional fresh fruits and vegetables are highlighted. Over three-quarters of the conventional produce contained residues (78%), while 16% of the organic samples contained residues. The odds of a conventional sample of fresh produce containing a residue were 4.9 times higher than an organic sample. Given that the odds were 3.9 times greater on average in the 1993–2004 period, the increase to 4.9 times in 2004 suggests that some progress has been made in the last decade in reducing inadvertent contamination of organic food with prohibited synthetic pesticides.

Further insight on the frequency of multiple residues is evident in Table 14.4, which shows the number of residues found per sample for selected foods tested in 2004 by the PDP. Conventional apples were found to contain, on average, 3.6 residues, while the one positive organic sample had a very minute level of the post-harvest fungicide thiabendazole. The level of residue found in the one positive organic apple sample was 0.0002 parts per million, while the mean thiabendazole residue found in 641 positive conventional samples was 0.43 ppm, over 2100 times higher than the level found in the organic sample.

Table 14.2 Frequency of pesticide residues in fresh fruits and vegetables by market claim, excluding the residues of banned organochlorines; PDP 1993–2004

	Organic			IPM/NDR			No market claim		
	Number of samples	Number of positives	Positive (%)	Number of samples	Number of positives	Positive (%)	Number of samples	Number of positives	Positive (%)
Total fruit	104	18	17	83	43	52	28,500	21,199	74
Total vegetables	310	53	17	188	80	43	38,273	23,274	61
Total fruits and vegetables for all years	414	71	17	271	123	45	67,389	44,728	66

Table 14.3 Frequency of pesticide residues in fruits and vegetables by market claim, excluding the residues of banned organochlorines; PDP 2004

	Organic			IPM/NDR			No market claim		
	Number of samples	Number of positives	Positive (%)	Number of samples	Number of positives	Positive (%)	Number of samples	Number of positives	Positive (%)
Fruits									
Cantaloupe	1			6	3	50	734	393	54
Grapes	3	1	33	1	1	100	732	569	78
Oranges	6	1	17				734	651	89
Pears	5						736	643	87
Strawberries	2						719	676	94
Apples	5	1	20	1	1	100	738	723	98
Total fruits	22	3	14	8	5	63	4393	3655	83
Vegetables									
Cucumbers	2						555	385	69
Winter squash	7	1	14				357	121	34
Green beans	2						543	384	71
Lettuce	5			1			737	654	89
Sweet bell peppers	11	1	9				552	534	97
Sweet potatoes	5	4	80				731	451	62
Tomatoes	5			2			733	356	49
Cauliflower	3	1	33				181	132	73
Total vegetables	40	7	18	3			4389	3017	69
All fresh produce: 2004 PDP	62	10	16	11	5	45	8782	6672	76

Table 14.4 Number of pesticide residues found by market claim and average number of residues in selected fruits and vegetables tested by the USDA's Pesticide Data Program; PDP 2004

		Number of samples tested	Number of positive samples	Number of unique residues found	Residues per sample tested	Residues per positive sample
Apples	IPM/NDR	1	1	2	2	2
	No market claim	738	723	2.614	3.54	3.61
	Organic	5	1	1	0.2	1
Cauliflower	IPM/NDR					
	No market claim	181	132	148	0.818	1.121
	Organic	3	1	1	0.333	1
Grapes	IPM/NDR	1	11	1	1	1
	No market claim	732	569	1.328	1.814	2.329
	Organic	3	1	1	0.333	1
Lettuce	IPM/NDR	1	0	0	0	0
	No market claim	737	654	1.879	2.55	2.873
	Organic	5	0	0	0	0
Oranges	IPM/NDR	0	0	0	0	0
	No market claim	181	90	91	0.503	1.011
	Organic	1	0	0	0	0
Sweet bell peppers	IPM/NDR	0	0	0	0	
	No market claim	552	534	2.251	4.078	4.215
	Organic	11	1	8	0.727	8
Strawberry	IPM/NDR	0	0	0	0	0
	No market claim	719	676	1.983	2.758	2.933
	Organic	2	0	0	0	0
Tomatoes	IPM/NDR	2	0	0	0	0
	No market claim	733	356	732	0.999	2.056
	Organic	5	0	0	0	0

This is a clear-cut example where movement of a post-harvest fungicide applied to conventional apples reached organic apples. This could have occurred in a cold storage facility, or it might have happened during trucking or even in a store if a box of treated conventional apples were placed too close to a box of organic apples. NOP rules governing the separation of conventional and organic produce are designed to prevent this sort of inadvertent cross-contamination and are, for the most part, working reasonably well, given that the vast majority of fresh organic fruits and vegetables lack post-harvest fungicide residues that are very common and indeed often ubiquitous on conventional produce. For example, 79.2% of the apples tested by PDP in 2004 had residues of the post-harvest fungicide diphenylamine and 86.6% had residues of thiabendazole. Obviously, over half of the apple samples tested had residues of both (Agricultural Market Service, 2006).

The data on sweet bell peppers in Table 14.4 is striking, one organic sample had eight residues! While there were 29 conventional bell pepper samples with nine or more residues, it is inconceivable that a truly organic sample could contain eight residues. In fact, this mislabeled organic sample contained 0.22 ppm of chlorpyrifos, a very high and dangerous level of this high-risk OP insecticide. The mean level of chlorpyrifos in the 95 conventional samples that also tested positive for this insecticide was 0.048 ppm, about one-fifth of the level in this exceptionally 'hot' organic sweet bell pepper sample. Curiously, one organic bell pepper sample in 2003 was also found to contain eight residues.

In response to these and other clearly mislabeled samples, The Organic Center has written to the USDA's National Organic Program and the PDP suggesting that a mechanism be put in place to flag samples with violative residues, based on NOP rules. According to NOP rules, any pesticide residue found at a level exceeding 5% of the published tolerance warrants investigation by the certifier.

Whenever the PDP finds a residue in an organic sample above 5% of the existing EPA tolerance, an e-mail could be sent to the NOP with the details of the sample: where and when it was collected and where it was produced. In many cases, this should allow the NOP to determine the certifier involved with the product. The certifier could then be alerted and future accreditation of that certifier could be based, in part, on how well the certifier responded to the incident and whether the source of violative residues was discovered and properly dealt with. Other countries conducting periodic government testing of organic foods, like the UK, could easily introduce a comparable system to alert certifiers of shipments of conventional food bearing organic labeling.

Table 14.5 provides a sense of how frequently the PDP might need to alert the NOP of a violative residue in an organic sample. The table shows all positive samples of fresh organic produce in 2004, the residue level found and the applicable EPA tolerance. The PDP would need to flag any value over one in the column 'Ratio of residue found to 5% of EPA tolerance'

Table 14.5 Overview of organic samples with positive residues; PDP 2004

Crop–pesticide data pairs		Residue level (ppm)	EPA tolerance (ppm)	Ratio of residue found to 5% of EPA tolerance	Mean residue level (all samples)	Ratio of residue found in organic samples to mean of all samples
Sweet bell pepper	Bifenthrin	0.096	0.5	3.84	0.0193189	4.97
Sweet bell pepper	Chlorpyrifos	0.22	1	4.40	0.0491	4.48
Sweet bell pepper	Permethrin *cis*	0.023	1	0.46	0.0165442	1.39
Sweet bell pepper	Oxamyl	0.033	3	0.22	0.0250789	1.32
Sweet potato	Chlorpyrifos	0.007	0.05	2.8	0.0055863	1.25
Sweet potato	Dieldrin	0.01	0.1	2.0	0.0085	1.18
Sweet bell pepper	Permethrin *trans*	0.026	1	0.52	0.0237	1.1
Winter squash	*o*-phenylphenol	0.017			0.016934	1.0
Oranges	Chlorpyrifos	0.007	0.5	0.28	0.007	1.0
Sweet bell pepper	Tetrahydrophthalimide	0.033	25	0.03	0.0328	1.0
Oranges	*o*-phenylphenol	0.017	10	0.03	0.0187042	0.91
Oranges	Imazalil	0.05	10	0.1	0.0909026	0.55
Sweet potato	Piperonyl butoxide	0.017	0.3	1.36	0.03541897	0.48
Sweet potato (4 samples)	*o*-phenylphenol	0.017	15	0.02	0.042125	0.40
Winter squash	Dieldrin	0.01	0.1	2.0	0.032423	0.31
Grapes	Imidacloprid	0.017	1	0.34	0.056383	0.30
Sweet bell pepper	Myclobutanil	0.005	1	0.1	0.0184816	0.27
Cauliflower	Imidacloprid	0.001	3.5	0.01	0.0051368	0.21
Sweet bell pepper	Methamidophos	0.002	1	0.04	0.0567585	0.04
Apples (3 samples)	Thiabendazole	0.0002	10	0.00	0.4220756	0.00

(because the residue level found was greater than 5% of the EPA tolerance). Note that, in 2004, PDP would only have to report six residues, two involving long-banned organochlorines and one the synergist piperonyl butoxide.

The next column reports the mean residue level found in conventional samples and the last column shows the ratio of the residue level in the organic sample compared to the mean of the residues in conventional samples. Residues in this table are ranked in descending order relative to this last column. Any value over about 0.5 in this last column suggests mislabeling, since the residue level in the organic sample was no less than one-half the mean level in all-positive conventional samples. Half to two-thirds of the residues in Table 14.5 are likely reflect mislabeling, while perhaps one-third are likely to be cases of inadvertent drift or cross-contamination.

14.3.1 Why organic food sometimes contains residues

Pesticides are ubiquitous and mobile across agricultural landscapes. Many positive organic samples contain low levels of pesticides used on nearby conventional crops that have moved across field boundaries by drift or through use of contaminated irrigation water. Soil-bound residues of persistent pesticides used years ago, before the farmer switched to organic methods, account for a large portion of the residues found in vegetables and root crops, especially squashes. Cross-contamination with post-harvest fungicides applied in storage facilities, or later along the food supply chain, is a major cause of low-level fungicide residues (Baker *et al.*, 2002).

Samples sold as organic and found to contain levels of residues comparable to conventional foods reflect laboratory error, inadvertent mixing of produce, or mislabeling, although some cases are likely to represent outright fraud. Consumers also need to understand that organic farmers are permitted to apply non-synthetic pesticides including sulfur, oils, several botanicals, *Bacillus thuringiensis* (*Bt*), soaps, certain microbial pesticides and pheromones. By volume, sulfur, horticultural oils, soaps and copper-based fungicides are among the most heavily used pesticides on both organic and conventional produce farms. These pesticides are used in similar ways for comparable reasons on organic and conventional fruit and vegetable farms.

While there were once several toxic botanical insecticides on the market and approved for organic production, only one remains in relatively common use – pyrethrum. Pesticides containing pyrethrums are highly toxic but they degrade rapidly after spraying and hence rarely leave detectable residues in harvested food. Plus, they are applied at very low rates, on the order of one to two one-hundredths of a pound per acre (0.0112 kilograms per hectare). In contrast, OP insecticides are applied at 50- to 100-times higher rates per acre. A survey of organic farmers carried out by the Organic Farming Research Foundation (OFRF) found that only 9% of 1045 farmers applied botanicals regularly (mostly pyrethrums and neem), and that 52% never use them, 21% use them rarely and 18% 'on occasion' (Walz, 1999).

Tables 14.6, 14.7, 14.8, 14.9 and 14.10 provide further insight into the comparative properties and toxicity of pesticides applied on organic and conventional farms to treat a given type of pest. Table 14.6 lists the primary pesticides approved for use on organic farms and their uses and target pests. Tables 14.7–14.10 again list the major organic pesticides, along with two or three conventional pesticide alternatives that are used by conventional farmers to manage the same pest problems. Tables 14.7 and 14.8 summarize the rates of application of these pesticides, while Tables 14.9 and 14.10 focus on relative measures of toxicity to mammals and other organisms.

In Tables 14.9 and 14.10, the last column reports the 'environmental impact points' (EIPs) for typical applications of organic and conventional pesticides derived from the Pesticide Environmental Assessment System, or PEAS. This model produces relative rankings of risks based on defined use rates and use patterns (the formulation used to apply a pesticide, timing, target of the application, spray equipment used, etc). PEAS scores reflect an equal balancing of acute pesticide risks to farm workers, chronic risks via dietary exposure and exposures to birds, Daphnia and bees.

Clearly, pesticides approved for use on organic farms are typically applied at a much lower rate (except for mineral and soap-based products), tend to be far less persistent in the soil and on plant tissues (data not reported) and are far less toxic. They rarely appear as residues in food and pose much lower overall risks to a diverse set of non-target organisms, as evidenced in the huge disparity in EIU scores between the organically approved pesticides and their common conventional alternatives. The high EIU value for spinosad is based almost entirely on its toxicity to bees. While the label directs growers to not spray spinosad when bees are actively foraging in the fields, it is not known whether and to what degree this label requirement is preventing bee mortality in the field.

14.4 Reducing exposure to the OP insecticides

Since passage of the FQPA in 1996, the US EPA has focused on reducing exposures to the organophosphate (OP) class of insecticides. This is the most widely used class of insecticides in the world. Of all classes of pesticides, the OPs pose by far the greatest risks of developmental effects during pregnancy and in the first years of a child's life as a result of dietary exposures and are also among the major causes of poisoning episodes among farm workers. The factors driving OP risk include when a person is exposed, levels of exposure in food, how much and how often contaminated foods are eaten, pesticide toxicity and general health status. Young people lacking developed immune systems and adults with compromised immune systems, for example, can be at heightened risk of diseases linked to proliferative cell growth, like cancer (National Research Council, 1993).

Major advances have been made in the last decade in the science of

Table 14.6 Primary pesticides approved in the United States for use in organic production

	Active ingredient	Approved product for organic production	Examples of pests controlled	Major use sites	Type of pesticide
Microbial	*Bacillus subtilus*	Rhapsody, Serenade	Downy mildew, bean rust, certain rots	Stone fruit, cucumbers, tomatoes	Fungicide
	Bacillus thuringiensis	Xentari, Dipel	Certain worms, borers, caterpillars, Lepidoptera	Various fruits and vegetables	Insecticide
	Spinosad	Entrust	Caterpillars, beetles, leafminers	Various fruits and vegetables	Insecticide
	Coniothyrium minitans	Contans	White mold, leafdrop	Cole crops, beans, tomatoes, carrots, lettuce	Fungicide
Biological	*Beuveria bassiana*	Mycotrol, Naturalis	Thrips, whitefly, aphid, Colorado potato beetle, caterpillars, ants	Various fruits and vegetables	Insecticide
	Pheromones	Multiple products	Codling moth, oriental fruit moth	Various fruits and vegetables	Insecticide
Mineral	Bicarbonate (K and NA)	Kaligreen	Powdery mildew, black spot, leaf spot	Grapes, strawberries	Fungicide
	Copper products	Champion, Clean Crop, Nordox	Powdery mildew, anthracnose, various blights	Various fruits and vegetables	Fungicide
	Kaolin clay	Surround	Apple maggot, leafhopper, pear psylla, plum curculio	Various fruits and vegetables	Insecticide, insect growth regulator
	Sulfur products	Multiple products	Powdery mildew, certain spot, scale, certain rust	Various fruits and vegetables	Fungicide
Botanical	Pyrethrum	Pyganic, Safer	Whitefly, aphid, leafhopper	Tomato, cucumber, apple, potato	Insecticide
	Rotenone	No longer approved	Caterpillar, beetles, aphids	Potato, apple	Insecticide
	Azadirachtin (Neem)	Aza-direct, Neemix	Whitefly, caterpillars, leafminer, aphid, DBmoth	Various fruits and vegetables	Insecticide
Petroleum	Oils	JMS Stylet oil	Whitefly, aphid, thrips, pysllids	Various fruits and vegetables	Insecticide
Soaps	Potassium salts of fatty acids	M-Pede	Aphid, whitefly, mealybug, powdery mildew	Various fruits and vegetables	Insecticide, fungicide

Table 14.7 Microbial, biological and botanical organic materials and their typical conventional alternatives: the average use rate in pesticides used in conventional farming compared to the pesticides used in organic farming

Organic pesticide conventional alternative	Trade name	Typical use rate (pounds active ingredient/acre)	Ratio of average conventional use rate to organic use rate
Bacillus thuriengensis	Xentari, Dipel	0.04	
Azinphos-methyl	Guthion	0.58	
Endosulfan	Thiodan	0.83	
Thiamethoxam	Platinum	0.062	
Average conventional		0.491	12.3
Bacillus subtilus	Serenade, Rhapsody	0.01	
Azoxystrobin	Abound	0.16	
Zoxamide	Gavel	0.16	
Captan	Captan	2.4	
Average conventional		0.907	90.7
Spinosad	Entrust	0.08	
Cypermethrin	Ammo, Cymbush	0.08	
Methomyl	Lannate	0.52	
Average conventional		0.3	3.8
Coniothyrium minitans	0.1		
Thiophanate methyl	Topsin M	0.58	
Iprodione	Rovral	0.73	
Chloropicrin	Chloropicrin	61.6	
Average conventional		20.97	209.7
Beauveria bassiana	Mycotrol, Naturalis	0.01	
Chlorpyrifos	Lorsban	1.25	
Imidacloprid	Admire	0.12	
Average conventional		0.685	68.5
Pheromones	Multiple products	0.001	
Pyriproxyfen	Esteem	0.0745	
Methoxyfenozide	Intrepid	0.25	
Average conventional		0.162	162.3
Pyrethrum	Pyganic, Safer	0.01	
Dimethoate	Dygon	0.55	
Carbofuran	Furadan	0.9	
Average conventional		0.725	72.5
Rotenone	Rotenone	0.04	
Acephate	Orthene	0.69	
Chlorpyrifos	Lorsday	1.25	
Average conventional		0.97	24.3
Azadirachtin (neem)	AZA-direct, Neemix	0.16	
Carbaryl	Sevin	1.58	
Phosmet	Imidan	1.43	
Average conventional		1.505	9.4

Notes: Rates based on the average use rate for all crops in the annual pesticide use surveys, carried out by the National Agricultural Statistics Service, USDA. Use rate for *Beauveria bassiana* is estimated.

Table 14.8 Mineral, petroleum and soap-base organic materials and their typical conventional alternatives: the average use rate in pesticides used in conventional farming compared to the pesticides used in organic farming

Organic pesticide Conventional alternative	Trade name	Typical use rate	Ratio of average conventional use rate to organic use rate
Copper products	Champion	2.69	
Chlorothalonil	Bravo	1.46	
Mancozeb	Manzate	1.43	
Average conventional		1.445	0.54
Bicarbonate (K and Na)	Kaligreen	2.15	
Maneb	Manex	1.33	
Metam sodium	Vapam	138.68	
Average conventional		70.005	32.56
Sulfur products	Multiple Products	9.37	
Maneb	Manex	1.33	
Captan	Captan	2.4	
Average conventional		1.87	0.2
Kaolin clay	Surround	25	
Methomyl	Lannate	0.6	
Esfenvalerate	Asana	0.05	
Average conventional		0.325	0.01
Petroleum oils	JMS Stylet Oil, Purespray	21.48	
Malathion	Fyfanon, Malixol	1.23	
Bifenthrin	Capture, Brigade	0.07	
Average conventional		0.65	0.03
Soaps	M-Pede	4.5	
Permethrin	Pounce, Ambush	0.13	
Lambda-cyhalothrin	Karate	0.03	
Average conventional		0.08	0.02

Notes: Rates based on the average use rate for all crops in the annual pesticide use surveys, carried out by the National Agricultural Statistics Service, USDA.

pesticide risk assessment – the process through which regulators integrate these factors into a single quantitative and/or probabilistic measure of risk. Quantitative measures of risk are needed to determine when regulatory restrictions must be applied to assure acceptable ‘margins of safety’.

14.4.1 EPA’s OP cumulative risk assessment

Early on in the FQPA implementation process, EPA decided to focus on OP insecticide risks. Residues of these insecticides were frequently found in foods consumed by children, these insecticides were widely used in the home and other urban environments and their toxicity was well established. In addition, EPA knew years before the passage of the FQPA that many then-

Table 14.9 Microbial, biological and botanical organic materials and their typical conventional alternatives: toxicity comparisons

Organic pesticide conventional alternative	Trade name	LD_{50} [a]	cPAD [b,e]	aPAD [c,e]	PEAS EIU [d]	EIU crop
Bacillus thuringiensis	Xentari, Dipel	5000	0.1	0.1	0.04	Peach
Azinphos-methyl	Guthion	16	0.00149	0.003	209.97	Peach
Endosulfan	Thiodan	80	0.00006	0.0015	93.29	Peach
Thiamethoxam	Platinum	1453	0.0006		0.09	Strawberry
	Average conventional	516.3	0.00072	0.00225	101.12	
Bacillus subtilus	Serenade, Rhapsody	5000	0.1	0.1	0.16	Grape
Azoxystrobin	Abound	5000	0.18	0.67	0.17	Grape
Zoxamide	Gavel	5000	0.48		NA	
Captan	Captan	5000	0.13	0.1	2.3	Grape
	Average conventional	5000	0.26	0.385	1.235	
Spinosad	Entrust	3738	0.268		100.14	Snap bean, proc
Cypermethrin	Ammo, Cymbush	86	0.01		13.92	Snap bean, proc
Methomyl	Lannate	17	0.008	0.02	57.02	Snap bean, proc
	Average conventional	51.5	0.009	0.02	35.47	
Coniothyium minitans		5000	0.1	0.1	NA	Strawberry
Thiophanate methyl	Topsin M	5000	0.27	0.067	0.76	Strawberry
Iprodione	Rovral	3500	0.073	0.067	10.3	Strawberry
Chloropicrin	Chloropicrin	250	NA		65.49	Strawberry
	Average conventional	2916.7	1.1715	0.067	25.53	
Beauveria bassiana	Mycotrol, Naturalis	5000	0.1	0.1	<1.0	Grape
Chlorpyrifos	Lorsban	135	0.00003	0.0005	270.82	Grape
Imidacloprid	Admire	450	0.019	0.14	1.94	Grape
	Average conventional	292.5	0.0095	0.07	136.38	
Pheromones	Multiple products	5000	0.1	0.1	0.0001	Peach
Pyriproxyfen	Esteem	5000	0.35		0.03	Peach

Table 14.9 Continued

Organic pesticide conventional alternative	Trade name	LD_{50}[a]	cPAD[b,e]	aPAD[c,e]	PEAS EIU[d]	EIU crop
Methoxyfenozide	Intrepid	5000	0.1		0.35	Peach
	Average conventional	5000	0.225		0.19	
Pyrethrum	Pyganic, Safer	500	0.064		3.27	Grape
Dimethoate	Dygon	150	0.0005	0.02	19.92	Grape
Carbofuran	Furadan	8	0.005		174.31	Grape
	Average conventional	79	0.00275	0.02	97.1	
Rotenone	Rotenone	1620	0.004		0.11	Strawberry
Acephate	Orthene	945	0.0012	0.005	122.08	Snap bean, proc
Chlorpyrifos	Lorsban	135	0.00003	0.0005	191.9	Strawberry
	Average conventional	540	0.0006	0.00275	156.99	
Azadirachtin (neem)	AZA-direct, Neemix	5000	0.1		0.07	Grape
Carbaryl	Sevin	300	0.014		17.62	Grape
Phosmet	Imidan	113	0.011	0.045	70.88	Grape
	Average conventional	206.5	0.0125	0.045	44.25	

Notes: a. LD_{50} s are measured in mg/kg of bodyweight and are the dose at which 50% of the experimental animals die after exposure to a chemical. The smaller the number, the more toxic the pesticide.

b. cPAD is the chronic population adjusted dose set by the US Environmental Protection Agency. cPAD equals the chronic 'Reference Dose' (RfD) for a chemical divided by any applicable additional safety factor triggered by the Food Quality Protection Act's 10-X provision.

c. aPAD is the acute population adjusted dose set by the US EPA.

d. EIU is the acronym for environmental impact units derived from the Pesticide Environmental Assessment System (PEAS). EIUs are based on acute and chronic mammalian exposure and toxicity, and risks to birds, daphnia and honeybees. EIUs reflect relative risk associated with a given pesticide use rate and use pattern. The higher the EIU, the greater the potential for adverse impacts on non-target organisms.

e. cPADs and aPADs for microbial and biological pesticides approved for organic production have not been set by the U.S. EPA because of the granting of exemptions from the requirement for tolerances. A default value of 0.1 is used for all untested microbial and biological pesticides approved for organic production.

Table 14.10 Mineral, petroleum and soap-base organic materials and their typical conventional alternatives: toxicity comparisons (see notes)

Organic pesticide conventional alternative	Trade name	LD_{50}[a]	cPAD[b,e]	aPAD[c,e]	PEAS EIU[d]	EIU crop
Copper products	Champion	1000	0.1		5.54	Tomato
Chlorothalonil	Bravo	5000	0.2		2.83	Tomato
Mancozeb	Manzate	5000	0.003		0.7	Tomato
	Average conventional	5000	0.1015		1.765	
Bicarbonate (K and Na)	Kaligreen	3358	0.1		0.45	Grape
Maneb	Manex	5000	0.005		1.51	Grape
Metam sodium	Vapam	285	0.01		1.99	Grape
	Average conventional	2642	0.0075		1.75	
Sulfur products	Multiple products	3000	0.1		2.94	Grape
Maneb	Manex	5000	0.005		1.51	Grape
Captan	Captan	5000	0.13	0.1	2.3	Grape
	Average conventional	5000	0.0675	0.1	1.905	
Kaolin clay	Surround	5000	0.1		1.87	Tomato
Methomyl	Lannate	17	0.008	0.02	8.85	Tomato
Esfenvalerate	Asana	67	0.02		17	Tomato
	Average conventional	42	0.014	0.02	12.925	
Petroleum oils	JMS Stylet oil, Purespray	5000	0.1		8.96	Winter Squash
Malathion	Fyfanon, Malixol	2100	0.02	0.5	59.06	Winter Squash
Bifenthrin	Capture, Brigade	55	0.015	0.01	33.23	Winter Squash
	Average conventional	1078	0.0175	0.255	46.145	
Soaps	M-Pede	5000	0.1		0.33	Grape
Permethrin	Pounce, Ambush	500	0.05		6.89	Pear
Lambda-cyhalothrin	Karate	56	0.001	0.0025	7.6	Pear
	Average conventional	278	0.0255	0.0025	7.245	

Table 14.10 Continued

Notes: a. LD_{50} s are measured in mg/kg of bodyweight and are the dose at which 50% of the experimental animals die after exposure to a chemical. The smaller the number, the more toxic the pesticide.

b cPAD is the chronic population adjusted dose set by the US Environmental Protection Agency. cPAD equals the chronic reference dose (RfD) for a chemical divided by any applicable additional safety factor triggered by the Food Quality Protection Act's 10-X provision.

c. aPAD is the acute population adjusted dose set by the US EPA.

d. EIU is the acronym for environmental impact units derived from the Pesticide Environmental Assessment System (PEAS). EIUs are based on acute and chronic mammalian exposure and toxicity, and risks to birds, Daphnia and honeybees. EIUs reflect relative risk associated with a given pesticide use rate and use pattern. The higher the EIU, the greater the potential for adverse impacts on non-target organisms.

e. cPADs and aPADs for microbial and biological pesticides approved for organic production have not been set by the US EPA because of the granting of exemptions from the requirement for tolerances. A default value of 0.1 is used for all untested microbial and biological pesticides approved for organic production.

current tolerances covering OP residues in food could not be defended as safe (National Research Council, 1993). The agency justified lack of action in such cases by pointing out that actual residues in food rarely approached the tolerance level. In other words, legal residues were not safe, but no action was needed to lower tolerances since residues rarely approach legal, but unsafe tolerance levels.

EPA released the first case study of cumulative risks from 24 OPs in food for scientific review in mid-2000. Public comments were solicited and several scientific panel (SAP) meetings were held on various aspects of EPA's quantitative methods. In December 2001 a preliminary OP-CRA (cumulative risk assessment) was released, this time encompassing 30 OPs, additional foods, more residue data and all major routes of exposure. Public comments were solicited again and another series of SAP meetings were held. The revised final OP-CRA was issued in June 2002 after more than 20 SAP meetings and four rounds of public comment (US Environmental Protection Agency, 2002). It is the most sophisticated and data-rich pesticide risk assessment ever carried out.

While EPA publicly downplayed the risks documented in the CRA, the results showed that infants aged one to two years old, at the 99.9th percentile of the exposure distribution, face at least twice the level of risk that EPA has traditionally found acceptable. Based on EPA's findings, several thousand children on any given day are likely to be exposed to OPs well in excess of levels consistent with the FQPA's 'reasonable certainty of no harm' standard.

In response to requests from interested parties, the EPA released detailed results of its CRA, allowing assessment of the distribution of risks across foods, pesticides and food–pesticide combinations. Some key insights emerge from the OP-CRA results:

- Eight of 30 OP insecticides accounted for 97% of total estimated risk;
- A single insecticide (dimethoate and its metabolite *o*-methoate) accounted for 47% of total risk, largely from residues in just two foods, grapes and apples;
- Grapes, apples and pears accounted for over three-quarters of total risk; and
- Fresh fruits and vegetables accounted for the vast majority of exposure and risk.

The June 2002 OP-CRA confirmed that a relatively small number of OP insecticide uses account for the majority of risks faced by infants and children. Grapes, in particular, emerged as a major risk driver, with imported grapes accounting for a major share of this risk. Of the 100 grape samples in the OP-CRA with the highest levels of dimethoate based on PDP testing, 94 were imports. The surprising dominance of imported grapes among the highest risk samples in the OP-CRA is not an isolated case. There is a pronounced shift in the distribution of OP residues and risk in certain foods from domestic production (decreasing) to imports (increasing) in the USDA's PDP dataset from 1993 through 2004. This trend warrants more attention and focused risk assessment.

Results may confirm the need for novel policy interventions to more fairly divide the risk reduction burden between US farmers and those growing crops for export to the USA. Recent trends in methamidophos residues in a variety of Mexican produce imported into the USA are particularly worrying, especially in contrast to the solid progress made by US growers in reducing the frequency and mean levels of methamidophos residues in the same foods.

14.4.2 Biomonitoring surveys show modest gains

The Centers for Disease Control (CDC) and National Institutes of Health have periodically monitored levels of OP metabolites in urine and blood across the population. CDC-NIH surveys were carried out in 1988–1994, 1999–2000 and 2001–2002; the first survey was before the FQPA, the later two well after passage, and after the only major actions taken to date by EPA targeting high-risk OPs (methyl parathion and chlorpyrifos). In CDC and private surveys of OP metabolites in urine and blood, 90% or more of children test positive for several OP metabolites (Adgate *et al.*, 2001; Centers for Disease Control and Prevention, 2001). Figure 14.1 shows trends in metabolites corresponding to the OPs methyl parathion (4-nitrophenol residues could reflect the use of other pesticides including nitrofen and EPN; residue data in food suggests that methyl and ethyl parathion are the primary pesticide sources of 4-nitrophenol in human urine, although other sources cannot be ruled out) and chlorpyrifos (Centers for Disease Control and Prevention, 2001; Hill *et al.*, 1995). While modest declines are evident, these data are not encouraging since all food uses of methyl parathion resulting in residues, according to PDP testing, were cancelled in 1999 and residues should have

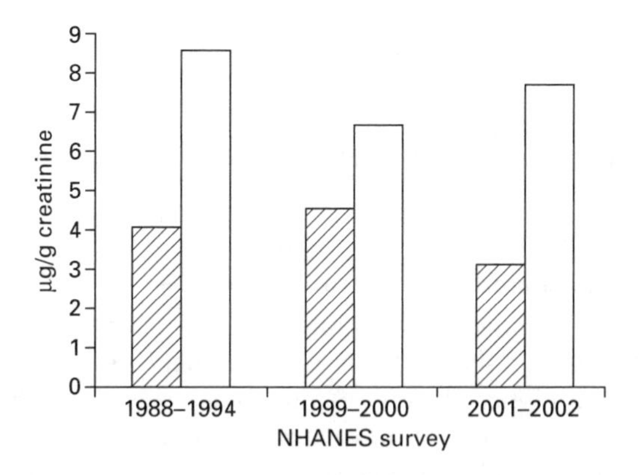

Fig. 14.1 95th percentile of urine concentrations (μg/g creatinine corrected) for the US population aged 20–59 years, three NHANES surveys. Shaded blocks show 4-nitrophenol (methyl parathion); unshaded blocks show 3,5,6-trichloro-*s*-pyridinol (chloropyrifos).

been out of the food supply by 2001. Likewise, major actions were taken in 2000 to end chlorpyrifos residential uses and reduce chlorpyrifos in the diet, yet the levels in urine actually went up from 1999/2000 to 2001/02 and have actually changed little since the 1988 sampling.

These data suggest that there are significant sources of exposure to these OPs other than those EPA identified as contributing most heavily to aggregate exposure. These might be additional crop uses in the USA or uses abroad, leading to exposures via imported foods.

A report entitled *Chemical Trespass* was issued in May 2004 by the Pesticide Action Network (Schafer *et al.*, 2006). It contained detailed analysis of 2000/01 National Health and Nutrition Examination Survey (NHANES) OP urinary metabolite data and used published methods to estimate exposure levels to parent compounds from creatinine corrected urinary metabolite levels. They focused on chlorpyrifos and its metabolite 3,4,6-trichloro-2-pyridinol (TCP), and found that chlorpyrifos exposures for children ages 6–11 and 12–19 exceeded EPA's chronic population-adjusted dose (cPAD) by surprisingly wide margins. Geometric mean TCP levels were 3 to 4.6 times higher than the EPA-estimated 'safe' dose, as shown in Fig. 14.2. The more heavily exposed children received daily doses more than ten times the 'safe' level.

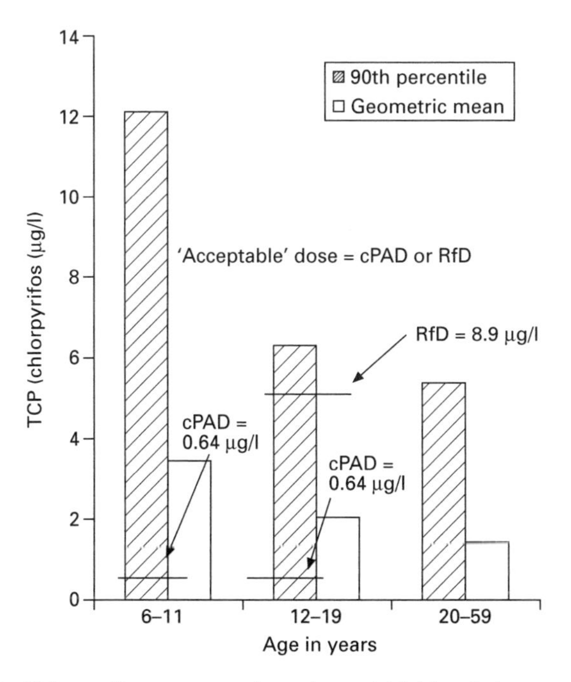

Fig. 14.2 Chlorpyrifos exposure above 'acceptable' levels in many children (2001–2002 NHANES). From Schafer *et al.*, 2004.

14.5 Need to reduce exposures further

Evidence linking low-dose exposures to pesticides to reproductive problems and developmental abnormalities has mounted steadily in the last decade. In June 2005, *Science* published the first study that showed that developmental changes triggered by pesticides can last multiple generations in experimental animals (Anway *et al.*, 2005). Fungicides were shown to cause decreased sperm counts and mobility, not just to animals exposed *in utero*, but for three subsequent generations. In other words, assuming that the same biological impacts occur in humans, exposures to a developing fetus might affect the health of the mother's great-grandchildren. During pregnancy, pesticides are readily transferred to the developing fetus via umbilical fluids (National Research Council, 1993).

A study involving more than 44,000 children measured pesticide residues in stored frozen blood samples from pregnancies in the early 1960s (Longnecker *et al.*, 1996, 2001, 2002, 2005; US Environmental Protection Agency, 2002; Walz, 1999). Children were divided into five groups based on levels of maternal pesticide exposure. Odds ratios were calculated for preterm birth and small-for-gestational-age babies across the five groups, and increased in a dose–response manner as shown in Table 14.11. Those in the group with the smallest exposure had a 50% increased chance of being born prematurely, compared to those with none. Those at the highest level had greater than a 200% increased chance of premature birth. The authors (Longnecker *et al.*) estimate that pesticide exposure was responsible for 15% of all infant deaths during the years of the study, the only such estimate this author is aware of.

A similar study was conducted jointly by investigators at the Center for Research on Women's and Children's Health, the Mount Sinai School of Medicine, the University of North Carolina, Chapel Hill, the Kaiser Permanent

Table 14.11 Maternal serum DDE concentration in relation to odds of preterm or small-for-gestational-age birth

	Serum DDE (μg/l)				
	<15	15–29	30–44	45–59	≥ 60
Preterm birth					
Number of cases	34	153	80	50	44
Number of controls	375	944	404	176	120
Adjusted odd ratio (95% confidence interval)	1	1·5	1·6	2·5	3·1
Small-for-gestational age					
Number of cases	20	106	47	22	26
Number of controls	389	991	436	204	138
Adjusted odd ratio (95% confidence interval)	1	1·9	1·7	1·6	2·6

Source: Longnecker *et al.*, 2001.

Division of Research and the University of California-San Francisco School of Medicine (Cohn *et al.*, 2003). They measured pesticide metabolites in preserved postpartum maternal serum samples from 1960 to 1963. They also recorded time to pregnancy in their eldest daughters 28–31 years later. The daughters' probability of pregnancy fell by 32% for each 10 mcg/l detected, three decades after the exposure.

A team of researchers at the University of California-Berkeley School of Public Health found that exposures to pesticides during pregnancy significantly heightened risk of children developing leukemia and that the more frequent the exposures and the earlier in life, the greater the increase in risk (Ma *et al.*, 2002). A team in the Department of Preventive Medicine, University of Southern California, found that exposure to pesticides in the home during fetal development and the early years of life increased the risk of non-Hodgkin's lymphoma, with odds ratios as high as 9.6 for Burkitt lymphoma (Buckley *et al.*, 2000).

A study in Ontario, Canada showed that exposures to pesticides three months prior to conception and during pregnancy increased the risk of spontaneous abortions (Arbuckle *et al.*, 2001). Pesticides can also alter fetal development and trigger life-long and/or long-delayed health problems from childhood exposures. Research by a team led Dr. Robin Wyatt has focused on the impacts of OP residential exposures during pregnancy and after birth among women from minority ethnic backgrounds in public housing projects in New York City. They found that chlorpyrifos exposures significantly reduced birthweight and length and the higher the exposures, the larger the impact (Whyatt *et al.*, 2004).

The team led by Whyatt used regression analysis to assess whether there was a difference in the association between chlorpyrifos exposure and birth outcome before and after the EPA's action in the summer of 2000 which had ended residential use of chlorpyrifos. Prior to 2001, chlorpyrifos clearly had an impact on birth outcome, but after the EPA action taken in June 2000, levels of exposure declined and there was no longer a statistically significant association between insecticide exposure and birth outcome (Whyatt *et al.*, 2004, 2005). This study provides encouraging evidence linking an action driven by the FQPA to a significant reduction in prenatal and infant exposures and risk.

14.5.1 Dietary intervention studies

EPA research investments since 1995 in pesticide exposure and risk assessment methods have helped pioneer novel approaches to quantify risk levels. A team at the University of Washington's School of Public Health and Community Medicine found that 2–5 year olds consuming predominantly organic foods over a 3-day period had 8.5-fold lower mean levels of OP insecticide metabolites in their urine than children eating mostly conventional (unlabeled) foods (Curl *et al.*, 2003). The study was carefully designed to minimize potentially confounding variables. The children came from similar socioeconomic backgrounds, households with recent use of pesticides in the home were excluded from the study and rigorous sampling and double-blind testing protocols were used. The research team concluded that:

> Dose estimates generated from pesticide metabolite data suggest that organic diets can reduce children's exposure levels from above to below EPA's chronic reference doses, thereby shifting exposures from a range of uncertain risk to a range of negligible risk. Consumption of organic produce represents a relatively simple means for parents to reduce their children's exposure to pesticides. (Curl *et al.*, 2003)

The Curl study provides the first direct empirical support for the conclusion that, for many children, pesticide exposures can probably be reduced to negligible risk levels during critical periods of development simply by seeking out organic fresh produce and fruit juices unlikely to contain significant OP residues. A wide array of reactions was triggered by the Curl findings (Curl *et al.*, 2003). Methodological issues were raised and addressed in a lengthy exchange of letters in *Environmental Health Perspectives*. One of the scientists on the Curl-led project conducted a second, more rigorous dietary intervention study in 2003–04 (Lu *et al.*, 2006). A cohort of 23 school-age children in the Seattle, Washington area was selected. This time, the study included three phases of testing for OP insecticide metabolites in urine. The first followed a period when the children consumed their normal diet containing conventionally grown foods. Phase two testing was carried out five days after the children switched to a predominantly organic diet and the third phase, after a return to a conventional diet for five days.

All 23 children had OP insecticide metabolites in their urine in phase one testing, while levels were below the limit of detection during phase two, following the consumption of mostly organic food for just five days. Once the children were back on their normal, conventional food in phase three, the levels of insecticide metabolites in urine returned to those found in phase one.

This carefully designed and conducted study confirmed the findings of the earlier study by Curl *et al.* (2003) and eliminated uncertainty regarding the identification of the OPs leading to specific metabolites in the children's urine. They accomplished this in the second study by only testing for two major OPs with distinct urinary metabolites – malathion and chlorpyrifos. Lu *et al.* (2006) concluded, however, that their findings on malathion and chlorpyrifos almost certainly apply to other major OPs in the diet. The Lu team concluded that their study shows that an organic diet '...provides a dramatic and immediate protective effect against exposures to OP pesticides'. Dr. Lu is currently conducting a third dietary intervention study, this time involving children in the Atlanta, Georgia area; it is expected that the results will be published in 2007.

14.6 Endnote

In February 2006 a symposium was sponsored by The Organic Center at the annual meeting of the American Association for the Advancement of Science (AAAS). The AAAS annual meeting is among the largest scientific gatherings in the country each year. The symposium was entitled '*Opportunities and Initiatives to Minimize Children's Exposures to Pesticides*' and included presentations by Greene (2006), the author (Benbrook, 2006), Lu (Lu *et al.*, 2006) and a paper jointly authored by Landrigan and Benbrook (2006). The four presenters issued a 'Joint Statement on Pesticides, Infants, and Children' summarizing their major findings and conclusions. It states:

> We believe that the scientific case supporting the need to significantly reduce prenatal and childhood exposures to pesticides has greatly strengthened over the last decade since passage of the Food Quality Protection Act (FQPA) in 1996. Evidence of the developmental neurotoxicity of several commonly used pesticides is particularly compelling. The FQPA provided the Environmental Protection Agency (EPA) with important new tools, ten years, and a mandate to address these sorts of risks and assure that there is a 'reasonable certainty of no harm' from government-approved pesticide uses, with special focus on pregnant women, infants and children. The EPA has acted decisively to eliminate most residential uses of the organophosphate (OP) insecticides. There is encouraging evidence that actions taken to date on residential pesticide uses are producing public health benefits. Equally decisive steps to reduce dietary exposures to high-risk OP pesticides have been regrettably few and far between. Human biomonitoring data shows that only modest progress has been made in reducing OP exposures since passage of the FQPA.
>
> Strong data point to a dramatic shift of pesticide dietary risks from fresh fruits and vegetables grown in the US to those imported from abroad. As a nation, we have more work to do, and contentious decisions ahead if we are to markedly reduce pesticide dietary risks.
>
> How can we best approach this task? In the last decade, significant public and private resources have been invested with the goal of reducing pesticide risks through:
>
> - the discovery and registration of safer pesticides
> - adoption of integrated pest management systems
> - ecolabel programs, including 'certified organic' and
> - regulation.
>
> We conclude that discovery of reduced risk pesticides has significantly facilitated the transition by many farmers away from high-risk pesticides. This transition has clearly helped reduce risks in some key children's foods. EPA policies put in place to expedite registration of reduced risk products should be strengthened.
>
> Adoption of integrated pest management (IPM) has had limited impacts

on pesticide use and risks. Most IPM systems are focused on using pesticides efficiently and lack even a secondary focus on dietary risk reduction.

Ecolabel programs have had modest impacts on pesticide risks because they collectively impact so few acres, and many programs do not require farmers to markedly change pest management systems. Organic farming is the clear exception, and offers one proven way to quickly and dramatically reduce children's exposures. Studies led by Dr. Chensheng Lu of Emory University have shown that a predominantly organic diet essentially eliminates evidence of exposure to certain widely used organophosphate insecticides.

Regulation, and the FQPA in particular, has advanced knowledge of pesticide risks and addressed residential risks reasonably well, but has done little to reduce pesticide dietary risks. The FQPA is fundamentally sound law, but it has not delivered fully on its promise to reduce children's pesticide risks because of the EPA's hesitancy to fully use the law's strong new provisions.

In the absence of more decisive action by EPA, significant near-term reductions in pesticide dietary risks are attainable, but only if farmers are provided support and incentives to change pest management systems, and only if consumers demand change.

We conclude that enhanced efforts by the government and food industry to increase both the supply and demand for organic food will deliver the most significant near-term public health gains, especially if the focus is on expanding consumption of fresh and processed organic fruits and vegetables, while reducing consumption of foods high in added sugar and added fat content. Building such requirements into the school lunch and WIC programs are obvious ways to start.

14.7 References

Adgate, J. L., Barr, D. B., Clayton, C. A., Eberly, L. E., Freeman, N. C., Lioy, P. J., Needham, L. L., Pellizzari, E. D., Quackenboss, J. J., Roy, A. and Sexton, K. (2001). 'Measurement of children's exposure to pesticides: analysis of urinary metabolite levels in a probability-based sample'. *Environ. Health Perspect*, **109**(6), 583–590.

Agricultural Market Service (2002). *Pesticide Data Program Annual Summary Calendar year 2000.* US Department of Agriculture, Washington, DC.

Agricultural Market Service (2004). *Pesticide Data Program Annual Summary Calendar Year 2002.* U.S. Department of Agriculture, Washington, DC.

Agricultural Market Service (2005). *Pesticide Data Program Annual Summary Calendar Year 2003.* US Department of Agriculture, Washington, DC.

Agricultural Market Service (2006). *Pesticide Data Program Annual Summary Calendar Year 2004.* US Department of Agriculture, Washington, DC.

Anway, M. D., Cupp, A. S., Uzumcu, M. and Skinner, M. K. (2005). 'Epigenetic transgenerational actions of endocrine disruptors and male fertility'. *Science*, **308**(5727), 1466–1469.

Arbuckle, T. E., Lin, Z. and Mery, L. S. (2001). 'An exploratory analysis of the effect of pesticide exposure on the risk of spontaneous abortion in an Ontario farm population'. *Environ. Health Perspect*, **109**(8), 851–857.

Baker, B. P., Benbrook, C. M., Groth, E., III and Benbrook, K. L. (2002). 'Pesticide residues in conventional, integrated pest management (IPM)-grown and organic foods: insights from three US data sets'. *Food Addit. Contam.*, **19**(5), 427–446.

Benbrook, C. (2006). 'The effectiveness of farm and private sector initiatives to reduce children's pesticide exposures'. Presented at the *2006 Annual Meeting of the AAAS Opportunities and Initiatives to Minimize Children's Exposure to Pesticides,* St. Louis, Missouri. 19 Feb 2006.

Buckley, J. D., Meadows, A. T., Kadin, M. E., Le Beau, M. M., Siegel, S. and Robison, L. L. (2000). 'Pesticide exposures in children with non-Hodgkin lymphoma'. *Cancer*, **89**(11), 2315–2321.

Centers for Disease Control and Prevention (2001). *National Report on Human Exposure to Environmental Chemicals.* Atlanta, Georgia.

Cohn, B. A., Cirillo, P. M., Wolff, M. S., Schwingl, P. J., Cohen, R. D., Sholtz, R. I., Ferrara, A., Christianson, R. E., van den Berg, B. J. and Siiteri, P. K. (2003). 'DDT and DDE exposure in mothers and time to pregnancy in daughters'. *Lancet*, **361**(9376), 2205–2206.

Consumers Union (2001). *A Report Card for the EPA: Successes and Failures in Implementing the Food Quality Protection Act.* Consumers Union of the United States, Yonkers, NY.

Curl, C.L., Fenske, R.A. and Elgethun, K. (2003). 'Organophosphorus pesticide exposure of urban and suburban preschool children with organic and conventional diets'. *Environ. Health Perspect*, **111** (3), 377–383.

Greene, A. (2006). 'Opportunities to reduce children's exposures to pesticides: a truly grand challenge'. Presented at the 2006 *Annual Meeting of the AAAS Opportunities and Initiatives to Minimize Children's Exposure to Pesticides,* St. Louis, Missouri, 19 February 2006.

Groth, E., Benbrook, C. M. and Benbrook K.L. (2000). *Pesticide Residues in Children's Food.* Consumers Union of the United States, Yonkers, NY.

Hill, R. K., Head, S. L., Baker, S., Gregg, M., Shealy, D. B., Bailey, S. L., Williams, C., Sampson, E. J., and Needham, L. (1995). 'Pesticide residues in urine of adults living in the United States: reference range concentrations'. *Environ. Res.*, **71**, 99–108.

Landrigan, P. and Benbrook, C. (2006). 'Impacts of the food quality protection act on children's exposures to pesticides'. Delivered at the 2006 *Annual Meeting of the AAAS Opportunities and Initiatives to Minimize Children's Exposure to Pesticides,* St. Louis, Missouri, 19 February 2006.

Longnecker, M. P., Bernstein, L., Bird, C. L., Yancey, A. K. and Peterson, J. C. (1996). 'Measurement of organochlorine levels in postprandial serum or in blood collected in serum separator tubes'. *Cancer Epidemiol. Biomarkers Prev.*, **5**(9), 753–755.

Longnecker, M. P., Klebanoff, M. A., Zhou, H. and Brock, J. W. (2001). 'Association between maternal serum concentration of the DDT metabolite DDE and preterm and small-for-gestational-age babies at birth'. *Lancet*, **358**(9276), 110–114.

Longnecker, M. P., Klebanoff, M. A., Brock, J. W., Zhou, H., Gray, K. A., Needham, L. L. and Wilcox, A. J. (2002). 'Maternal serum level of 1,1-dichloro-2,2-bis(*p*-chlorophenyl)ethylene and risk of cryptorchidism, hypospadias, and polythelia among male offspring'. *Am. J. Epidemiol.*, **155**(4), 313–322.

Longnecker, M. P., Klebanoff, M. A., Dunson, D. B., Guo, X., Chen, Z., Zhou, H. and Brock, J. W. (2005). 'Maternal serum level of the DDT metabolite DDE in relation to fetal loss in previous pregnancies'. *Environ. Res.*, **97**(2), 127–133.

Lu, C., Toepel, K., Irish, R., Fenske, R. A., Barr, D. B. and Bravo, R. (2006). 'Organic diets significantly lower children's dietary exposure to organophosphorus pesticides'. *Environ. Health Perspect.*, **114**(2), 260–263.

Ma, X., Buffler, P. A., Gunier, R. B., Dahl, G., Smith, M. T., Reinier, K. and Reynolds,

P. (2002). 'Critical windows of exposure to household pesticides and risk of childhood leukemia'. *Environ. Health Perspect.*, **110**(9), 955–960.

National Research Council (1987). *Regulating Pesticides in Food: The Delaney Paradox.* National Academy Press, Washington, DC.

National Research Council (1993). *Pesticides in the Diets of Infants and Children.* National Academy Press, Washington, DC.

Pesticide Residue Committee (2001). *Annual Report of the Pesticide Residue Committee 2001.* United Kingdom Food Standards Agency.

Schafer, K. S., Reeves, M., Spitzer, S. and Kegley, S. E. (2006). *Chemical Trespass: Pesticides in Our Bodies and Corporate Accountability.* Pesticide Action Network North America, San Francisco, California.

US Environmental Protection Agency (20002). *OPP Revised OP Risk Assessment – Cumulative Risk From Pesticides in Foods.* 1.C.1–1.C.24.

Walz, E. (1999). *Final Results of the Thrid Biennial National Organic Farmers' Survey.* Organic Farming Research Foundation, Santa Cruz, California.

Whyatt, R. M., Rauh, V., Barr, D. B., Camann, D. E., Andrews, H. F., Garfinkel, R., Hoepner, L. A., Diaz, D., Dietrich, J., Reyes, A., Tang, D., Kinney, P. L. and Perera, F. P. (2004). 'Prenatal insecticide exposures and birth weight and length among an urban minority cohort'. *Environ. Health Perspect.*, **112**(10), 1125–1132.

Whyatt, R. M., Camann, D., Perera, F. P., Rauh, V. A., Tang, D., Kinney, P. L., Garfinkel, R., Andrews, H., Hoepner, L. and Barr, D. B. (2005). 'Biomarkers in assessing residential insecticide exposures during pregnancy and effects on fetal growth'. *Toxicol. Appl. Pharmacol.*, **206**(2), 246–254.

15

Levels and potential health impacts of nutritionally relevant phytochemicals in organic and conventional food production systems

Eduardo A. S. Rosa, Richard N. Bennett and Alfredo Aires, Universidade de Trás-os-Montes e Alto Douro, Portugal

15.1 Introduction

Modern agriculture has to feed approximately 6 billion people and for the next 50 years it is expected, based on population growth, that there will be a doubling in global food demand. The continuous population growth and reduction of global famine are challenges for the sustainability of the ecosystems involved in the crop and livestock production systems. Thus there is the inevitable need to increase food production, without increasing negative environmental impacts and also to answer public health demands. Intensive agricultural systems using conventional practices have been detrimental to the environment and production pressure of some foods may increase the rate of environmental deterioration. The world's ecosystems have been simplified and homogenized through monocultures, replacing natural ecosystems that once contained hundreds if not thousands of plant species, thousands of insect species and many species of vertebrates in ecological balance.

Improvements required in the production system include more efficient use of fertilizers and water and integrated pest management strategies to minimize the need for toxic pesticides and fungicides. Meeting these requirements is a great scientific challenge for humankind because of the trade-offs among competing economic and environmental goals and inadequate knowledge of the key biological, biogeochemical and ecological processes (Tilman *et al.*, 2002).

Increasing public awareness that agricultural practices are determinants of the health properties and attributes of foods requires changes to current systems of agriculture, particularly ones that are more intensive and far from the natural ecosystems. To meet these requirements new policies and regulations have been defined (EU Regulation 2092/91/EE) and incentives given for the production of foods under a more sustainable production system. Organic farming can substantially reduce the environmental impact of crop and livestock production and address the concerns about environment and food plant quality and subsequent health effects. Thus, individual countries and the EU have defined and set goals for the near future regarding the proportion of organic farming.

During the last decade there was a rapid expansion in the organic sector of European agriculture (Fig. 15.1). By the end of 2002, organic farming in Europe accounted for 5.8 million hectares on 190 000 holdings. This represents 4% of European agricultural land area, with 10% or more in some countries, and an annual retail market currently valued at more than €10 billion (Lampkin, 2004). The current scale and future potential of the sector is enormous.

Surveys on consumer choices for organic products rely on parameters such as effects on the environment, safety, freshness, visual quality, better taste and consistent quality (Bollinger, 2001). Some of these parameters are associated with food free from pesticide and fungicide residues and also animal products free of antibiotics; these are major concerns of a large proportion of consumers. Indeed, plant products from organic cultivation should be grown without the aid of chemical–synthetic pesticides and largely without the use of readily soluble mineral fertilizers within a diverse range of crop rotation and extensive soil tillage (Woese *et al.*, 1997). Maintenance of genetic diversity of the production systems and attention to the wider social and economic impact of the food production and processing systems, whilst producing food of high quality in sufficient quantity are also other attributes of organic production (Bourn and Prescott, 2002).

Despite the wide variety of organic products and continuous incentives to convert to organic farming many questions are still being raised concerning the quality of the organic products and how they compare with conventionally grown crops. Since fertilization and tillage practices and pest and disease control are the major changes needed for organic food production there are questions about how these changes might affect quality. Very few comparative studies are available on the concentrations of nutrients and bioactive compounds in foods conventionally and organically grown; in addition the number of foods examined has been small. Although there is a tendency to consider that organically produced plant foods have a better health effect than products from conventional production systems, available data and information seem to be controversial and inconclusive, based on the poor quality and wide variations in style of the available evidence. In this chapter we address these questions, selecting the major natural compounds in common food plants and how these can affect human health.

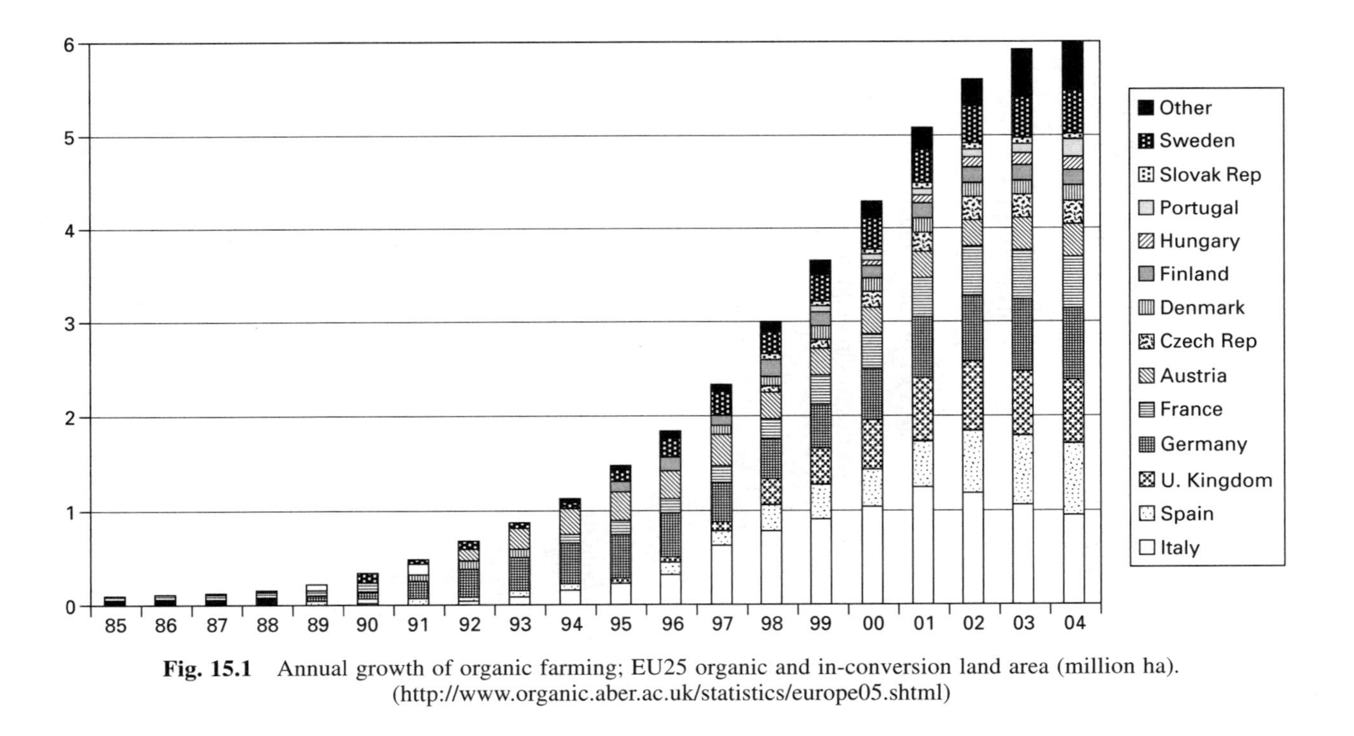

Fig. 15.1 Annual growth of organic farming; EU25 organic and in-conversion land area (million ha). (http://www.organic.aber.ac.uk/statistics/europe05.shtml)

15.2 Plants as sources of phytochemicals

Plants products are usually seen as a good source of vitamins, minerals, proteins and carbohydrates and minimum levels of each must be present in the diet. Levels of these compounds have been reported in food composition tables. In balanced diets, if consumption greatly exceeds the recommended daily allowance (RDA), normally no benefit is provided and the contents of these nutrients in fruits and vegetables are, in practice, never so high as to become harmful. Plants also have the characteristic ability and capacity to synthesize a large number of compounds of low to medium molecular weight, the so-called phytochemicals (i.e. plant secondary metabolites), which can be grouped into two major classes: the nitrogen-free, such as phenolics and polyphenolics, terpenes, saponins and polyacetylenes and the nitrogen-containing compounds such as the alkaloids, cyanogenic glycosides, non-protein amino acids and glucosinolates.

Plants are thus complex mixtures of primary and secondary metabolites of more than one class, depending on plant species, a characteristic that tends to be used in taxonomy studies to screen plant material. Within the plant, phytochemicals might be transported via the xylem, phloem or the apoplast to different plant organs, since sites of synthesis are not necessarily the sites of storage (Wink, 1999). Hydrophilic compounds are often stored in the vacuoles or the apoplast, whereas lipophilic substances are deposited in resin ducts, laticifers, trichomes, oil cells or in the cuticle (Wink, 1999). Although stored in different plant organelles and organs, they can be metabolized and catabolized, and this commonly occurs when plants are subjected to stress conditions of various kinds. Thus, phytochemical levels vary, within a relatively wide range, according to the plant organ but also with the conditions of stress in a dynamic and complex system still not completely understood.

15.2.1 Major food plants and their bioactive constituents

Class of plants

Globally, humans consume and use a very large number of plant species from diverse plant families including Gymnosperms and Angiosperms (see Table 15.1 for examples of the most commonly used plant families and example food species in these families). These plants are used as major dietary crops, for oil production (seed and essential oils), as herbs and spices, as sources of medicinal compounds and also as animal feed (and thus indirectly affecting human health). Many of these species are cultivated commercially using both conventional (herbicide, pesticide, fungicide and chemical fertilizer input) and organic and sustainable farming practices. Within the scope of this review only the most commonly used crops and their major phytochemicals are compared in relation to the different cultivation practices and the external (genotype-independent) and internal (genotype-dependent) factors that affect the levels of the phytochemicals important for health. This includes effects

Table 15.1 List of common vegetable, herb and spice plant families and example species

Plant family	Example species		
Agaricaceae (fungi)	*Agaricus bisporus* (cultivated mushroom)	*Agaricus campestris* (field mushroom)	
Alliaceae	*Allium cepa* (onion)	*Allium porrum* (leek)	*Allium sativum* (garlic)
Anacardiaceae	*Anacardium occidentale* (cashew nut)	*Pistacia vera* (pistachio nut)	
Asteraceae/Compositae	*Cichorium endiva* (endive)	*Cynara scolymus* (artichoke)	*Lactuca sativa* (lettuce)
	Porophyllum ruderale (papalo)	*Scorzonera hispanica* (scorzonera)	*Tragopogon porrifolius* (salsify)
Betulaceae	*Corylus avellana* (hazel nuts)		
Brassicaceae	*Brassica napus* var. *napobrassica* (turnip)	*Brassica oleracea* var. *acephala* (kale)	*Brassica oleracea* var. *botrytis* subvar. *italica* (broccoli)
	Brassica oleracea var. *cauliflora* (cauliflower)	*Brassica oleracea* var. *capitata* (white cabbage)	
	Eruca sativa (salad rocket)	*Lepidium sativum* (cress)	*Nasturtium officinale* (watercress)
Chenopodiaceae	*Atriplex hortensis* (red and green orach)	*Beta vulgaris* var. *conditiva* (red beet)	*Beta vulgaris* ssp. *vulgaris* var. *cicla* (Swiss chard)
	Chenopodium quinoa (quinoa)	*Spinacia oleracea* (spinach)	
Cucurbitaceae	*Citrullus vulgaris* (watermelon)	*Cucumis melo* (melon)	*Cucumis sativus* (cucumber)
	Cucurbita pepo (pumpkin)		
Fagaceae	*Castanea sativa* (sweet chestnut)	*Fagus sylvatica* (beech; nuts)	*Quercus robur* (oak; acorns)
Lamiaceae/Labiatae	*Melissa officinalis* (lemon balm)	*Mentha* species (mints)	*Ocimum basilicum* (basil)
	Origanum vulgare (oregano)	*Rosmarinus officinalis* (rosemary)	*Thymus vulgaris* (thyme)
Lauraceae	*Cinnamomum zeylandicum* (cinnamon)	*Laurus nobilis* (laurel)	*Persea americana* (avocado)
Leguminoseae	*Cicer arietinum* (chick pea)	*Glycine max* (soya bean)	*Phaseolus vulgaris* (french bean)
	Vicia faba (faba bean)	*Vigna radiata* (mung bean)	

Table 15.1 Continued

Plant family	Example species		
Myrtaceae	*Eugenia uniflora* (Brazil cherry)	*Psidium guajava* (guava)	
Piperaceae	*Piper longusm* (Indian long pepper)	*Piper nigrum* (green, white, red and black pepper)	
Poaceae (cereals)	*Avena sativa* (oats)	*Hordeum vulgare* (barley)	*Oryza sativa* (rice)
	Secale cereale (rye)	*Sorghum bicolor* (sorghum)	*Triticum aestivum* (wheat)
	Zea mays (maize)		
Rosaceae	*Eriobotrya japonica* (loquat)	*Fragaria* x *annassa* (cultivated strawberry)	*Malus domestica* (apple)
	Prunus armeniaca (apricot)	*Prunus avium* (sweet cherry)	*Prunus domestica* (plum)
	Pyrus communis (pear)	*Rubus fruticosus* (blackberry)	*Rubus ideaus* (raspberry)
Rutaceae	*Citrus limon* (lemon)	*Citrus paradisi* (grapefruit)	*Citrus sinensis* (sweet orange)
Saxifragaceae	*Ribes nigrum* (blackcurrant)	*Ribes rubrum* (redcurrant)	
Solanaceae	*Capsicum annum* (sweet and bell peppers)	*Lycopersicum esculentum* (tomato)	*Solanum melongena* (eggplant)
	Solanum tuberosum (potato)		
Sterculiaceae	*Theobroma cacao* (cacao)		
Umbelliferae/Apiaceae	*Daucus carota* (carrot)	*Foeniculum vulgare* (fennel)	*Petroselinum crispum* (parsley)
Zingerberaceae	*Alpinia officinarum* (galangal)	*Curcuma longa* (turmeric)	*Zingiber officinale* (ginger)

that cause changes in health beneficial and health negative (anti-nutritional and toxic) components of the foods.

Class of compounds and their function in plants
Because of the great diversity of phytochemicals in fruits, vegetables, herbs and spices, only the major structural classes in the most common food plants, have been selected for discussion (see below and Figs. 15.2 and 15.3, for example, phytochemicals). These are also the phytochemicals for which there is generally the greatest information on food composition and health effects.

It must always be remembered that, when we consume food, it is as a 'whole food' and not as individual compounds, i.e. food is a mixture of primary (proteins, lipids, carbohydrates, vitamins and minerals) and secondary (phytochemicals) plant compounds. It is the balance of agonistic and antagonistic compounds that create overall health effects upon consumption. Therefore, subtle changes in the concentrations of various phtyochemicals could have great effects on health, for example, consider the hormetic dose–response model and effects such as synergism. There is also increasing evidence that some plant defence molecules, elicited in response to microbial and fungal infection, are immune system stimulators and may also be toxic, for example, the PR (pathogenesis-related)-proteins can elicit allergic reactions in some people and some phytoalexins may be directly toxic to humans (Beier and Nigg, 1992; Hoffmann-Sommergruber, 2002 Asenio *et al.*, 2004). This is a consideration when comparing the health and cost benefits of organic versus conventional cultivation. The human body has effective processes for xenobiotic metabolism of both health-beneficial and many toxic phytochemicals. However, for humans there are very few reports on the long-term effects of exposure to low doses of anti-nutritional substances, for example, potato glycoalkaloids. Therefore, using appropriate agricultural practices that ensure very low levels of these anti-nutritional and toxic compounds are synthesized in the food plants is very important. It should also be noted that many compounds that were traditionally considered as anti-nutritional are now known as potent anti-cancer or health-promoting agents such as certain alkaloids, coumarins, *Allium* sulfur compounds, glucosinolate hydrolysis products (isothiocyanates and indoles) and many types of simple phenolics and flavonoids. So, it needs to be made clear which compounds should now be considered anti-nutritional and/or toxic and those consumed in the diet that exert beneficial effects. Once again, the hormetic dose–response is a very important consideration: a phytochemical at a low dose exerts a positive health effect, at a higher dose becomes toxic. A good example, recently considered, is indole-3-carbinol, which is considered to be a 'Janus' molecule, i.e. it shows a positive face (anti-carcinogenic) at low concentrations and a negative face (potentially pro-carcinogenic) at high concentrations.

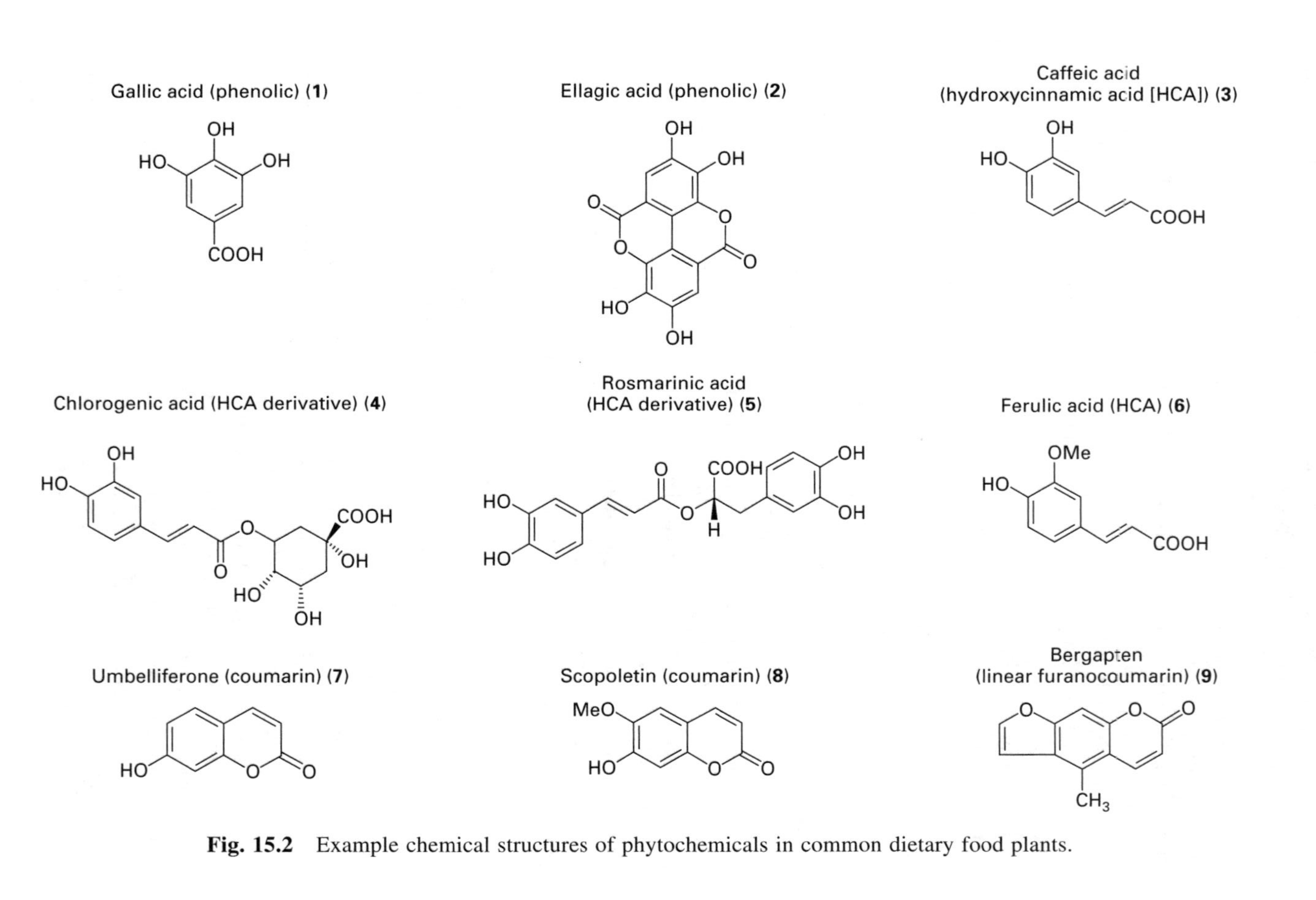

Fig. 15.2 Example chemical structures of phytochemicals in common dietary food plants.

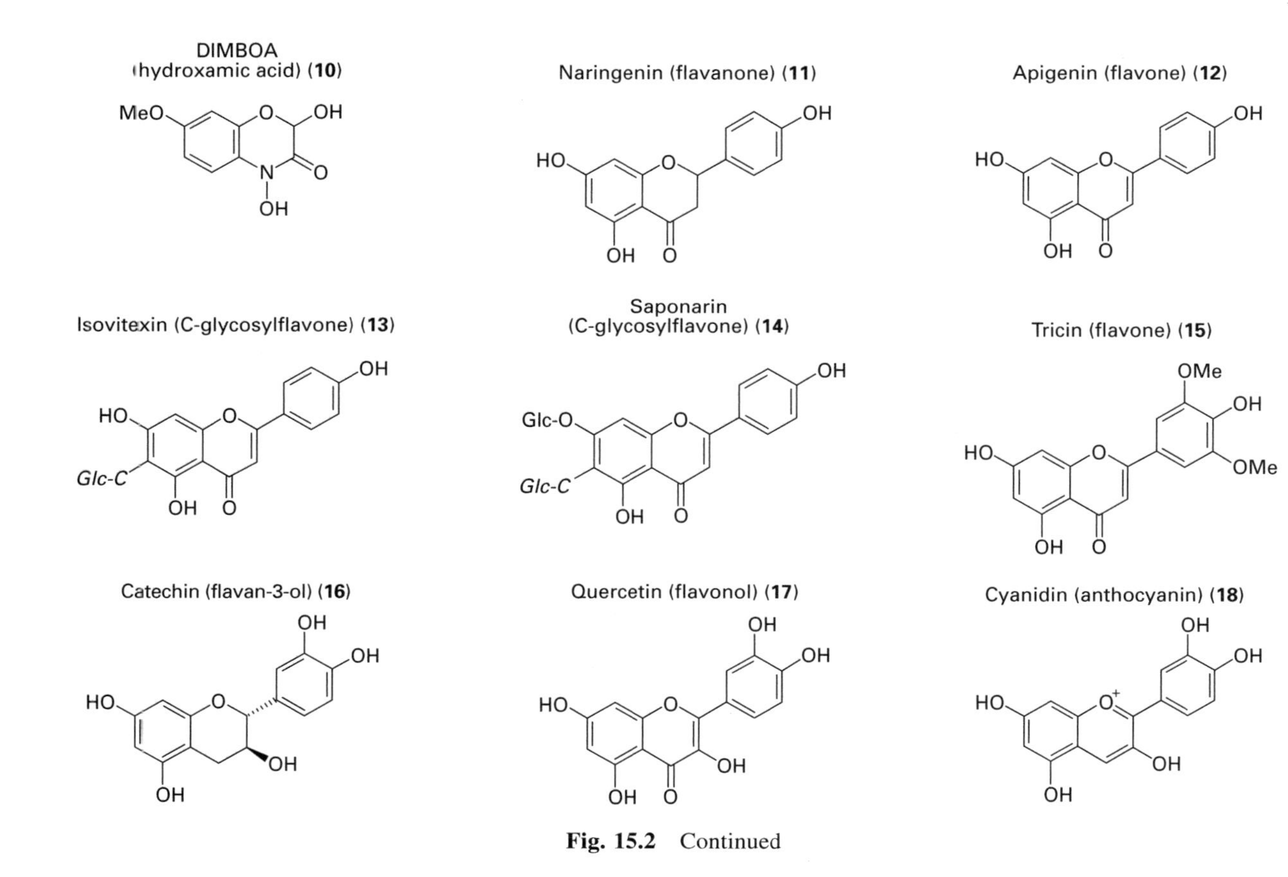

Fig. 15.2 Continued

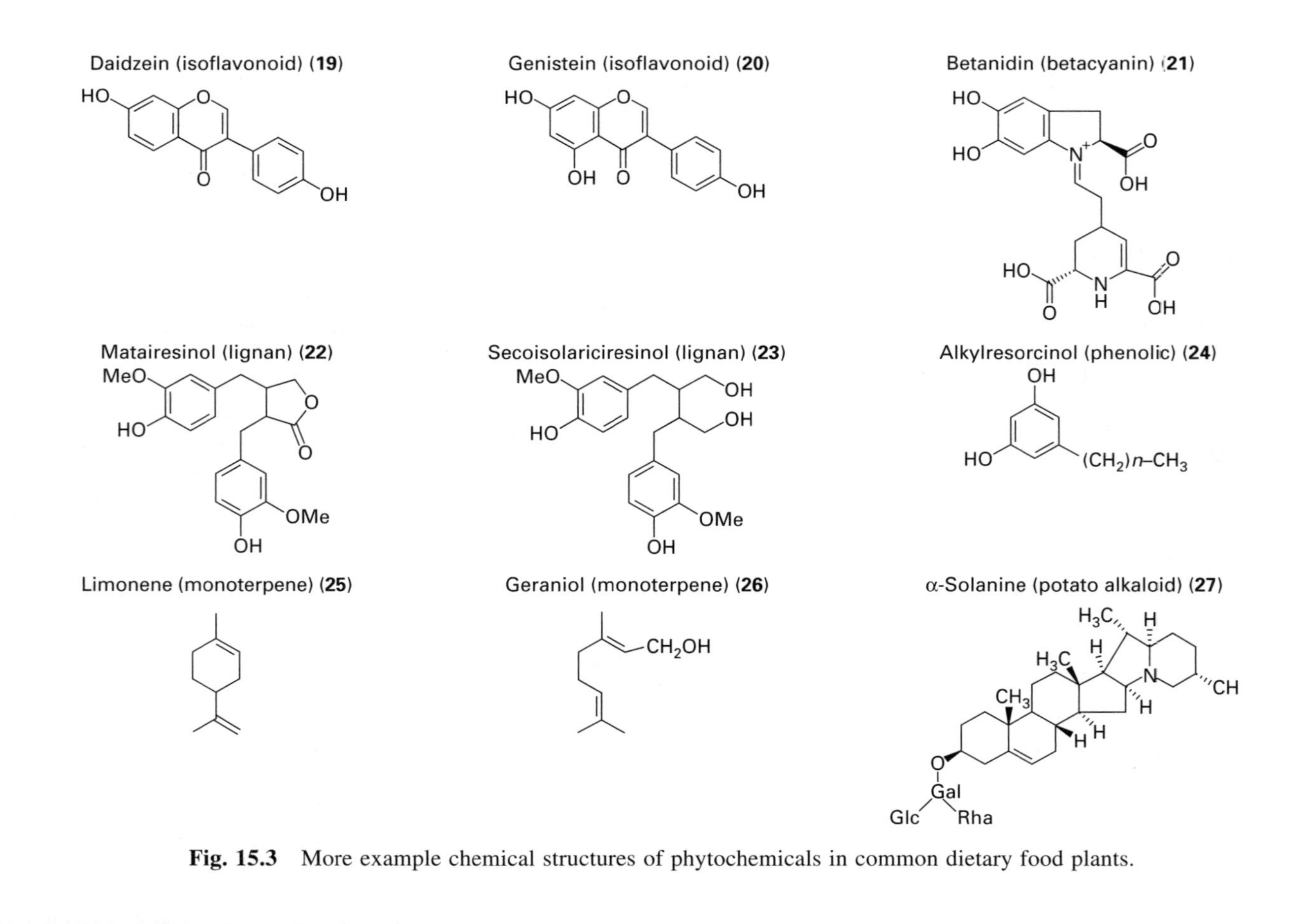

Fig. 15.3 More example chemical structures of phytochemicals in common dietary food plants.

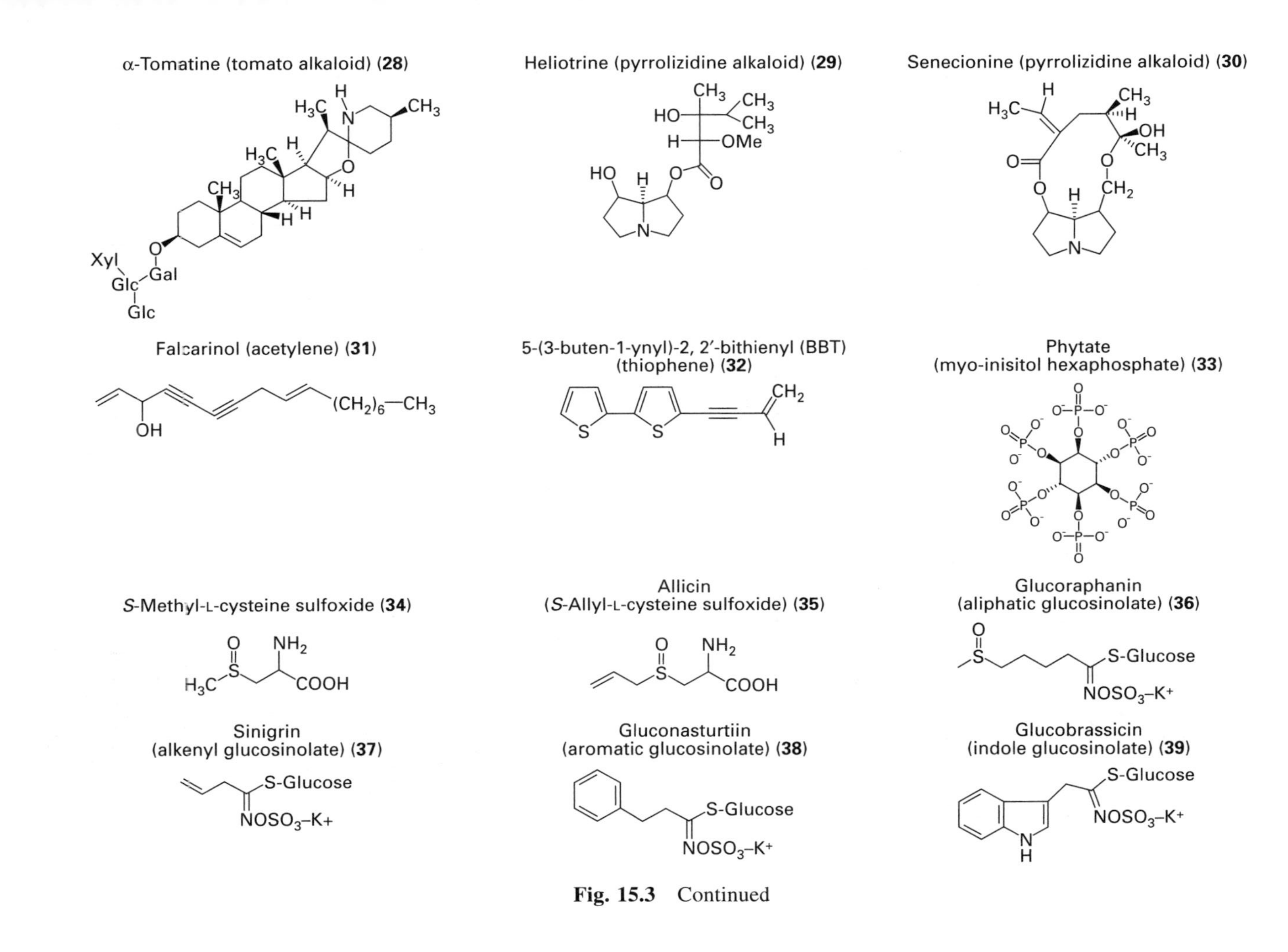

Fig. 15.3 Continued

Simple phenolics and their derivatives (benzoic, gallic, ellagic, gallotannins and ellagitannins), flavonoids and condensed tannins

Simple phenolics (Fig. 15.2, structures **1–6**) and polyphenolics (Fig. 15.3: flavonoids, structures **11–18**; isoflavonoids, structures **19–20**) are present in the majority of plant foods. Tannins are also widely distributed in food plants. They are divided into two classes (i) hydrolysable tannins (derived from gallic acid (**1**) and ellagic acid (**2**) producing gallo- and ellagi-tannins, respectively: Niemetz and Gross, 2005) and (ii) condensed tannins (non-hydrolysable tannins, also known as proanthocyanidins or procyanidins; commonly derived from catechin (**16**) and its isomer epicatechin). In plants many of these compounds function as antimicrobial, antifungal and insect deterrent compounds. Flavonoids and anthocyanins also contribute to flower colour, acting as attractants for pollinating insects (Harborne, 1994). Many phenolics are also involved in cell wall structure. Simple phenolics and polyphenolics have received interest because of their direct antioxidant activities and also the abilities to regulate xenobiotic detoxification mechanisms and exert anticancer effects in humans (Scalbert *et al.*, 2005; Halliwell *et al.*, 2005). Isoflavonoids are of interest because they exert estrogenic activity in humans and may be important anti-carcinogens in hormone-related cancers, for example breast cancer. In cereals C-glycosylflavonoids such as vitexin, derivatives such as saponarin and the polymethoxyflavone tricin (Fig. 15.2, structures **13–15**, respectively) and flavolignan derivatives of tricin are common (Harborne, 1994). Vitexin and related flavonoids have been investigated for their anti-oxidant activity and tricin is being evaluated as an anti-cancer compound (Lin *et al.*, 2002a; Verschoyle *et al.*, 2005). Low molecular weight phenolics are readily absorbed and metabolized in humans and animals. Several metabolic routes have been characterized, for example, phenolic acid *o*-methylation and conjugation to glycine producing hippuric acids, and flavonoid *o*-methylation and conjugation to sulfate and glucuronic acid. These metabolites are generally excreted within 48 h of consumption (Nielsen *et al.*, 2002; Kroon *et al.*, 2004; Manach *et al.*, 2004). Higher molecular weight compounds, such as proanthocyanidins, not absorbed in the small intestine are metabolized by colonic microorganisms to low molecular weight acids that can then be absorbed from the colon. Anti-nutritional effects reported for tannins include a reduction in amino acid release from dietary proteins, i.e. tannins can cause the precipitation of proteins and also form covalent complexes with proteins thus making them harder to digest by proteases (Griffiths, 1991).

Hydroxycinnamic acids and their derivatives (free, soluble conjugated, cell wall bound)

Hydroxycinnamic acids are common in the majority of plant species and certain fruits and vegetables (e.g. plums and artichoke) and cereal brans (e.g. whole grains) are a good source (Clifford, 1999). They occur less often as the free acids (e.g. *p*-coumaric, caffeic, ferulic, sinapic) but usually occur as either soluble conjugated compounds (chlorogenic acids in many species, tartaric acid conjugates in lettuce, sinapine/sinapoyl-malate in the Brassicaceae, gentiobiose esters in the Brassicaceae, and less widely distributed derivatives such as rosmarinic acid in the Lamiaceae herbs such as *Rosmarinus*, *Ocimum*, *Thymus* and *Salvia*) (Fig. 15.2, structures **3–6**). They are generally bioavailable in livestock and humans. Their function in plants appears to be anti-oxidant, anti-microbial, anti-fungal and structural/defence as precursors of lignans and lignin.

Coumarins (derived from hydroxycinnamates; Fig. 15.2, structures **7–9**) are also distributed in many plant species but are especially common in the Asteraceae, Apiaceae and Rutaceae. The linear furanocoumarins are well known as photoactive stress-induced defence compounds that can cause extreme dermatitis, skin damage and also cancer (Beier and Nigg, 1992). Several herbs of the Asteraceae and many species in the food plants in the Umbelliferae (Apiaceae) contain coumarins, and in some situations can be induced in very high concentrations (Beier and Nigg, 1992). A good example is celery (which contains psoralen, bergapten (**9**), xantotoxin and isopimpinellin) for which there are several epidemiological reports on photophytodermatitis (Beier and Nigg, 1992; Diawara *et al.*, 1995).

There are also more complex hydroxycinnamic acids in cereals and other members of the Poaceae. These are cell wall-bound: mono-, di- and tri-ferulates. Although they appear to have significant antioxidant activity, their bioavailability (release from cell walls and subsequent absorption in humans) is very low.

Hydroxamic acids

The hydroxamic acids, a group of phytochemicals found in wheat, are known for their fungistatic, insect anti-feedant and phytotoxic properties. They have been reported to be toxic to a broad range of insects, bacteria and fungi, including northern leaf blight, *Agrobacterium tumefaciens* (root gall bacteria), the European, Asian and southwestern corn borers and various aphid species (Melanson *et al.*, 1997). These compounds are DIMBOA (2,4-dihydroxy-7-methoxy-(2*H*)-1,4-benzoxazin-3(4*H*)-one) (Fig. 15.2, structure **10**), DIBOA (2,4-dihydroxy-(2*H*)-1,4-benzoxazin-3(4*H*)-one) and their transformation products 6-methoxybenzoazolin-2-one (MBOA), 2-hydroxy-7-methoxy-1,4-benzoazin-3-one (HMBOA), 2-hydroxy-1,4-benzoazin-3-one (HBOA) and benzoazolin-3-one (BOA). Natural variation in these phytochemicals can be large and total levels of DIMBOA and DIBOA vary between 250 and 1650 $\mu g\ g^{-1}$ dry weight in the whole plant (growth stage 9–10) when grown under organic conditions, whilst levels can be half of this under conventional growing (Mogensen *et al.*, 2004). Such differences were partly attributable to soil types. These hydroxamic acids are not present in wheat grain but are synthesized early in seed germination, with the highest concentrations occurring at the seedling stage (Oleszek, *et al.*, 2004). There is commonly a low impact of these compounds on human health since the non-germinated grain is predominantly used for consumption, for example, for the production of flour. However, there is an increasing trend to eat seedlings (sprouts) of some plants and this includes wheat, therefore consumption of wheat seedlings could lead to a high exposure to hydroxamic acids. When DIMBOA and DIBOA are leached out into the soil, they are chemically transformed to MBOA and BOA, respectively. The degradation of both BOA and MBOA at the lowest concentration (400 ng g^{-1}) occurs quickly with a half life of less than 1 day and none of the metabolites 2-aminophenoxazin-3-one (APO), 2-acetamidophenoxazin-3-one (AAPO), 2-amino-7-methoxyphenoxazin-3-one (AMPO) and 2-acetamido-7-methoxyphenoxazin-3-one (AAMPO) detected (Fomsgaard *et al.*, 2004). The same authors performed studies with MBOA at 400 $\mu g\ g^{-1}$ and BOA at 400 and 4000 $\mu g\ g^{-1}$ (covering by far the highest levels observed in the whole plants grown under organic farming) and found that the half-life is still below 90 days, which is the limit that is set for a half-life for synthetic pesticides in the European registration procedure.

Lignans and alkylresorcinols

Cereals are good sources of lignans and specifically whole grain foods (Liggins *et al.*, 2000; Milder *et al.*, 2005). They are bioavailable and upon consumption the dietary lignans matairesinol and secoisolariciresinol (Fig. 15.3, structures **22** and **23**, respectively) are converted to enterolactone and enterodiol, compounds with phytoestrogenic activity (Nicolle *et al.*, 2002). Alk(en)yl-resorcinols are phenolic lipids found in many cereal species (Ross *et al.*, 2004; Mattila *et al.*, 2005). In humans they are readily absorbed, are present in plasma, and excreted in urine (Ross *et al.*, 2004). They also appear to have antimutagenic activity (Gąsiorowski *et al.*, 1996).

Betalains

The betalains (including the yellow betaxanthins and the red–violet betacyanins) are a class of nitrogenous phytochemicals found in the 13 families of the plant order Caryophyllales and some Cactaceae (Stintzing *et al.*, 2002; Kanner *et al.*, 2001; Cai *et al.*, 2001; Kugler *et al.*, 2004). They give the characteristic red and yellow colours to beet, Swiss chard and *Amaranthus* species, as well as many of the Chenopodiaceae (e.g. red orach, *Atriplex hortensis*) and also *Opuntia ficus-indica* fruits. Their role in plants is thought to be primarily as photoprotectants, i.e. anti-oxidants for preventing ROS (reactive oxygen species) formation. The core structure of many betalains is the betacyanin betanidin (Fig. 15.3, structure **21**). In humans betalains are readily absorbed and excreted (Tesoriere *et al.*, 2004). There is considerable interest in using betalains and anthocyanins as safe, natural, food colours as replacements for the older synthetic colour chemicals (many of which have been shown to be carcinogenic).

Terpenoids

Terpenoids are widely distributed in the plant kingdom, often because they have an essential role in plant defence (insect deterrents, anti-microbial and anti-fungal) and in insect attraction for pollination (volatile terpenoids and non-volatile pigments such as carotenoids), and also as part of cell membranes. The terpenoids include a wide range of structures from simple monoterpenes and sesquiterpenes to the more complex triterpenoids (saponins), carotenoids, vitamin E (tocopherols and tocotrienols) and phytosterols. Limonene and geraniol (Fig. 15.3, structures **25** and **26**, respectively) are commonly occurring monoterpenes present in many essential oil producing plants (e.g. herbs from the Lamiaceae such as *Rosmarinus* and *Salvia*) and *Citrus* species (Rutaceae). The structural diversity means that there is a wide range of both positive (health-promoting) and negative (irritant and toxic) effects dependent on the specific terpenoid. Monoterpenes and herb essential oils, especially limonene, have recently received interest owing to their potential as anti-cancer compounds, both dietary and for chemotherapy (Crowell, 1999).

Alkaloids

Alkaloids are present in a wide range of plant families and have a variety of biological effects (Roberts and Wink, 1998). They are divided into two major classes: the proto-alkaloids (nitrogen-containing but not heterocyclic in structure, e.g. phenylethylamines) and the true alkaloids (nitrogen-containing heterocyclic compounds). Their function in plants is primarily as defence compounds against insect pests and herbivorous animals. Other than the known medicinal plant species, there are few food plants that contain significant levels of alkaloids. However, members of the Solanaceae have the potential under stress conditions, such as potato tubers exposed to light, to accumulate high concentrations of alkaloids. Examples include α-solanine (Fig. 15.3, structure **27**), the related α-chaconine and also α-tomatine in tomatoes (Fig. 15.3, structure **28**) (Beier and Nigg, 1992). Alkaloid exposure can also occur from consumption of some herbal preparations that have been contaminated with weed species that have high levels of alkaloids. This is a possible risk with both organic and conventional cultivation, i.e. weed species could inadvertently be harvested and incorporated into the food material. The alkaloids causing greatest concern are the highly toxic/mutagenic pyrrolizidine alkaloids (Prakash *et al.*, 1999; Lin *et al.*, 2002b). Examples include heliotropine from *Heliotropus* species and senecionine from *Senecio* (groundsel) species (Fig. 15.3, structures **29** and **30**, respectively).

(Poly)acetylenes and thiophenes

These phytochemicals are common in the Asteraceae (Compositae) and Apiaceae (Umbelliferae) (Chitwood, 1992; Zidorn *et al.*, 2005). There is considerable interest in these compounds because of their potential as anti-cancer agents and therapeutic agents (e.g. Zheng *et al.*, 1999; Kobæk-Larsen *et al.*, 2005). Falcarinol (Fig. 15.3, structure **31**) and its derivative falcarindiol are phytoalexins and are induced by fungal infection (Kuc, 1992). Many members of the Asteraceae also contain sulfur-containing polyacetylene derivatives known as thiophenes (Tosi *et al.*, 1991; Margl *et al.*, 2001). These compounds have a wide range of biological activities including anti-nociceptive (pain-killing) properties and most are potent insecticidal and insect deterrent compounds (Goncales *et al.*, 2005). A common example is 5-(3-buten-1-ynyl)-2, 2′-bithienyl (BBT) (Fig. 15.3, structure **32**). Both acetylenes and thiophenes can be absorbed and metabolized by livestock and humans.

Phytate

Phytate (myo-inositol hexaphosphate; Fig. 15.3, structure **33**) is found in many food species and can be considered as a phytochemical. Its role in the plant is primarily as a phosphate store in seeds, but it is found in other tissues as well, for example, tubers (Harland *et al.*, 2004). Phytate and its hydrolysis products are anti-nutrients that chelate metal ions and thus reduce their bioavailability (Persson *et al.*, 1998; House, 1999). This is particularly a problem with cereal grains, but pre-processing can improve mineral absorption from these foods (Agte and Joshi, 1997). There is some concern that high phytate foods could also contain higher levels of toxic heavy metals caused by natural accumulation. Plants also contain phytate-degrading enzymes that can also influence metal ion bioavailability (Viveros *et al.*, 2000).

S-Alk(en)yl-L-cysteines and *S*-alk(en)yl-L-cysteine sulfoxides

The most common dietary sources of these compounds are the Alliaceae (*Allium* species such as onion, garlic, leeks and chives) and Brassicaceae (specifically *Brassica* species such as cabbage, Brussels sprouts and broccoli) (Marks *et al.*, 1992; Kubec *et al.*, 2000). There are diverse structures with different alk(en)yl substitutions, but the core is the cysteine molecule. In garlic the *S*-alkenyl-L-cysteines are predominant, for example, alliin (Fig. 15.3, structure **35**), whereas in onion *S*-alkyl-L-cysteines are predominant. Common to both the Alliaceae and Brassicaceae is *S*-methyl-L-cysteine sulfoxide (SMCSO; Fig. 15.3, structure **34**). When the plant tissues are crushed or cooked, the *S*-alk(en)yl-L-cysteines are catabolized by endogenous *C-S* lyases to pyruvate, ammonia and unstable sulfur intermediates. In the case of SMCSO this intermedediate is sulfenic acid (CH_3-S-OH) which dimerizes to form methylmethane-thiosulfinate (MMTSO; CH_3-SO-S-CH_3), which can further react non-enzymatically to form various sulfur compounds including di- and trisulfides (CH_3-S-S-CH_3 and CH_3-S-S-S-CH_3) and dimethylthiosulfonate (CH_3-SO_2-S-CH_3) (Marks *et al.*, 1992). The flavour compounds released from the *S*-alk(en)yl-L-cysteines are generally considered to be health beneficial and there is a very large amount of literature on their effects (e.g. Tapiero *et al.*, 2004; Herman-Antosiewicz and Singh, 2004).

The most clear-cut effects are anti-bacterial, anti-oxidant, anti-atherogenic, anti-cancer and immunostimulatory.

Glucosinolates

As with the *S*-alk(en)yl-L-cysteines there is a very large amount of literature on glucosinolates. There have been many extensive reviews on the glucosinolate composition of common Brassicaceae and related vegetables and thousands of papers on individual species from almost every family within the *Capparales*. The key reviews on common *Brassica* crops are those of Fenwick *et al.* (1983; with some additional data on salad and herb species) and Rosa *et al.* (1997). For an overall glucosinolate review of other data and all *Capparales* species, see Fahey *et al.* (2001). There is increasing information on the health effects of secondary metabolites in *Capparales* species on animals and humans. Several decades ago the focus was on the negative, anti-nutritional effects of glucosinolates and their hydrolysis products (e.g. see Fenwick *et al.*, 1983). However, it is becoming clear that many of the hydrolysis products from dietary crucifers have beneficial effects in humans – specifically the anti-carcinogenic activity of isothiocyanates such as sulforaphane (e.g. Johnson, 2002) and indoles (e.g. Bonnesen *et al.*, 2001). Examples of common glucosinolates found in food plants include glucoraphanin (found in *Brassica* species and *Eruca sativa* (salad rocket) which is the precursor of sulforaphane (SFN), sinigrin (found in many *Brassica* species), which is the precursor of allylisothiocyanate (AITC), gluconasturtiin (found in watercress (*Nasturtium officinalis*) and land cress (*Barbarea praecox*), which is the precursor of 2-phenylethylisothiocyanate (PEITC) and glucobrassicin (found in all *Brassica* species) which is the precursor of indole-3-carbinol (I3C) and 3, 3′-di-indolylmethane (DIM) (Fig. 15.3, structures **36–39**, respectively).

15.3 Assessment and bioavailability of phytochemicals

Most studies comparing organically and conventional grown foods have only focused on the major nutrients (e.g. proteins, lipids, fibre, minerals, etc.) and there is less data for phytochemicals. This is clearly a limited approach and only recently have researchers begun to address the potential differences between food plants grown using organic and conventional methods, that is in terms of the metabolism and bioavailability of the various nutrients and phytochemicals. Thus, bioavailability studies are urgently needed. Bourn and Prescott (2002) also raised the question of whether the nutrient concentration is expressed on a fresh or dry weight basis, since organically grown products tend to have higher dry matter content than conventionally grown products. Another key aspect to consider is that diets are not likely to be made entirely of organic products.

15.4 Potential positive and negative effects of phytochemicals on livestock and human health

Phytochemicals have been the subject of many studies evaluating their effects in relation to common chronic human illnesses such as cancer and cardiovascular diseases. These studies encounter difficulties in using this information to influence the dietary patterns of consumers because in the past they have used models or experiments with animals. However, in the last decade, researchers have moved away from animal studies in favour of human cell models or human intervention studies. Scientists still need to determine the likely incidence of illness from exposure to known amounts of a given natural compound in the diet and specifically in relation to the complex matrices of whole foods. Therefore, it is inevitable that some animal studies have to be continued for toxicological studies.

At normal levels of consumption, although large variations might occur according to type of population and special groups of consumers, it is generally accepted that phytochemicals are not as harmful as would be expected from their LD_{50} values (Brandt and Mølgaard, 2001). This in part is due to extrapolation of data from toxicology models that are not always the best indicators of toxicity *in vivo*, for example, the large amount of erroneous data derived from the Ames test. The other aspect is that humans are capable of metabolizing and efficiently excreting a wide range of xenobiotics including dietary phytochemicals.

There is also the additional concern of using transgenic plants and their environmental and genetic impact, and more specifically their effects in the human diet. Although a detailed discussion on transgenic crops is not within the scope of this review, it is worth mentioning a recent commentary on bioactives in transgenic plants by Finley (2005).

In this section a brief reference can be made to xenobiotics. Organic products are usually found to contain no pesticide residues and, if present, they are typically of significantly lower incidence and levels than those found in non-organic products, as a result mostly of environmental pollution from non organic agriculture (Heaton, 2002; see also Ch 14). Heaton (2002) also commented on food additives and found that, whilst more than 500 are permitted for use in processed foods, only around 30 are permitted in organic processing meeting the expectation of consumers regarding foreign compounds in foods.

15.5 Impact of phytochemicals on crop resistance to pests and diseases

The specific function of many phytochemicals is still unclear; however, a considerable number of studies have shown that they are involved in the interaction of plants/pests/diseases. Most plants produce phytochemicals as anti-microbial, antibiotic, insecticidal and hormonal agents either as part of their normal programme of growth and development (inbuilt chemical barriers), or in response to pathogen attack or stress (phytoalexins). They may play multiples roles and could interact synergistically. There are many examples that illustrate the enormous impact of these compounds in ecological, biological and agronomic fields, particularly relating to plant–herbivore and plant–pests and pathogens interactions.

Based on the tendency for higher levels of phenolics in organically grown plants, these are potentially less susceptible to pests and diseases. Indeed, Schultz and Baldwin (1982) and Rossiter *et al.*, (1986) observed that high phenolic contents reduced the size of female gypsy moth pupae, leading to a decline in reproductive production of female gypsy moths.

Organically grown *Brassica* seem to have higher levels of glucosinolates, which act as defensive compounds against the cabbage white butterfly (*Pieris rapae*) and form volatiles that attract predators and parasites of the butterfly, reduce the nutritional quality of the plant for generalist herbivores and increase the extra-floral nectar in ant–plant systems (Agrawal, 2000). This general behaviour could result in a decrease in herbivore damage owing to induced resistance or induced defence, as previously reported by Karban and Baldwin (1997). The effect of glucosinolates and their derivatives on pests and diseases was reviewd by Rosa *et al.* (1997) and Brown and Morra (1997). If more precise studies can confirm the general tendency for higher glucosinolate levels in organically grown Brassicaceae, this might result in reduced susceptibility to most of their natural pests. However, since plants contain other phytochemicals which might interact and change the response to a given pest, and because several attempts to correlate the levels of glucosinolates with resistance to specific pathogens have failed (Mithen, 2001), there must be caution in considering that high levels of glucosinolates, their derivatives and other phytochemicals are synonymous with resistance. This probably reflects the complex interplay between co-occurrence and possible co-variation of numerous other defence compounds in plants, for example, alkaloids (Macel *et al.*, 2002) or terpenoids (Langenheim, 1994; Fassbinder *et al.*, 2002) and other interacting organisms (Moyes *et al.*, 2000).

In plants grown without pesticides, there is likely to be some degree of infestation and this, apart from inducing the formation of volatiles which might act as plant protectants, as previously described, may also alter the host plant physiology and chemistry, as shown by Mayer *et al.* (2002) in a study with the pest silverleaf whitefly (*Bemisia argentifolii*) (Bellows & Perring). The silver whitefly induces a number of host plant defences, including pathogenesis-related (PR) protein accumulation, for example, chitinases, β-1,3-glucanases, peroxidases, and so on. (Konstantopoulou *et al.* 2004), studying the effect of volatile oils [(7,11,15-tetramethyl- 2-hexadecen-1-ol (phytol); (*Z*)-3-hexenol nonanal, pentadecanal, neophytadiene, (*Z*)-3-hexenyl acetate and an analogue of 2,4-dihydroxy-7-methoxy-(2H)-1,4-benzoxazin-3-(4H)-one (DIMBOA)], from leaves of seven corn hybrids on the oviposition of *Sesamia nonagrioides* (corn borer) females, verified that those compounds affected the rate of oviposition, confirming the results obtained in a previous work (Konstantopoulou *et al.*, 2002). Both studies concluded that these volatile oils deter females from ovipositing, suggesting that corn genotypes with high quantities of C9–C14 aldehydes can deter females from ovipositing and could be largely used as pest management tools for development of effective and environmentally safe control methods for *S. nonagrioides*.

Another field of research is the possibility offered by phytochemicals in protecting plants against diseases and pathogens (fungus, bacteria and nematodes). Numerous studies have suggested that plant–pathogen interactions are partially mediated via plant secondary metabolite production, despite the inconsistency revealed by some works on the ability of particular compounds to provide resistance to a specific pathogen.

According to several authors, even healthy plants produce a huge number of compounds by secondary metabolism that can protect plants against attack by a wide range of potential pathogens. For example, Lüning and Schlösser (1976), Price *et al.* (1987) and Osbourn (1996), report anti-fungal activity for saponins (glycosylated triterpenoids, steroid and steroid alkaloid molecules). Défago and Kern (1983) and Défago *et al.* (1983), reported an anti-fungal activity of the glycoalkaloid α-tomatine for tomato and pea plants, when green tomatoes were submitted to the action of *Fusarium solani*, a common fungus in tomato soil. They found that green tomatoes, with high levels of α-tomatine, were more resistant to *Fusarium solani* than the α-tomatine-deficient plants.

The Solanaceae plants such as potato, tomato, eggplant and peppers are a good example of the mechanism of host–plant resistance. They accumulate a variety of secondary metabolites including phenelic compounds, phytoalexins, protease inhibitors, carotenoids, lycopenes and glycoalkaloids (atropine, α-chaconine, α-solanine, dehydromatine and α-tomatine among others), which serve as natural defences against plant phytopathogens including fungi, bacteria and viruses; (Valkonen *et al.*, 1996; Roddick, 1996; Friedman and McDonald, 1997; Lachman *et al.*, 2001; Friedman, 2002; and more recently Friedman 2004).

Other examples include the glucosinolates and their hydrolysis products, the isothiocyanates. Isothiocyanates possess a range of anti-fungal, anti-bacterial and anti-microbial activities and thus inhibit microorganisms and repel insects and molluscs (Fenwick *et al.*, 1983, Glen *et al.*, 1990; Fahey *et al.*, 2001). Isothiocyanates are highly effective, volatile, general biocides generally owing to their reactions with glutathione and proteins (Brown and Morra, 1997). The toxicity of several isothiocyanates to certain nematode species is well documented (Mojtahedi *et al.*, 1991; Johnson *et al.* 1992; Lazzeri *et al.*, 1993; Heaney and Fenwick, 1995; Griffiths *et al.*, 1998; Ferris and Zheng, 1999; Chitwood, 2002; Buskov *et al.*, 2002; and Zasada and Ferris, 2003). These bioactive compounds could intervene in the biological activity of the soil, depending, however, on the chemical class and individual glucosinolate concentrations of incorporated *Brassica* material (Gardiner *et al.*, 1999; Dandurand *et al.*, 2000; Warton *et al.*, 2001). Many more examples can illustrate the real impact and importance of phytochemicals in crop resistance to pests and diseases. The last four decades have witnessed a rising consciousness of the functions of these compounds, as well as interest in the biochemistry and molecular biology underlying their synthesis and control mechanisms.

15.6 Factors that modulate differences in phytochemical levels and other major constituents between organic and conventional farming

Apart from genotype variation, biotic and abiotic factors account for the large differences found between the same cultivars. The use of fertilization, pesticides and weed control agents are the major differences between organic and conventional farming. Several studies have compared both cultivation systems but results have been inconsistent and they did not evaluate each potential modulating factor on the product composition. Comparisons of differences in all farm management practices between organic and conventionally grown crops were found to be even more difficult with very few differences and high variability between the results. This was attributed to the interaction of a large number of variables affecting nutritional value (Bourn and Prescott, 2002). Thus, in this section we are going to separate the effect of each input on the quality of fruits and vegetables, particularly with respect to the previously discussed major phytochemcials.

15.6.1 Genotype variation

If all biotic and abiotic factors can be controlled, genetic influence is one of the major features responsible for differences in phytochemical composition. Several reviews on diverse compounds have reported the genetic influence on phytochemicals, sometimes to a surprising order of magnitude. The influence of cultivar can be even greater than the influence of growth conditions, thus complicating the data analyses for plant composition.

The phytochemical contents of cultivars of the same species can be significantly different, even within the same field. Evidence for this statement has been given in several recent papers and reviews: glucosinolates (Rosa *et al.*, 1997), phenolic compounds (Osier and Lindroth, 2001; Kalt *et al.*, 2001; Howard *et al.*, 2002; Scalzo *et al.*, 2005; Dykes *et al.*, 2005; Pandjaitan *et al.*, 2005; Mpofu *et al.*; 2006) and alkaloids and terpenoids (Theis and Lerdau, 2003).

Although disease-resistant cultivars are used in organic farming, some landraces and local ecotypes of grown plants are well adpated to the abiotic conditions being used, with adavantages in organic farming being due to less susceptibility to pests and diseases. Good examples of these can be seen in Brassicaceae and Fabaceae. Studies with potatoes have shown that resistant cultivars tend to have higher levels of defence compounds like glycoalkaloids (Sanford *et al.*, 1992), and similar observations were noted in *Brassica rapa* and *Brassica oleracea* var. *acephala*, but not for broccoli with regard to the glucosinolates (Rosa, unpublished data). Based on this, Brandt and Mølgaard (2001) stated that fruits and vegetables grown organically could provide greater health-promoting benefits than conventionally grown crops. However, it would be more prudent to say that they are safer, with respect to xenobiotic inputs, since we do not yet know the phytochemical variation compared with conventional farming. We believe that, with the development of organic farming, particularly composting methods and biological control of pests, diseases and weeds, there would be significant ecological and health benefits, and that differences in yields would be negligible.

15.6.2 Fertilization

There are several studies on the effect of fertilization on product composition. However, they were of limited scope because they do not provide clear answers on the effect of farming systems on composition. Bourn and Prescott (2002) made a comprehensive review on this subject and, overall, the studies suggested that the use of organic fertilizers may result in lower nitrate concentration for some crops and some cultivars than when using more soluble mineral fertilizers. Based on the observed large variations, they emphasized the strong influence of climatic conditions on nitrogen, nitrate and mineral content, as well as fertilizer treatments.

Nitrogen

The use of nitrogen fertilization results in higher content of N-containing compounds, including free amino acids, and also increases in terpene content in wood plants, whilst starch, total carbohydrates, phenylpropanoids and total carbon-based phytochemicals decreased (Koricheva *et al.*, 1998). Higher levels of nitrogen favoured its uptake and increased the nitrate content of the crop, which is critical for salad vegetables and baby foods.

On organic farms, the supply of required nutrients for plant growth is predominantly from manure from different sources, usually composted, and without any contamination from synthetic chemicals, in combination with crop rotation. The release of plant-available nutrients from manure is gradual and, on average, is estimated as 50% for the first year and 35% and 15% for the second and third years, respectively. Thus, even applied in large amounts, it is not likely that there will be a rapid release of nitrogen which would influence the biosynthesis of several classes of phytochemicals. For instance, the low nitrogen availability caused by manure application in organic farming reduces the synthesis of protein and the nitrogen content of grain of wheat and other cereals. This situation was seen in several studies reviewed by

Woese *et al.* (1997), in which levels of protein in organically produced wheat were lower than in conventionally grown wheat. This fact seems to be a general feature of the food products produced in organic farming systems. In the case of wheat this leads to undesirable consequences for baking properties. However, this effect can be overcome by selecting suitable cultivars for this purpose. The low availability of nitrogen also induces a lower synthesis of amino acids, the precursors of many phytochemcials (e.g. phenolics, glucosinolates, *Allium* flavour compounds), potentially leading to a reduction in phytochemical content. Regular use of composted manure will help to overcome the low availability of nitrogen in the soil, particularly in situations where the use of organic fertilization was not practised for many years. Thus, in these circumstances, nitrogen levels will be closer to conventional farming levels and this factor would smooth the differences between both growing systems.

The acceptable level of nitrates is regulated for only a small group of fruits and vegetables (EU Reg. 466/2001). Thus, commodities should not be traded with levels exceeding the fixed values. However, situations might occur where recommended levels are unclear leaving uncertainty and major concern for consumers. Obviously, such concerns are extended to non-regulated commodities. Recommended fertilization practices in organic farming clearly reduce the nitrate concentrations of food products as reviewed by Woese *et al.* (1997) and Brandt and Mølgaard (2001). Recent comprehensive studies of potatoes confirm the lower nitrate concentration in organic production, in which 40-45 t ha^{-1} of manure was applied as a nitrogen source (Hajslová *et al*, 2005).

Phosphorus, potassium and other minerals

There is some evidence that organic farming methods and the use of manure leads to a lower soil phosphorus content leading to a lower carbohydrate content in the crops (Heldt *et al.*, 1977), but there have been few recent studies to support this data. Potassium levels do not seem to affect the composition of plant material systemically (Brandt and Mølgaard, 2001) and well-balanced compost providing major and minor nutrients is a guarantee of negligible effects on the quality of organically produced fruits and vegetables.

15.6.3 Pesticides and fungicides

Mechanical injury to plants caused by agricultural practices and management, or from pests or diseases, induces a reaction in the plant by activation of the plant secondary metabolism. Increases in glucosinolates, phenolics, particularly chlorogenic acid and the activation of polyphenyloxidases and peroxidases are examples of the consequence of herbivore damage and infection (Rosa *et al.*, 1997; Lattanzio *et al.*, 2001; Du and Bramlage, 1995), which is more likely to occur in organic cultivation. Studies of plant disease severity in conventional and organic systems have shown that overall disease incidence is not much higher in organic than in conventional farming, even when the same cultivars are used (Van Bruggen, 1995) (cultivars known to be very susceptible were not included in these comparisons). Ferulic acid and *p*-coumaric acids are also part of the defence mechanism since they are able to form esterified bonds with the cell wall's polysacharides, increasing fruit firmness and consequently mechanical resistance (Lombardi-Boccia *et al.*, 2004). Thus, plant damage that induces the metabolism of these anti-fungal agents and increases the activity of some enzymatic systems, can lead to organically grown plants having a higher intrinsic resistance than the conventional ones, a possible explanation of why they can cope so relatively well in the absence of pesticides (Brandt and Mølgaard, 2001).

When plants are exposed to pathogens or certain chemicals, they react by producing phytoalexins. Significant induction of phytoalexins is usually a consequence of resistant reactions and is often restricted to the site of infection and thus a small percentage of the total crop tissue, whereas in susceptible plants there is a lower or negligible induction of phytoalexins. Therefore, in terms of potential negative health effects, an organic crop that has low levels of infection will not be less healthy than one produced by conventional methods using high chemical input. It is not likely that a product heavily infected with pathogens, with extreme tissue damage and subsequent high phytoalexin content, would be brought to market since external appearance is a key factor for consumers. Many fungicides are also of concern since they can accumulate in plant foods. In addition, it has been shown that fungicides can induce toxic furanocoumarins in celery (*Apium graveolens*) and thus there is a direct negative correlation with human health and the use of these chemicals (Nigg *et al.*, 1997). Pesticides are also known to affect the biosynthesis of other phytochemcials (Lydon and Duke, 1989).

15.6.4 Overall growth management effects

In many studies it was not possible to separate the overall effects of the management practices of organic farming and so we include this section to describe the overall effects on important plant food components. We have tried to separate the effect for each plant compound when possible.

Minerals

Marginal differences and inconclusive findings were recently reported for organic versus conventionally grown plums (Lombardi-Boccia *et al.*, 2004). Regarding trace elements, these seem to be more influenced by soil type and the respective levels in the soil than by cultivation systems (Woese *et al.*, 1997).

Vitamins

Vitamins have been one of the focus compounds in comparisons between organic and conventional food plants. Organic practices have been reported to increase vitamin C levels in potatoes (Hajslová *et al.*, 2005), in corn and strawberries (Asami *et al.*, 2003), in other vegetables (Mozafar, 1993) and in plums grown in grass-covered soils (Lombardia-Boccia *et al.*, 2004). The α- and γ-tocopherols and β-carotene were also higher in organic plums grown in grass-covered soil (Lombardi-Boccia *et al.*, 2004). In this study, the authors attributed the major differences in fruit composition to the type of soil management within the whole agriculture system. A decrease in β-carotene levels seems to be a feature of conventionally grown crops where high nitrogen fertilization levels were used as well as some pesticides (Leclerc *et al.*, 1991; Mercadante and Rodriguez-Amaya, 1991).

Phenolics

When comparing organic versus conventional grown plums, Lombardi-Boccia *et al.* (2004) found that total polyphenols and quercetin were higher in the conventional system whilst two other flavonols, myricetin and kaempferol, showed the highest levels in the organic plums. The highest phenolic acid content was detected in plums grown on soil covered with *Trifolium* species (clover). Previous studies (Andersen, 2000) also showed that high nitrogen fertilization reduced the levels of phenolic compounds associated with susceptibility to pests and diseases. Organic potatoes and apples were also reported to have higher contents of polyphenols and flavonols, respectively (Hamouz *et al.*, 1999; Weibel *et al.*, 2000). These findings are in agreement with a recent review on phenolics (Manach *et al.*, 2004) supported by other studies on strawberries, blackberries and corn (Asami *et al.*, 2003; Hakkinen and Torronen, 2000) and pears and peaches (Carbonaro *et al.*, 2002). In peaches and corn, levels in organic samplers were 33% and 58% higher than in conventional samples, respectively. Thus, the intake of major health-promoting flavonoids, such as quercetin and kaempferol, is higher for organic fruits and vegetables, as shown in an intervention study (Grinder-Pedersen *et al.*, 2003). Resveratrol, a grape (*Vitis vinifera*) phytoalexin, content in organic wine was 26% higher than in conventional wines in paired comparisons of the same grape variety (Levite *et al.*, 2000). A Japanese study with vegetables revealed an anti-oxidant capacity of organic spinach of 2.2-fold compared to conventionally grown spinach (Ren *et al.*, 2001a). In organic qing-gen-cai, Chinese cabbage and Welsh onions, the antioxidant capacities were 20–50% higher (Ren *et al.*, 2001a). The organic vegetables also displayed higher antimutangenic (anti-cancer) activity, whilst the juices contained 1.3 to 10.4 times the flavonoid concentrations of comparable conventionally grown vegetables. In a further study with 11 vegetables, the same team studied anti-mutagenicity by evaluating 33 different bioassays. Eight of these bioassays showed a higher anti-mutagenic activity in the organic than in the conventional vegetables (Ren *et al.*, 2001b).

Glutathione

Levels of glutathione, as a measure of anti-oxidant capacity, were shown to be higher in strawberry plants grown in compost than in plants grown with mineral fertilizers, and the same applies to the juices from these plants (Wang and Lin, 2003).

Taste

Organic fruits and vegetables may also have superior taste and longer shelf-life (Ren *et al.*, 2001b) as a result of the changes in dry matter and individual composition (see above). This may be correlated with the way nutrients are delivered to the plant, that is more slowly and changing in relation to the pace of the crop's physiological development, and often the size of the fruit, grain or leaf (Benbrook, 2005). Further explanations may involve more complex plant–microbe interactions, higher levels of nutrient cycling, changes in plant metabolism and biochemistry.

15.7 Gaps in knowledge – future research evaluations

- Research is needed to evaluate the interactions of all cultivation variables that influence plant product composition. Climate induces a large variation in levels of phytochemicals and so data from more than one year are needed. This is reinforced by the global climatic changes we are currently facing and also based on future predictions of climate shifts.
- More data are needed on the effects of organic practices on phytochemicals and their precursors. How does modification of precursor pools affect biosynthesis? For example, see studies on the effects of lower levels of free amino acids caused by a reduction in nitrogen input.
- More data on the effects of low level infestation by pests and diseases, which usually occurs in organic crops, on phytochemical induction and levels in harvested food material.
- More data are needed on post-harvest storage diseases of organic crops, because chemical controls cannot be used on organic products. This is important in relation to fungal toxins, e.g. aflatoxin.
- Bioavailability studies need to be performed to evaluate the differences between conventional and organic crops, e.g. human intervention studies.

15.8 References

Agrawal A (2000), 'Benefits and costs of induced plant defenses for *Lepidium virginicum* (*Brassicaceae*)', *Ecology*, **87**, 804–813.

Agte V V and Joshi S R (1997), 'Effect of traditional food processing on phytate degradation in wheat and millets', *J Agric Food Chem*, **45**,1659–1661.

Andersen J-O (2000), 'Farming, plant nutrition and food quality', in *Proceedings of the UK Soil Association Conference*, Jan 2000, Bristol, UK.

Asami D K, Hong Y J, Barret D M, and Mitchell, A E (2003), 'Comparison of the total phenolic and ascorbic acid content of freeze-dried and air.dried marionberry, strawberry, and corn grown using conventional, organic, and sustainable agricultural practices'. *J Agric Food Chem*, **51**, 1237–1241.

Asenio T A, Crespo J F, Sanchez-Monge R, Lopez-Torrejon G, Somoza M L, Rodríguez J and Salcedo G (2004), 'Novel plant pathogenesis-related protein family involved in food allergy', *J Allergy Clin Immunol*, **114**, 896–899.

Beier R C and Nigg H N (1992), 'Natural toxicants in foods', in Nigg H N and Seigler D, *Phytochemical Resources for Medicine and Agriculture*, Plenum Press, UK, 247–368.

Benbrook C M (2005), 'Elevating antioxidant levels in food through organic farming and food processing', *The Organic Center for Education & Promotion*, 78 pp.

Bollinger H (2001), 'Consumer expectations and eating behavior over time', *Food Market and Technol*, 10–13.

Bonnesen C, Eggleston I M and Hayes J D (2001), 'Dietary indoles and isothiocyanates that are generated from cruciferous vegetables can both stimulate apoptosis and confer protection against DNA damage in human colon cell lines', *Cancer Rese*, **61**, 6120–6130.

Bourn D and Prescott J (2002), 'A comparison of the nutritional value, sensory qualities, and food safety of organically and conventionally produced foods', *Crit Rev Food Sci Nutr*, **42**, 1–34.

Brandt K and Mølgaard J P (2001), 'Featured article, Organic agriculture: does it enhance or reduce the nutritional value of plant foods?', *J Sci Food Agric*, **81**, 924–931.

Brown P D and Morra M J (1997), 'Control of soil-borne plant pests using glucosinolate-containing plants', *Adv Agronomy*, **61**, 167–231.

Buskov S, Serra B, Rosa E, Sorensen H and Sorensen J C (2002), 'Effects of intact glucosinolates and products produced from glucosinolates in myrosinase-catalyzed hydrolysis on the potato cyst nematode (*Globodera rostochiensis* Cv Woll)', *J Agric Food Chem*, **50**, 690–695.

Cai Y, Sun M and Corke H (2001), 'Identification and distribution of simple and acylated betacyanins in the *Amaranthaceae*', *J Agric Food Chem*, **49**, 1971–1978.

Carbonaro M, Mattera M, Nicoli S, Bergamo P and Cappelloni M (2002), 'Modulation of antioxidant compounds in organic vs conventional fruit (peach, *Prunus persica* L., and pear, *Pyrus communis* L.)', *J Agric Food Chem*, **50**, 5458–5462.

Chitwood D J (1992), 'Nematicidal compounds from plants', in Nigg H N and Seigler D, *Phytochemical Resources for Medicine and Agriculture*, Plenum Press UK, 185–204.

Chitwood D J (2002), 'Phytochemical based strategies for nematode control', *Annu Rev Phytopathol*, **40**, 221–249.

Clifford M N (1999), 'Chlorogenic acids and other cinnamates – nature, occurrence and dietary burden', *J Sci Food Agric*, **79**, 362–372.

Crowell PL (1999), 'Prevention and therapy of cancer by dietary monoterpenes', *J Nutr*, **129** (Suppl), 775S–778S.

Dandurand L M, Mosher R D and Knudsen G R (2000), 'Combined effects of *Brassica napus* seed meal and *Trichoderma harzianum* on two soilborne plant pathogens', *Can J Microbiol*, **46**, 1051–1057.

Défago G and Kern H (1983), 'Induction of *Fusarium solani* mutants insensitive to tomatine, their pathogenicity and aggressiveness to tomato fruits and pea plants', *Physiol Plant Pathol*, **22**, 29–37.

Défago G, Kern H and Sedlar L (1983), 'Genetic analysis of tomatine insensitivity, sterol content and pathogenicity for green tomato fruits in mutants of *Fusarium solani*', *Physiol Mol Plant Pathol*, **22**, 39–43.

Diawara M M, Trumble J T, Quiros C F and Hansen R (1995), 'Implications of distributions of linear furanocoumarins within celery', *J Agric Food Chem*, **43**, 723–727.

Du Z and Bramlage W J (1995), 'Peroxidative activity of apple peel in relation to development of post-storage disorders', *Hortic Sci*, **30**, 205–209.

Dykes L, Rooney LW, Waniska RD, and Rooney WL (2005), 'Phenolic compounds and

antioxidant activity in sorghum grains of varying genotypes', *J Agric Food Chem*, **53**, 6813–6818.
Fahey J W, Zalcmann A T and Talalay P (2001), 'The chemical diversity and distribution of glucosinolates and isothiocyanates among plants', *Phytochemistry*, **56**, 5–51.
Fassbinder C, Grodnitzky J and Coats J (2002), 'Monoterpenoids as possible control agents for *Varroa destructor*', *J Agric Resources*, **41**, 83–88.
Fenwick G R, Heaney R K and Mullin W J (1983), 'Glucosinolates and their breakdown products in food and food plants', *Crit Rev Food Sci Nutr*, **18**, 123–301.
Ferris H and Zheng L (1999), 'Plant sources of Chinese herbal remedies: effects on *Pratylenchus neglectus* and *Meloidogyne javanica*', *J Nematol*, **31**, 241–263.
Finley J W (2005), 'Bioactive compounds in designer plant foods: the need for clear guidelines to evaluate potential benefits to human health', *Chronica Horticulturae*, **45**, 6–11.
Fomsgaard I S, Mortensen A G, Gents M B and Understrup A G (2004), 'Time-dependent transformation of varying concentration of the hydroxamic acid metabolites MBOA and BOA in soil', *2nd European Allelopathy Symposium*, 3–5 June 2004, Poland, Pulawy, 61–63.
Friedman M (2002), 'Tomato glycoalkaloids: role in the plant and in the diet', *J Agric Food Chem*, **50**, 5751–5778.
Friedman M (2004), 'Analysis of biologically active compounds in potatoes (*Solanun tuberosum*), tomatoes (*Lycopersicum esculentum*), and jimson weed (*Datura stramonium*) seeds', *J Chromatogr A*, **1054**, 143–155.
Friedman M and McDonald G M (1997), 'Potato glycoalkaloids: chemistry, analysis, safety and plant physiology', *Crit Rev Plant Sci*, **16**, 55–132.
Gardiner J, Morra M J, Eberlein C V, Brown P D and Borek V (1999), 'Allelochemicals released in soil following incorporation of rapeseed (*Brassica napus*) green manures', *J Agric Food Chem*, **47**, 3837–3842.
Gąsiorowski K, Szyba K, Brokos B and Kozubek A (1996), 'Antimutagenic activity of alkylresorcinols from cereal grains', *Cancer Lett*, **106**, 109–115.
Glen D M, Jones H and Fieldsend J K (1990), 'Damage to oilseed rape seedlings by the filed slug *Deroceras reticulatum* in relation to glucosinolate concentration', *Ann Appl Biol*, **117**, 197–207.
Goncales C E P, Araldi D, Panatieri R B, Rocha J B T, Zeni G and Nogueira C W (2005), 'Antinociceptive properties of acetylenic thiophene and furan derivatives: Evidence for the mechanism of action', *Life Sciences*, **76**, 2221–2234.
Griffiths D W (1991), 'Condensed tannins', in D'Mello J P F, Duffus C M and Duffus J H, *Toxic Substances in Crop Plants*, The Royal Society of Chemistry, Cambridge, UK, 180–201.
Griffiths D W, Birch A N E and Hillman J R (1998), 'Antinutritional coumponds in the *Brassicaceae*: Analysis, biosynthesis, chemistry and dietary effects-Review article', *J Hort Sci Biotech*, **73**, 1–18.
Grinder-Pedersen L, Rasmussen S E, Bugel S, Jorgensen L V, Dragsted L O, Gundersen V and Sandstrom B (2003), 'Effect of diets based on foods from conventional versus organic production on intake and excretion of flavonoids and markers of antioxidative defense in humans', *J Agric. Food Chem*, **51**, 5671–5676.
Hajslová J, Schulzová V, Slanina P, Janné K, Hellenäs K E and Andersson C. (2005), 'Quality of organically and conventionally grown potatoes: four-year study of micronutrients, metals, secondary metabolites, enzymic browning and organoleptic properties', *Food Addit Contam*, **22**, 514–534.
Hakkinen S H and Torronen A R (2000), 'Content of flavonols and selected phenolic acids in strawberries and *vaccinium* species: influence of cultivar, cultivation site and technique', *Food Res Int*, **33**, 517–524.
Halliwell B, Rafter J and Jenner A (2005), 'Health promotion by flavonoids, tocopherols, tocotrienols, and other phenols: direct or indirect effects? Antioxidant or not?', *Am J Clin Nutr*, **81** (Suppl), 268S–276S.

Hamouz K, Lachman J, Vokal B and Pivec V (1999), 'Influence of environmental conditions and way of cultivation on the polyphenol and ascorbic acid content in potato tubers', *Rostlinna Vyroba*, **45**, 293–298.

Harborne J B (1994), *The Flavonoids: Advances in Research since 1986*, Chapman and Hall, London.

Harland B F, Smikle-Williams S and Oberleas D (2004), 'High-performance liquid chromatography of phytate (IP6) in selected foods', *J Food Comp Anal*, **17**, 227–233.

Heaney R K and Fenwick G R (1995), 'Natural toxins and protective factors in *Brassica* species, including rapeseed', *Natural Toxins*, **3**, 233–237.

Heaton S (2002), 'Assessing organic food quality: is it better for you?', in Owell P, *Proceedings of. COR Conference – UK Organic Research 2002*, 26–28 March 2002, Aberystwyth, UK, 55–60.

Heldt H W, Chon C J, Maronde D, Harold A, Stankovic Z S, Walker D A, Kraminer A, Kirk M R and Heber U (1977), 'Role of orthophosphate and other factors in the regulation of starch formation in leaves and isolated chloroplasts', *Plant Physiol*, **59**, 1146–1155.

Herman-Antosiewicz A and Singh S V (2004), 'Signal transduction pathways leading to cell cycle arrest and apoptosis induction in cancer cells by *Allium* vegetable-derived organosulfur compounds: a review', *Mutation Res*, **555**, 121–131.

Hoffmann-Sommergruber K (2002), 'Pathogenesis-related (PR) proteins identified as allergens', *Biochem Soc Trans*, **30**, 930–935.

House W A (1999), 'Trace element bioavailability as exemplified by iron and zinc', *Field Crops Res*, **60**, 115–141.

Howard L R, Pandjaitan N, Morelock T, and Gil M L (2002), 'Antioxidant capacity and phenolic content of spinach as affected by genetics and growing season', *J Agric Food Chem*, **50**, 5891–5896.

Johnson A W, Golden A M, Auld D L and Sumner D R (1992), 'Effect of rapeseed and vetch as green manure crops and fallow on nematodes and soil-borne pathogens', *J Nematol*, **24**, 117–126.

Johnson I T (2002), 'Anticarcinogenic effects of diet-related apoptosis in the colorectal mucosa', *Food Chem Toxicol*, **40**, 1171–1178.

Kalt, W, Ryan D A, Duy J C, Prior R L, Ehlenfeldt M K, and Vander Kloet S P (2001), 'Interspecific variation in anthocyanins, phenolics, and antioxidant capacity among genotypes of highbush and lowbush blueberries (*Vaccinium* section *Cyanococcus* spp.)', *J Agric Food Chem*, **49**, 4761–4767.

Kanner J, Harel, S, Granit and R (2001), 'Betalains – a new class of dietary cationized antioxidants', *J Agric Food Chem*, **49**, 5178–5185.

Karban R and Baldwin I T (1997), Induced Responses to Herbivory, University of Chicago Press, Chicago, IL, USA.

Kobæk-Larsen M, Christensen L P, Vach W, Ritskes-Hoitinga J and Brandt K (2005), 'Inhibitory effects of feeding with carrots or (–) falcarinol on development of azomethane-induced preneoplastic lesions in the rat colon', *J Agric Food Chem*, **53**, 1823–1827.

Konstantopoulou M A, Krokos F D and Mazomenos B E (2002), 'Chemical stimuli from corn plants affect host selection and oviposition behavior of *Sesamia nonagrioides* (Lepidoptera: *Noctuidae*)', *J Econ Entomol*, **95**, 1289–1293.

Konstantopoulou M A, Krokos F D and Mazomenos B E (2004), 'Chemical composition of corn leaf essential oils and their role in the oviposition behaviour of *Sesamia nonagrioides* female', *J Chem Ecol*, **30**, 2243–2256.

Koricheva J, Larsson S, Haukioja E and Keinänen M (1998), 'Regulation of woody plant secondary metabolism by resource availability: hypothesis testing by means of meta-analysis', *Oikos*, **83**, 212–226.

Kroon P A, Clifford M N, Crozier A, Day A J, Donovan J L, Manach C and Williamson G (2004), 'How should we assess the effects of exposure to dietary polyphenols in vitro', *Am J Clin Nutr*, **80**, 15–21.

Kubec R, Svobodová M and Velíšek J (2000), 'Distribution of *S*-alk(en)ylcysteine sulfoxides in some Allium species. Identification of a new flavour precursor: *S*-ethylcysteine sulfoxide (ethiin)', *J Agric Food Chem*, **48**, 428–433.

Kuc J (1992), 'Antifungal compounds in plants', in Nigg H N and Seigler D, *Phytochemical Resources for Medicine and Agriculture*, Plenum Press UK, 159–184.

Kugler F, Stintzing F C and Carle R (2004), 'Identification of betalains from petioles of differently coloured swiss chard (Beta vulgaris L. ssp. cicla [L.] Alef. Cv. Bright Lights) by high-performance liquid chromatography–electrospray ionisation mass spectrometry, *J Agric Food Chem*, **52**, 2975–2981.

Lachman J, Hamouz K, Orsak M and Pivec V (2001), 'Potato glycolakaloids and their significance in plant protection', *Rostlinna Vyroba*, **47**, 181–191.

Lampkin N (2004), 'Introduction to the European Information System for Organic Markets and the aims of the 1st EISfOM European Seminar, Berlin, April 2004', in *Proceedings of the 1st EISfOM European Seminar* held in Berlin, Germany, 26–27 April,17–22.

Langenheim J H (1994), 'Higher plant terpenoids-a phytocentric overview of their ecological roles', *J Chem Ecol*, **20**, 1223–1280.

Lattanzio V, Di Venere D, Linsalata V, Bertolini P, Ippolito A and Salerno M (2001), 'Low-temperature metabolism of apple phenolics and quiescence of *Phyllyctaena vagabonda*', *J Agric Food Chem*, **49**, 5817–5821.

Lazzeri L, Tacconi R and Palmieri S (1993), '*In vitro* activity of some glucosinolates and their reaction products toward a population of the nematode *Heterodera schachtii*', *J Agric Food Chem*, **41**, 825–829.

Leclerc J, Miller M L, Joliet E and Rocquelin G (1991), 'Vitamin and mineral contents of carrot and celeriac under mineral or organic fertilization', *Biol Agric Hort*, **7**, 349–361.

Levite D, Adrian M and Tamm L (2000), 'Preliminary results of resveratrol in wine of organic and conventional vineyards', in Willer H and Meier U, *Proceedings 6th International Congress on Organic Viticulture*, 25–26 August 2000, Ackerstrasse, Germany, 256–257.

Liggins J, Grimwood R and Bingham S A (2000), 'Extraction and quantification of lignan phytoestrogens in food and human samples', *Adv Biochem*, **287**, 102–109.

Lin C M, Chen C T, Lee H H and Lin J K (2002a), 'Prevention of cellular ROS by isovitexin and related flavonoids', *Planta Med*, **68**, 365–367.

Lin G, Cui Y-Y, Liu X-Q and Wang Z-T (2002b), 'Species differences in the in vitro metabolic activation of the hepatotoxic pyrrolizidine alkaloids clivorine', *Chem Res Toxicol*, **15**, 1421–1428.

Lombardi-Boccia G, Lucarini M, Lanzi S, Aguzzi A and Cappelloni M (2004), 'Nutrients and antioxidant molecules in yellow plums (*Prunus domestica* L.) from conventional and organic productions: a comparative study', *J Agric Food Chem*, **52**, 90–94.

Lüning H U and Schlösser E (1976), 'Role of saponins in antifungal resistance. VI. Interactions *Avena sativa–Drechslera avenacea*', *J Plant Dis Protect*, **83**, 317–327.

Lydon J and Duke S D (1989), 'Pesticides effects on secondary metabolism of higher plants', *Pest Sci,* **25**, 361–374.

Macel G L, Klinkhamer P, Vrieling K and van der Meijden E (2002), 'Diversity of pyrrolizidine alkaloids in Senecio species does not affect the specialist herbivore *Tyria jacobaeae*', *Oecologia*, **133**, 541–550.

Manach C, Scalbert A, Morand C, Rémésy C and Jiménez L (2004), 'Polyphenols: foods sources and bioavailability', *Am J Clin Nutr*, **79**, 727–747.

Margl L, Eisenreich, W, Adam, P, Bacher, A and Zenk M H (2001), 'Biosynthesis of thiophenes in *Tagetes patula*', *Phytochemistry*, **58**, 875–881.

Marks H S, Hilson J A, Leichtweis H C and Stoewsand G S (1992), '*S*–Methylcysteine sulfoxide in Brassica vegetables and formation of methyl methanethiosulfinate from Brussels sprouts', *J Agric Food Chem*, **40**, 2098–2101.

Mattila P, Pihlava J-M and Hellstrøm J (2005), 'Contents of phenolic acids, alkyl- and alkenylresorcinols, and avenanthramides in commercial grain products', *J Agric Food Chem*, **53**, 8290–8295.

Mayer R T, Inbar M, McKenzie C L, Shatters R and Borowicz V (2002), 'Multitrophic interactions of the silverleaf whitefly host plants, competing herbivores, and phytopathogens', *Arch Insect Biochem Physiol*, **51**, 151–169.

Melanson D, Chilton M D, Masters-Moore D, Chilton WS (1997), 'A deletion in an indole synthase gene is responsible for the DIMBOA-deficient phenotype of bxbx maize', *Proc. Natl. Acad. Sci. USA*, **94**, 13345–13350.

Mercadante A Z and Rodriguez-Amaya D B (1991), 'Carotenoid composition of a leafy vegetable in relation to some agricultural variables', *J Agric Food Chem*, **39**, 1094–1097.

Milder I E, Arts I C, van de Putte B, Venema D P and Hollman P C (2005), 'Lignan contents of Dutch plant foods: a database including lariciresinol, pinoresinol, secoisolariciresinol and matairesinol', *Br J Nutr*, **93**, 393–402.

Mithen R (2001), 'Glucosinolates – biochemistry, genetics and biological activity', *Plant Growth Regul*, **34**, 91–103.

Mogensen B B, Mathiassen S, Krongaard T, Eljarrat E, Villagrasa M, Guillamón M, Taberner A and Barceló D (2004), 'Quantification of hydroxamic acid allelochemicals in wheat varieties grown under varying conditions', *2nd European Allelopathy Symposium*, Pulawy, Poland, 50–53.

Mojtahedi H, Santo G S, Hang A N and Wilson J H (1991), 'Suppression of root-knot nematode populations with selected rapeseed cultivars as green manure', *J Nematol*, **23**, 170–174.

Moyes C L, Collin H A, Britton G and Raybould A E (2000), 'Glucosinolates and differential herbivory in wild populations of *Brassica oleracea*', *J Chem Ecol*, **26**, 2625–2641.

Mozafar A (1993), 'Nitrogen fertilizers and the amount of vitamins in plants – a review', *J Plant Nutr*, **16**, 2479–2506.

Mpofu A, Sapirstein HD, and Beta T (2006), 'Genotype and environmental variation in phenolic content, phenolic acid composition, and antioxidant activity of hard spring wheat', *J Agric Food Chem*, **54**, 1265–1270.

Nicolle C, Manach C, Morand C, Mazur W, Adlercreutz H, Rémésy C and Scalbert A (2002), 'Mammalian lignan formation in rats fed a wheat bran diet', *J Agric Food Chem*, **50**, 6222–6226.

Nielsen S E, Freese R, Kleemola P and Mutanen M (2002), 'Flavonoids in human urine as biomarkers for intake of fruits and vegetables', *Cancer Epidemiol Biomark Prevent*, **11**, 459–466.

Niemetz R, Gross GG (2005), 'Enzymology of gallotannin and ellagitannin biosynthesis', *Phytochemistry*, **66**, 2001–2011.

Nigg H N, Strandberg J O, Beier R C, Petersen H D and Harrison J M (1997), 'Furanocoumarins in Florida celery varieties increased by fungicide treatment', *J Agric Food Chem*, **45**, 1430–1436.

Oleszek W, Stochmal A and Kus J (2004), 'Concentration of hydroxamic acids in Polish wheat varieties', *2nd European Allelopathy Symposium*, Pulawy, Poland, 54–57.

Osbourn A E (1996), 'Saponins and plant defence – a soap story', *Trends Plant Sci*, **1**, 4–9.

Osier T L, and Lindroth R L (2001), 'Effects of genotype, nutrient availability, and defoliation on aspen phytochemistry and insect performance', *J Chem Ecol*, **27**, 1289–1313.

Pandjaitan N, Howard L R, Morelock and Gil MI (2005), 'Antioxidant capacity and phenolic content of spinach as affected by genetics and maturation', *J Agric Food Chem*, **53**, 8618–8623.

Persson H, Türk M, Nyman M and Sandberg A-S (1998), 'Binding of Cu^{2+}, Zn^{2+} and Cd^{2+} to inositol tri-, tetra-, penta- and hexaphosphates', *J Agric Food Chem*, **46**, 3194–3200.

Prakash A S, Pereira T N, Reilly P E B and Seawright A A (1999), 'Pyrrolizidine alkaloids in the human diet', *Mutat Res*, **443**, 53–67.

Price K R, Johnson I T and Fenwick G R (1987), 'The chemistry and biological significance of saponins in food and feeding stuffs', *Crit Rev Food Sci Nutr*, **26**, 27–133.

Ren H, Endo H and Hayashi T (2001a), 'Antioxidative and antimutagenic activities and polyphenol content of pesticide-free and organically cultivated green vegetables using water-soluble chitosan as a soil modifier and leaf surface spray', *J Sci Food Agric*, **81**, 1426–1432.

Ren H, Endo H and Hayashi T (2001b), 'The superiority of organically cultivated vegetables to general ones regarding antimutagenic activities', *Mutat Res*, **496**, 83–88.

Roberts M F and Wink M (1998), *Alkaloids: Biochemistry, Ecology, and Medicinal Applications*, in Roberts M F and Wink M, Kluwer Academic Publishers, UK.

Roddick J G (1996), 'Steroidal glyolakaloids: nature and consequences of bioactivity', *Adv Exp Med Biol*, **404**, 277–297.

Rosa E A S, Heaney R K, Fenwick R G and Portas C A M (1997), 'Glucosinolates in crop plants', *Hort Rev*, **19**, 99–215.

Ross A B, Aman P, Andersson R and Kamal-Eldin A (2004), 'Chromatographic analysis of alkylresorcinols and their metabolites' *J Chromatogr A*, **1054**, 157–164.

Rossister M, Schultz J C and Baldwin I T (1986), 'Relationships among defoliation, red oak phenolics, and gypsy moth growth and reproduction', *Ecology*, **69**, 267–277.

Sanford L L, Deahl K L, Sinde S L and Ladd Jr T L (1992), 'Glycoalkaloid content in tubers from *Solanum tuberosum* populations selected for potato leafhopper resistance', *Am Potato J*, **69**, 693–703.

Scalbert A, Johnson I T and Saltmarsh M (2005), 'Polyphenols: antioxidants and beyond?', *Am J Clin Nutr*, **81** (Suppl.), 215S–217S.

Scalzo J, Politi A, Pellegrini N, Mezzetti B, and Battino M (2005), 'Plant genotype affects total antioxidant capacity and phenolic contents in fruit', *Nutrition*, **21**, 207–213.

Schultz J C and Baldwin I T (1982), 'Oak leaf quality declines in response to defoliation by gypsy moth larvae', *Science*, **217**, 149–50.

Stintzing F C, Schieber A and Carle R (2002), 'Identification of betalains from yellow beet (*Beta vulgaris* L.) and cactus pear (*Opuntia ficus-indica* (L.) Mill.) by high-performance liquid chromatography – electrospray ionisation mass spectrometry', *J Agric Food Chem*, **50**, 2302–2307.

Tapiero H, Townsend D M and Tew K D (2004), 'Organosulfur compounds from the *Alliaceae* in the prevention of human pathologies', *Biomed Pharmacother*, **58**,183–193.

Tesoriere L, Allegra M, Butera D and Livrea M A (2004), 'Absorption, excretion, and distribution of dietary antioxidant betalains in LDLs: potential health effects of betalains in humans', *Am J Clin Nutr*, **80**, 941–945.

Theis N and Lerdau, M (2003), 'The evolution of function in plant secondary metabolites', *Int. J. Plant Sci.*, **164** (3 Suppl.), S93–S102.

Tilman D, Cassman K G, Matson P A, Naylor R and Polasky S (2002), 'Agricultural sustainability and intensive production practices', *Nature*, **418**, 671–677.

Tosi B, Bonora A, Dall'Ollio G and Bruni A (1991), 'Screening for toxic thiophene compounds from crude drugs of the family *Compositae* used in Northern Italy', *Phytother Res*, **5**, 59–62.

Valkonen J P T, Keskitalo M, Vasara T and Pietila L (1996), 'Potato glycoalkaloids: a burden or a blessing?', *Crit Rev Plant Sci*, **15**, 1–20.

Van Bruggen A H C (1995), 'Plant disease severity in high-input compared to reduced-input and organic farming systems', *Plant Dis*, **79**, 976–984.

Verschoyle R D, Greaves P, Cai H, Borkhardt A, Broggini M, D'Incalci M, Ricci E, Doppalapudi R, Kapetanovic I M, Steward W P and Gescher A J (2005), 'Preliminary safety evaluation of the putative cancer chemopreventive agent tricin, a naturally occurring flavone', *Cancer Chem Pharmacol*, **57**, 1–6.

Viveros A, Centeno C, Brenes A, Canales R and Lozano A (2000), 'Phytase and acid phosphatase in plant feedstuffs', *J Agric Food Chem*, **48**, 4009–4013.

Wang S Y and Lin H S (2003), 'Compost as a soil supplement increases the level of antioxidant compounds and oxygen radical absorbance capacity in strawberries', *J Agric Food Chem*, **51**, 6844–6850.

Warton B, Mattiessen J N and Shackleton M A (2001), 'Glucosinolate content and isothiocyanate evolution-two measures of the biofumigation potential of plants', *J Agric Food Chem*, **49**, 5244–5250.

Weibel F P, Bickel R, Leuthold S and Alfoldi T (2000), 'Are organically grown apples tastier and healthier? A comparative field study using conventional and alternative methods to measure fruit quality', *Acta Horticulturae*, **517**, 417–426.

Wink M (1999), 'Introduction: biochemistry, role, and biotechnology of secondary metabolites', in Michael Wink, *Functions of Plant Secondary Metabolites and their Exploitation in Biotechnology, Annual Plant Reviews*, Academic Press, UK, **3**, 1–16.

Woese K, Lange D, Boess C and Bögl K W (1997), 'A comparison of organically and conventionally grown foods – results of a review of the relevant literature', *J Sci Food Agric*, **74**, 281–293.

Zasada I A and Ferris H (2003), 'Sensitivity of *Meloidogyne javanica* and *Tylenchulus semipenetrans* to isothiocyanates in laboratory assays', *Phytopathology*, **93**, 747–750.

Zheng G, Lu W, Aisa H A and Cai J (1999), 'Absolute configuration of falcarinol, a potent antitumor agent commonly occurring in plants', *Tetrahedron Lett*, **40**, 2181–2182.

Zidorn C, Jöhrer K, Ganzera M, Schubert B, Sigmund E M, Mader J, Greil R, Ellmerer E P and Stuppner H (2005), 'Polyacetylenes from the *Apiaceae* vegetables, carrot, celery, fennel, parsley, and parsnip and their cytotoxic activities', *J Agric Food Chem*, **53**, 2518–2523.

16

Improving the quality and shelf life of fruit from organic production systems

Franco P. Weibel and Thomas Alföldi, FiBL, Switzerland

16.1 Introduction

Since the mid-1990s important key markets like the EU, USA and Japan have demonstrated a growth in the organic sector of around 20% per year. This strong growth of organics is underpinned by consumer concern over genetic engineering, food safety, health and nutrition (is it safe and healthy to eat?) and greater awareness of environmental issues (Torjusen and Sangstad, 2004; Granatstein and Kirby, 2006). For farmers this provides an important opportunity in the 'search for' and 'development of' niche, new and non-traditional premium earning markets.

However on-going improvements in conventional food production systems in terms of visible produce quality, but also with respect to environmental criteria, require a clear demonstration of the additional values that organic food can provide for the consumers and for the sustainability of food production and the environment (see Chapter 2). Sound scientific studies quantifying 'added values' play a key role in the marketing of organic produce if appropriately communicated to consumers, but are also important in gaining continued political support for an expansion of organic agriculture.

Organic fruit production is one of the most important and challenging areas of organic production for a variety of reasons. These include:

- Fruits are still sold and consumed mainly fresh, essentially as they are picked from the tree. On the tree, however, fruit (e.g. apples) remain exposed to the attacks of diseases, pests and climatic stress factors during 130–160 days of the year. Given the often low tolerance among consumers for even minor blemishes on fruit, it is a complex and laborious task for fruit growers to keep the fruit in a visibly attractive and immaculate state until maturity and harvest. Additionally, fruit are very thin-skinned and therefore require gentle handling post-harvest and often specific storage conditions.

- As tree fruit species are perennial crops, year-to-year influences are often detected. For example, factors in the previous year(s) (e.g. water or nutrient deficiency, hail storm damage, shoot deformation caused by aphids, too high or too low crop load) strongly influence the tree's performance in the next year (Tromp and Wertheim, 2005). Thus, a major objective of agronomic practices used is to 'buffer' the orchard from stress and to keep trees in a balance/equilibrium between vegetative and generative activity.
- Owing to the more limited range, and relatively less efficient inputs (e.g. for pest and disease control, crop regulation, weed control and fertilisation) available in organic fruit production, yields are still considerably lower than in conventional fruit production, although exceptions have been reported (Reganold *et al.*, 2001). However, the (usually around 15–30%) reduced yields compared to conventional fruit farms are currently compensated for by higher farm gate prices (Weibel *et al.*, 2002). Nevertheless, premiums may decrease and/or be more difficult to achieve, the more mature and larger the organic food market becomes.
- The investment costs to convert/plant orchards are very high (40–80 k€ per ha) with duration of amortisation usually longer than 12 years (Zürcher and Leumann, 2006). To achieve high specifications (with respect to visual, but increasingly also organoleptic qualities) requires a high level of professional skill. However, since many organic fruit farms are mixed farms, it is getting harder and harder to remain up to date with technology development/innovations and to provide the same level of visual quality as the specialised, stockless conventional fruit farms.

16.2 Reasons for varying fruit quality: interactions between site conditions and management factors

For the early pioneers of organic agriculture in the beginning of the twentieth century, achieving a high and natural product quality was always the central concern and motivation (Weibel *et al.*, 2002). Indeed, there are several principles in organic agriculture that can (but not necessarily always do) lead to a high produce quality. Quality determining factors can be divided into factors that can be influenced by the farmer (e.g. agronomic practices) and those that cannot be manipulated (e.g. climatic conditions) (Fig. 16.1). Those that can be influenced can be further subdivided into (i) 'conservative' factors, that once chosen cannot be changed easily (e.g. choice of cultivar), and (ii) 'dynamic' factors that can be changed during the season (e.g. irrigation, fertilisation and disease control protocols used).

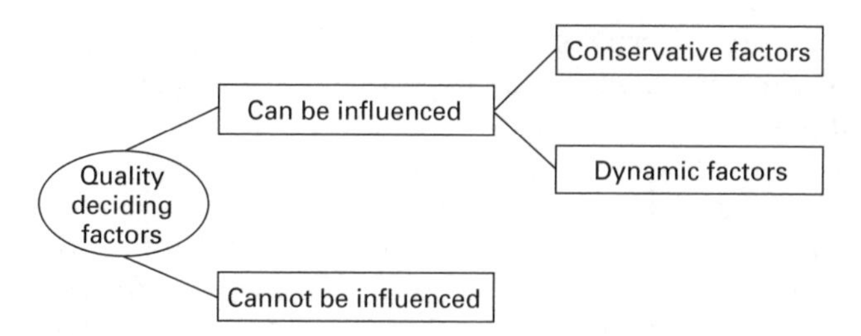

Fig. 16.1 Divisions of fruit quality-determining factors.

16.2.1 Quality affecting factors that cannot be influenced

There are two major quality-deciding factors that cannot be influenced by farmers:

- Climatic conditions and certain environmental conditions (e.g. hours of sunshine, day and night temperatures, elevation, ozone stress), etc.
- Requirements, specifications and distance to retail markets. For example, retailers increasingly apply the same size and visual quality requirements to organic and conventional fruit. Under supermarket specifications, even slight deviations in size or minor skin defaults that do not negatively affect the eating quality or hygiene of the fruit are usually rejected. The often substantial effort required to achieve these mainly 'cosmetic' quality characteristics is often considered to detract attention and resources from important nutritional and sensory-related qualities.

16.2.2 Quality determining factors that can be influenced

Conservative factors

'Conservative' is used to define orchard and management factors that once introduced cannot, or are extremely difficult to, change or require substantial additional investment to do so. Among these are soil and other site-specific conditions, choice of variety/cultivar and rootstock, planting density and tree canopy formation, system stabilisation measures at orchard set-up and installations to buffer extreme events.

Soil and other site-specific conditions

Before the establishment of organic orchards, it is essential to consider practical experience and local knowledge about the most suitable sites and microclimatic conditions (reviewed bv Webster, 2005; Barden and Neilson, 2003) as well as species and cultivar suitability. These factors should be considered before planting orchards, because it is more difficult than in conventional production to overcome problems associated with poor choice of site and variety/cultivar. However, some site-specific problems may be addressed within certain limits (e.g. suboptimal soil conditions) and should be improved before planting a new orchard (e.g. installation of drainage to prevent temporary water logging, deep ploughing to break up compacted soil layers and/or lime applications to neutralise a too low pH value of the soil).

Choice of variety/cultivar and rootstock

Market demands must be the main focus for every grower when deciding on the choice of cultivar. However, a major challenge for organic fruit growing is that the preferred current or new cultivars established in the conventional market are highly disease sensitive and therefore have too low a yield security in the more 'input use restricted' organic production systems. The use of cultivars with higher levels of resistance to the main pests and diseases (e.g. scab) would therefore in many geographic regions essentially contribute to obtaining satisfactory yields and comply with supermarket quality specifications if these niche cultivars were to be accepted by retailers and consumers. The marketing of new and relatively unknown 'resistant' varieties can, however, be extremely difficult, especially with respect to 'conservative' supermarket customers used to conventional varieties. A solution for introducing 'unknown' scab-resistant apple cultivars more easily is the so-called 'Flavour Group Concept' developed and successfully introduced in Switzerland (Weibel and Häseli, 2003; Weibel and Leder, 2004), which is marketing new varieties by communicating the new cultivar's taste character rather than its variety name.

Important intrinsic quality criteria currently determining the market potential of new apple cultivars are related to the sensory quality such as fruit firmness (crispness) and the sugar and acidity contents. On the other hand, the nutritional composition (e.g. the vitamin or antioxidant contents) is currently not used as a criterion in the choice of cultivars, neither in conventional nor in organic fruit production. The difference in the content of such components between fruit species is in most cases more relevant than between cultivars of the same species (e.g. vitamin C content of oranges versus apples).

The choice of rootstocks was until recently thought to primarily affect the growth and vigour of the tree and to only have a minor effect on intrinsic fruit quality. However, recent research has shown that, under 'low input' organic production practice, the choice of rootstock can have a significant influence on tree fitness and tree nutrient acquisition, and thereby also on fruit quality (Weibel *et al.*, 2006a).

Planting density and tree canopy formation

Planting density and tree canopy formation influence the amount of light that will reach the tree canopies. Densely planted orchards create a high yield potential, but also create a higher proportion of fruit with low light exposure, which may reduce the supply of assimilates (sugars, nutrient elements) to individual fruits and consequently lower fruit quality. Today, there is a substantial body of knowledge about optimum planting densities and how different tree pruning/training systems influence light distribution within the orchard and individual trees (Robinson, 2003).

In organic fruit production, in particular in more humid climates, a lower planting density with well light-penetrated tree forms is recommended in order to favour quick drying of the leaves and fruits after rain and dew, which is essential as an indirect measure to prevent fungal and bacterial diseases (Tamm *et al.*, 2004). As mentioned before, this measure at the same time increases the potential for higher fruit quality through a better nutrition/assimilate supply to the single fruit.

In commercial practice, many organic orchards that were originally designed and used for intensive conventional production have been converted to organic production. In such orchards the higher tree density often causes severe problems with regard to plant protection and fruit quality and it is regularly necessary to thin out such orchards by removing up to 50% of the trees in order to achieve a more favourable micro-climate necessary for organic production.

System stabilisation measures in the orchard set-up

System stabilisation measures, like hedges, wild-flower strips and other ecological compensation areas, are set up in and around the orchard to enhance beneficial antagonists for pest control and the general floral and faunal biodiversity of the plantation. These areas are expensive to install and maintain, but were shown to increase the population density of beneficial invertebrates (Wyss *et al.*, 1999) and are thought to thereby minimise pest damage in orchards. Ecological areas should be integrated when orchards are established, because it is difficult to introduce them later on when trees are fully established. However, apart from indirect effects on invertebrate pest damage, ecological compensation areas and are not thought to affect intrinsic (sensory or nutritional) quality of fruit. Nevertheless, these measures are important for a well functioning organic orchard and for the credibility of the growing method, and support the holistic quality image of organic fruit.

Installations to buffer extreme events

To achieve a good intrinsic visible fruit quality, biotic and abiotic stress to the trees has to be controlled. Therefore, conventional fruit orchards are often fitted with relatively expensive installations such as wind break hedges, hail nets, irrigation systems and wind machines for frost prevention, irrigation and liquid fertilisation facilities and wild animal fences. Apart from liquid fertilisation systems, most of these installations can also be installed in organic orchards and help to assure yield and quality security.

Dynamic factors affecting fruit quality

Most dynamic factors that affect fruit quality are agronomic practices that can be changed by farmers over relatively short time spans. They are mainly related to fertility management and crop protection and other husbandry interventions and inputs such as pesticides, fertilisers, herbicides, thinning agents and so on. The permitted tools for these activities, however, differ greatly between conventional and organic fruit production. The consequences of practices permitted in organic farming systems that relate to fruit quality are described in separate paragraphs below.

Crop load of the trees

The fruit load of a tree (which is expressed as the number of fruits per tree or per orchard area, or as number of leaves per fruit), is one of the most important factors affecting fruit quality (Schumacher, 1989; Tromp and Wertheim, 2005). Effects on quality – especially for apple – are related to within-tree competition for nutrients and assimilates between shoots, fruits and flower bud formation for the next year (an excellent overview of the physiological crop load interactions in apple is given by Wünsche and Ferguson, 2005). For example, when fruit set is very low, the few fruits present are oversupplied with assimilates and nutrients (esp. nitrogen and potassium) and become too big, showing thin cell walls, a low firmness of the fruit flesh, a low storability and poor sensory/eating quality (Schumacher *et al.*, 1980; Friedrich *et al.*, 1997). In contrast, when the number of fruits per tree is too high, nutrient supply to individual fruit is poor and they remain small and underdeveloped resulting in inferior eating/sensory quality. Additionally, trees with a supra-optimal fruit load over cropping cannot develop sufficient flower buds for the next season, resulting in alternate or bi-annually bearing trees and long-term problems with poor fruit quality.

For these reasons, fruit growers need to focus on adjusting fruit load to optimum levels that can be sustainably maintained year after year. The methods available for fruit, or more precisely flower blossom thinning, in organic fruit growing are very limited compared to conventional production where synthetic chemical hormones and desiccants are permitted (see Byers, 2003 for a review). In organic fruit growing, mechanical and hand thinning and some natural agents (which may vary between countries and organic sector/certification bodies) are permitted for use. These include lime sulphur, rape oil, soap or molasses. For example, Weibel *et al.* (2006b) found good thinning effects with an ecofriendly and economic combination of a mechanical rope thinner and molasses applications.

In comparative studies of organic and conventional fruit quality it is important that the fruit samples compared originate from trees with an optimum or equivalent fruit load (e.g. see Weibel *et al.*, 2000), to avoid the confounding effects of fruit load on quality. Unfortunately, in most comparative studies on fruit quality this factor was not addressed (Bourn and Prescott, 2002; Harker, 2004).

Fertilisation

The level of plant mineral nutrients available to trees is known to affect fruit quality, but its relative effect is often overestimated compared to other factors such as fruit load and light (and associated assimilate supply to fruit) (see sections above) (for review see Neilsen and Neilsen, 2003). The mineral nutrition of trees and fruits is complex. Uptake of the macronutrients nitrogen and potassium is closely related to their concentrations in the soil solution and oversupply can occur relatively easily, especially when fertilisers with a high water soluble nitrogen or potassium content are used. However, other mineral nutrients that are also known to affect fruit quality, such as calcium, phosphorus and micro-elements, rely on more complex physiological processes in the roots (Marschner, 1995). Their uptake is also known to be antagonised by high levels of other mineral nutrients (in particular, nitrogen and potassium). Deficiency in the supply of these elements cannot therefore simply be corrected by application of fertilisers.

In organic agriculture highly soluble chemosynthetic nitrogen and phosphorus fertilisers and KCl are not permitted. This restriction clearly decreases the risk of nutrient oversupply, in particular for nitrogen and potassium. For apple, this was shown to have a quality assuring effect and this may be related to the calcium uptake pattern, which is known to be one of the most important quality affecting factors. In the case of calcium deficiency, cell walls are weak and fruits show lower fruit flesh firmness and storability. In a farm comparison study and a replicated field trial in Switzerland, it was shown that microbial activity (expressed as microbial bound carbon and nitrogen, or the relation between them) in soil of orchards under organic cultivation was increased significantly (Weibel *et al.*, 2004a; Widmer *et al.*, 2004; Fig. 16.2, Fig. 16.3). Interestingly, in both cases close correlations with phosphorus content of the fruit and calcium availability in the soil solution, respectively, could be found (Fig. 16.2a).

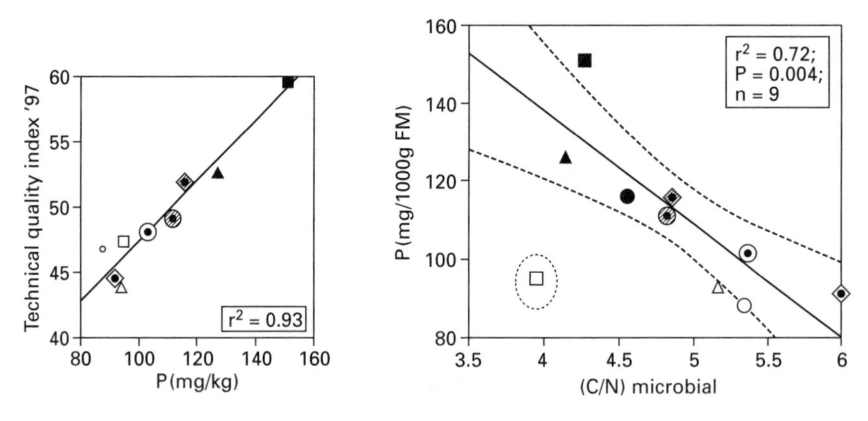

Fig. 16.2 A 3-year on-farm comparison study in Switzerland where microbial activity in orchard soils (expressed as the ratio of microbial bound carbon:nitrogen) was higher (low values indicate higher bacterial presence and activity) in organic (filled symbols) than in integrated (open symbols) managed orchards (same symbol shape = orchards in the same village). (a) A clear correlation between the C_{mic}:N_{mic} ratio to phosphorus content of the fruit flesh was found which (b) in turn was correlated with the fruit quality index (including sugar and acidity content and fruit flesh firmness) *P* = phosphorus content of the fruit fresh in mg/kg dry matter.

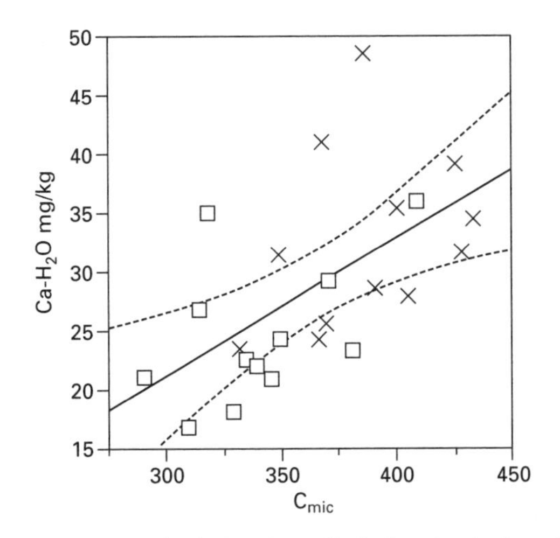

Fig. 16.3 Replicated field trial in Wädenswil, Switzerland where higher bacteria biomass (expressed as microbial bound carbon (C_{mic}) was found in the organically managed orchard soils (crosses = organic; squares = conventional, integrated) than in orchards managed according to integrated farming practice. Higher bacterial biomass was correlated with increased content of water-extractable calcium in soil samples ($n = 24$, $R^2 = 0,34$).

These results support the hypothesis that organic management methods with regular organic matter inputs, but without synthetic chemical input such as pesticides, herbicides and fertilisers, increase the microbial activity of the soil and that this in turn improves the mineralisation capacity of the soil biota resulting in a more balanced availability of mineral nutrients to the plant. Similar conclusions are drawn by Mäder *et al.* (2002) when interpreting their findings in the DOK long-term system comparison trial with arable crops. Also Reganold *et al.* (2001) found improved soil fertility parameters in organically managed plots of a major comparison trial with apple in the USA. Increases in soil microbial activity related to organic fertilisation protocols and increased calcium and phosphorous availability and contents in apple were also reported (Weibel, 1997; Weibel *et al.*, 2004a) (Fig. 16.2).

Tree nutrition can also affect the synthesis of polyphenolic compounds. For example, Leser and Treutter (2005) and Strissel *et al.* (2005) reported that apple trees receiving increased levels of nitrogen inputs contained significantly lower levels of polyphenolic compounds in their leaves. With the scab-(*Venturia inaequalis*) sensitive cultivar Golden Delicious, the lower polyphenol contents were also shown to lead to higher disease susceptibility. Raised levels of polyphenolics in fruit are linked to potential self-defence benefits, indicating that organic fertility management practices may therefore have beneficial effects on both crop health and nutritional quality (see also Section 16.2.2).

On the other hand, there is also evidence that polyphenolic synthesis in trees is reduced under nutrient stress. For example, in an organic/conventional comparison study with two apple cultivars, trees in the organic plots suffered from nutrient deficiency (detected by leaf analyses) caused by root damage after too harsh tillage (Weibel *et al.*, 2004b). While apples from the organic plots in this study showed similar polyphenol content to that of fruit from conventional plots, the content of polyphenols was significantly positively correlated with leaf nutritional status. Independently of the production method, fruits from plots with a low nutritional status showed lower polyphenol contents and vice versa (Fig. 16.4).

These results clearly indicate that fruit growers should not lower fertility inputs below the level required for satisfactory tree performance and intrinsic fruit quality, and also content of polyphenols. Thus, balanced nutrition without over or under supply of nutrients should be targeted in organic fruit production.

Weed competition for water and nutrients

Weed competition for water and nutrients can have similar effects on fruit quality as described above for fertilisation. For example, if weed competition is completely prevented by chemosynthetic herbicides in conventional production, this can lead to excess supply of certain mineral nutrients, in particular nitrogen and potassium, which in turn results in reduced sensory quality and shelf-life (Section 16.2.2). On the other hand, excessive weed competition, in particular, during the pre-bloom phase and the end of the first shoot growth period (Gut and Weibel, 2005), can induce nutrient and/or water deficiency and a risk of quality loss.

Because herbicides are not permitted in organic agriculture and mechanical tillage, the most widespread alternative is less radical; the tree rows are often

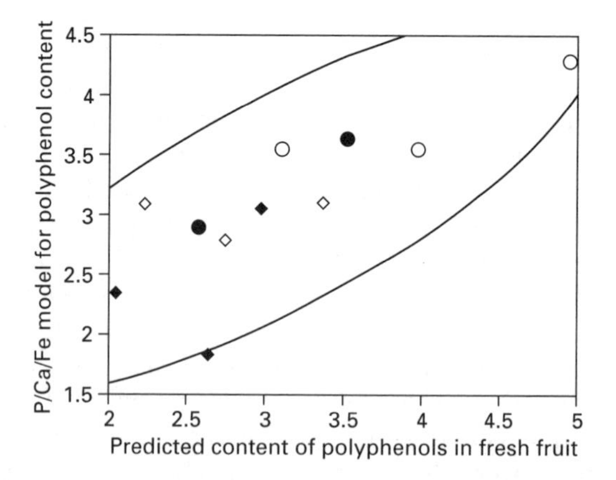

Fig. 16.4 Multiple linear model containing phosphorus, calcium and iron contents of the leaves correlated significantly with the polyphenol contents of fruit. Organic (filled symbols) and conventional fruits (open symbols) did not differ significantly in their average polyphenol contents; however, trees with a low leaf nutrient status showed a lower polyphenol contents in fruit ($R^2 = 0.64$). Circles = cultivar 'Glockenapfel', squares = cultivar 'Idared'; area between the curves = space where 95% of the modelled values can be expected.

weedier than in conventional production systems. The fruit quality-related effect of a managed weed cover in the root zone of the trees is thought to be partially due to the prevention of excess nutrient and/or water supply, but also to the positive effects of weed cover on soil stability and microbial activity. Weeds, by their substantial release of carbon-rich root exudates, serve as a primary energy source for the soil biota (Ingham *et al.*, 1985). This increase in soil microbial activity can then have positive effects on fruit quality (see also Section 16.2.2).

Disease and pest control

Owing to the prohibition of chemosynthetic pesticides under organic farming standards, there is a greatly reduced availability of intervention/treatment-based methods for disease and pest control in organic fruit production systems. The efficacy of the permitted biological control, extract or mineral element (e.g. S and Cu)-based crop protection products is also usually lower than of chemosynthetic pesticides. Permitted plant protection products show efficacies of between 60 and 80% while chemosynthetic fungicides and pesticides often have efficacy levels >95% (Tamm *et al.*, 2004).

As a result, organic fruit growers have to make maximum use of preventative or indirect plant protection measures, some of which were already described in terms of their effect on fruit quality (e.g. lower planting densities, see sections above). However, there remains an increased risk for pests and diseases causing stress, decreased photosynthetic activity and capacity. Also in certain sensitive cultivars, the application of permitted plant protection products (such as lime sulphur) can result in phytotoxic effects or latent stress in trees (Palmer *et al.*, 2002). The decrease in photosynthetic activity by both pest/disease attack and pesticides may, in turn, result in reduced fruit quality. Both long-term latent and short-term acute side effects of organic plant protection products have not yet been sufficiently investigated.

Effect on content of polyphenolic compounds: Complete protection from stress, however, appears to reduce plant synthesis of polyphenolic compounds, which are known to be produced as part of the plant's inducible resistance response to fungal and pest attack, but are also produced in response to certain abiotic stress factors such as mechanical injury (Feucht and Treutter, 1999).

In a recent field study five commercial orchards with both conventional and organic apple production units of the cultivar Golden Delicious were monitored for three years. The fruit content of flavonols, the major polyphenolic compounds in apples, was significantly (up to 23%) higher in organic than in conventional apples (Table 16.1; Weibel *et al.*, 2004a). As indicated above, this was probably due to the combined effects of different pest and disease pressure and management methods within the organic orchards.

In the few studies that have been carried out to compare polyphenol content in organic and conventional crops, many, but not all, show cases

Table 16.1 Mean content (mg/g DM) of polyphenols in organically produced Golden Delicious apples in three successive years (analyses were carried out at the beginning of the storage period in December). The percentage differences compared to conventional fruit from the same production unit, the level of significance between conventional and organic fruit (** = $P > 0.01$; * = $p \geq 0.05$) or trends (exact p-value listed if $0.05 < p < 0.13$) are indicated

Year and month of analysis	Flavanols bio vs. IP	Cinnamon acids bio vs. IP	Phloretin glycosides bio vs. IP	Quercetin glycosides bio vs. IP	Total polyphenols bio vs. IP
1997, Dec.	** 3.19 (+22.9%)	* 0.94 (+12.4%)	* 0.24 (+21.6%)	ns 0.29	* 4.66 (+18.7%)
1998, Dec	** 2.34 +15.8%)	ns 0.64	ns 0.25	* 0.22 (−11.6%)	ns 3.45
1999, Dec.	($P = 0.07$) 2.70 (+8.0%)	($P = 0.09$) 1.23 (+6.4%)	($P = 0.06$) 0.22 (+9.9%)	*0.24 (−18.1%)	($P = 0.13$) 4.41 (+6.3%)

IP = integrated production. ns = not statistically significant.

with higher levels of polyphenols in organic compared to conventional fruit, vegetables and grapevine (Levite *et al.*, 2000; Afssa, 2003; Asami *et al.*, 2003; Benbrook, 2005).

Pesticide residues: Another important quality aspect of organic fruit is the absence (or greatly reduced) risk of residues of plant protection products being present, because chemosynthetic pesticides are prohibited under organic farming standards. It has also been shown that, except for rare cases of fraud, historic use of very persistent pesticides (e.g. organochlorine pesticides) prior to conversion of orchards or pre- or post-harvest accidents (e.g. spray drift from neighbouring farms, mix-up of fruit during processing or storage), organic fruit are virtually free of detectable pesticide residues (e.g. European Commission, 2004 and CVUA, 2005). In the CVUA study the average content of pesticide residues on organic fruit was, on average, 550 times lower than on conventional fruit. A recent study in Switzerland in which 835 conventional and 99 organic fruit samples were analysed, found that 68% of conventional fruit samples had detectable residues below the legal threshold value, 25% were without pesticide residues and 7% had pesticide levels above the legal threshold value (Klein, 2004; Table 16.2). Among the 99 organic fruit samples, 92% had no detectable pesticide contents and in the remaining 8% of samples only traces far below the legal threshold were found.

Optimal harvest date, storage method

Picking fruit at an optimal harvest date is also an important factor affecting the nutritional and sensory quality and storability of fruit. Regardless of the production method, a fruit grower needs precise information on the seasonal (e.g. climate) and site-specific parameters that determine the optimal harvest

Table 16.2 Pesticide residues found in organic and conventional fruits in an inter-cantonal study 2002/03 in Switzerland (Klein, 2004)

Production method (no. of samples)	No detectable residues	Residues below legal threshold value	Residues above legal threshold value
Conventional (835)	212 (25%)	567 (68%)	56 (7%)
Organic (99)	91 (92%)	8 (traces, 8%)	0

date. There is widespread use among both organic and conventional growers of the so-called 'Streif-Ripeness Index', which is based on a starch test with iodide and measurements of sugar and firmness of the fruit to indicate the optimal picking date as a function of the desired storage duration (Watkins, 2003). For organic versus conventional comparison trials and trials focused on detecting effects of production system parameters (e.g. irrigation, fertility and/or crop health management methods), it is essential that the ripeness index at harvest is determined for all samples to identify potential confounding effects of the ripening stage of fruit (e.g. Weibel *et al.* 2004a). However, in many comparative studies that have previously been published this has not been done or reported.

Although no pre- or post-harvest pesticide/fungicide treatments are applied to top fruit in organic production systems prior to storage, their storability is similar to conventional apples or pears, as long as care is taken during harvest and only clean, healthy fruit is put into storage (Tschabold, 2003). A specific post-harvest problem that occurs in organic apples is *Gloeosporium* sp., a fungal disease that does not pose problems in conventional fruit production, owing to the use of chemosynthetic fungicides. Dipping fruit into hot water (49–52 °C) was shown to be an effective treatment to stop epidemics of the fungus during storage (Weibel *et al.*, 2004c), but there is still a need to reduce pre-harvest infections by this fungus in organic production systems. Zingg and Weibel (2004) achieved interesting results with applications of acidified clay powder.

Use of ethylene blockers (1-MCP): In conventional apple storage a new, revolutionary invention (1-MCP = 1-methylcyclopropene) has recently been introduced into commercial practice. 1-MCP is a volatile compound, which acts as an ethylene blocker (ethylene is the main ripening hormone for apple fruit). Applied at very low dose rates into the storage atmosphere, it is extremely efficient in slowing down the ripening of climacteric fruit such as apples (Watkins, 2003). The molecule also has a residual effect after fruit is removed from cold stores; thus it extends shelf-life and maintains sensory quality (crispness) of fruit until reaching the consumer's household. In organic fruit storage 1-MCP is not allowed, mainly because it is a chemosynthetic molecule that does not occur in nature (Speiser *et al.*, 2005). Although 1-MCP does not work for all top fruit cultivars, it can significantly reduce post-harvest losses and thereby can increase profitability in conventional fruit production and the price differential between organic and conventional fruit of similar quality.

Sorting and grading

Fruit with visual blemishes, poor sensory quality and/or measurable lesions should be eliminated in the pack house and not reach the end consumer. However, the threshold limits for cosmetic blemishes are currently very high and it has been questioned whether visual perfection leads to a solely 'cosmetic quality' focused increase in use of permitted plant protection products, which is incompatible with the principles of organic farming. To address this dilemma, in Switzerland organic fruit specific sorting and grading rules (which are slightly more tolerant towards purely cosmetic defaults) have been established between the market partners (Lösch, 2006). These organic-specific sorting prescriptions create useful and ecologically friendly degrees of freedom in production without affecting the interior or hygienic quality of the fruit. Unfortunately, in most countries organic fruit usually needs to achieve the same appearance criteria as conventional fruit.

16.3 Comparison of quality parameters between organic and conventional fruit

16.3.1 Standard assessments of quality parameters

In a three-year comparison trial involving ten commercial orchards, a wide range of quality parameters were assessed including fruit weight, dry matter, vitamin C and vitamin E (Weibel *et al.*, 2004a). For several yield, visual and nutritionally relevant quality parameters, no significant differences between fruit from organic and conventional orchards could be detected. However, for some nutritional and sensory quality parameters, significant differences were detected (Table 16.3). This included certain mineral nutrients (e.g. in the first year phosphorus content was 32% higher in organic fruit; Fig. 16.2a). Also, fruit firmness, nutritional fibres and the quality index ((3 × malic acidity g/L + 2 x firmness kg/cm^2 + Brix)/3) differed significantly in the first year (Table 16.3) and Brix content (soluble solids, sugar) was significantly different in the third year. Food preference tests with rats revealed a tendency ($P = 0.08$) for the rats to preferred conventional to organic apples (173.0 g of fruit eaten per animal vs. 123.5 g per animal, respectively). However, rat preference tests may not reflect consumer preferences determined in taste panels, because feed preference in rats correlated positively with ripeness ($R^2 = 0.28$), whereas sensory scores by humans correlated negatively with ripeness ($R^2 = -0.48$). Neither the self-decomposing test according to Samaras (1978) nor the feeding test with laboratory rats, revealed significant differences between production methods.

Table 16.3 Standard fruit quality parameters measured at the beginning (December) and towards the end (February) of storage period over three years. Means are of five organic orchards (bio, upper figure) and five conventional orchards (IP, lower figure) working to integrated farming standards (* = $P \leq 0.05$)

Year and month of analysis	Sugar (Brix) % bio vs. IP	Malic acid (g/l) bio vs. IP	Fruit flesh firmness bio vs. IP	Quality index*) bio vs. IP	Other parameters bio vs. IP
1997, Dec.	ns 14.03 > 13.58	ns 4.3 > 3.9	* 6.35 > 5.57	* 40.5 > 36.4	Vitamin C: ns Fibres: + 9.6 % P-content: + 31.9 % Selenium: ns
1998, Feb	ns 13.5 > 13.3	ns 2.7 > 2.1	ns 4.16 < 4.07	ns 29.9 >27.8	Vit. C: ns
1998, Dec	ns 14.6 < 15.2	ns 4.4 > 4.2	ns 5.30 > 5.15	ns 38.2 > 37.9	Fibres: ns
1999, Feb.	ns 13.9 < 14.7	ns 2.7 > 2.1	ns 4.55 < 4.92	ns 31.2 > 30.9	
1999, Dec.	* 14.5 > 13.5	ns 4.3 > 4.2	ns 5.72 > 5.38	ns 38.5 > 36.7	Fibres: ns
2000, Feb.	ns 13.7 > 13.0	ns 2.7 > 2.8	ns 4.50 = 4.50	ns 30.7 > 30.5	

In six 'blind' sensory panel tests with apples from the same group of orchards (as used in the rat preference tests) organic apples achieved significantly higher scores (up to 15%) in three of the six tests, with no significant difference between conventional and organic fruit being detected in the other three tests. In all three years the correlation between the technical quality index and the sensorial score was high and significant (Fig. 16.5). This emphasises that, independent of the production method used, good standard values in terms of sugar and acidity content and flesh firmness are the basis for sensorial perception by consumers.

Similar results were found in a replicated field trial involving four cultivars at Wädenswil, Switzerland (Weibel and Widmer, 2004) and in a similar study carried out in the USA (Reganold *et al.*, 2001) with organic fruit repeatedly achieving higher taste panel scores for fruit firmness, acidity, sugar content and overall sensory score.

16.3.2 Alternative assessments of fruit quality

Picture forming methods

Among all holistic methods proposed for quality assessments, image or 'picture' forming methods are the most frequently used. In 10 out of 11 studies they were shown to differentiate accurately between samples of products from organic and conventional farming systems (Alföldi *et al.*, 2006). The

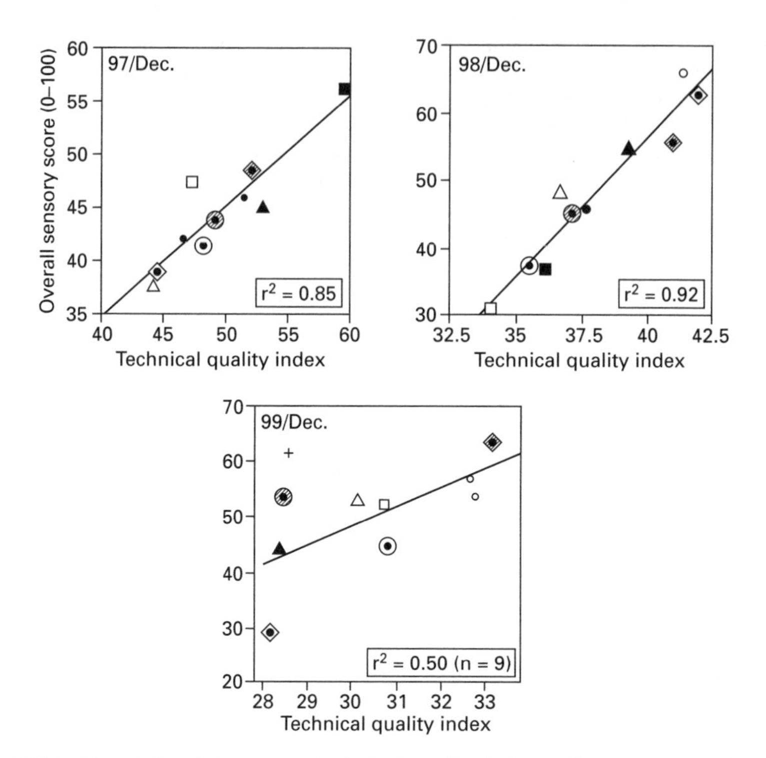

Fig. 16.5 Correlation between the technical quality index and sensory score (overall judgement) in blind sensory panel tests.

image forming methods were originally intitated by Rudolf Steiner in the early twentieth century. He suggested making life forces visible by putting an extract of the undisturbed food substance into a solution of copper chloride, creating a specific crystallisation picture that can be interpreted for 'vitality'. Pfeiffer, WALA, Ursula Balzer (Balzer-Graf and Balzer, 1991) and many others then developed further the so-called image forming method, also including circular chromatography ('Chroma') and vertical capillary dynamolysis ('Steigbild') over a silver nitrate solution.

In a study comparing organic and conventional apples (variety Golden Delicious) from a range of commercial orchards it was found that reproducible differences between apples from organic and conventional systems can be detected (Fig. 16.6). Additionally, the 'vitality index' produced based on the picture forming method was found to be correlated with the technical standard index based on chemical quality tests and the sensory test results (Fig. 16.7, left) (Weibel *et al.*, 2004a).

However, while it has become more generally accepted that the 'picture forming method' can be a suitable tool for identifying products from organic and conventional production systems, there is controversy about the suitability of using these data to detect or extrapolate specific additional or non-material 'quality' differences between samples of the same crop.

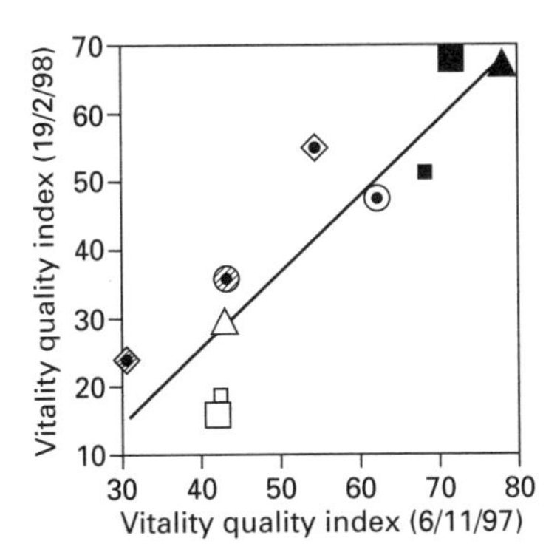

Fig. 16.6 Reproducibility of the picture forming methods. Fruits from the same sample were analysed twice (before and after storage) and blind coded. The quality-interpretation in the form of a 'vitality index' correlated significantly and well between the two analyses ($R^2 = 0.72$; filled symbols = organic; open symbols = conventional; same symbol form = same village).

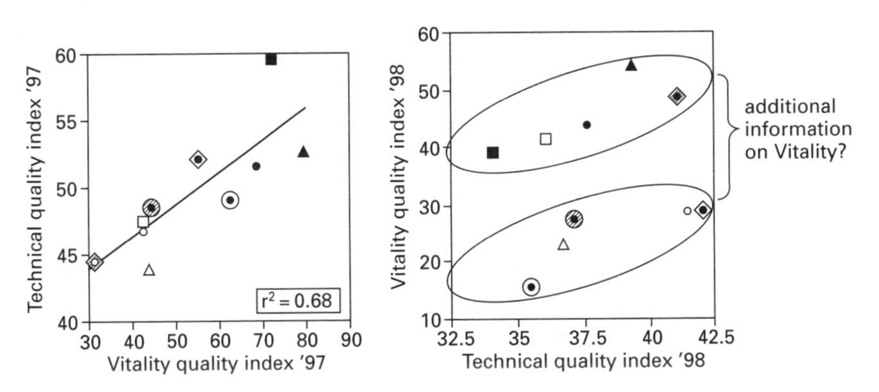

Fig. 16.7 Correlations between the 'vitality quality index' (based on data from picture forming methods and the index of technical quality (based on classical fruit quality tests, see text) in the first year (left) and the second year (right) of the study (filled symbols = organic, open symbols = conventional; same symbol form = same village).

Fluorescence excitation spectroscopy

Fluorescence excitation spectroscopy (FES; Strube and Stolz, 2002) is the second most frequently used alternative or complementary food quality assessment method. Seven out of eight studies identified differences between production systems (Alföldi *et al.*, 2006). Currently, several evaluation studies are carried out with FES on conventional versus organic apples of yields from several years (Strube, 2005; Weibel and Bigler, 2005; personal communications). Roughly, the assumption behind the method is 'life is light'. The hypothesis is that a vital plant or animal product has incorporated more light (in form of photons) than a less vital foodstuff. After excitation with light, the emission of photons from food samples can be measured in light-tight dark-chambers with ultra-sensitive light detectors. Strube and Stolz

(2002) carried out separate measures with nine different light wavelengths. Based on numerous calibration curves, they could show that between different foods species (e.g. seeds versus fruit) specific emissions patterns exist allowing the results to be interpreted as being 'typical' or 'not typical' for a certain species or stage of development.

Kirlian photography or gas discharge visualisation
The gas discharge visualisation method was applied in food quality research only very recently (preliminary studies of Sadikov *et al.*, 2004 and Weibel *et al.*, 2005). 'Kirlian photography', later called gas discharge visualisation GDV (Korotov, 2004) is a method where a fresh food sample is exposed to a high voltage and high frequency electrical field. While the current is on, a cloud of electrons and photons forming a spectacular light emission is released. This 'corona' is then digitally recorded, sequenced into defined image parameters, which then can be statistically analysed (Fig. 16.8). One advantage of this method would be that measurements are easier and faster to carry out (including statistical analyses of the data) compared with, for example, the picture forming method. To date, the method is still in the phase of validation and not ready for application to interpret quality differences.

16.4 Conclusions and future challenges

In a recent report (Alföldi *et al.*, 2006) seven recent literature reviews (Woese *et al.*, 1997; Worthington, 1998; Heaton, 2001; Bourn and Prescott, 2002; Velimirov and Müller, 2003; Tauscher *et al.*, 2003; Afssa, 2003) on the quality comparison of organic and conventional food were examined, compared and summarised. Major conclusions were that organic products:

- contain markedly fewer value-reducing constituents (pesticides, nitrates); this enhances their physiological nutritional value;
- are as safe as conventional products regarding pathogenic microorganisms (mycotoxins, *E. coli* bacteria);
- tend to have a higher vitamin C content;
- tend to have higher than average scores for taste;
- have a higher content of certain health-promoting secondary plant compounds;
- have a lower total protein content (most studies focused on cereals); this can mean that grain produced for bread is less suitable for baking.

Therefore, many research findings appear to confirm one of the fundamental hypotheses/recommendations of the pioneers of organic farming: 'Don't apply chemical-synthetic inputs to soil, plant or animals because it will directly or indirectly affect their well being and quality'. However, while it is generally accepted that organic food is virtually free from agrochemical residues, is as safe as conventional food in terms of microbial pathogens and mycotoxins

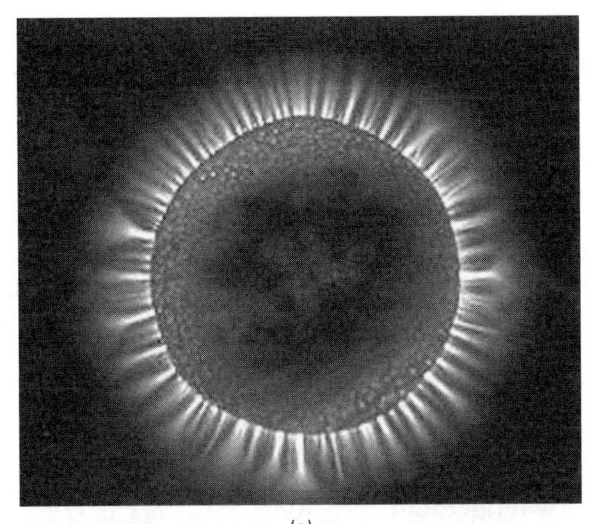

(a)

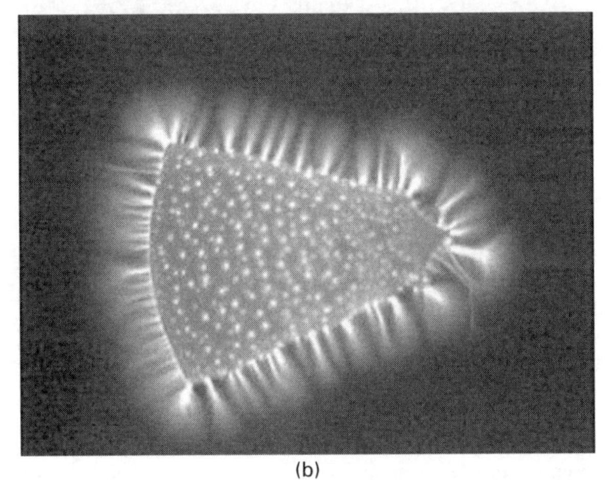

(b)

Fig. 16.8 Pictures of the gas discharge 'corona' of (a) an apple fruitlet at T-stage and (b) of a slice of a ripe apple. The digitally recorded picture is described by 23 parameters that undergo statistical analyses with learning tool algorithms or methods of discriminate analysis.

(see Chapter 17) and has a tendency to be tastier and higher in nutritionally desirable compounds, there are virtually no scientifically acknowledged studies that demonstrate beneficial effects of organic food consumption on human health. Dietary intervention and/or cohort studies with animal models and/or humans are therefore urgently needed to address this issue. With respect to methodologies used in food quality assessments, alternative or holistic methods have been shown to have the potential to deliver complementary information to standard chemical analysis methods. However, while proven to provide consistent, reproducible and objectively interpretable results, which can be used to distinguish between organic and conventional foods, more work is required to calibrate such methods against standard chemical analyses to identify if, and to what extent, they can be used to identify nutritional qualities that are potentially linked to human health.

In this chapter we have explained several system-inherent factors of organic fruit growing that can improve fruit quality. However, with the intensification of organic fruit production currently under way worldwide (e.g. more intensive nitrogen application on horticultural crops), there is a risk of quality decrease. Therefore, technical progress in organic farming should be closely and scientifically monitored for (side) effects on food quality, possibly in a holistic view that also includes environmental, social and human health criteria.

16.5 Acknowledgement

Our particular thanks go to the NATURAplan team of Coop Switzerland (F. Wehrle, R. Isella, K. Rapp) who supported the apple studies used as examples in this chapter.

16.6 References

AFSSA (Agence Française de Sécurité Sanitare des Aliments) (2003). *Evaluation Nutritionelle et Sanitaire des Aliments Issus de L'agriculture Biologique*, Agence Française de Sécurité Sanitaire des Aliments, 236 pp.

Alföldi, T., Granado, J., Kieffer, E., Kretzschmar, U., Morgner, M., Niggli, U., Schädeli, A., Speiser, B., Weibel, F.P. and Wyss, G. (2006). *Quality and Safety of Organic Producs*. FiBL-Dossier No. 4. FiBL, Frick 24 pp.

Asami, D.K., Hong Y.-J., Barret D.M. and Mitchell A.E. (2003). 'Comparison of the total phenolic and ascorbic acid content of freeze-dried and air-dried marionberry, strawberry, and corn grown using conventional, organic, and sustainable agricultural practices', *Journal of Agricultural Food Chemistry*, **51**, 1237–1241.

Balzer-Graf, U. and Balzer, F. (1991). 'Steigbild und Kupferchloridkristallisation – Spiegel der Vitalaktivität von Lebensmitteln', in Meier-Ploeger A. and Vogtmann H. *Lebensmittelqualität–Ganzheitliche Methoden und Konzepte*, Verlag C.F. Müller, Karlsruhe, 163–210.

Barden, J.A. and Neilsen, G.H. (2003). 'Selecting the orchard site, site preparation and orchard planning and establishment', in Ferree D.C. and Warrington I.J. *Apples: Botany, Production and Establishment*, CABI Publishing, Oxon and Cambridge, MA, 237–266.

Benbrook, C.M. (2005). 'Elevation of antioxidant levels in food through organic farming and food processing', *Organic Center State of Science Review*, Foster, RI, 78 pp.

Bourn, D. and Prescott, J. (2002). 'A comparison of the nutritional value, sensory qualities, and food safety of organically and conventionally produced foods', *Critical Reviews in Food Science and Nutrition*, **42**(1), 1–34.

Byers, R.E. (2003). 'Flower and fruit thinning', in Ferree D.C. and Robinson T.L. *Apple: Botany, Production and Uses*, CABI Publishing, Oxon and Cambridge, MA, 409–436.

CVUA Stuttgart (2005). (Eco-Monitoring 2004) *Ökomonitoring 2004*, Die Chemischen und Veterinäruntersuchungsämter in Baden Württemberg, Deutschland.

European Commission, (2004). *Monitoring of Pesticide Residues in Products of Plant Origin in the European Union, Norway, Iceland and Liechtenstein*, 2001 Report, Sanco/17/04–Final/EC.

Feucht, W. and Treutter, D. (1999). 'The role of flavan-3-ols and proanthocyanidins in plant defense', in Inderjit, K. and Dakshini M.M F.C.L., *Principles and Practices of Plant Ecology*, CRC Press, London, 307–338.

Friedrich, G., Neumann, D. and Vogl, M. (1997). *Physiologie der Obstgehölze*, Springer, Berlin, 2nd extended edition, 601 pp.

Granatstein, D. and Kirby, E. (2006). 'The changing face of organic tree fruit productions', *Proceedings of the First International Symposium on Organic Apple and Pearl* in Wolfsville (CAN). Submitted to *Acta Horticulturae* in Feb. 2006, 8 pp.

Gut, D. and Weibel, F.P. (2005). 'Integrated and organic weed control in pome and stone fruit', in Tromp J., Webster A.D. and Wertheim S.J. *Fundamentals of Temperate Zone Tree Fruit Production*, Backhuys Publishers, Leiden, The Netherlands, 372–377.

Harker, F.R. (2004). 'Organic food claims cannot be substantiated through testing of samples intercepted in the marketplace: a horticulturalist's opinion'. *Food Quality and Preference*, **15**(2), 91–95.

Heaton, S. (2001). *Organic Farming, Food Quality and Human Health. A review of the evidence*, Soil Association, Bristol, Great Britain, 87 pp.

Ingham, R.E., Trofymow, J.A., Ingham, E.R. and Coleman, D.C. (1985). 'Interactions of bacteria, fungi, and their nematode grazers: effects on nutrient cycling and plant growth'. *Ecological Monographs*, **55**, 119–140.

Klein, B. (2004). (Inter-cantonal monitoring on pesticide residues in fruit and vegetables 2002–2004) *Rapport Inter-cantonal sur des Résidus de pesticides en Fruits et légumes 2002–2003*, Rapport des Laboratoires Cantonales GE et VD of Switzerland.

Korotkov, K. (2004). *Measuring Energy Fields: State of the Science*, Backbone Publishing, Fair Lawn, USA.

Leser, C. and Treutter, D. (2005). 'Effects of nitrogen supply on growth, contents of phenolic compounds and pathogen (scab) resistance of apple trees', *Physiologia Plantarum*, **123**, 49–56.

Levite, D., Adrian, M. and Tamm, L. (2000). 'Preliminary results on contents of resveratrol in wine of organic and conventional vineyards', *Proceedings of the 6th International. Congress on Organic* Viticulture, Basel, 256–257.

Lösch, R., Weibel, F.P., J.Y., C., Schmid, C. and Vogt, C. (2006). (Sorting prescriptions for organic fruit) *Sortiervorschriften für Bioobst. Bio-Suisse leaflet*; http://www.biosuisse.ch/de/produkte/fruechtegemuese/index.php: 8 pp.

Mäder, P., Fliessbach, A., Dubois D., Gunstl L., Fried P. and Niggli U. (2002). 'Soil fertility and biodiversity in organic farming', *Science*, **296**, 1694–1697.

Marschner, H. (1995). *Mineral Nutrition of Higher Plants*, Academic Press, London 566–595.

Neilsen, G.H. and Neilsen, D. (2003). 'Nutritional requirements of apple', in Ferree D.C. and Warrington I.J. *Apples: Botany, Production and Uses*, CABI Publishing, Oxon-UK and Cambridge (MA) 267–302.

Palmer, J.W., S.B., D., Shaw, P. and Wünsche, J.N. (2002). 'Growth and fruit quality of 'Braeburn' apple (*Malus domestica*) trees as influenced by fungicide programmes suitable for organic production', *New Zealand Journal of Crop and Horticultural Science*, **31**, 169–177.

Reganold, J.P., Glover, J.D., Andrews, P.K. and Hinman, H.R. (2001). 'Sustainability of three apple production systems', *Nature*, **410**, 926–930.

Robinson, T.L. (2003). 'Apple orchard planting systems', in Ferree D.C. and Warrington I.J. *Apples: Botany, Production and Uses*, CABI Publishing, Oxon and Cambridge-MA, 345–408.

Sadikov, A., Kononenko, I. and Weibel, F.P. (2004). 'Analyzing coronas of fruits and leaves', in Korotkov K. *Measuring Energy Fields: State of the Science*, Backbone Publishing, Fairlown USA 143–154.

Samaras, L. (1978). *Nachernteverhalten unterschiedlich gedüngter Gemüsearten mit besonderer Berücksichtigung physiologischer und mikrobiologischer Parameter*, Schriftenreihe 'Lebendige Erde', Darmstadt.

Schumacher, R. (1989). *Die Fruchtbarkeit der Obstgehölze*, Ulmer Verlag, Stuttgart, 242 pp.

Schumacher, R., Fankhauser, F. and Stadler, W. (1980). 'Influence of shoot growth, average fruit weight and daminozide on bitter pit', in Atkinson D., Jackson J.E., Sharples R.O. and Waller W.M. *Mineral Nutriton of Apple Trees*, Butterworth, London, 83–91.

Speiser, B., Tamm, L. and Weibel, F.P. (2005). *Evaluation of 1-MCP for use in Organic Farming*, FiBL-Statement, Research Institute of Organic Farming (FiBL), Switzerland: http://www.fibl.org/english/news/single-article.php?id=486.

Strissel, T., Halbwirth, H., Hoyer, U., Zistler, C., Stich, K., and Treutter, D. (2005). 'Growth-promoting nitrogen nutrition affects flavonoid biosynthesis in young apple (*Malus domestica* Borkh.) leaves', *Plant Biology*, **7**, 677–685.

Strube, J. and Stolz, P. (2002). (Fluorescence-Excitation-Spectroscopy for determination of quality of apples from organic farming) 'Fluoreszenz-Anregungs-Spektroskopie zur Bestimmung der Qualität von Äpfeln aus ökologischem Anbau', *Vortragstagung der Deutschen Gesellschaft für Qualitätsforschung*, Hannover, 04.03.2002, pp. 209–214.

Tamm, L., Häseli, A., Fuchs, J.G., Weibel, F.P. and Wyss, E. (2004). 'Organic fruit production in humid climates of Europe: Bottlenecks and new approaches in disease and pest control', *Acta Horticulturae (ISHS)*, **638**, 333–339.

Tauscher, B., Brack, G., Flachowsky, G., Henning, M., Köpke, U., Meier-Ploeger, A., Münzing, K., Niggli, U., Pabst, K., Rahmann, G., Willhöft, C. and Maywer-Miebach, E. (2003). (Appraisal of food from different production systems) *Bewertung von Lebensmitteln verschiedner Produktionsverfahren*, Status-Bericht 2003, Landwirtschaftsverlag, Münster-Hiltrup, 166 pp; http:www.bmvel-forschung.de.

Torjusen, H. and Sangstad, L. (2004). *European Consumers' Conceptions of Organic Food: A Review of Available Research*', Professional Report no. 4–2004, National Institute for Consumer Research, Oslo. 150 pp.

Tromp, J. and Wertheim, S.J. (2005). 'Fruit growth and development', in Tromp J., Webster A.D. and Wertheim, S.J. *Fundamentals of Temperate Zone Tree Fruit Production*, Backhuys Publishers, Leiden, 240–262.

Tschabold, J.-L. (2003). *Storage Capability of Organic and Conventional Fruit in the Western Part of Switzerland*, Internal Report of FiBL, Frick, 12.

Velimirov, A. and Müller, W. (2003). (Quality of organically grown food. A broad literature search on potential advantages of organically produced food) *Die Qualität biologisch erzeugter Nahrungsmittel. Umfassende Literaturrecherche zur Ermittlung potenzieller Vorteile biologisch erzeugter Lebensmittel*, Bio Ernte Austria/Wien, 59 pp.

Watkins, C.B. (2003). 'Principles and practices of postharvest handling and stress', in Ferree D.C. and Warrington I.J., *Apples: Botany, Production, and Uses*, CABI Publishing, Oxon and Cambridge, MA, 585–614.

Webster, A.D. (2005). 'Sites and soils for temperate tree-fruit production: their selection and amelioration', in Tromp J., Webster A.D. and Wertheim S.J. *Fundamentals of Temperate Zone Tree Fruit Production*, Backhuys Publishers, Leiden, 12–23.

Weibel, F.P. (1997). 'Enhancing calcium uptake in organic apple growing', *Acta Horticulturae*, **448**, 337–343.

Weibel, F.P. and Häseli, A. (2003). 'Organic apple production – with emphasis on European systems', in Ferree D.C. and Warrington I.J. *The CABI Apple Book*, CABI Publishing, Wallingford Oxon, 551–583.

Weibel, F.P. and Leder, A. (2004). 'Consumer reaction to the "Flavour Group Concept" to introduce scab resistant apple varieties into the market. "Variety Teams" as further development of the concept', in Weinsberg FÖKO, *ECO-FRUIT; 11th International Conference on Cultivation Technique and Phytopathological Problems in Organic Fruit-Growing*, LVWO, Weinsberg/Germany, 196–201.

Weibel, F.P. and Widmer, A. (2004). (System comparison. Integrated and organic apple production: Part III: Inner quality – substances of content and sensory tests) 'Systemvergleichsversuch: Integrierte und biologische Apfelproduktion; Teil III: Innere Qualität – Inhaltsstoffe und Sensorik', *Schweiz. Zeitschrift für Obst- und Weinbau*, **140**(7), 10–13.

Weibel, F.P., Bickel, R., Leuthold, S. and Alföldi, T. (2000). 'Are organically grown apple tastier and healthier? A comparative field study using conventional and alternative methods to measure fruit quality', *Acta Horticulturae*, **517**, 417–427.

Weibel, F.P., Häseli A. and Schmid O. (2002). 'Organic fruit production in Europe (overview and farm economy)', *The Compact Fruit Tree*, **35**(3), 77–82.

Weibel, F., Treutter, D., Häseli, A. and Graf, U. (2004a). 'Sensory and health-related quality of organic apples: a comparative field study over three years using conventional and holistic methods to assess fruit quality', in Weinsberg F.Ö.O.e.V. *ECO-FRUIT; 11th International Conference on Cultivation Technique and Phytopathological Problems in Organic Fruit Growing*, LVWO, Weinsberg/Germany, 185–195.

Weibel, F.P., Widmer, A. and Treutter, D. (2004b). (System comparison. Integrated and organic apple production: Part III: Inner quality – contents of polyphenolic substances) 'Systemvergleichsversuch: Integrierte und biologische Apfelproduktion. Teil IV: Innere Qualität: Gehalt an antioxodativen Stoffen (Pflanzenphenolen)', *Schweiz. Zeitschrift für Obst- und Weinbau*, **140**(19), 6–9.

Weibel, F.P., Suter, F. and Zingg, D. (2004c). (Control of *Gloeosprium* sp. on apple with different post-harvest treatments) *Bekämpfung von Gloeosporium auf Apfel mit Nacherntebehandlungen*, FiBL-Bioobstbautagung, Forschungsinstitut für biologischen Landbau (FiBL), CH-Frick, 35–37.

Weibel, F.P., Sadikov, A. and Bigler, C. (2005). 'First results with the gas discharge visualisation (GDV) method (Kirlian photography) to assess the inner quality of apples', *1st Scientific FQH Conference*, FQH, Research Institute of Organic Agriculture (FiBL), CH-5070 Frick, Switzerland, p. 77.

Weibel, F.P., Ladner, J., Monney, P. and Sutter, F. (2006a). 'Improved organic and low-input apple production by weed competition tolerant rootstocks', *Acta Horticulturae (ISHS)*, Submitted Sept. 2006.

Weibel, F.P., Chevillat, V., Tschabold, J.-L. and Stadler, W. (2006b). Fruit thinning in organic apple growing with optimized timing and combinations strategies including (new) natural spray products and mechanical rope-divices', in Weinsberg F.Ö.O.e.V. *12th International Conference on Cultivation Technique and Phytopathological Problems in Organic Fruit Growing*, Weinsberg/Germany, 183–197.

Widmer, A., Husistein, A., Bertschinger L., Weibel, F.P., Fliessbach, A. and Käser, M. (2004). (System comparison integrated and organic apple production. Part II: Growth, yield, sorting results and soil parameters). 'Systemvergleichsversuch: Integrierte und biologische Apfelproduktion. Teil II: Wachstum, Ertrag, Kalibrierung, Boden', *Schweiz. Zeitschrift für Obst- und Weinbau*, **140**, 6–9.

Woese, K., Lange, D., Boess, C. and Bögl, K.W. (1997). 'A comparison of organically and conventionally grown foods – Result of a review of the relevant literature', *Journal of the Science of Food and Agriculture*, **74**, 281–293.

Worthington, V. (1998). 'Effect of agricultural methods on nutirtional quality: A comparison of organic with conventional crops', *Alternative Therapies*, **4**(1), 58–69.

Wünsche, J.N. and Ferguson, I.B. (2005). 'Crop load interaction in apple', in Janick, J., *Horticultural Reviews*, John Wiley and Sons, Chichester, 231–290.

Wyss, E., Villiger, M. and Müller-Schärer, H. (1999). 'The potential of three native insect predators to control the rosy apple aphid, *Dysaphis plantaginea*', *Entomologia Experimentals et Applicata*, **44**, 171–182.

Zingg, D. and Weibel, F.P. (2004). (Pre-harvest control of *Gloesporium* sp. on apple: results of the trials in 2003) *Vorernte-Bekämpfung von Gloeosporium auf Apfel: Versuchsresultate FiBL-Bioobstbautagung 2004*, Forschungsinstitut für biologischen Landbau (FiBL), CH-Frick, 38–39.

Zürcher, M. and Leumann, M. (2006). 'Wie hoch sind die Produktionskosten im Obstbau unter Schweizer Produktionsbedingungen und mit deutschen Produktions-und Betriebsmittelpreisen?' *Schweizer-Obstkulturtage* 12–13 Jan. 2006 at Flawil. Agroscope Changings-Wädenswil, http://www.db-acw.admin.ch/pubs/wa_arb_06_div_1914_d.pdf; pp. 8.

17

Strategies to reduce mycotoxin and fungal alkaloid contamination in organic and conventional cereal production systems

Ulrich Köpke and Barbara Thiel, University of Bonn, Germany and Susanne Elmholt, University of Aarhus, Denmark

17.1 Introduction

Mycotoxins are toxic metabolites synthesized by specific fungi that grow on living plants and their residues under favourable conditions. They are undesirable ingredients of food and feed. Risks are also posed by the spores and toxin-contaminated aerosols. Contamination by mycotoxins is a severe problem in food security. More than 300 species of fungi with the ability to form mycotoxins have been identified and more than 400 metabolites are assigned to the group of mycotoxins. Fortunately, only about 20 mycotoxins produced by five genera of fungi (*Fusarium*, *Penicillium*, *Claviceps*, *Alternaria* and *Aspergillus*) are found regularly or occasionally in food and feed at levels which might have an impact on human and animal health (Gareis, 1999a,b).

Since synthetic fungicides are not allowed in organic agriculture, fungal populations on crops are often presumed to be higher in this farming system and a higher frequency of mycotoxin contamination of organic food and feed has frequently been postulated (e.g. Tinker 2001; Trewavas, 2001, 2004). Others argued that omitting fungicides might lead to higher diversification of microbial populations and therefore limit growth of specific mycotoxin producers. There are controversial reports about mycotoxin contamination levels in grains produced in organic, 'low input' and conventional farming (Marx *et al.*, 1995; Berleth *et al.*, 1998; Birzele *et al.*, 2002; Meier *et al.*, 1999; Stähle *et al.*, 1998; Döll *et al.*, 2002). Thus, the first aim of this paper was to analyse, based on currently available data, whether there is a higher occurrence of fungal pathogens on organically grown crops and, as a result, a higher risk of mycotoxin contamination in grain produced according to organic as well as other low input farming methods. The second aim was to describe factors that are known to influence mycotoxin contamination levels in the different cereal production systems. Finally, strategies to reduce mycotoxin contamination are described. The recommendations given are limited to the conditions of the temperate European climate.

17.2 Mycotoxin- and alkaloid-producing fungi

Food safety risks in cereals grown in Europe are posed mainly by mycotoxin-producing species of the fungal genera *Fusarium*, *Penicillium*, *Claviceps*, *Alternaria* and *Aspergillus*: deoxynivalenol (DON), nivalenol (NIV), zearalenone (ZEA), fumonisins, ochratoxin A (OTA), alternariol, ergot alkaloids and – less important in temperate climatic regions – aflatoxins (Table 17.1). Phytopathologically relevant fungi that inhabit seeds and their products can be divided into field and storage pathogens.

Field pathogens invade and damage seeds almost exclusively during the growing season. Infection and growth are normally related to high relative air humidity and comparatively high moisture contents in seeds of at least 19%. The main toxin-producing fungi in central Europe belong to the genus *Fusarium*, *Claviceps* and *Alternaria*.

Table 17.1 Specific mycotoxins that are produced by different genera of fungi important for food safety

Fusarium	*Alternaria*	*Claviceps*	*Penicillium*	*Aspergillus*
Deoxynivalenol (DON)	Alternariol (AOH)	Ergot alkaloids	Ochratoxin A (OTA)	Aflatoxin B1
Nivalenol	Alternariol-methylether (AME)		Patulin	Aflatoxin G1
Zearalenone (ZEA)	Tenuazonic acid (TeA)		Citrinin	Aflatoxin M1
T-2	Altertoxins		Penitrem A (PA)	Ochratoxin A (OTA)
Fumonisin	Altenuen		Cyclopiazonic acid	Sterigmatocysteine
Moniliform				Cyclopiazonic acid
Diacetoxy-scripenol				

Source: Derived from information in Bennett and Kilch (2003) and Steyn (1995).

Storage pathogens invade and damage seeds predominantly during storage. Storage fungi can grow at relatively low relative humidities (>65%) and at comparatively low temperatures (usually >5 °C). *Aspergillus*, and in particular *Penicillium* species, are storage pathogens, some of which producing mycotoxins.

17.2.1 *Fusarium* species and toxins

The fungal genus *Fusarium* encompasses some hundreds of species that are widely distributed both in temperate and tropical regions, where they can be isolated from soil and plant materials. These species are important pathogens of cereals causing diseases such as the seedling blight, *Fusarium* root rot and *Fusarium* head blight (FHB), one of the most important fungal diseases affecting a wide range of cereals including wheat, maize and barley. *Fusarium culmorum*, *Fusarium avenaceum* and *Fusarium poae* are more often associated with FHB in cooler regions, such as central and northern Europe, whereas *Fusarium graminearum* is considered as the major pathogen found worldwide (Nicholson *et al.*, 2004). Fortunately, only a few *Fusarium* species produce a wider range of mycotoxins and levels of mycotoxins that can pose food safety risks (Moss, 2002), but more than 100 *Fusarium* toxins are known. The most important *Fusarium* toxins from the point of view of human health are the trichothecenes, zearalenones, moniliformins and fumonisins (D'Mello *et al.*, 1997, Table 17.2). Not much is known of the species attacking fodder grass species, and their mycotoxin production. A few publications and more research are needed in this field (e.g. Engels and Krämer, 1996).

Trichothecenes

Of four basic trichothecene groups (A, B, C and D), types A and B represent the most important mycotoxins. Type A includes the T-2 toxin that can suppress the immune system and can cause damage to bone marrow. The T-2 toxin is about 14 times more toxic than the most widespread *Fusarium*-based type B mycotoxin, deoxynivalenol (DON), previously called vomitoxin. DON is a

Table 17.2 Mycotoxins produced by different *Fusarium* species (after Desjardins and Proctor 2001)

Fusarium species	Trichothecenes	Zearalenone	Fumonisins	Moniliformin
F. avenaceum	–		–	+
F. culmorum	+	+		
F. cerealis	+	+		
F. equiseti	+	+		+
F. graminearum	+	+	–	+
F. poae	+		–	
F. sporotrichioides	+		+	
F. tricinctum	–			

low-molecular weight inhibitor of protein synthesis with cell membrane and haemolytic activity and is found in wheat, barley, maize and safflower as well as in compound animal feed based on these crops. The synthesis of the two types appears to be a characteristic of particular *Fusarium* species (Table 17.2). Type A trichothecenes predominate in *Fusarium sporotrichoides* and possibly *F. poae*, whereas type B occurs principally in *F. graminearum* and *F. culmorum* (D'Mello, 2003).

Zearalenone

Zearalenone (ZEA) is synthesised by many *Fusarium* species, which also produce DON, and is found worldwide in maize and maize products. Young sows seem to be the most susceptible animals and may show symptoms of hyperoestrogenism even with low ZEA intakes (0.06 mg/kg body weight intake). In contrast, 300 ppm are reported to be tolerated by hens and up to 2000 ppm by young heifers (Fink-Gremmels, 2005a).

ZEA resembles the human 17β-oestradiol hormone produced by the ovaries. Although almost non-toxic, in very small doses it has oestrogenic effects that can disrupt the human endocrine system (Benbrook, 2005). It is important to note that transformation products of ZEA can have three to four times higher endocrine disrupting activity than ZEA.

Fuminosins and moniliformin

Fusarium moniliforme syn. *verticillioides* causes the so-called 'ear rot' disease in maize and produces fumonisin B_1 (FB_1), one of the most frequently detected mycotoxins in the food supply chain worldwide (Steyn, 1995). FB_1 can inhibit lipid formation, particularly in the liver. Fumonisins have been detected and investigated only relatively recently. Several structurally related forms of fumonisins (FB_s) have been associated with human cancer (e.g. FB_1 with oesophageal cancer) as well as with a host of problems in livestock fed with FB_1-contaminated feed (D'Mello, 2003; Benbrook, 2005).

17.2.2 *Penicillium* species and toxins

Most mycotoxin-producing *Penicillium* species are primarily known as storage pathogens; however, some may be found in the field environment and these will be addressed here.

Ochratoxin

Ochratoxin A (OTA) is a mycotoxin produced by some species of *Penicillium* and *Aspergillus*. It is nephrotoxic to all animal species tested and the causal agent of mycotoxic porcine nephropathy (Krogh, 1978). It was previously associated with the human renal disorder, Balcan endemic nephropathy (BEN), and tumours of the urinary tract (Pfohl-Leszkowicz *et al.*, 2002). Recently, another endemic kidney disease (Tunisian chronic interstitial nephropathy, CIN) was linked to OTA-contaminated food (Creppy, 1999; Wafa *et al.*,

2004) and it has also been linked with testicular cancer (Schwartz, 2002). OTA acts through several molecular pathways and, apart from being nephrotoxic, it is reported to be teratogenic, immunotoxic, mutagenic, carcinogenic and genotoxic (Boorman, 1989; WHO, 1990; Smith *et al.*, 1994; Creppy, 2005). Two major OTA producing *Penicillium* species are currently recognized: *Penicillium nordicum*, which is mainly found on meat and cheese, and *Penicillium verrucosum,* which is found to contaminate grain (Larsen *et al.*, 2001; Seifert and Levesque, 2004). Some *Aspergillus* species are also known to produce OTA (Krogh, 1978; Frisvad and Thrane, 2000; Seifert and Levesque, 2004) including *Aspergillus niger* (not all strains), which is thought to be responsible for OTA production in wine and *Aspergillus ochraceus*, the main OTA producer in coffee and cocoa (Frank, 1999; Frisvad and Thrane, 2000).

17.2.3 *Claviceps* and other fungi-producing ergot alkaloids

Ergot alkaloids are produced by *Claviceps purpurea*, which has contaminated rye flour for centuries and is still a problem in many areas of the world. Poisoning from *C. purpurea* dates back to antiquity. Several epidemics have been reported since the Middle Ages that were linked to the ingestion of mouldy rye bread. *C. purpurea* is a species complex that, besides rye, can infect wheat, barley, triticale and more than 600 grasses, especially under cool and rainy weather conditions. Thus, infection can also originate from grasses growing off-site or as weeds in the stand (Agrios, 1997).

17.2.4 Aflatoxin and other important mycotoxins and their producers

Although poisoning by mycotoxins has been common over the centuries, the mass deaths of turkeys in modern intensive livestock farming in Great Britain (turkey-X disease) at the beginning of the 1960s triggered the first systematic research into mycotoxins (Gareis, 1999a, b). Deaths were caused by the use of aflatoxin (a mycotoxin produced by *Aspergillus flavus*) contaminated Brazilian peanuts, which led to carcinoma of the liver. Two species of *Aspergillus* are regarded as the main producers of aflatoxins: *A. flavus* (Group I producing aflatoxins B and Group II producing aflatoxins B and G (Geiser *et al.*, 2000)) and *Aspergillus parasiticus*, which produces aflatoxins B and G (Frisvad and Thrane, 2000; Seifert and Levesque, 2004). Aflatoxins mainly cause cancer of the liver, kidneys and bile duct and are carcinogenic, mutagenic, nephrotoxic and immunosuppressive. Aflatoxins have the highest toxicity of all known mycotoxins. This is the reason why maximum tolerance levels for aflatoxins were first to be set in the 1970s and 1980s. However, as carry-over rates of other mycotoxins into edible products of animals were for a long time considered too low to constitute a public health risk, aflatoxins remained until recently the only group of mycotoxins for which a harmonized European feed legislation was established. Contamination with aflatoxins is usually found in crops imported as animal feed from tropical and subtropical countries. Regulations on European levels (2003/2174/EC of 12 December 2003 amending 2001/466/EC; EC, 2003) have been made for groundnuts, nuts and dried fruits, for cereals, milk and their products as well as for spices. Separate limits were set for baby foods, infant formulae and dietary foods (2004/683/EC; EC, 2004a). Since aflatoxin B_1 is metabolised and excreted by cows as aflatoxin M_1 in milk, upper limits for aflatoxin B_1 in feed have been set extremely low (cows: 10 ppb; pigs: 20 ppb; poultry: 20 ppb). Since the highest levels of toxicity are attributed to aflatoxin B_1, regulations include separate maximum levels for aflatoxin B_1, aflatoxin M_1 and the sum of aflatoxins B_1, B_2, G_1 and G_2.

In cereal stands that have experienced wet weather and extended dew periods (especially during the grain ripening phase), saprophytic, dark pigmented fungi such as *Alternaria* spp., *Cladosporium* spp., *Drechslera* and *Epicoccum* also occur. Those low-input or organic farming systems that do not use fungicides to protect the ears from diseases are more susceptible. These fungi may cause allergic reactions of the human pulmonary and immunological system. Mycotoxins produced by *Alternaria* spp. are currently considered to be of minor toxicological relevance in the concentrations that have been detected in cereals (Griffin and Chu, 1983; Thalmann, 1990; Davis and Stack, 1994; Schrader *et al.*, 2001). However, high toxicity of cereals infested with *Cladosporium* has been reported, although the specific toxins involved are not known. Close interactions between *Fusarium*, *Alternaria* and *Cladosporium* spp. with respect to grain colonization and mycotoxin production can be assumed as they are similarly located on the grains.

17.3 Problems associated with dietary mycotoxins/alkaloid intake in livestock and humans

Mycotoxins can have carcinogenic, teratogenic, nephrotoxic, hepatotoxic and immunosuppressive properties (Table 17.3). They have also been linked to damage of several organs, especially the liver and kidneys and negative impacts on the pulmonary, neurological, gastrointestinal and immunological systems of humans and livestock (Fung and Clark, 2004).

Feeding of *Fusarium*-contaminated cereals to livestock can result in adverse effects on animal health and performance. In addition to acute poisoning symptoms in livestock, mycotoxins can cause subclinical effects, which are considered to be economically more important than acute poisoning (Fink-Gremmels, 1999, 2005a). These include severe long-term consequences, including immunosuppression, neurotoxicity and changes in nutrient uptake (Rotter and Prelusky, 1996; Gilbert and Tekauz, 2000; Miller *et al.*, 2001; Walker *et al.*, 2001). In livestock, particularly in pigs, even extremely low DON levels can induce protracted feed refusal and at higher levels vomiting

Table 17.3 Effects of some important mycotoxins on humans

	Deoxynivalenol	Nivalenol	Zearalenone	T-2 toxin	Moniliformin	Ochratoxin A	Aflatoxins
Carcinogenic						×	×
Hepatotoxic						×	×
Nephrotoxic						×	×
Mutagenic			×			×	×
Skin irritating		×		×			
Emetic	×	×		×			
Immunosuppressive	×	×		×		×	×
Necrotising				×	×		
Oestrogenic			×				

is commonly observed. Reduction in weight gain has been recorded for all monogastric animals when the DON concentration in feed is higher than 1000 μg/kg (Fink-Gremmels, 2005a).

Fumonisin B_1 in corn fed to pigs causes pulmonary oedema (Becker *et al.*, 1995). OTA was shown to affect the kidneys of most monogastric livestock species, but severe ochratoxicosis in agricultural practice has, to date, only been described in pigs and fowl. Ruminants are able to inactivate OTA (Fink-Gremmels, 2005a, b). The long half-life of OTA residues in pigs contributes to cumulative exposure in humans (Fink-Gremmels, 2005a).

Intoxications with ergot alkaloids are among the earliest described mycotoxicoses in humans and animals. Ergot alkaloids are formed by *Claviceps* species and comprise ergometrine, which is best known, ergotamine, ergosine, ergocristine, ergocryptine and ergocornine and their related -inines (Fink-Gremmels, 2005b). *Claviceps* species that have invaded monocots can easily be identified by the formation of sclerotia, replacing the normal seeds or caryopses. These sclerotia contain ergot alkaloids (ergolines) and exert toxic effects in all animal species, even in low concentrations (WHO, 1990). The toxins interact with adrenergic, serotinergic and dopaminergic receptors in the target organisms leading to neurotoxic and vascular symptoms such as vasoconstriction, sometimes progressing into vaso-occlusion and gangrenous changes, as well as abortions. Neurotoxic symptoms of ergotism are feed refusal, dizziness and convulsions (Fink-Gremmels 2005a).

Potential impacts of mycotoxins on human health were often deduced from animal studies, epidemiological investigations and studies with people working in grain storage and/or processing facilities (Benbrook, 2005). Impacts of mycotoxins can be exacerbated by many (stress-) factors such as other toxins, microbial fodder quality or unfavourable housing conditions. Several authors have shown that antioxidants have a highly protective effect with respect to the impact of DON on laboratory animals (e.g. Rizzo *et al.*, 1994; Atroshi *et al.*, 1998). Antioxidant-rich cabbage seed extracts and garlic had a protective effect with regard to the negative impacts of fumonisins (Abdel-Wahhab *et al.*, 2004).

17.4 Mycotoxin regulation and monitoring

Recently harmonized European regulations have set comprehensive maximum levels for the different aflatoxins that are allowed in feeds and foodstuffs. The maximum levels for OTA in raw cereal grains (including raw rice and buckwheat) are set at 5 μg/kg for grain and 3 μg/kg for all products derived from cereals by Commission Regulation (EC) No 472/2002 (EC, 2002a) and Commission Regulation (EC) No 123/2005 (EC, 2005a). Commission Regulation (EC) 683/2004 sets maximum levels for food intended for infants and young children. Commission Directive 2002/5/EC (EC, 2005b) defines the sampling and analysis methods for the official control of the levels of certain contaminants in foodstuffs. The design of appropriate sampling procedures is extremely important, because distribution of mycotoxins can be very heterogeneous. For instance, the number of contaminated peanut kernels in a lot is usually low, but the mycotoxin level in a single peanut can be very high (FAO, 2004). The more recent directives 98/53/EC (EC, 1998), 2002/27/EC (EC, 2002b), 2004/43/EC (EC, 2004b) and 2005/5/EC (EC, 2005b) set limits and regulate aflatoxins and ochratoxins. Sampling methods and methods of analysis to control *Fusarium* toxins in foodstuffs have also been adopted just recently (2005/38/EC of 6 June 2005, EC, 2005c) and maximum concentrations for *Fusarium* toxins have just been regulated by regulation 2005/856/EC of 6 June 2005 (EC, 2005d). For cereal products as consumed and other cereal products at the retail stage, less than 500 μg/kg of DON are permitted by the EU and 750 μg/kg for flour used as raw material in food products (Tables 17.4 and 17.5). Aflatoxins and ergot content are currently still regulated in feedstuff regulations at the national level (e. g. Germany: Futtermittel-Verordnung, 2003).

Compared with other regions of the world, Europe has the most extensive and detailed regulations for mycotoxins in food. Harmonized regulations exist for aflatoxins, aflatoxin M1 in milk, OTA in cereals and dried wine fruits, patulin in apple juice and apple products, and for aflatoxin B1 in various feedstuffs, and from 1 July 2006 also for *Fusarium* toxins (2005/856/EC, EC, 2005d). Upper EU limits for DON and ZEA are listed in Table 17.4. The German threshold values as an example of national regulations that apply to edible cereals but not to the crude harvested grains are listed in Table 17.5. It is to be expected that the number of regulations for mycotoxins other than aflatoxins will increase significantly in the next few years, both for food and feed.

Table 17.4 Regulations for deoxynivalenol (DON) and zearalenone (ZEA) in the European Union according to the Commission Regulation (EC) No 856/2005 of 6 June 2005 (EC, 2005d)

Product	DON (µg/kg)	ZEA (µg/kg)
Unprocessed cereals other than durum wheat, oats and maize	1250	
Unprocessed cereals other than maize		100
Unprocessed durum wheat and oats	1750	
Unprocessed maize	–	–
Cereal flour, including maize flour, maize grits and maize meal	750	
Cereal flour, except maize flour		75
Maize flour, maize grits and refined maize oil		–
Bread, pastries, biscuits, cereal snacks and breakfast cereals	500	50
Maize snacks and maize-based breakfast cereals	–	–
Pasta (dry)	750	–
Processed cereal-based food for infants and young children and baby food	200	20

Table 17.5 Regulations for deoxynivalenol (DON) and zearalenone (ZEA) in Germany according to the German 'Ordinance laying down maximum levels for mycotoxins in foodstuffs' and 'Ordinance on dietetic foodstuffs' of 4 February 2004 (BMVEL, 2004)

Product	DON (µg/kg)	ZEA (µg/kg)
Cereal grains ready to eat and processed cereal products except durum wheat products, bread, cookies and fine pastries	500	50
Bread, cookies and fine pastries	350	50
Cereal grains ready to eat and processed cereal products used for dietary foods for nurslings and babies	100	20

17.5 Factors affecting mycotoxin/alkaloid contamination of cereal grains

A review of the available literature allows a hierarchical ranking of the importance of different factors that influence the pre-harvest infection of cereal grains by *Fusarium*: weather/climate (year) > inoculum pressure/tillage/previous crop > fungicides > available plant nitrogen. The individual factors related to other fungi are described in separate sections below.

17.5.1 Effect of environmental conditions on fungal infection and mycotoxin loads

Fusarium

Temperature and humidity are the main climate factors influencing the development of *Fusarium* diseases of cereals. These factors show strong interactions and modulation with different environmental and host factors (Doohan *et al.*, 2003). High incidences of *Fusarium* contamination in wheat have frequently been reported as being caused by high rainfall during the vegetative period and especially the flowering season (e.g. Birzele *et al.*, 2002; Schauder, 2003). Obst and Bechtel (2000) showed that temperatures have to be above 16 °C and rainfall above 4 mm between growth stages (GS) 39/41 and 61 for ascospores of *F. graminearum* to be released. Ear infection can take place when, for at least two days in growth stages GS 55-69, temperatures exceed 17 °C and rainfall 2 mm. Infection by means of conidia can occur at lower temperatures.

Fungi in general are able to grow under a broad spectrum of environmental conditions. However, synthesis of mycotoxins is more dependent on specific weather and climate conditions (Drusch and Ragab, 2003). The impact of grain moisture levels at time of harvest and during storage on DON synthesis was determined by Birzele *et al.* (2000) during two years with different weather conditions during the flowering period. In the rainy 1998 season, DON levels were almost three times higher (280 μg/kg) than in 1997 (111 μg/kg) when precipitation was lower. Accordingly, about 10% of the samples exceeded the German threshold level of 500 μg/kg (BMVEL, 2004) in 1998. Reutter (1999) reported an average of 2700 μg/kg DON in wheat of the same year, for example in conventionally produced wheat from Schleswig–Holstein in Northern Germany.

Earlier studies by Schauder (2003) of seeds derived from organic farms in Germany during 1991 to 1993 showed that more than 80% of wheat samples were infected by *Microdochium nivale* and *Fusarium* spp. Seasons like 1993 with very high rainfall resulted in 97% of wheat samples with seed infection by these pathogens. Grain infection with *M. nivale* and *Fusarium* spp. was correlated with rainfall during the flowering stage.

Penicillium

Variations in OTA findings in cereals and incidences of 'mycotoxic porcine nephropathy' have often been correlated with years and geographical areas with wet harvest conditions (Büchmann and Hald, 1985; Czerwiecki *et al.*, 2002a; Jørgensen and Jacobsen, 2002), for example, the very wet harvest year of 1987 (Holmberg at al., 1990a, b; Jørgensen *et al.*, 1996). Barley harvested in 1987 contained on average 25% moisture, the highest in ten years and the average harvest time was delayed by $3^1/_2$ weeks compared to the 'driest' year in the study period. As a result, 35% of blood samples from swine herds were OTA-positive compared to less than 20% in the other tested years (Holmberg *et al.*, 1990a, b). In conclusion, moist weather conditions at harvest increase the risk of OTA problems and the need for quick and effective grain drying as shown also by Elmholt (2003) and Elmholt and Rasmussen (2005).

17.5.2 Inoculum pressure/tillage/previous crop effects on fungal infection and mycotoxin loads

Fusarium

While grain infection of most *Fusarium* spp. including all mycotoxin-producing species originates from the soil and/or plant residues or stubble on the soil surface, infection with *M. nivale* (which is not thought to form mycotoxins) is seedborne (Birzele *et al.*, 2002). Infection of grains from soil-borne fungi and mycotoxin production is the result of complex interactions between climatic conditions (see above), inoculum levels in the soil and plant litter on the soil surface, as well as tillage methods and crop rotational sequence, especially the immediate previous crop. In the case of wheat, a previous crop of wheat and particularly maize, and minimum tillage, were shown to increase significantly the inoculum pressure and *Fusarium* grain infection and mycotoxin levels. However, a high density of inoculum does not necessarily lead to high crop infestation. Soil suppressiveness, that is the complex microbial interaction between plant pathogens and all or a part of the saprophytic microflora, may diminish infestation of the pathogen (Alabouvette, 1990). Benbrook (2005) linked the tendency of organically managed soils to sustain more diverse microbial communities to the low levels of grain diseases observed on many organic farms. Investigations into the isolation and development of fungal antagonists as biological control products for seedborne diseases caused by *F. culmorum*, confirmed this assumption (Knudsen *et al.*, 1995). These authors determined that the occurrence of *F. culmorum* was 1.7 times higher in the organically managed compared with the conventionally managed fields, while the occurrence of antagonistic strains was about 4.6 times higher in the organically cultivated compared to the conventionally cultivated fields. A subsequent study with five soils, two of which were under organic or low input management, did not reveal differences in suppression of brown foot rot of barley. This was thought to be due to low saprophytic competitiveness of the *F. culmorum* isolate used (Knudsen *et al.*, 1999).

Penicillium

Penicillium verrucosum is mainly considered a storage pathogen. However, several authors highlighted a lack of knowledge about the pre-harvest ecology of storage fungi and argued that such knowledge would be essential to improve forecasting and prevention of *Penicillium* mycotoxins (Lillehøj and Elling, 1983; Miller, 1995; Wicklow, 1995). A limited number of studies have contributed to elucidating the role of *P. verrucosum* in the field ecosystem. Elmholt (2003) successfully isolated *P. verrucosum* in 14 of 76 Danish fields, using a new selective isolation/culture medium (dichloran yeast extract sucrose glycerol; DYSG), which is recommended for isolating *P. verrucosum* from foods, feeds (Frisvad *et al.*, 1992) and soil (Elmholt *et al.*, 1999). The low abundancies indicated that *P. verrucosum* was an ephemeral invader brought into the soil during soil management. However, it was shown to survive in soil for many months and proliferate even without addition of nutrient resources,

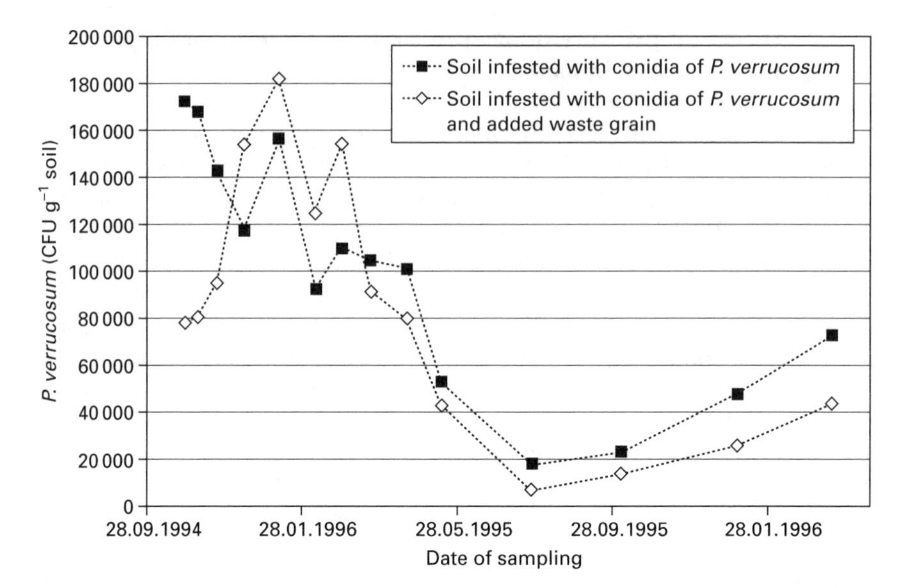

Fig. 17.1 Survival of *P. verrucosum* in soil (Elmholt and Hestbjerg, 2000).

as shown in Fig. 17.1. At one of the two sites with high abundancies of *P. verrucosum* in the soil, distinct spatial variation supported the hypothesis that the fungus had established and proliferated in the soil environment (Elmholt, 2003). The toxin itself, on the other hand, decomposes quickly in soil (Mortensen *et al.*, 2003).

Alternaria

Saprophytic, dark pigmented fungi such as *Alternaria* spp. can infect a wide range of plant species, especially tissues that are exposed to other biotic or abiotic stressors and older and senescing plant tissues. Also, wet weather conditions favour attack by *Alternaria* spp. Inoculum of *Alternaria* and potentially production of alternariol is further enhanced when cereal straw and stubble is left on the soil surface and not sufficiently incorporated into the soil after harvest (direct seeding and minimum tillage systems).

17.6 Agronomic strategies to reduce mycotoxin grain infection and mycotoxin levels

Although there is no close correlation of crop infection and mycotoxin production, all efforts to reduce mycotoxin contamination of cereal grains start with agronomic factors that can be controlled to minimize crop infection.

17.6.1 Rotational design and pre-crop effects

Fusarium

The life cycle of *Fusarium* spp. involves long periods of saprophytic growth on crop residues and the main source of *Fusarium* inoculum is crop residues that remain on the soil surface after harvest (Pirgozliev *et al.*, 2003). Maize, especially when grown to full maturity as a pre-crop to wheat, was shown to result in particularly high levels of residue-based inoculum levels. Silage maize, which is harvested earlier in a physiologically immature stage, is thought to result in lower levels of inoculum being produced. *F. graminearum* develops perithecia-containing ascospores on maize residues left on the soil surface in early spring. Ascospores are then actively released from dried and ripened perithecia during rainfall and can infect the newly planted cereal crops. Beck *et al.* (1997) found approximately 500 μg/kg DON on wheat after maize grown for grain compared with 300 μg/kg DON after silage-maize, but less than 100 μg/kg DON on wheat grown after wheat and barley.

One of the most important methods of reducing *Fusarium* infection risk by minimizing inoculum levels is to establish diversified crop rotations which avoid growing maize and cereals before cereals. Since maize as a pre-crop for cereals (especially wheat) is rarely used in organic systems, this risk is applied mainly to conventional farming systems where this practice has increased over the last 20 years.

Diversified rotations (with a lower proportion of cereals and a higher percentage of fodder crops), in combination with regular organic matter-based fertility inputs, were also linked to soils with (i) higher contents and turnover of soil organic matter and (ii) greater soil biodiversity and biological activity. These characteristics are also thought to increase the suppressiveness of soils and thereby reduce pathogen inocula (Alabouvette, 1990; Knudsen *et al.*, 1995, 1997,1999; Gattinger *et al.*, 2002; Mäder *et al.*, 2002; Alabouvette *et al.*, 2004).

Penicillium

Elmholt (2003) found no effects of previous crops or crop sequences on *P. verrucosum* abundance in soil. However, she suggested that some agronomic practices used in organic farming systems may increase the risk of post-harvest production of mycotoxins by *P. verrucosum*, for example, by delaying the time of harvest, increasing the amount of moist impurities in the bulk of grain and thus prolonging the drying time in some drying systems (Elmholt, 2003).

17.6.2 Tillage/crop residue management

Fusarium

Incorporation of plant residues using an inverting plough is known to be the most efficient tillage method of reducing levels of *Fusarium* inoculum on the soil surface, the incidence of FHB and DON levels in wheat, especially when wheat is grown after maize. Although traffic and tillage may cause problems with subsoil compaction in organic agriculture (Schjønning *et al.*, 2002), direct seeding and reduced tillage systems are currently rarely used, owing to the problems of controlling weeds in systems that do not permit herbicides. However, minimum tillage is increasingly used in conventional agriculture, resulting in higher incidences of *Fusarium* diseases and DON levels (Krebs *et al.*, 2000).

The problems associated with leaving infected plant residues on the soil surface after harvest are thought to be particularly acute with maize and in areas with high precipitation during anthesis (see Section 17.6.1 above and Beck *et al.*, 1997).

Claviceps
The risk of infection with *C. purpurea* can also be diminished by ploughing plant residues to bury the sclerotia and enhance their decay.

17.6.3 Choice of crop species

Fusarium
There are significant differences between cereal species in their susceptibility to mycotoxin-producing fungi and/or mycotoxin contamination levels. For example, wheat is generally more frequently contaminated with DON than rye (Tanaka *et al.*, 1988; Lepschy *et al.*, 1989; Döll *et al.*, 2002), but rye usually has higher levels of ergot contamination than wheat.

Penicillium
Cereal species also differ in their susceptibility to OTA contamination. Abramson (1998) and Abramson *et al.* (1990) ranked cereals grown in Canada in the following order: no risk – oats; low risk – HY-320 wheat, hard red spring wheat, 2-row barley; moderate risk – 6-row barley, corn; high risk – amber durum wheat. Most Canadian comparisons were made under controlled environmental conditions, for example, equal moisture additions to the tested grain. A range of European studies, as discussed below, also report species differences in OTA susceptibility, but these results should be interpreted with some care as the samples originate from 'real life', that is the grain had been subjected to the prevailing weather conditions at harvest (Jørgensen *et al.*, 1996; Czerwiecki *et al.*, 2002a, b; Elmholt and Rasmussen, 2005). As harvest time differs from species to species, this will inevitably lead to interactions between weather conditions and species sensitivity.

The Danish OTA surveillance programme has particularly focused on rye (Jørgensen *et al.*, 1996; Jørgensen and Jacobsen, 2002). From 1986–1992, the OTA content was determined in both cultivated rye and wheat (kernels): 78% of the rye samples contained OTA, but only 40% of the wheat samples did. Lower frequency of OTA contamination was also reported by Elmholt (2003) and Czerwiecki *et al.* (2002b) suggesting that rye is more sensitive than wheat.

17.6.4 Cultivar choice

Fusarium

In general, crops resistant to infection by fungi during the growing season are less likely to become contaminated by mycotoxins. Currently, no highly resistant cultivars are available and cultivars with some level of resistance/tolerance that are available in Europe would best be described as 'less susceptible'. Nevertheless, the percentage of 'resistant' cultivars listed in the official European cultivar rankings has been increased during the last 10 years. Both crop morphology and physiology related components of resistance have been identified. Cultivars with less susceptibility to *Fusarium* diseases are characterized by morphological features such as: (i) tall types with long culms, (ii) long distances between flag leaf and ear, (iii) awnless with not too dense ears (Mesterházy, 1995) as well as (iv) closed flowering and short flowering phase (Aufhammer *et al.*, 1999).

Physiological traits/components including: (i) resistance against penetration by fungal hyphae, (ii) production of anti-fungal compounds and/or (iii) prevention of mycotoxin production, are more complicated to identify, measure and use as selection criteria. Increasing the resistance to *Fusarium* spp. must not necessarily be based on the introgression of genes from exotic wheat genotypes. A successful use of a European genotype has been reported for the Swiss winter wheat cultivar *Arina* (Miedaner *et al.*, 2000).

Distinct differences in the susceptibility of cultivars to *Fusarium* infection and DON content have been reported by Döll *et al.* (2002). These authors compared 12 winter wheat cultivars amounting to a total of 120 accessions and at least five samples per cultivar that were grown conventionally on different sites in Germany during the 1998 growing season, which was characterized by high rainfall. Maximum DON contents of cultivars ranged from 180 μg/kg in *cv. Bussard* to higher than 11 660 μg/kg (*cv. Ritmo*). Although results may have been confounded by differences in agronomy practices between sites, since there were no clear morphological differences between cultivars included in the trial, results suggested a clear effect of cultivar on mycotoxin production and confirmed other studies showing differences in the susceptibility of wheat cultivars to *Fusarium* infection and mycotoxin contamination (Lauren *et al.*, 1996; Langseth and Stabbetorp 1996; Obst *et al.*, 1997; Klingenhagen and Frahm, 1999; Lienemann, 2002; Meier, 2003). Morphological features such as longer culms were also shown to reduce the risk of ear infection by *Fusarium* species in barley (Buerstmayr *et al.*, 2004). However, no morphological, biochemical or molecular markers, which correlate with reduced mycotoxin levels that could be used in breeding programmes, have been identified so far (Ruckenbauer *et al.*, 1999). Lemmens *et al.* (1997) concluded that, for practical breeding purposes, visual scores of several disease symptoms can be used, but that this again does not allow precise predictions of mycotoxin content.

Penicillium

There is very little information on cultivar differences with respect to (i) resistance to infection by *P. verrucosum* and (ii) OTA formation (Hökby *et al.*, 1979; Axberg *et al.*, 1997). The latter performed a laboratory experiment with six artificially inoculated cultivars of barley and three of wheat. After incubation for 23 weeks at approximately 19% moisture, the barley cultivars contained between 6 and 350 and the wheat cultivars between 25 and 890 ng OTA g^{-1} grain. In another study, Elmholt and Rasmussen (2005) reported that two of four cultivars of spring spelt from the same field contained 18 and 92 ng OTA g^{-1}, respectively, while the other two contained less than 0.5 ng OTA g^{-1}. As differences in OTA neither correlated with moisture content at harvest nor with the level of *P. verrucosum*, they were probably caused by differences in cultivar sensitivity. This should, however, be further studied.

Claviceps

Differences in susceptibility to ergot have been described for spring wheat (Platford and Bernier, 1976; Bretag and Merriman, 1981), durum (Platford and Bernier, 1976) and barley (Cunfer *et al.*, 1975; Pageau *et al.*, 1994) cultivars. It was suggested that, in order to increase resistance to *C. purpurea*, plants with bigger anthers and a high amount of pollen should be selected. For rye, hybrid cultivars have been shown to have higher susceptibility than non-hybrid cultivars, with the latter still more frequently used in organic agriculture.

17.6.5 Optimization of soil-nitrogen status

Fusarium

It has often been shown that sites with higher yield potential and/or fertilization regimes/soil conditions resulting in higher soil-nitrogen availability show higher infection levels of *Fusarium* spp. and *M. nivale*. A trend towards higher levels of grain contamination by DON was also found in soils with high supplemental nitrogen fertilization (e.g. Schauder, 2003). High levels of readily soluble mineral nitrogen were shown to increase the risk of mycotoxin loads in conventional agricultural systems (Buschbell and Hoffmann, 1992) and supplementary nitrogen applied as mineral and organic fertilizers with a high content of water soluble N (e.g. chicken pellets, manure slurry) was shown to increase *Fusarium* disease (Lemmens *et al.*, 2004; Heier *et al.*, 2005) and DON-levels (Aufhammer *et al.*, 2000).

This may be due to (i) an increased physiological susceptibility in crops grown with high nitrogen inputs and/or (ii) an extended vegetative growth and longer flowering period and later ripening of crops, which prolongs the period of time that wheat is susceptible to infection (Weinert and Wolf, 1995). Increased water soluble nitrogen-input levels may also increase FHB by increasing crop density, which is known to maintain a more humid microclimate in the crop canopy for longer periods of time after dew or rainfall (Martin and MacLeod, 1991). Since formation of both microconidia and ascospores needs moist conditions, inoculum production on crop residues remaining on the soil surface might also be enhanced by denser canopy structures (Fauzi and Paulitz, 1994; Benbrook, 2005). Thus, under the conditions of restricted nitrogen availability in organic agriculture, the probability of mycotoxin formation caused by an excess of nitrogen uptake is lower when compared with current mainstream farming.

Precision farming techniques and split applications adapted to micro-sites and specific to the demand of the crop might be a quality assurance tool for conventional farming in the future. Since input rate has shown significant effects on FHB development in wheat, lower amounts of nitrogen, both applied or available from the soil, may generally result in reduced FHB as well as DON contamination.

Penicillium

Studying barley that had been fertilized with 0, 90 or 240 kg N ha^{-1}, respectively, Häggblom and Ghosh (1985) found that strains of *Aspergillus* and *Penicillium* produced more OTA with increasing nitrogen-input levels and resultant barley protein contents. This was ascribed to a stimulatory effect of the amino acids glutamic acid and proline. The observed correlation is not surprising when considering the chemical structure of OTA, in which the amino acid phenylalanine is a component. Chemically, OTA is a chlorine-containing dihydroisocumarin linked as an amide of its 7-carboxyl group to L-β-phenylalanine.

Claviceps

The influence of nitrogen fertilizer source on the germination rate of ergots placed on the ground in rye culture has been studied. Compared with calcium ammonium nitrate, application of calcium cyanamide reduced the germination of ergots and formation of perithecia by 40–50% (Mielke, 1993).

17.6.6 Seed grading

Fusarium

Since *Fusarium* head blight (FHB) often results in a higher percentage of small/shrivelled grains, grading of seeds is considered an important tool in improving the quality of infected seed lots, especially in the production of organic seed, where the use of chemical seed coatings is not permitted (Piorr, 1990, 1992). Nevertheless, seed grading is neither sufficient nor precise enough to remove all infected grains and/or DON.

Some authors suggest that there is no clear relationship between grain size and DON levels (Lepschy, 1992). However, Perkowski (1998) reported that the fraction of smallest barley kernels contained more than 75% of the total DON. On the other hand, Chelkowsky and Perkowski (1992) determined higher DON contents in grains bigger than 2.5 mm grading compared to smaller grains. Sinha and Savard (1997) found less DON (< 5 mg/kg) in shrivelled grains compared with typically *Fusarium*-damaged grains (1–600 mg/kg). It was also shown that normal size grains may be infected, even if they are from ears without FHB symptoms (Rintelen, 1995). Visual scoring for FHB symptoms may therefore be relatively imprecise for prediction of FHB and DON production (Herrmann *et al.*, 1998; Aufhammer *et al.*, 1999; Lemmens *et al.*, 2004).

17.6.7 Seed treatments

Fusarium

Seed treatments aim to eliminate/reduce inoculum levels and/or disease development from seedborne fungi that are present on or within seeds, without affecting the viability and germination/emergence capacity of seed. Thermal seed treatments have been one of the earliest seed treatments used in agriculture. They may be based on warm water (45 °C, 2 h), hot-water treatments (52 °C, 10 min) and/or hot air/steam treatments. Warm or hot water treatments were shown to prevent/reduce *Fusarium* disease development effectively (Jahn, 2002), often showing similar efficiency to chemical seed dressing (Winter *et al.*, 1997; Forsberg *et al.*, 2005). On the other hand, thermal treatments based on hot air and hot steam have not yet found their way into agricultural practice (Jahn, 2002).

A seed dressing method based on 'electrons' was developed and tested by Lindner *et al.* (1996) and Schauder (2003) as a direct method to improve seed quality and remove/reduce seedborne disease inocula. 'Electron' seed dressing effectively removed common bunt spores *(Tilletia caries)* and reduced bunt levels compared to untreated seeds in field trials. However, its efficacy against *M. nivale* has not, as yet, been confirmed. Since this technique has potentially negative effects on germination rates of seed, it is limited to surface treatments (Jahn, 2002; Jahn *et al.*, 2005).

Seed coating with extracts of mustard (*Sinapis alba)* and horse radish (*Amoracia lapathifolia)* was shown to have high efficacy against *T. caries* spores but not *Fusarium* spp. (Spieß and Dutschke, 1991).

Claviceps

Chemical seed treatments of seed lots contaminated with *C. purpurea* sclerotia have been shown to delay ascocarp emergence and significantly reduce ascocarp numbers above the soil surface at rye anthesis (Dabkevicius and Semaskiene, 2002). A range of biological control agents (e.g. *Trichoderma lignorum* and *Pseudomonas aureofaciens*) were also investigated for their ability to provide control of *C. purpurea*, but did not significantly affect germination and ascocarp formation of sclerotia.

17.6.8 Use of 'organic', 'home-grown' and/or 'farm-saved' seed

Fusarium

Several studies in the early 1990s have shown lower *Fusarium* infection in seed grown on organically certified farms compared to conventionally produced seeds (Piorr, 1992; Dornbusch *et al.*, 1993). This was attributed to the relatively low efficacy of the fungicide treatments used in conventional seed production and a range of practices used in organic systems such as the (i) non-use of maize as a pre-crop for cereals in the rotation, (ii) fertilization regimes which result in lower plant-available nitrogen levels and (iii) soil management practices which result in higher biological activity and disease suppression (Knudsen *et al.*, 1995, 1999). However, the efficient use of better designed fungicide treatments may overcome the problems observed in conventional seed production systems (Stähle *et al.*, 1998).

Penicillium

The low abundance of *P. verrucosum* in most soils (Elmholt, 2003) indicates that seed contamination may be an important infection source. However, only a few studies have investigated the importance of seed inocula on pre-harvest contamination by *P. verrucosum.* Elmholt (2003) reported examples of much lower contamination levels in commercially produced compared to home-grown/farm-saved seed. However, it should be noted that many of the home-grown samples originated from farms with insufficient drying facilities, where up to 100% of the kernels were infested by *P. verrucosum.* Future studies should determine whether rapid and efficient drying of seed will avoid the risk of *P. verrucosum* contamination of 'home-grown' seed.

17.6.9 Plant growth regulator and fungicide application

Fusarium

Dwarf or semi-dwarf cultivars (which have a shorter straw length) have been widely introduced into conventional agriculture, to increase yield and reduce the risk of lodging. However, it has been suggested that the higher density of tillers/stems within the crop canopy, often achieved with short-straw cultivars, also results in a humid microclimate that can enhance *Fusarium* infection. To increase yields further and reduce lodging risk, it is common practice in many European countries to treat cereal crops (in particular, wheat) with plant growth regulators (e.g. chlormequat/CCC) that further shorten the straw length. In short straw cultivars treated with growth regulators, ears and grains are closer to the soil surface and are thought (during periods of high humidity) to facilitate a more rapid progress of infection from primary *Fusarium* inocula on the soil surface to lower leaves, from leaf to leaf and eventually from leaves (especially the flag leaf) to the grain. The combined use of short straw varieties and straw shortening growth regulators is therefore thought to increase the risk of *Fusarium* and mycotoxin load in grain.

Until the 1990s, no effective fungicides to control *Fusarium* diseases were available to conventional farmers. During the last decade significant efforts have focused on identifying or developing fungicides that would be effective in preventing FHB and/or mycotoxin production in cereals. However, studies on the effects of fungicide applications on *Fusarium*, FHB and/or the proportion of *Fusarium*-infected grains have produced variable and often contradictory results (D'Mello *et al.*, 1998; Pirgozliev *et al.*, 2003, Heier *et al.*, 2005). For example, D'Mello *et al.* (1998) concluded that 'the overall evidence concerning the effectiveness of fungicides is contradictory and in certain cases somewhat unexpected'. In fact, some studies indicated that mycotoxin production by *Fusarium* pathogens may increase if sub-lethal doses of fungicides such as carbendazim, tridemorph, difenoconazole and tebuconazole with triadimenol are applied. Miconazole and fenpropimorph have also been shown to increase aflatoxin production from *Aspergillus parasiticus* (D'Mello *et al.*, 1998). Two other studies described complex interactions occurring between fungicide type, fungal isolates and environmental factors, in relation to the level of control achieved by fungicides and DON production (Meier, 2003; Mesterhàzy *et al.*, 2003). In one of the most recent studies, fungicides such as tebuconazole and metconazole were described as providing efficient control of both FHB and DON production, while others such as *azoxystrobin* and related fungicides were less effective (Henriksen and Elen, 2005).

In fact, one of the strobilurin-based fungicides (azoxystrobin) developed primarily for powdery mildew control has repeatedly been found to result in increased concentrations of mycotoxins in wheat (Edwards *et al.*, 2001; Simpson *et al.*, 2001; Magan *et al.*, 2002). This indicates that the latest generation of fungicides introduced into cereal spray regimes may have increased the risk of mycotoxin contamination. Early application of fungicides (before anthesis), which targets the control of other cereal pathogens, was also shown to increase the level of natural *Fusarium* infection. For example, Liggitt *et al.* (1997) observed that the inhibitory effect of saprophytic flora on the grains can suppress the growth of *F. culmorum* and Henriksen and Elen (2005) concluded that early fungicide application may suppress competition by the saprophytic microflora (or other pathogens) on leaves, while having little effect on *Fusarium* species. Based on the currently available published information it can be concluded that currently available fungicide treatments are relatively ineffective in controlling mycotoxin-producing fungal species belonging to *Fusarium* and *Aspergillus*. Since fungicides were designed to control disease (but may in fact increase mycotoxin production caused by the stress imposed on the pathogen), D'Mello *et al.* (1998) suggested the inclusion of monitoring of mycotoxin levels in future evaluation protocols for candidate fungicides.

Penicillium

There is virtually no information on the sensitivity of *P. verrucosum* to fungicides. Banks *et al.* (2002) found no effect of different fungicide applications on post-harvest development of *P. verrucosum*.

Claviceps

It has long been known that breeding strategies and/or applications of growth regulators, which lead to an extended or more open flowering of cereals, increase the risk of infection with *C. purpurea*. Evans and Jenkyn (2000) observed that 23 out of 34 fungicides that showed evidence or at least some activity against *C. purpurea in vitro* did not show any effect in the field. Azole fungicides displayed highest levels of activity, while application of strobilurins resulted in a significant increase in the number and weight of ergots. The authors concluded that, since the agrochemical industry does not consider ergot to be a major crop protection target, there is currently no available economically feasible fungicide control method for *C. purpurea*.

17.6.10 Competition between fungi and potential for biological control on mycotoxin-producing fungi

Fungal–fungal interactions may also affect grain colonization and toxin production. Ramakrishna *et al.* (1996b) reported that competing fungal species did not affect spore germination of *F. sporotrichioides* in the phyllosphere, but almost completely inhibited growth and grain colonization. However, mycotoxin (T-2) production by *F. sporotrichioides* was, in general, stimulated by the presence of *A. flavus* and *P. verrucosum*, but slightly decreased by the yeast *Hyphopichia burtonii*. In a similar experiment with *P. verrucosum*, Ramakrishna *et al.* (1996a) found that germination and sporulation were not, or only slightly, inhibited by other species. However, seed infection was generally decreased, as was OTA production, in the presence of *A. flavus* and *H. burtonii*, but not in the presence of *F. sporotrichioides*. In search of potential biocontrol agents, Petersson *et al.* (1998) examined the interaction between *P. verrucosum* and two yeast species. They found that *Pichia anomala* suppressed growth and OTA production of one of the tested *P. verrucosum* isolates in wheat. Elmholt *et al.* (1999) found that *P. verrucosum* suppressed some of the indigenous soil fungi when grown together in Petri dishes. From the currently available information it must be concluded that efficient commercial biological control methods for mycotoxin-producing fungi are unlikely to become available in the near future.

17.6.11 Weed management

Fusarium

Compared to the preventative strategies described above, weed control is thought to have a relatively minor effect on *Fusarium* infection and mycotoxin levels in grain, but there are few studies available. However, it should be noted that weed density is often higher in organic farming systems and that certain weeds can act as alternative hosts for *Fusarium*. For example, *Fusarium* contamination has been detected by Meier (2003) especially on climbing weeds such as *Galium aparine* and *Convolvulus arvensis*. These weeds may therefore act as vectors that transport conidia to the upper parts of cereals, thus increasing the chance of infection of ears and grains. Meier (2003) also reported that infection rates of *G. aparine* with *Fusarium* were higher than those of the surrounding wheat plants, when overall infection pressure with *Fusarium* was high.

Claviceps

Many grasses are known to act as reservoirs and sources of primary *C. purpurea* inoculum (Hoffmann and Schmutterer, 1983; Agrios, 1997). Since beetle banks and non-crop field margins are commonly used in organic farming to increase biodiversity, it is possible that these areas may also become sources of *Claviceps* inoculum. The goal for farmers must be to achieve a balance between the benefits of on-farm biodiversity (e.g. improved habitat for pest predators) and the risk of increased infection by fungi such as *Claviceps*.

17.6.12 Management of invertebrate and vertebrate animals

Damage from rodents, birds and, in particular, insect pests, can break the outer seed coat in cereals and facilitate fungal infection and mycotoxin production (Twiddy, 1994; Beti *et al.*, 1995; Blank *et al.*, 1995). The ecological implications of these animal–fungal interactions are discussed by Wicklow (1995) and Janzen (1977). For maize, insects were shown to act as vectors for the dispersal of fungal conidia. For example, *Sitophilus zeamais* acts as a vector for conidia of *A. flavus* (McMillian *et al.*, 1980). However, for wheat there is little published information on the importance of insect vectors in the transmission of mycotoxin-producing fungi, but they are widely thought to play a minor role. Several authors also hypothesize that the mycotoxin-producing capacity in fungi is not merely 'a metabolic accident', but provides a competitive advantage for the fungus with respect to excluding other fungi (and possibly invertebrate and vertebrate animals) from colonizing (using as a food source) the same plant tissues.

In storage, insect activity may increase grain moisture and thereby improve conditions for fungal growth and toxin production. For example, the rice weevil *Sitophilus oryza* increased the moisture content in stored wheat, leading to *A. flavus* infection and aflatoxin production (Sinha and Sinha, 1991). There is, however, also some evidence that insects feeding on mouldy grain may metabolize and reduce mycotoxin loads (Wicklow, 1995). The impact of insect (and potentially other animal) vectors on fungal growth and mycotoxin loads in organic and conventional crop production systems certainly warrants further investigation.

17.7 Effect of harvest conditions and post-harvest handling on mycotoxin contamination levels

Grain moisture and impurities with a high moisture content harvested together with the grain (e.g. moist weed seed) may increase mycotoxin contamination levels if grains are not immediately dried after harvest. Fortunately, the critical moisture level for toxin production is often higher than for growth. Impurities in the harvested bulk grain may prolong the drying time (Elmholt, 2003) and make drying more heterogeneous (Hill and Lacey, 1983a), which is again thought potentially to increase mycotoxin levels. For example, it was shown that both DON and OTA levels can increase within a few weeks when moisture conditions are favourable for mycotoxin formation (e.g. Langseth *et al.*, 1993; Backes and Krämer, 1999; Birzele *et al.*, 2000; Abramson *et al.*, 1980; Abramson *et al.*, 1990).

Early harvest increases the number of immature, wet kernels in the bulk, making drying more difficult (Hill and Lacey, 1983a). Harvest time also affects fungal abundance, as more propagules are present on late compared to early harvested grain (Dickinson, 1973; Hill and Lacey, 1983b; Elmholt, 2003; Elmholt and Rasmussen, 2005; Kristensen *et al.*, 2005).

The time between harvest and reduction of grain moisture to a safe level is of vital importance. The higher the moisture level of grain at harvest and the higher the temperature, the more critical it becomes to reduce the time of interim storage prior to drying. If the capacity of the dryers is exceeded and damp grain has to be stored before drying, priority must be given to the wettest grain and the grain that is intended for human consumption or feeding to sensitive animals (Banks *et al.*, 2002).

For both conventional and organic grains, appropriate drying, storage and processing are essential to limit mycotoxin formation after harvest. Elevated moisture of 17% and 20% was shown to increase the DON level of naturally infected grains (Birzele *et al.*, 2000; Birzele and Prange, 2003); however, the increase in DON content varied in degree and rate, independent of cultivar.

Post-harvest infection by typical storage pathogens (e.g. *Penicillium* spp.) also leads to increased mycotoxin loads and further loss of quality (e.g. deterioration in colour, texture and taste, reduced seed viability and germination rates, loss of nutrients, and development of fungal odours and allergenicity) (Abramson, 1998). Physical, chemical and biological factors – including moisture, temperature, time, substrate, O_2, CO_2, mechanical damage, microflora and fauna – interact during drying and storage and affect the risk of fungal growth and mycotoxin formation in a very complex manner. This is not described in detail in this review and the reader is referred to reviews by, for example, Lillehøj and Elling (1983), Frisvad and Samson (1991), Abramson (1998) and Wicklow (1995). Some general aspects must nevertheless be enumerated here.

Cleaning of combines, trailers, augers, dryers and stores is needed to prevent carryover of fungal propagules from year to year. Conventional farmers can use pesticides. This option is not available to organic farmers and the problem is most severe where drying and storage facilities are made of materials that are difficult to clean. Elmholt (2003) and Haase (2003) studied a farm with a home-made natural air-drying system. The main and side ducts were made of wood and the side ducts covered by old hessian sacks, some being 20 years old and virtually impossible to clean. Both ducts and sacks contained large amounts of viable *P. verrucosum* conidia. Drying took several weeks and during this process the grain was constantly infested by *P. verrucosum* conidia. Banks *et al.* (2002) concluded that grain may become contaminated with *P. verrucosum* during or shortly after harvest through the presence of the fungus in handling equipment or the store environment. They state 'Scrupulous hygiene in handling equipment, ventilation systems and stores are vital to reduce the infection of fresh grain'.

The fungal biomass level, measured as ergosterol concentration, is generally several times higher in small kernels and impurities than in the original lots (Regner *et al.* 1994). Removal of these fractions did not significantly influence the average ergosterol content owing to the relatively low weight of the small kernels. However, as this material often separates during transport and bin loading (cf. on-floor drying), zones with very high ergosterol concentrations and fungal biomass may arise in certain parts of the grain and serve as starting points for mould growth. As small kernels have been shown to contain high amounts of certain toxins, removal would seem worth considering.

Winnowing is recommended at high levels of impurities especially if these impurities hold much water, the drying system is slow and/or if the technical facilities are conducive to the formation of damp pockets. Sclerotia of *C. purpurea* can be separated with the fraction of light grains, but this treatment brings a high percentage of grain loss. Small ergots and similar sized and shaped cereal grains may pass through the cleaning equipment and be sown alongside the seed.

On-floor drying using ambient or low-heated air is common (Scudamore, 1999; Kristensen *et al.*, 2005). The system is cheap and attractive to small farmers, including many organic growers. On-floor systems have a high intake capacity, as the bin can be filled at harvest rate, but their drying capacity is low and depends on, for example, moisture content, air flow rate, heating of the drying air and weather conditions. On-floor drying may therefore take several weeks or even months, increasing the risk of fungal growth if the grain is harvested with high moisture content. The use of heated air will, if handled correctly, decrease the drying time and the risk of mycotoxin formation (Holmberg *et al.*, 1990a, b). Fan capacity, construction of main and side ducts and uniformity of the grain bed are extremely important in on-floor drying. If facilities are not properly dimensioned and operated, if areas contain a particularly large number of impurities, and/or if the grain bed is non-uniform, the air flow will be poorly distributed and 'damp pockets' may arise (Lai, 1980). The drying air frequently does not reach these pockets properly and these therefore dry much slower than the rest of the grain bed. The risk of mycotoxin production is much higher. If moisture and temperature are sufficiently high and aeration is unrestricted, this may cause spontaneous heating, so-called 'hot spots' (Lillehøj and Elling, 1983; Sinha and Wallace, 1965).

In the study by Elmholt (2003), results from three consecutive years at two farms with slow drying systems (on-floor and in-bin) were very consistent and it was concluded that these drying facilities were insufficient and conducive to growth of *P. verrucosum*. Further examples of on-floor drying being conducive to *P. verrucosum* were given by Elmholt and Rasmussen (2005).

In contrast to on-floor drying, batch dryers have a high drying capacity and the high drying air temperature enables drying of grain with high moisture contents. They have, however, a much lower intake capacity, resulting in a risk of bottleneck problems at harvest and fungal growth prior to drying. Furthermore, batch drying leaves the grain with a considerable moisture gradient, which may lead to damp pockets with a very high risk of mould growth if the grain is insufficiently mixed (Hellevang, 1994). This problem is overcome in recirculation batch dryers. The recirculation dryer described by Elmholt (2003) showed no problems with *P. verrucosum* during three consecutive years.

To avoid bottleneck problems in years with wet harvest conditions, a combination of high intake capacity and high drying capacity is essential. These demands are fulfilled by continuous flow dryers, which are quite common but more expensive than on-floor dryers and therefore less attractive to small farms (Pabis *et al.*, 1998). Continuous flow dryers can be very efficient with a high degree of automation and the grain is generally dried uniformly. A disadvantage of both batch and continuous flow dryers is the combination of relatively high drying air temperatures and a long retention time of the grain in the drier (0.5–1 h). This may heat the grain to a point where baking qualities and germination ability are negatively affected (Brooker *et al.*, 1974).

Drum drying also fulfils the demand of a high intake capacity and a high drying capacity. It is a continuous drying process but differs in having very high drying air temperatures and a short treatment duration. The drying capacity depends on the temperature of the drying air, the volume of air, the retention time, the moisture content of the grain and the acceptable maximal grain temperature. Drum dryers also maintain their capacity at high moisture contents at intake. They were developed for feed grain and until recently it was not possible to control the drying process precisely enough for damage caused by the very high temperatures to be avoided (Kristensen, 1998; Kristensen and Søgaard, 1995). The use of very high temperatures enables simultaneous heat treatment of the grain, which very efficiently reduces the number of fungal propagules on the grain, including *P. verrucosum*, *F. avenaceum*, *F. culmorum*, *F. poae*, *Fusarium sporotrichioides* and *Fusarium tricinctum* (Kristensen *et al.*, 2005). Both temperature and retention time in the drum affected the survival rate of the fungi and, using the optimal drying regime, 99% of the yeast propagules and 98% of the filamentous fungi were killed. Moisture contents were reduced to about 12%. The combination of high drying capacity and short but efficient heat treatment reduces the risk of mould deterioration to almost zero, when the grain is properly stored afterwards. At the same time, high quality baking is maintained.

17.8 Do organic and 'low input' systems present a particular risk for mycotoxin contamination?

Critics of organic and low input farming systems have frequently advanced the idea that the non-use of fungicides increases the risk of mycotoxin contamination in organic foods (Avery 1998; Trevewas, 2001, 2004). However, given the wide range of factors affecting fungal infection levels and mycotoxin production (see Sections 17.1 to 17.7 above) it is difficult to make a general conclusion about the relative risks of mycotoxin contamination in organic and conventional farming systems. A literature-based study by Paulsen and Weissmann (2002) identified the main factors that may influence mycotoxin formation and contamination of food and feed in organic and conventional crop production systems. Three of 13 factors known to affect mycotoxin contamination (climatic conditions, site and storage conditions) were described as affecting all farming systems in a similar way. Another three factors were considered potentially to increase the risk in organic systems. These were prohibition of the use of appropriate modern fungicides, higher weed densities and the use of livestock bedding on litter straw, which the authors believe might lead to higher exposures to pathogenic fungi and their mycotoxins. However, seven factors were considered to reduce the mycotoxin contamination risk in organic compared to conventional farming systems. These included (i) the trend towards the use of taller cultivars, (ii) non-use of growth regulators, (iii) lower nitrogen input levels, (iv) lower crop densities resulting from (i), (ii) and (iii) and (v) non-use of no or minimum tillage systems. For wheat, the lower density of wheat in organic rotations (vi) and the fact that maize is very rarely used as a previous crop (vii) is also thought to decrease the risk of mycotoxin (especially *Fusarium* mycotoxin) contamination levels.

Organic farming systems do not, as such, facilitate higher contamination levels of OTA than do conventional farming systems, but as exemplified in the sections above, certain management practices are susceptible to OTA problems and these management practices may be more prevalent in organic than in conventional farming (e.g. use of home-grown/farm-saved seed, large amount of impurities in harvested grain and slow and heterogeneous drying systems (Elmholt, 2003). Krogh (1976) showed that high incidence of mycotoxic porcine nephropathy was correlated to farm size: the smaller the farm, the higher the frequency. Jørgensen *et al.* (1996) speculated that the higher OTA contents in organically than in conventionally cultivated rye could be due to organic farms being smaller with less efficient drying facilities. Small farmers, some of whom farm as a sideline, may be less willing and able to invest in machinery and expensive drying facilities.

Surveys of mycotoxin contamination levels in organic and conventional crops often give conflicting results and will therefore not be described in detail here. They often poorly describe the management history of samples included in the survey and may therefore be misleading with respect to the causes of differential mycotoxin loads. Also, the contribution made by primary production practices/factors and storage conditions to overall mycotoxin loads were unclear for most of these studies.

17.9 Conclusions

Based on current knowledge about the agronomic factors that affect mycotoxin contamination levels in crops, the following recommendations can be made.

17.9.1 Harvest and processing

- Harvest equipment should be properly cleaned and the combine harvester adjusted so that kernels are minimally damaged.
- Interim storage should be avoided, especially at moisture contents above 18%, as OTA formed during this phase will not be degraded by drying. If dryer capacity is exceeded and damp grain has to be stored before drying, priority should be given to the wettest lot and the batch that is intended for human consumption.
- Cool products during interim storage (< 17 °C).
- Immediate separation of grain visibly infested with saprophytic, dark pigmented fungi spores is suggested.
- Winnowing/sieving separation before drying is recommended at elevated contents of weeds, unripe grain, straw, soil, etc. Winnowing or other cleaning techniques will reduce the risk of damp pockets. This is especially important in slow dryers.
- Quick and effective drying is crucial if the grain is harvested with moisture contents above 18%. The need increases with increasing moisture content.

17.9.2 *Fusarium*

- Seed quality depends heavily on the climatic characteristics of site locations. Thus, in order to limit seedborne disease inocula, seed production should be in areas characterized by low rainfall during flowering and ripening stages and by soils with low soil-nitrogen levels.
- Primary tillage based on inverting ploughs is recommended to bury

plant residues and minimize plant residue-based inoculum on the soil surface; ploughing is considered to be essential if cereals are grown after maize (maize as a previous crop for wheat is extremely rare in organic systems).

- Efficient control of stinging and sucking insects prevents the creation of pathways for fungal penetration (e.g. *Ostrinia nubilalis* in maize, aphids in cereals).
- Efficient weed control is necessary, especially for climbing weeds that can carry fungal spores vertically upwards to ears.
- High seeding density as well as high nitrogen availability should be avoided because it results in dense crop stands with a favourable microclimate for fungal growth.
- All measures that extend the period of flowering and growth (e.g. excess of nitrogen applied in the growth phase, in conventional agriculture: growth regulators and fungicide applications, especially *azoxystrobin*) should be avoided, because they increase the risk of mycotoxin production.
- Grain yields of resistant cultivars are often lower. Thus, these cultivars are currently less attractive for conventional agriculture, but interesting options for organic producers.
- Tall cultivars with greater distances between ears and leaves as well as the upper leaf distances should be considered for use, because they are less susceptible to splash dispersal of spores.

17.9.3 *Penicillium*

- *P. verrucosum* is present in some soils, and soil contamination during harvest should be avoided as much as possible.
- Cooling of grain has many well-established advantages relating to the prevention of grain pests and grain respiration. It will also inhibit the growth rate of many fungi including *P. verrucosum*. But it is not sufficient to prevent *P. verrucosum* from growing and forming OTA if the grain is wet and time sufficient! At low temperatures, *P. verrucosum* even has a competitive advantage compared with other storage fungi like *Aspergillus* spp.

17.9.4 *Claviceps* and other fungi producing ergot alkaloids

- Only sclerotia-free seeds should be used. Contaminated farm-saved seed should be cleaned or not used!
- Generally, rye is the most susceptible cereal. Since ergots are frequently higher in hybrid rye compared to population cultivars, higher yields of hybrid cultivars have to be balanced with the high risk of ergots being produced.
- Spring wheat, durum and barley show different susceptibilities to ergot.

In areas where ergot is endemic, the use of resistant cultivars can reduce the amount of sclerotia at harvest. Cultivars with quickly fading blossoms should be selected.

- Avoid agronomic practices that result in high crop densities and late tillering.
- Grass weeds pose a risk for infections with *C. purpurea.*
- Ploughing and diversified crop rotation counteracts enrichment of sclerotia on the soil surface.
- For conventional agriculture: efficient fungicides for the control of *C. purpurea* are not yet available. When fungicides are used, application has to be timed before infection has occurred. Strobilurins can lead to an increase in the number and weight of ergots.
- Pearl calcium cyanamide is able to reduce germination of ergots and formation of perithecia better than calcium ammonium nitrate in conventional farming.

17.10 Acknowledgements

The work by Susanne Elmholt was financed by the project PREMYTOX, performed in the context of the Danish Research Centre for Organic Farming (DARCOF). The investigations of Barbara Thiel, Ulrich Köpke and their colleagues were financially supported by Deutsche Forschungsgemeinschaft (DFG, German Research Foundation). The authors are indebted to Julia Cooper and Carlo Leifert for the critical review of the manuscript and to Anja Schneider for the vigilant text processing.

17.11 Sources of further information and advice

A comprehensive recent survey on mycotoxins in organic and conventional agriculture is given by Benbrook (2005), accessible via internet: www.organic-center.org.

Current literature concerning organic agriculture is accessible via www.orgprints.org.

17.12 References

Abdel-Wahhab M A, Hassan A M, Armer H A and Naguib K M (2004), 'Prevention of fumonisin-induced maternal and developmental toxicity in rats by certain plant extracts', *J. Appl. Toxicol.*, **24** (6), 469–474.

Abramson D, Sinha R N and Mills J T (1980), 'Mycotoxin and odor formation in moist cereal grain during granary storage', *Cereal Chem.*, **57**, 346–351.

Abramson D, Sinha R N and Mills J T (1990), 'Mycotoxin formation in HY-320 wheat during granary storage at 15 and 19% moisture content', *Mycopathologia*, **111**, 181–189.

Abramson D (1998), 'Mycotoxin formation and environmental factors', in Sinha K K and Bhatnagar D, *Mycotoxins in Agriculture and Food Safety*, Marcel Dekker, New York, 255–277.

Agrios G N (1997), *Plant Pathol.*, Academic Press, London.

Alabouvette C (1990), 'Biological control of *Fusarium* wilt pathogens in suppressive soils', in Homby D, *Biological Control of Soil Borne Pathogens*, CAB International, Wallingford, 27–45.

Alabouvette C, Backhouse D, Steinberg C, Donovan N J, Edel-Hermann V and Burgess L W (2004), 'Microbial diversity in soil – effects on crop health', in Schjønning P, Elmholt S and Christensen B T, *Managing Soil Quality – Challenges in Modern Agriculture*, CAB International, Wallingford, 121–138.

Atroshi F, Rizzo A F, Westermarck T and Ali-Vehmas T (1998), 'Effects of tamoxifen, melatonin, coenzyme Q10 and L-carnitine supplementation on bacterial growth in the presence of mycotoxins', *Pharmacol. Res.*, **38** (4), 289–295.

Aufhammer W, Hermann W, Kübler E, Lauber U and Schollenberger M (1999), '*Fusarium* (*F. graminearum*) infection of ears and toxin concentration of grains of winter wheat, triticale and rye depending on cultivars and production intensity', *Pflanzenbauwissenschaften*, **3** (1), 32–39.

Aufhammer W, Kübler E, Kaul H P, Hermann W, Höhn D and Cuilin Yi (2000), 'Ährenbefall mit Fusarien (*F. graminearum, F. culmorum*) und Deoxynivalenolgehalt im Korngut von Winterweizen in Abhängigkeit von der N-Düngung', *Plfanzenbauwissenschaften*, **4** (2), 72–78.

Avery D T (1998), 'The hidden dangers of organic food', *American Outlook*, Fall, 19–22.

Axberg K, Jansson G, Svensson G and Hult K (1997), 'Varietal differences in accumulation of ochratoxin A in barley and wheat cultivars after inoculation of *Penicillium verrucosum*', *Acta Agric. Scand.*, **47**, 229–237.

Backes F and Krämer J (1999), 'Mikrobiologische und mykotoxikologische Qualität von Winterweizen aus Organischem Landbau als Rohstoff für Lebensmittel', *Getreide Mehl und Brot*, **53**, 197–201.

Banks J, Scudamore K A, Norman K and Jennings P (2002), 'Practical guidelines to minimise mycotoxin development in UK cereals, in line with forthcoming EU legislation, using the correct agronomic techniques and grain storage management', *Home Grown Cereals Authority Project Report*, **289**, UK.

Beck R, Lepschy J and Obst A (1997), 'Gefahr aus der Maisstoppel', Deutsche Landwirtschafts Gesellschaft (*DLG*) *Mitteilungen*, **5**, 34–38.

Becker B A, Pace L, Rottinghaus G E, Shelby R, Misfeldt M and Ross P F (1995), 'Effects of feeding fumonisin B1 in lactating sows and their suckling pigs', *Am. J. Vet. Res.*, **56** (9), 1253–1258.

Benbrook, C M (2005), 'Breaking the mold – impacts of organic and conventional farming systems on mycotoxins in food and livestock feed', *An Organic Center State of Science Review*, 58 pp, www.organic-center.org.

Bennett J W and Klich M (2003). 'Mycotoxins', *Clin. Microbiol. Rev.*, **16**, (3), 497–516.

Berleth M, Backes F and Krämer J (1998), 'Schimmelpilzspektrum und Mykotoxine (Deoxynivalenol und Ochatoxin A) in Getreideproben aus ökologischem und integriertem Anbau', *Agribiol. Res.*, **51**, 369–376.

Beti J A, Phillips T W and Smalley E B (1995), 'Effects of maize weevils (Coleoptera: Curculionidae) on production of aflatoxin B1 by *Aspergillus flavus* in stored corn', *J. Economic Entomol.*, **88**, 1776–1782.

Birzele B and Prange A (2003), *Fusarium* spp. and storage fungi in suboptimally stored wheat: mycotoxins and influence on wheat gluten proteins. *Mycotoxin Res.*, **19**, 162–170.

Birzele B, Prange A and Krämer J (2000), 'Deoxynivalenol and ochratoxin A in German wheat and changes of level in relation to storage parameters', *Food Addit. Contam.*, **17** (12), 1027–1035.

Birzele B, Meier A, Hindorf H, Krämer J and Dehne H W (2002), 'Epidemiology of Fusarium infection and Deoxynivalenol content in winter wheat in the Rhineland, Germany', *Europ. J. Plant Pathol.* **108**, 667–673.

Blank G, Goswami N, Madrid F, Marquardt R R and Frohlich A A (1995), 'Evaluation of *Trifolium castaneum* (Herbst) (Coleoptera: Tenebrionidae) excreta on ochratoxin production in stored wheat', *J. Stored Products Res.*, **31**, 151–155.

BMVEL (Bundesminterium für Verbraucherschutz, Ernährung und Landwirschaft) (2004), 'Verordnung zur Änderung der Mykotoxin-Höchstmengen-Verordnung und der Diätverordnung', *Bundesgesetzblatt*, Teil **I**, 151–152.

Boorman G A (1989), 'Toxicology and carcinogenesis of ochratoxin A', (CAS NO. 303–47–9) in *F344/N Rats (Gavage studies)*, 358. US Department of Health and Human Services, National Toxicology Programme, Technical Report Series, Research Triangle Park, NC 27709.

Bretag T W and Merriman P R (1981), 'Epidemiology and cross-infection of *Claviceps purpurea*', *Trans. Br. Mycol. Soc.*, **77**, 211–213.

Brooker D B, Bakker-Arkema F W and Hall C V (1974), *Drying Cereal Grains*, AVI Publishing, Westport, CT.

Büchmann N B and Hald B (1985), 'Analysis, occurrence and control of ochratoxin A residues in Danish pig kidneys', *Food Addit. Contam.*, **2**, 193–199.

Buerstmayr H, Legzdina L, Steiner B and Lemmens M (2004), 'Variation for resistance to *Fusarium* head blight in spring barley', *Euphytica*, **137**, 279–290.

Buschbell T and Hoffmann G M (1992), 'The effects of different nitrogen regimens on the epidemiological development of pathogens on winter wheat and their control', *Z. Pflanzenkrankh. Pflanzenschutz*, **99**, 381–403.

Chelkowski J and Perkowski J (1992), 'Mycotoxins in cereal grain (Part 15). Distribution of deoxynivalenol in naturally contaminated wheat kernels', *Mycotoxin Research*, **8**, 27–31.

Creppy E E (1999), 'Human ochratoxicosis', *J. Toxicol.*, *Toxin Rev.*, 277–293.

Creppy E E (2005), 'Update of survey, regulation and toxic effects of mycotoxins in Europe', *Toxicol. Lett.*, **127**, 19–28.

Cunfer B, Mathre D E and Hockett E A (1975), 'Factors influencing the susceptibility of male-sterile barley to ergot', *Crop Sci.*, **15**, 194–196.

Czerwiecki L, Czajkowska D and Witkowska-Gwiazdowska A (2002a), 'On ochratoxin A and fungal flora in Polish cereals from conventional and ecological farms. Part 2: Occurrence of ochratoxin A and fungi in cereals in 1998', *Food Addit. Contam.*, **19**, 1051–1057.

Czerwiecki L, Czajkowska D and Witkowska-Gwiazdowska A (2002b), 'On ochratoxin A and fungal flora in Polish cereals from conventional and ecological farms. Part 1: Occurrence of ochratoxin A and fungi in cereals in 1997', *Food Addit. Contam.*, **19**, 470–477.

Dabkevicius Z and Semaskiene R (2002), 'Control of ergot (*Claviceps purpurea* (Fr.) Tul.) *ascoscarpus* formation under the impact of chemical and biological seed dressing', *Plant Protect. Sci.*, **38** (2), 681–683.

Davis V M and Stack M E (1994), 'Evaluation of alternariol and alterariol methyl ether for mutagenic activity in *Salmonella thyphimurium*', *App. Environ. Microbiol.* **60**, 3901–3902.

Desjardins A E and Proctor R H (2001), 'Biochemistry and genetics of *Fusarium* toxins', in Summerell B A, Leslie J F, Backhouse D, Bryden W L and Burgess L W, *Fusarium – Paul E. Nelson Memorial Symposium*, APS Press, American Phytolpathological Society, St. Paul, USA, 50–69.

Dickinson C H (1973), 'Interactions of fungicides and leaf saprophytes', *Pesticide Sci.*, **4**, 563–574.

D'Mello J P F (2003), 'Mycotoxins in cereal grains, nuts and other plant products', *Food Safety: Contaminants and Toxins*, CAB International, Wallingford, 65–90.

D'Mello J P F, Porter J K and Macdonald A M C (1997), '*Fusarium* mycotoxins', in D'Mello J P F, *Handbook of Plant and Fungal Toxicants*, CRC Press, Boca Raton, Florida, 287–301.

D'Mello J P F, Macdonald A M C, Postel D, Dijksma W P T and Desjardins A (1998), 'Pesticide use and mycotoxin production in *Fusarium* and *Aspergillus* phytopathogens', *Europ. J. Plant Pathol.*, **104**, 741–751.

Döll S, Valenta H, Danicke S and Flachowsky G (2002), '*Fusarium* mycotoxins in conventionally and organically grown grain from Thuringia/Germany', *Landbauforschung Volkenrode*, **52**, 91–96.

Doohan F M, Brennan J and Cooke B M (2003), 'Influence of climatic factors on *Fusarium* species pathogenic to cereals', *Europ. J. Plant Pathol.*, **109**, 755–768.

Dornbusch C, Schauder A, Piorr H P and Köpke U (1993), 'Qualitätsbeeinflussende Parameter von Saatgutpartien aus dem Organischen Landbau', in VDLUFA (Hrsg.): *VDLUFA-Schriftenreihe*, Kongreßband 1993, Hamburg. Vorträge zum Generalthema "Qualität und Hygiene von Lebensmitteln in Produktion und Verarbeitung", 305–308.

Drusch S and Ragab W (2003), 'Mycotoxins in fruits, fruit juices and dried fruits', *J. Food Prot.*, **66** (8), 1514–1527.

EC (European Commission) (1998), 'Commission directive No 98/53/EC of 16 July 1998 laying down the sampling methods and the methods of analysis for the official control of the levels for certain contaminants in foodstuffs', *Official J. Europ. Communities*, **L201**, 93–101.

EC (European Commission) (2002a), 'Commission Regulation No 472/2002 of 12 March 2002 amending regulation (EC) No 466/2001 setting maximum limits for certain contaminants in food stuffs', *Official J. Europ. Union*, **L77**, 1–13.

EC (European Commission) (2002), 'Commission directive 2002/27/EC of 13 March 2002 amending directive 98/53/EC laying down the sampling methods and the methods of analysis for the official control of the levels for certain contaminants in foodstuffs', *Official J. Europ. Communities*, **L75**, 44–45.

EC (European Commission) (2003), 'Commission regulation (EC) No 2174/2003 of 12 December 2003 amending regulation (EC) No 466/2001 as regards aflatoxins', *Official J. Europ. Union*, **L326**, 12–15.

EC (European Commission) (2004a), 'Commission regulation (EC) No 683/2004 of 13 April 2004 amending regulation (EC) No 466/2001 as regards aflatoxins and ochratoxin A in foods for infants and young children', *Official J. Europ. Union*, **L106**, 3–5.

EC (European Commission) (2004b), 'Commission directive 2004/43/EC of 13 April 2004 amending directive 98/53/EC and directive 2002/26/EC as regards sampling methods and methods of analysis for the official control of the levels of aflatoxin and ochratoxin A in food for infants and young children', *Official J. Europ. Union*, **L113**, 14–16.

EC (European Commission) (2005a), 'Commission regulation (EC) No 123/2005 of 26 January 2005 amending regulation (EC) No 466/2001 as regards ochratoxin A', *Official J. Europ. Union*, **L25**, 3–5.

EC (European Commission) (2005b), 'Commission directive 2005/5/EC of 26 January 2005 amending directive 2002/26/EC as regards sampling methods and methods of analysis for the official control of the levels of ochratoxin A in certain foodstuffs', *Official J. Europ. Union*, **L27**, 38–40.

EC (European Commission) (2005c), 'Commission directive 2005/38/EC of 6 June 2005 laying down the sampling methods and the methods of analysis for the official control of the levels of *Fusarium* toxins in foodstuffs', *Official J. Europ. Union*, **L143**, 18–26.

EC (European Commission) (2005d), 'Commission regulation (EC) No 856/2005 of 6

June 2005 amending regulation (EC) No 466/2001 as regards *Fusarium* toxins', *Official J. Europ. Union*, **L143**, 3–8.

Edwards S G, Pirgozliev S R, Hare M C and Jenkinson P (2001), 'Quantification of trichothecene-producing *Fusarium* species in harvested grain by competitive PCR to determine efficacies of fungicides against *Fusarium* head blight of winter wheat', *Appl. Environ. Microbiol.*, **67**, 1575–1580.

Elmholt S (2003), 'Ecology of the ochratoxin A producing *Penicillium verrucosum:* Occurrence in field soil and grain with special attention to farming system and on-farm drying practices', *Biol. Agric. Horticul.*, **20**, 311–337.

Elmholt S and Hestbjerg H (2000), 'Field ecology of the ochratoxin A-producing *Penicillium verrucosum*: Survival and resource colonisation in soil', *Mycopathologia*, **147**, 67–81.

Elmholt S and Rasmussen P H (2005), '*Penicillium verrucosum* occurrence and Ochratoxin A contents in organically cultivated grain with special reference to ancient wheat types and drying practice', *Mycopathologia*, **159**, 421–432.

Elmholt S, Labouriau R, Hestbjerg H and Nielsen J M (1999), 'Detection and estimation of conidial abundance of *Penicillium verrucosum* in soil by dilution plating on a selective and diagnostic agar medium (DYSG,') *Mycol. Res.*, **103**, 887–895.

Engels R and Krämer J (1996), 'Incidence of Fusaria and occurrence of selected *Fusarium* mycotoxins on *Lolium* spp. in Germany', *Mycotoxin Res.*, **12**, 31–40.

Evans V J and Jenkyn J F (2000), 'Fungicides for control of ergot in cereal crops', *Pest and Diseases*, Proceedings of an international conference held at the Brighton Hilton Metropole Hotel, UK, 13–16 November 2000, 511–514.

FAO (2004), 'World wide regulations for mycotoxins in food and feed in 2003', *Food and Agriculture Organization of the United Nations*, Rome.

Fauzi M T and Paulitz T C (1994), 'The effect of plant growth regulators and nitrogen on *Fusarium* head blight of the spring wheat cultivar Max', *Plant Dis.*, **78**, 289–292.

Fink-Gremmels J (1999), 'Mycotoxins: their implications for human and animal health', *Vet. Q.*, **21** (4), 115–120.

Fink-Gremmels J (2005a), 'Mycotoxicosity and animal health', *European Mycotoxin Seminar Series*, Alltech, 9–25 February 2005, 30–43.

Fink-Gremmels J (2005b), 'Mycotoxins in animal feeds: Current issues in Europe', *XV Animal Science Congress*, Zootec I and Zootec D, Vila Real, Portugal, November 3–5, 2005.

Forsberg G, Johnsson L and Lagerholm J (2005), 'Effects of aerated steam seed treatment on cereal seed-borne diseases and crop yield', *J. Plant Dis. Protect.*, **112** (3), 247–256.

Frank M (1999), 'Mycotoxin prevention and decontamination. HACCP and its mycotoxin control potential: an evaluation of ochratoxin A in coffee production', *Third Joint FAO/WHO/UNEP International Conference on Mycotoxins*, Tunis, Tunisia, 3-3-0099.

Frisvad J C and Samson R A (1991), 'Filamentous fungi in foods and feeds: ecology, spoilage, and mycotoxin production', in Arora D K, Mukerji K G and Marth E H, *Handbook of Applied Mycology, Vol. 3, Food and Feeds,*. Marcel Dekker, New York, 31–68.

Frisvad J C and Thrane U (2000), 'Mycotoxin production by common filamentous fungi', in Samson R A, Hoekstra E S, Frisvad J C and Filtenborg O, *Introduction to Food- and Airborne Fungi*, Centraalbureau Voor Schimmelcultures, Utrecht, 321–332.

Frisvad J C, Filtenborg O, Lund F and Thrane U (1992), 'New selective media for the detection of toxigenic fungi in cereal products, meat and cheese', in Samson R A, Hocking A D, Pitt J I and King A D *Modern Methods in Food Mycology*, Elsevier Science Publishers, Amsterdam, 275–285.

Fung F and Clark R F (2004), 'Health effects of mycotoxins: a toxicological overview', *J. Toxicol. Clin. Toxicol.*, **42** (2), 217–234.

Futtermittel-Verordnung (2003), German Federal Animal Feed Regulation, Futtermittel-VO vom 23. Nov. 2000 i.d.F. vom 11. April 2003 (BGBl I S. 534), Anlage 5, unerwünschte Stoffe.

Gareis M (1999a), 'Mykotoxine und Schimmelpilze', *Forschungs Report (Ernährung, Landwirtschaft, Forsten)*, **2**, 4–5.

Gareis M (1999b), 'Occurrence of the mycotoxins ochratoxin A and B in German malting barley and from manufactured malt', *Archiv fur Lebensmittelhygiene*, **50**, 83–87.

Gattinger A, Embacher A, Emmerling C, Fliessbach A and Schloter M (2002), 'Microbial community analyses in organically and conventionally managed soil ecosystems', in *Organic Production and Environmental Responsibilities*, Proceedings of the 14th IFOAM Organic World Congress, Victoria, Canada, 41.

Geiser D M, Dorner J W, Horn B W, and Taylor J W (2000), 'The phylogenetics of mycotoxin and sclerotium production in *Aspergillus flavus* and *Aspergillus oryzae*', *Fungal Genet. Biol.*, **31**, 169–179.

Gilbert J and Tekauz A (2000), 'Recent developments in research on *Fusarium* head blight of wheat in Canada', *Can. J. Plant Pathol.*, **22**, 1–8.

Griffin G F and Chu F S (1983), 'Toxicity of the alternaria metabolites alternariol, alternariol methyl ether, alternuene, and tenuazonic acid in the chicken embryo assay', *Appl. Environ. Microbiol.*, **46**, 1420–1421.

Haase M S (2003), *Prevention of Mycotoxin Formation In Organically Cultivated Bread Grain – with Special Emphasis* on Penicillium verrucosum *and Formation of Ochratoksin* A (in Danish). Biologisk Institut, Århus Universitet, Århus.

Häggblom P E and Ghosh J (1985), Postharvest production of ochratoxin A by *Aspergillus ochraceus* and *Penicillium viridicatum* in barley with different protein levels. *Appl. Environ. Microbiol.*, **49**, 787–790.

Heier T, Jain S K, Kogel K H and Pons-Kuehnemann J (2005), 'Influence of N-fertilization and fungicide strategies on *Fusarium* head blight severity and mycotoxin content in winter wheat', *J. Phytopathol.*, **153**, 551–557.

Hellevang K J (1994), *Grain Drying*, NDSU Extension Service, Fargo, ND.

Henriksen B and Elen O (2005), 'Natural *Fusarium* grain infection level in wheat, barley and oat after early application of fungicides and herbicides', *J. Phytopathol.*, **153**, 214–220.

Herrmann W, Kübler E and Aufhammer W (1998), 'Ährenbefall mit Fusarien und Toxingehalt im Korngut verschiedener Wintergetreidearten', *Pflanzenbauwissenschaften* **3**, 97–107.

Hill R A and Lacey J (1983a), 'Factors determining the microflora of stored barley grain', *Annal. Appl. Biol.*, **102**, 467–483.

Hill R A and Lacey J (1983b), 'The microflora of ripening barley grain and the effects of pre-harvest fungicide application', *Annal. Appl. Biol.*, **102**, 455–465.

Hoffmann G M and Schmutterer H (1983), Parasitäre Krankheiten und Schädlinge an landwirtschaftlichen Kulturpflanzen. Ulmer, Stuttgart.

Hökby E, Hult K, Gatenbeck S and Rutqvist L (1979), 'Ochratoxin A and citrinin in 1976 crop of barley stored on farms in Sweden', *Acta Agricu. Scand.*, **29**, 174–178.

Holmberg T, Breitholtz A, Bengtsson A and Hult K (1990a), 'Ochratoxin A in swine blood in relation to moisture content in feeding barley at harvest', *Acta Agric. Scand.*, **40**, 201–204.

Holmberg T, Hagelberg S, Lundeheim N, Thafvelin B and Hult K (1990b), Ochratoxin A in swine blood used for evaluation of cereal handling procedures. *J. Vet. Med.*, **37**, 97–105.

Jahn M (2002), 'Saatgutbehandlung im Ökologischen Landbau', *Forschungsreport*, 1/2002, 12–15.

Jahn M, Röder O and Tigges J (2005), 'Die Elektronenbehandlung von Getreidesaatgut – Zusammenfassende Wertung der Freilandergebnisse', *Mitteilungen aus der Biologischen Bundesanstalt für Land- und Forstwirtschaft Berlin-Dahlem*, pp. 63.

Janzen D H (1977), 'Why fruits rot, seeds mold, and meat spoils', *Am Naturalist*, **111**, 691–713.

Jørgensen K and Jacobsen J S (2002), 'Occurrence of ochratoxin A in Danish wheat and ryes, 1992–99', *Food Addit. Contam.*, **19**, 1184–1189.

Jørgensen K, Rasmussen G and Thorup I (1996), 'Ochratoxin A in Danish cereals 1986–1992 and daily intake by the Danish population', *Food Addit. Contam.*, **13**, 95–104.

Klingenhagen G and Frahm J (1999), 'Fusariumbefall in Getreide', *Getreide* 5/2, 74–76.

Knudsen I M B, Elmholt S, Hockenhull J and Jensen D F (1995), 'Distribution of saprohytic fungi antagonistic to *Fusarium culmorum* in two differently cultivated field soils with special emphasis on the genus *Fusarium', Biol. Agric. Hort.*, **12**, 61–79.

Knudsen I M B, Hockenhull J, Jensen D F, Gerhardson B, Hökeberg R, Tahvonen R, Teperi E, Sundheim L and Henriksen B (1997), 'Selection of biological control agents for controlling soil and seed-borne diseases in the field', *Europ. J. Plant Pathol.*, **103**, 775–784.

Knudsen I M B, Debosz K, Hockenhull J, Jensen D F and Elmholt S (1999), 'Suppressiveness of organically and conventionally managed soils towards brown foot rot of barley', *Appl. Soil Ecol.*, **12**, 61–72.

Krebs H, Dubois D and Külling C (2000), 'Effects of preceding crop and tillage on the incidence of *Fusarium* spp. and mycotoxin deoxynivalenol content in winter wheat grain', *AGRARForschung*, **7** (6), 264–268.

Kristensen E F (1998), New drying method for improvement of the quality of malting barley and bread grain. *Proceedings of the European Association of Agricultural Engineers,* Ag Eng '98, Norway. Abstract: 98-F-011, p. 324.

Kristensen E F and Søgaard H T (1995), '*Production of High Quality Flour from Danish Grain*' (in Danish). Statens Husdyrbrugsforsøg. Intern Rapport, no. 52. Danish Institute of Agricultural Sciences, Foulum.

Kristensen E F, Elmholt S and Thrane U (2005), 'High-temperature treatment for efficient drying of bread rye and reduction of fungal contaminants', *Biosyst. Eng.*, **92**, 183–195.

Krogh P (1976), 'Epidemiology of mycotoxic porcine nephropathy', *Nordisk Veterinaermedicin*, **28**, 452–458.

Krogh P (1978), 'Causual association of mycotoxic nephropathy', *Acta Pathol. Microbiol. Scandi., Section A*, Supplement **269**, 28 pp.

Lai F S (1980) 'Three-dimensional flow of air through nonuniform grain beds', *Trans. ASAE*, **23**, 729–734.

Langseth W and Stabbetorp H (1996), 'The effect of lodging and time of harvest on deoxynivalenol contamination in barley and oats', *J. Phytopathol.*, **144**, 241–245.

Langseth W, Stenwig H, Stogn L and Mo E (1993), 'Growth of moulds and production of mycotoxins in wheat during drying and storage', *Acta Scand., Section B, Soil Plant Sci.* **43**, 32–37.

Larsen T O, Svendsen A and Smedsgaard J (2001), 'Biochemical characterization of ochratoxin A producing strains of the genus *Penicillium*', *Appl. Environ. Microbiol.*, **67**, 3630–3635.

Lauren D R, Jensen D J, Smith W A and Dow W (1996), 'Mycotoxins in New Zealand maize: A study of some factors influencing contamination levels in grain', *New Zealand J. Crop Hort. Sci.*, **24**, 13–20.

Lemmens M, Josephs R, Schuhmacher R, Grausgruber H, Buerstmayr H, Ruckenbauer P, Neuhold G, Fidesser M and Krska R (1997), 'Head blight (*Fusarium* spp.) on wheat: investigations on the relationship between disease symptoms and mycotoxin content', Fifth European Fusarium Seminar, *Cereal Res. Commun.* **25** (3/1), 495–465.

Lemmens M, Haim K, Lew H and Ruckenbauer P (2004), 'The effect of nitrogen fertilization on Fusarium head blight development and deoxynivalenol contamination in wheat', *J. Phytopathol.*, **152**, 1–8.

Lepschy J, Dietrich R, Märtlbauer E, Schuster M, Süß A and Terplan G (1989), 'A survey on the occurrence of *Fusarium* mycotoxins in Bavarian cereals from the 1987 harvest', *Lebensm. Unters. Forsch.* **188**, 521–526.

Lepschy J (1992), 'Fusarientoxine in Getreide – ihre Entstehung und Vorbeugungsmaßnahmen', *Gesunde Pflanze* **44**, 35–39.

Lienemann K (2002), *Auftreten von Fusarium-Arten an Winterweizen im Rheinland und Möglichkeiten der Befallskontrolle unter besonderer Berücksichtigung der Weizensorte*, Dissertation, Rheinische Friedrich-Wilhelms-Universität, Bonn.

Liggitt J, Jenkinson P and Parry D W (1997), 'The role of saprophytic microflora in the development of Fusarium ear blight of winter wheat caused by *Fusarium culmorum*', *Crop Protect.*, **16**, 679–685.

Lillehøj E B and Elling F (1983), 'Environmental conditions that facilitate ochratoxin contamination of agricultural commodities', *Acta Agric. Scand.*, **33**, 113–128.

Lindner K, Burth U and Roeder O (1996), 'Einführung der Saatgutbehandlung von Winterweizen mit niederenergetischen Elektronen in die landwirtschaftliche Praxis', *Mitt. aus der Biol. Bundesanstalt H.*, **321**, 50.

Mäder P, Fliessbach A, Dubois D, Gunst L, Fried P, and Niggli U (2002), 'Soil fertility and biodiversity in organic farming', *Science*, **296**, 1694–1697.

Magan N, Hope R, Colleate A and Baxter E S (2002), 'Relationship between growth and mycotoxin production by *Fusarium* species, biocides and environment', *Europ. J. Plant Pathol.*, 108, 685–690.

Martin R A and Mac Leod J A (1991), 'Influences of production inputs on incidence of infection by *Fusarium* species on cereal seed', *Plant Dis.*, **75**, 784–788.

Marx H, Gedek B and Kollarczik B (1995), 'Vergleichende Untersuchungen zum mykotoxikologischen Status von ökologisch und konventionell angebautem Getreide', *Lebensm. Unters. Forsch.*, **201**, 83–86.

McMillian W W, Widstrom N W, Wilson D M and Hill R A (1980), 'Transmission by maize weevils of *Aspergillus flavus* and its survival on selected corn hybrids', *J. Economic Entomol.*, **73**, 793–794.

Meier A (2003), *Zur Bedeutung von Umweltbedingungen und pflanzenbaulichen Maßnahmen auf den Fusarium-Befall und die Mykotoxinbelastung von Weizen*, Dissertation, Rheinische Friedrich-Wilhelms-Universität, Bonn.

Meier A, Birzele B, Oerke E C and Dehne H W (1999), 'Auftreten von *Fusarium* spp. und Mykotoxingehalte von Winterweizen und Möglichkeiten zur Bekämpfung', *Proccedings 21. Mykotoxinworkshop*, Jena, pp 19–25.

Mesterházy A (1995), 'Types and components of resistance to *Fusarium* head blight of wheat', *Plant Breeding*, **114**, 377–386.

Mesterházy A, Bartók T and Lamper C (2003), 'Influence of wheat cultivar, species of *Fusarium* and isolate aggressiveness on the efficacy of fungicides for control of *Fusarium* head blight', *Plant Dis.*, **87**, 1107–1115.

Miedaner T, Reinbrecht C, Schollenberger M and Lauber U (2000), 'Vorbeugende Maßnahmen gegen Befall mit ährenfusariosen und mykotoxinbelastung des Getreides', *Mühle Mischfutter*, **137** (15), 485–489.

Mielke H (1993), 'Investigations on the control of ergot', *Nachrichtenbl. Deut. Pflanzenschutzd.*, **45**, 97–102.

Miller J D (1995), 'Fungi and mycotoxins in grain: implications for stored product research', *J. Stored Products Res.*, **31**, 1–16.

Miller J D, ApSimon J W, Blackwell B A K Greenhalgh R and Taylor A (2001), 'Deoxynivalenol: a 25 year perspective on a trichothecene of agricultural importance', in Summerell B A, Leslie J F, Backhouse D, Bryden W L and Burgess L W, *Fusarium: Paul E. Nelson Memorial Symposium*, The American Phytopathological Society Press, St. Paul, Minnesota, 310–320.

Mortensen G K, Strobel B W and Hansen H-C B (2003), 'Determination of zearalenone and ochratoxin A in soil', *Anal. Bioanal. Chem.*, **376**, 98–101.

Moss M O (2002), 'Mycotoxin review – 2. *Fusarium', Mycologist* 16, Part 4.

Nicholson P, Simpson D R, Wilson A H, Chandler E and Thomsett M (2004), 'Detection and differentiation of trichothecene enniatin-producing *Fusarium* species on small-grain cereals', *Europ. J. Plant Pathol.*, **110**, 503–514.

Obst A and Bechtel A (2000), 'Weather conditions conducive for wheat head blight, caused by *Fusarium graminearum*', *Bodenkultur und Pflanzenbau*, **3**, 81–88.

Obst A, Lepschy J and Beck R (1997), 'On the etiology of fusarium head blight of wheat in south Germany – preceding crops, weather conditions for inoculum production and head infection, proneness of the crop to infection and mycotoxin production', *Cereal Res. Commun.*, **25**, 699–703.

Pabis S, Jayas D S and Cenkowski S (1998), *Grain Drying – Theory and Practice*. John Wiley and Sons, NY, USA.

Pageau D, Collin J and Wauthy J M (1994), 'Evaluation of barley cultivars for resistance to ergot fungus, *Claviceps purpurea* (Fr.) Tul.', *Can. J. Plant Sc.*, **74**, 663–665.

Paulsen H M and Weissmann F (2002), 'Relevance of mycotoxins to product quality and animal health in organic farming', in *Redesigning Food Systems*, Proceedings of the 14th IFOAM Organic World Congress, Victoria, Canada, 212.

Perkowski J (1998), 'Distribution of deoxynivalenol in barley kernels infected by *Fusarium*', *Nahrung*, **42**, 81–83.

Petersson S, Hansen M W, Axberg K, Hult K and Schnürer J (1998), 'Ochratoxin A accumulation in cultures of Penicillium verrucosum with the antagonistic yeast *Pichia anomala* and *Saccharomyces cerevisiae*', *Mycol. Res.*, **102**, 1003–1008.

Pfohl-Leszkowicz A, Petkova-Bocharova T, Chernozemsky I N and Castegnaro M (2002), 'Balkan endemic nephropathy and associated urinary tract tumours: a review on aetiological causes and the potential role of mycotoxins', *Food Add. Contam.*, **19**, 282–302.

Piorr H P (1990), 'Saatgutqualität im Organischen Landbau', in Finke L and Linscheid J, *Vorträge der 42. Hochschultagung der Landwirtschaftlichen Fakultät der Universität Bonn*, 221–236.

Piorr H P (1992), 'Phytopathological advantages and risks of organic farming systems: Future perspectives to improve organic cropping systems', in Altman J, *Pesticide-interactions in Crop Production: Beneficial and deleterious effects*, CRC Press, Cleveland, 461–472.

Pirgozliev S R, Edwards S G, Hare M C and Jenkinson P (2003), 'Strategies for the control of *Fusarium* head blight in cereals', *Europ. J. Plant Pathol.*, **109**, 731–742.

Platford G R and Bernier C C (1976), 'Reaction of cultivated cereals to *Claviceps purpurea*', *Can. J. Plant Sci.*, **56**, 51–58.

Ramakrishna N, Lacey J and Smith J E (1996a), 'Colonization of barley grain by *Penicillium verrucosum* and *ochratoxin* A formation in the presence of competing fungi'. *J. Food Protect.*, **59**, 1311–1317.

Ramakrishna N, Lacey J and Smith J E (1996b), 'The effects of fungal competition on colonization of barley grain by Fusarium sporotrichioides on T-2 toxin formation'. *Food Addit. Contam.*, **13**, 939–948.

Regner S, Schnürer J and Jonsson A (1994), 'Ergosterol content in relation to grain kernel weight'. *Cereal Chem.*, **71**, 55–58.

Reutter M (1999), 'Zearalenon und Deoxynivalenol in Getreide und Futtermitteln Schleswig-Holsteins: Untersuchungen aus dem Erntejahr 1998', *Proceedings 21st Mykotoxin-Workshop*, Jena, 5–9.

Rintelen J (1995), 'Zum Infektionszeitpunkt von Fusarien an Weizenkörnern', *Gesunde Pflanze*, **47** (8), 315–317.

Rizzo A F, Atroshi F, Ahotupa M, Sankari S and Elovaara E (1994), 'Protective effect of antioxidants against free radical-mediated lipid peroxidation induced by DON or T-2 toxin', *Zentralb.Veterinarmed. A*, **41** (2), 81–90.

Rotter B A and Prelusky D B (1996), 'Toxicology of deoxynivalenol (vomitoxin)', *J. Toxicol. Environ. Health*, **48**, 1–34.

Ruckenbauer P, Bürstmayr H, Grausgruber H and Lemmens M (1999), 'Breeding research on resistance to *Fusarium* head blight in wheat', in Scarascia Mugnozza G T, Porceddu E and Pagnotta M A, *Genetics Breeding Crop Qual. Resistance*, 51–59.

Schauder A (2003), 'Saatgutvermehrung im Organischen Landbau unter besonderer Berücksichtigung der Schaderreger *Microdochium nivale* und der Gattung *Fusarium', Diss. agr., Schriftenreihe Institut für Organischen Landbau*, Verlag Dr. Köster, Berlin, Hrsg. Prof. Dr. Ulrich Köpke.

Schjønning P, Elmholt S, Munkholm L J, and Debosz K (2002), 'Soil quality aspects of humid sandy loams as influenced by different long-term management'. *Agric., Ecosystems Environ.*, **88**, 195–214.

Schrader T J, Cherry W, Soper K, Langlois I and Vijay H M (2001), 'Examination of *Alternaria alternata* mutagenicity and effects of nitrosylation using the Ames *Salmonella* Test', *Teratogenesis, Carcinogenesis, and Mutagenesis*, **21**, 261–274.

Schwartz G G (2002), 'Hypothesis: does ochratoxin A cause testicular cancer?' *Cancer Causes and Control*. **13**, 91–100.

Scudamore K A (1999), *A Study to Determine Whether On-floor Ambient Drying Systems are Conducive to the Formation of Ochratoxin A in Grain*, Home Grown Cereals Authority Project Report 196, UK.

Seifert K A and Levesque C A (2004), 'Phylogeny and molecular diagnosis of mycotoxigenic fungi', *Europ. J. Plant Pathol.*, **110**, 449–471.

Simpson D R, Weston G E, Turner J A, Jennings P and Nicholson P (2001), 'Differential control of head blight pathogens of wheat by fungicides and consequences for mycotoxin contamination of grain', *Europ. J. Plant Pathol.*, **107**, 421–431.

Sinha R C and Savard M E (1997), 'Concentration of deoxynivalenol in single kernels and various tissues of wheat heads', *Can. J. Plant Pathol.*, **19**, 8–12.

Sinha K K and Sinha A K (1991), 'Effect of *Sitophilus oryzae* infestation on *Aspergillus flavus* infection and aflatoxin contamination in stored wheat', *J. Stored Products Res.*, **27**, 65–68.

Sinha R N and Wallace H A H (1965), 'Ecology of a fungus-induced hot spot in stored grain', *Can. J. Plant Sci.*, **45**, 48–59.

Smith J E, Lewis C W, Anderson J G and Solomons G L (1994), *Mycotoxins in Human Nutrition and Health*, European Commission Directorate-General XII, Agro-industrial division, E–2, Brussels.

Spieß and Dutschke (1991), 'Bekämpfung des Weizensteinbrandes (*Tilletia caries*) im Biologisch-Dynamischen Landbau unter experimentellen und praktischen Bedingungen', *Gesunde Pflanzen*, **43**, 8, 264–270.

Stähle A, Birzele B, Schütze A, Krämer J and Dehne H W (1998), 'Optimierungsstrategien im Organischen Landbau: Auftreten von *Fusarium* spp. und DON-Konzentrationen in Winterweizen aus Organischem Landbau im Jahr 1997', in Wolff J, Betsche T, *Proceedings 20. Mykotoxinworkshop*, Bundesanstalt für Getreide-, Kartoffel- und Fettforschung, Detmold, 262–266.

Steyn P S (1995), 'Mycotoxins, general view, chemistry and structure'. *Toxicol. Lett.*, **82–83**, 843–851.

Tanaka T, Hasegawa A, Yamamoto S, Lee U, Sugiura Y and Ueno Y (1988), 'Worldwide contamination of cereals by *Fusarium* mycotoxins nivalenol, deoxynivalenol and zearalenone', 1. Survey of 19 countries, *J. Agric. Food Chem.*, **36**, 979–983.

Thalmann A (1990), 'Mykotoxine in Getreide', *Angew. Botanik*, **64**, 167–173.

Tinker P B (2001), *Shade of Green: A review of UK Farming System*, Royal Agricultural Society of England, Stoneleigh Park, UK.

Trewavas A (2001), 'Urban myths of organic farming', *Nature*, **410**, 409–410.

Trewavas A (2004), 'A critical assessment of organic farming-and-food assertions with particular respect to the UK and the potential environmental benefits of no-till agriculture', *Crop Protect.*, **23**, 757–781.

Twiddy D R (1994) Volatiles as indicators of fungal growth on cereal grains. *Trop. Sci.*, **34**, 416–428.

Wafa H, Abid-Essafi S, Abdellatif A, Noureddine G, Abdelfettah Z, Farielle E, Creppy E E and Hassen B (2004), 'Karyomegaly of tubular kidney cells in human chronic interstitial nephropathy in Tunisia: respective role of Ochratoxin A and possible genetic predisposition', *Human Experimen. Toxicol.*, **23**, 339–346.

Walker S L, Leath S, Hagler W M and Murphy J P (2001), 'Variation among isolates of *Fusarium graminearum* associated with *Fusarium* head blight in North Carolina', *Plant Dis.*, **85**, 404–410.

Weinert J and Wolf G A (1995), 'Gegen Ährenfusarien helfen nur resistente Sorten', *Pflanzenschutz-Praxis*, **2**, 30–32.

WHO (1990), 'Selected mycotoxins: ochratoxins, trichothecenes, ergot', *Environ. Health Criteria* 105, Geneva, Switzerland, WHO, 1–163.

Wicklow D T (1995), 'The mycology of stored grain: an ecological perspective', in Jayas D S, White N D G and Muir W E, *Stored-Grain Ecosystems*, Marcel Dekker, New York, 107–249.

Winter W, Bänziger I, Krebs H and Rüegger A (1997), 'Warm- und Heisswasserbehandlung gegen Auflaufkrankheiten', *Agrarforschung* **4** (11–12), 467–470.

18

Reducing copper-based fungicide use in organic crop production systems

Reza Ghorbani, Ferdowsi University of Mashhad, Iran and Steve Wilcockson, Newcastle University, UK

18.1 Introduction

Fungal and bacterial diseases causing losses of crop yield and quality create challenging practical and economic problems for producers. These are generally more acute in organic and 'lower-input' than in conventional production systems because the scope for use of fungicides is much more limited. Whilst several synthetic active ingredients are available to conventional growers, these are not allowed in organic agriculture, except for certain copper products, uses of which are 'considered to be traditional organic farming practices' (Anonymous, 2002). In most countries, copper-based fungicides can be used in organic crops where needed, with the permission of the relevant certifying authority. However, there are restrictions and national legislation or the organic certifying authorities' standards may either limit or forbid their use, or some growers may elect not to use them for various reasons (Anonymous, 2002; IFOAM, 2000; Tamm *et al.*, 2004). This reflects concerns of producers, consumers, environmentalists and health professionals about the potential toxic effects on plants, beneficial soil and other organisms, biodiversity and human health (Madge, 2005). It also recognizes the widespread view that the use of copper fungicides in organic farming is contentious and incompatible with the underlying principles.

In the European Union (EU) replacement of copper-based fungicides with other methods of disease control is a priority in organic farming policy (Anonymous, 2002). They were due to be prohibited by law for use in organic farming in the EU from March 2002 but the ban was delayed because of the increased risk of crop diseases and associated economic losses for organic producers in the medium to long term, until effective alternative control methods are developed. A maximum application of 8 kg of elemental copper/ha/year was imposed for annual crops until the end of 2005, decreasing to 6 kg per year thereafter. More complex rules apply for perennial crops such as grapevines and apples, but the overall aim is to reduce inputs further and phase them out completely. However, the regulations may be changed at any time in light of the development of viable alternatives. Moreover, any proposals to withdraw approval for the use of copper-based fungicides under the EC Review programme for existing active substances based on safety grounds, will take precedence.

Research in organic crop production must endeavour to develop techniques and products, including organically based fungicides, plant 'strengtheners' and biocontrol agents, and formulate effective management strategies that can replace or minimize the need for copper-based fungicides. The results of this endeavour may also have beneficial applications in both conventional and lower input systems leading to a more sustainable agriculture.

18.2 Effects of diseases on crop yield and quality in organic systems

Crop species are susceptible to a multitude of diseases caused by fungi, bacteria and viruses. The magnitude of losses of yield and quality depends on several factors. These include origin, amount and characteristics of the disease inoculum, weather and soil conditions, nutrient availability, site topography, varietal resistance, crop management and protection. Growers should take all these factors into account to develop an effective disease management strategy. The range of crop losses caused by incomplete control of diseases is the same in both organic and conventional crop production systems. At one extreme, a crop may fail completely because of an attack by disease. The devastating effects of late blight on the potato crop caused by the fungus *Phytophthora infestans* in the 1840s in Ireland is the best known example. At the other extreme, a crop may show symptoms of a disease but suffer little or no significant damage. Levels of infection that restrict growth by causing premature leaf death, root damage and impaired photosynthetic efficiency, decrease total yields because dry matter accumulation is restricted: dry matter distribution may also be adversely affected. The fraction of total yield which is marketable will be further reduced because of undersized, blemished or damaged crop products and keeping quality may also be limited because of the risk of the development of diseases in storage. In some cases, the pathogen may contaminate the product with toxins that may threaten human health. Contamination of cereal by mycotoxins produced by several species of *Fusarium*, which cause the early blight complex in wheat and barley, is a current high-profile issue. All of these undesirable effects occur in organic, low-input and conventional crop production systems to a greater or lesser extent and various approaches are taken to control them.

18.3 Crop protection with copper-based fungicides in organic production systems

As mentioned previously, the main group of fungicides currently permitted for use in organic crop production systems contains copper. The fungicidal properties of copper compounds have been recognized for many years and some compounds also kill bacteria. Copper sulphate was used as a seed treatment for control of bunt in wheat (*Tilleta caries*) as early as 1761. Later, from about 1885 onwards, Bordeaux mixture (a complex of copper sulphate and lime) was used in grapevines to control downy mildew caused by *Plasmopara viticola* (Copper Development Association, 2003) and this was the first fungicide to be used on a large scale worldwide (Schneiderhan, 1933). Even today, the only fungicides allowable under organic standards that are effective against downy mildew are based on copper hydroxide and copper sulphate (Madge, 2005).

Since 1920, many mineral and organic chemical copper fungicides have been developed and there are about 2000 registered products for control of numerous crop diseases (Gianessi and Puffer, 1992). Active ingredients include copper carbonate, copper ammonium carbonate, copper hydroxide, copper naphthenate, copper octanoate, copper oleate, copper oxide, copper oxychloride, copper oxychloride sulphate, copper 8-quinolinolate, copper sulphate, copper salts of fatty and rosin acids, copper sodium sulphate, phosphate complex and copper–zinc sulphate complex. Copper compounds are often used in combination with other active ingredients (chlorothalonil, streptomycin, maneb, mancozeb or sulphur) to enhance control, but few are permitted in organic production systems and in the EU are restricted to those included on Annex 2 of Regulation (EEC) 2092/91. Specifically, these are copper hydroxide, copper oxychloride (tribasic), copper sulphate and cuprous oxide. These are protectants with no systemic or eradicant action, which must be applied to the crop before infection occurs in order to be effective. They have multi-site activity and so there is a low risk of pathogens developing resistance to them (Van-Zwieten *et al.*, 2005). However, efficacy of fungicides containing copper varies and sole reliance upon them is unlikely to result in complete disease control: additional measures are usually required (Kuepper and Sullivan, 2004).

Copper compounds are recommended for disease control in a wide range of annual and perennial crops throughout the world: they are relatively cheap and widely available. Some examples are shown in Table 18.1. Citrus, rice, almonds, walnuts and tomatoes together account for 70% and 60% of the total weight and the area treated with copper compounds, respectively (Gianessi and Puffer, 1992). In Europe, the majority of copper-based fungicides used in organic farming are applied to potatoes for the control of late blight, grapevines for the control of downy mildew (*P. viticola*) and apple for the control of scab (*Venturia inaequlais*). These are extremely problematic diseases that result in severe economic penalties where control is unsuccessful.

Table 18.1 Examples of crop diseases treated with copper-based fungicides (based on Gianessi and Puffer, 1992 and British Crop Protection Council, 2004)

Crop species	Disease	Causal organism	Recommended copper-based product
Fruits			
Citrus fruits	Brown rot	*Alternaria alternata*	Copper oxychloride, copper sulphate
Tree nuts			
Almonds	Brown rot	*Alternaria alternata*	Copper ammonium carbonate, copper sulphate
	Greasy spot	*Xanthomonas arboricola*	
Walnuts			
Pome fruits			
Apples	Canker	*Nectria galligena*	Bordeaux mixture, copper oxychloride
Pears	Canker	*Nectria galligena*	Bordeaux mixture
Apples	Scab	*Venturia inaequalis*	Bordeaux mixture
Pears	Scab	*Venturia inaequalis*	Bordeaux mixture
Stone fruits			
Cherries	Bacterial canker	*Pseudomonas syringae*	Bordeaux mixture, copper oxychloride
Plums	Bacterial canker	*Pseudomonas syringae*	Copper oxychloride
Apricots	Leaf curl	*Taphrina deformans*	Bordeaux mixture
Nectarines	Leaf curl	*Taphrina deformans*	Bordeaux mixture
Peaches	Leaf curl	*Taphrina deformans*	Bordeaux mixture, copper ammonium carbonate,
		copper oxychloride	
Berries and small fruit			
Raspberries	Cane spot	*Elsinoe veneta*	Bordeaux mixture, copper ammonium carbonate,
		copper oxychloride	
Raspberries	Spur blight	*Didymella applanata*	Bordeaux mixture, copper oxychloride
Grapevine vines	Powdery mildew	*Uncinula necator*	Copper sulphate + sulphur
Grapevine vines	Downy mildew	*Plasmopara viticola*	Bordeaux mixture, copper oxychloride
Blackcurrants	Currant leaf spot	*Pseudopeziza ribis*	Bordeaux mixture, copper ammonium carbonate

Table 18.1 Continued

Crop species	Disease	Causal organism	Recommended copper-based product
Vegetables			
Swedes	Powdery mildew	*Erysiphe cruciferarum*	Copper sulphate + sulphur
Turnips	Powdery mildew	*Erysiphe cruciferarum*	Copper sulphate + sulphur
Fruiting vegetables			
Tomatoes	Blight	*Phytophthora infestans*	Bordeaux mixture, copper ammonium carbonate, copper oxychloride, copper sulphate + sulphur
Cucumbers	Powdery mildew	*Erysiphe cichoracearum*	Bordeaux mixture, copper ammonium carbonate
Peppers	Phytophthora	*Phytophthora capsici*	Copper oxychloride
Legume vegetables			
Beans	Powdery mildew	*Erysiphe polygoni*	Bordeaux mixture
Stem vegetables			
Celery	Celery leaf spot	*Septoria apiicola*	Bordeaux mixture, copper ammonium carbonate, copper oxychloride
Potatoes	Blight	*Phytophthora infestans*	Bordeaux mixture, copper oxychloride, copper sulphate + sulphur
Sugar beet	Powdery mildew	*Erysiphe polygoni*	Copper sulphate + sulphur
Hops			
Hops	Downy mildew	*Pseudoperonospora humuli*	Bordeaux mixture, copper oxychloride, copper sulphate + sulphur
Hops	Powdery mildew	*Sphaerotheca macularis*	Copper sulphate + sulphur

18.3.1 Effects of copper-based fungicides on crop growth, the environment and food safety

Copper is an essential micronutrient required in the growth of both plants and animals. In humans, it helps in the production of blood haemoglobin. In plants, copper is an important component of proteins found in the enzymes that regulate the rate of many biochemical reactions in plants. Plants would not grow without the presence of these specific enzymes. Research projects show that copper promotes seed production and formation, plays an essential role in chlorophyll formation and is essential for proper enzyme activity, disease resistance and regulation of water in plants (Rehm and Schmitt, 2002).

As the Ministry of the Environment Programs and Initiatives in Canada (2001) reported, copper is relatively abundant and varies widely from soil to soil depending on the parent rock type and proximity to manufactured sources. Available copper can vary from 1 to 200 ppm (parts per million) in both mineral and organic soils as a function of soil pH and soil texture. The finer textured mineral soils generally contain the highest amounts of copper and the lowest concentrations are associated with the organic or peat soils. Availability of copper is related to soil pH. As soil pH increases, the availability of this nutrient decreases. Copper is not mobile in soils and it is attracted to soil organic matter and clay minerals (Rehm and Schmitt, 2002). Acidification of the soil through the application of certain fertilizers, such as ammonium nitrate or organic materials such as peat moss and pine needles, can increase the uptake of copper into plants. Alternatively, the addition of non-acidic matter (compost or manure) or lime to the soil can reduce the uptake of copper into plants.

Copper is an essential element required by all organisms; however, it is toxic at high levels to humans, animals, plants, fungi, algae and microorganisms. Generally, a copper soil test above 0.4 ppm is adequate for crop needs. Karamanos *et al.* (1986) developed a critical level of 0.4 ppm for spring wheat and 0.35 ppm for canola grown in northern prairie soils. Alberta Agriculture, Food and Rural Development (1999) also used 0.4 ppm as a critical level for copper deficiency without any specific reference to plant species. Whilst the fungicides can scorch sensitive vegetable crops under some environmental conditions (Koike *et al.*, 2000), they may improve the nutritional status of both crop and soil when copper is in short supply.

Major concerns surround the potential detrimental effects of a build-up of copper in agricultural soils. This is less of a problem in annual crops, but in cropping systems with a long history of regular copper fungicide application – in perennial crops such as grapevines, for example – significant accumulations in soils' surface horizons have been recorded (Gallagher *et al.*, 2001; Chaignon *et al.*, 2003). Prolonged use in Europe has led to high levels in soil, for example 200–500 ppm in France over a large area of agricultural land (Brun *et al.*, 1998). In Australia, up to 250 ppm total copper were present in a 20–30-year-old vineyard soil, while 8–14 vineyards studied exceeded 60 ppm (Pietrzak and McPhail, 2004). Avocado orchards in northern Australia had even higher residues of copper in the top 2 cm of soil (280–340 ppm) than in a nearby reference site under natural vegetation (13 ppm) (Merrington *et al.,* 2002).

Residues are likely to remain indefinitely in most soils and have a continuous impact on soil ecology (Giller *et al.*, 1998; Van-Zwieten *et al.,* 2005). Even relatively low concentrations of copper in the soil may result in long-term effects, including reduced microbial and earthworm activity and subsequent loss of fertility (Dumestre *et al.*, 1999). Merrington *et al.* (2002) showed that the copper residues in avocado orchards significantly affected soil microorganisms. Biomass carbon (C_{mic}) was significantly lower in the orchard soils but levels of total organic carbon (C_{org}) were similar or higher than in the untreated site under natural vegetation. Concerns about potential impacts of copper contamination in agricultural soils are a major reason for restrictions on the use of fungicides containing this element. Whilst management strategies/ remediation technologies need to be developed to reduce the bioavailability of existing residues, reduction of further inputs, through the use of alternative fungicides and control methods, will help to prevent the situation from worsening.

Although humans are exposed to copper from many sources, including coins, cooking utensils, drinking water, soil and dust, 75–99% of total copper intake is from food. However, possible undesirable effects of copper-based fungicides on the health of workers exposed to the chemicals and consumers of crop products treated with them are a major concern. In humans, acute ingestion of copper sulphate has been reported to cause gastrointestinal injury, haemolysis, methemoglobinaemia, hepatorenal failure, shock, or even death, although the toxicity of organo-copper compounds remains largely unknown (Yang *et al.*, 2004). Risks should be minimal if the fungicides are used according to the conditions of approval laid down by pesticide regulations and decline even further as the need for them is reduced or eliminated.

18.4 Crop protection without copper-based fungicides

Measures that can be used to reduce reliance on copper-based fungicides include the use of new varieties, diversification, agronomic management and novel alternative treatments. Used independently, they are unlikely to be as adequate, but when combined into an integrated disease control system they might achieve a similar level of control. Approaches to reducing the inputs of copper-based fungicides in organic cropping systems can be both direct and indirect. Alternative methods of suppressing diseases caused by pathogens that are susceptible and routinely treated with copper-based fungicides, is a direct approach. Use can be indirectly reduced by adopting practices which result in healthier, more vigorous plants that are more resilient to diseases in general with less need for such fungicides. Examples of both will be mentioned in the following sections to illustrate general principles.

18.4.1 Prevention and monitoring

Prevention of diseases by adopting strict hygiene measures is a traditional method. The starting point is to plant healthy seed or transplants into clean soil as there are few 'chemical' pre-planting treatments that are permitted in organic farming. Regular removal and destruction of diseased plant residues and volunteers or ground keepers after harvest, in both field and protected crops, minimizes sources of infection and spread to healthy plants. There are many examples of this approach. In apples, an ecological approach to scab (*V. inaequalis*) control is to accelerate breakdown of scab-infected leaves on the orchard floor over winter by encouraging microbes and earthworms responsible for decomposition. In glasshouse production of tomato and cucumbers, removal of senescent plants might reduce (though not prevent) the spread of *Botrytis* spores. In potatoes, control of volunteer tubers, which are the host of the late blight (*P. infestans*) pathogen and may survive over winter to infect newly emerging crops in spring, is a prerequisite.

During the growing season, assessing the potential for disease, coupled with accurate timely diagnosis and continuous monitoring, is an essential component of integrated disease control (Koike *et al.*, 2000). Decision support systems (DSS) based on predictive models that use weather data and detailed epidemiological information, provide an invaluable service to growers (Madge, 2005). By confirming when protection from infection is required, spray timing can be optimized and unnecessary treatments avoided. This approach is far more effective than relying on a calendar-date based spray schedule and several DSS are in use for a range of crops. The RIMpro scab warning system developed in the Netherlands in the 1990s is used in apples (Bouma, 2003). Dacom's PLANT-plus system, also developed in the Netherlands, can be used for a range of crops. DSS are used widely in late blight control in the potato crop including PLANT-plus, the British Potato Council's Blight-Watch scheme and Phyto-pre from Switzerland (Madge, 2005). By optimizing spray programmes, use of copper-based fungicides in organic crop production systems and synthetic chemicals in conventional systems can be minimized. This is highly desirable as potatoes receive the highest quantities of pesticides applied to arable crops and growers are under increasing pressure from consumers and retailers to reduce inputs.

18.4.2 Choice of variety

A major component of any integrated disease control programme within a cropping system is to exploit varietal resistance to predominant diseases wherever possible. Resistant varieties suppress or retard a pathogen's activity and show little or no symptoms of infection. Tolerant varieties do not significantly inhibit the pathogen: they may show severe disease symptoms but without significant losses in yield or quality. Resistance can be race-specific (highly effective resistance based on R-genes) or race non-specific (partial resistance/tolerance). Growing the same varieties year after year in an area can lead to evolution of new races or more aggressive strains of a fungus and breakdown of resistance. As partial resistance/tolerance is based on several genes rather than a single major one, it slows the spread of the disease, imposes less selection pressure on the fungus and is more durable.

Disease-resistant varieties are attractive because they should pose little or no risk to the environment and enable growers to reduce and in some cases eliminate the need for pesticides. In some host–pathogen systems, resistance may persist for many years, but in others it may be short-lived (Koike *et al.*, 2000). Unfortunately, resistance is not available to counter every disease and for some of the most damaging ones, such as tomato late blight (*P. infestans*) and white rot (*Sclerotium cepivorum*) of alliums, no acceptable resistant varieties are currently available.

Most crop varieties currently grown have been bred for production in intensive, conventional systems and may not have all the characteristics which are suitable in lower input or organic systems. Nevertheless, some of the new, resistant varieties emerging from breeding programmes, are already available in some crops in most countries and can give effective prevention of soilborne and foliar diseases when grown to organic standards. (Of course, these exclude genetically modified varieties which are not permitted under any organic farming standards.) Speiser *et al.* (2006) demonstrated this for potatoes in several European countries, but the resistant varieties did not necessarily outyield more susceptible ones. However, they did decrease the source of inoculum significantly and in the UK the most resistant 'Sarpo' varieties of potato bred in Hungary did not benefit much from copper fungicide treatment. Varieties are certainly available with greater resistance than some of the most popular ones grown organically, but their uptake is very much dependent on their market acceptability and the cost of seed (Speiser *et al.*, 2006). For all crops, market demands may make it necessary for organic producers to grow some varieties that lack disease resistance. Whilst the changeover to new varieties can be achieved relatively quickly and easily for annual crops, the process is much more challenging in perennials such as grapevines and apples. Despite the challenges of developing resistant cultivars and the problems of breakdown, resistance is one of the most important weapons for disease control in organic systems.

18.4.3 Diversification strategies

Evidence suggests that disease epidemics can be delayed, if not prevented, by employing diversification strategies. At the system level, crop rotation is often the foundation of an integrated disease control programme. At the individual crop level, diversification can be achieved by growing several varieties with different forms or levels of resistance. Usually, each variety is grown as a separate monoculture, but growing them in random mixtures or alternating rows is an alternative approach. Another diversification method is intercropping, where rows or beds of the crop are separated by other crop species that provide physical barriers to the spread of the disease.

Crop rotation

Crop rotation using a range of contrasting crops (including cover crops or a period of fallow) can decrease soilborne pathogen inocula, increase diversity in soil microflora over time and be unfavourable for the of spread foliar diseases in the shorter term. This reduces the need for pesticides.

Recent research has shown that certain cash and catch crops also have a suppressive effect on diseases. For example, incorporated brassica crop debris decomposes releasing natural chemicals that can significantly reduce the number of *Verticillium dahliae* microsclerotia, but there is a risk of encouraging the clubroot fungus *(Plasmodiophora brassicae)* (Koike *et al.,* 2000). Unfortunately, others, such as vetch, can greatly increase the number of infective sclerotia of *Sclerotinia minor* if planted into a field with a history of lettuce crops and might also increase soil populations of *Pythium* and *Rhizoctonia* damping-off fungi. Vetch is also a known host of root-knot nematode *(Meloidogyne* species) (Koike *et al.,* 2000).

Crop rotation is the keystone of organic cropping systems and its importance is emphasized by all organic crop production standards, not only for its role in crop protection, but also for its contribution to soil fertility. Soil fertility may influence the severity of disease infection or offset its effects by accelerating the build-up of yield before the disease takes its toll. This aspect is considered further in Section 18.4.4.

Crop variety mixtures and intercropping

Monocultures, based on a single variety of a single species and often grown on a large scale, predominate in agriculture. The success of this approach in conventional farming owes much to the use of synthetic pesticides for crop protection. With the limited pesticides available for use in organic systems, more diverse configurations within the same field may be an advantage and stands composed of variety mixtures are a prime example (Wolfe, 1985). In potatoes, Andrivon *et al.* (2003) and Garrett and Mundt (1999) reported significant decreases in late blight infection (*P. infestans*) where a resistant variety was mixed with a susceptible variety. The latter authors also reported yield increases compared with pure stands of the constituent varieties. Some work suggests that diversification strategies could be useful against slow developing epidemics and/or combined with other measures with partial effects such as reduced fungicide sprays. Clearly, success will depend on the choice and compatibility of varieties grown in the mixture, and disease severity, but are worth considering as part of an integrated disease management system. In practice, growing variety mixtures may increase establishment, harvest and handling costs and there may be marketing issues, offsetting any disease prevention or yield advantages. Furthermore, it may not be feasible in certain crops, for example, grapevines and those grown for seed, where varietal integrity is of paramount importance. Intercropping with different species is an alternative and may be an easier system to manage but may be less effective.

18.4.4 Soil and fertility management

Physical, chemical and biological soil conditions influence occurrence and severity of diseases and the ability of plants to withstand infection as well as directly supplying water and nutrients. It is particularly important to manage and exploit suppressive effects of the soil environment on diseases as part of an integrated control strategy, because of the potentially significant contribution to agricultural sustainability and environmental quality (Quimby *et al.*, 2002). Major differences exist between conventional and organic production systems in their approaches to soil fertility and pest management. In organic systems, a greater reliance is placed on chemical and biological processes to release nutrients in plant available forms in soil solution (Stockdale *et al.*, 2002). Experiments and on-farm surveys conducted by Brown *et al.* (2000) have suggested that organic farm management gives better soil conditions and more balanced fertility than in conventional systems, leading to fewer problems in disease management (Lampkin, 1999). Literature also suggests that soil in organically managed farms is more fertile (higher total N, total P, humic acid, exchangeable nutrient cations, water-holding capacity and microbial biomass) than in conventional systems (Wells *et al.*, 2000). However, excessive fertilizer applications can increase a crop's susceptibility to diseases if they cause prolific growth of the leaf canopy (Davies *et al.*, 1997).

Organic matter is the main fertility input in organic systems. As well as supplying nutrients and improving soil structure and water-holding capacity and hence plant vigour, it has beneficial crop protection effects. These are associated with increased microbial activity leading to reduced aggressiveness and infestation of pathogens. Various chemicals in the soil are known to contribute to the anti-phytopathogenic potential of the soil. The release of carbon dioxide, which is harmful to some pathogens in high concentrations and toxins available in the plant residue (allelochemicals), may also act against them. Plant uptake of other breakdown products, including phenols, phenolics and other compounds such as salicylic acid, has an antibiotic effect on pathogens such as the fungi causing damping-off diseases (*Rhizoctonia*, *Fusarium* and *Pythium*) (Lampkin, 1999). Typically, the level of disease suppressiveness is related to the level of total soil microbiological activity (Sullivan, 2001).

In many cases, the incorporation of organic matter in the soil may be a beneficial alternative to chemicals for plant disease control. For example, Viana *et al.* (2000) reported that matured cattle manure and sugarcane husks are efficient alternatives for control of bean damping-off (*Sclerotinia sclerotiorum*). Farmyard manure applied at 5 t/ha once every three years reduced dry root rot (*Macrophomina phaseolina*) to 32% in groundnut (*Arachis hypogaea* L.) compared with untreated plants (Harinath and Subbarami, 1996). Whilst it is clear that soil organic matter contributes to the suppression of plant diseases, the level of understanding of the mechanisms involved is still limited (Bailey and Lazarovits, 2003) but enhanced soil microbiological activity is a major contributor.

Many fungi are hyperparasites of other fungi (Adams, 1990). Species of *Trichoderma* that secrete lytic enzymes are active against fungal cell walls (Sivan and Chet, 1989). Sullivan (2001) reported a direct correlation between general microbial activity and amount of microbial biomass, and the degree of *Pythium* suppression. These activities not only influence the general nutrition, health and vigour of higher plants (which affects disease susceptibility), but also determine the competitive behaviour of root-infecting fungi and their microbial antagonists (Curl, 1988). Streptomycetes are common filamentous bacteria that are effective, persistent, soil saprophytes. They are often associated with plant roots and have the ability to colonize plants and decrease damage from a broad range of pathogens. They are well-known producers of antibiotics and extra-cellular hydrolytic enzymes with potential to contribute significantly to an integrated disease management system that includes alfalfa and other crops such as potato, maize and soybeans (Samac *et al.*, 2003).

Compost is a particular form of organic matter recommended for use in organic farming systems. It encourages the proliferation of antagonists that compete with plant pathogens, organisms that prey on and parasitize pathogens, and beneficial microorganisms that produce antibiotics (Sullivan, 2001), including those mentioned above. Disease suppression in compost is very much related to the degree of decomposition; as the compost matures, it generally becomes more suppressive. However, readily available carbon compounds found in low-quality immature compost suppressed *Pythium* and *Rhizoctonia* in turf-grass (Nelson *et al.,* 1994). The moisture content of compost is also critical to the range of organisms inhabiting the finished product. Compost with at least 40–50% moisture will be colonized by both bacteria and fungi that suppress *Pythium* (Hoitink *et al.,* 1997). Some researchers recommend inoculating composts with specific beneficial agents such as certain strains of *Trichoderma* and *Flavobacterium* that produce anti-fungal exudates which suppress *Rhizoctonia solani* in potatoes (Sullivan, 2001). Applications of non-composted organic matter including animal manures and crop by-products may also be beneficial.

18.4.5 Crop management

The way that the crop is managed can have a significant effect on the timing and severity of disease incidences. It may also help to avoid some of the consequences of infection by advancing the achievement of acceptable yields of quality crop products before infection becomes serious. In Denmark, copper fungicides are not permitted by national legislation and organic growers employ cultural plant protection methods such as crop rotation, crop selection, time of planting and field sanitation extensively, while direct control methods are rarely applied (Langer, 1995).

In annual crops, planting may be timed to avoid periods of high risk and produce a marketable crop before serious infection occurs, largely avoiding the need for fungicides. The planting date may also directly influence susceptibility to certain pathogens because the crop's physiological and chronological age when infection occurs will be affected and will have implications for disease management. For example, the changing susceptibility of potato crops to late blight (*P. infestans*) throughout growth is well documented: plants are most susceptible when very young, most resistant when of intermediate age, becoming more susceptible again as they senesce (Mooi, 1965; Lapwood, 1977; Stewart, 1990; Carnegie and Colhoun, 1982) and the age distribution of leaves may have an effect at the individual plant level (Visker *et al.,* 2003).

Other treatments have similar effects to those of early planting and include priming of true seed, presprouting of tubers as in potatoes and the use of transplants. Generally, there is some flexibility in the use of such treatments in annual crops but it may be limited by local agro-meteorological conditions or the need to deliver a crop to the market at a predetermined time. Of course, the limitations are even greater in perennial crops.

Planting different populations in different configurations by changing spacing between row and/or within-rows affects a crop's microclimate because of changes in canopy structure and time of closure. Consequently, the duration of periods of high humidity and leaf wetness that encourage certain fungal and bacterial pathogens can be managed by altering row spacing. Lower plant populations with a wider spacing and a more open canopy should improve air-flow and accelerate the drying of wet foliage, making it less susceptible to some diseases than denser crops. However, choice of planting configurations may be constrained by effects on yield and size grading of produce (particularly important in vegetable crops), as well as quality and crop mechanization requirements.

Irrigation has implications for disease control as well as for yield. Management of irrigation to optimize crop yield and quality, minimize disease risks and attempt to schedule restricted fungicide applications, is extremely challenging. Water should be pathogen-free and applied to meet the crop's demand. Excess water from over-irrigation or heavy rainfall soon after overhead irrigation may create conditions, such as extended periods of soil and leaf wetness, that favour most soilborne and foliar disease pathogens, respectively, especially those that produce motile zoospores such as *Pythium* and *Phytophthora* (Madge, 2005). Despite the potentially large responses of crop yield to irrigation, overhead application may need to be limited or stopped if there is a high risk of foliar diseases (Koike *et al.,* 2000). Poor drainage and waterlogging may increase diseases because plant resistance may be lowered from lack of oxygen and lower soil temperatures. Cavity spot disease of carrot caused by the genus *Pythium* (e.g. *Pythium violae* and *Pythium sulcatum*) was more common in badly drained, poorly structured soils (Hiltunen and White, 2002). High soil moisture levels promoted the development of root rot (*Phytophthora citrophthora*) of citrus (Feld *et al.*, 1990). However, many soil fungi (e.g. *Phytophthora*, *Rhizoctonia*, *Sclerotinia*) and some bacteria (e.g. *Erwinia* and *Pseudomonas*) cause most severe plant symptoms when the soil is at field capacity and not saturated. Water stress may predispose a crop to infection. Drought enhances infection of groundnuts by *Aspergillus flavus* (Wotton and Strange, 1987), *Fusarium solani* which causes dry rot in beans and *Fusarium roseum* which causes seedling blight (Agrios, 1997). On the other hand, lack of moisture coupled with high air temperatures often causes fungal lesions in leaves to dry up, arresting development of the disease and decreasing the need for fungicides.

18.4.6 Alternative treatments

Copper-based fungicides are not the only 'chemical' fungicides permitted in organic farming. Other such substances traditionally used in organic farming include sulphur and lime sulphur (calcium polysulphide) for use on all crops, and mineral oils and potassium permanganate, but only in fruit trees, olive trees and grapevines. These exhibit some activity against several fungi: sulphur and lime sulphur are particularly effective against powdery mildew and potato late blight. Use of these active ingredients is more limited and seems to be less controversial than that of copper-containing materials, but there is growing interest in the development of biologically based alternatives. Some are already permitted under organic production standards, for example, microorganisms (bacteria, viruses and fungi) such as *Bacillus thuringiensis*, *Granulosis* virus and so on, provided they are not genetically modified (Anonymous, 2002).

Various alternative 'biological' treatments including microbial antagonists and plant or compost extracts have been developed. Effective ones offer the opportunity to reduce or eliminate the need not only for copper-based fungicides in organic cropping systems but also for pesticides, in general, in conventional and low-input systems. They work via direct antifungal effects, or stimulate competitor microorganisms, and/or induce plant resistance. Jindal *et al.*, (1988) reported that maximum inhibition (93%) *in vitro* against *P. infestans*, was caused by *Penicillium aurantiogriseum* when spores of *P. aurantiogriseum* were applied 12 hours prior to inoculation by the disease agent, followed by *Fusarium equiseti*, *Mucor hiemalis*, *Trichoderma koningii*, *Epicoccum purpurascens* and *Stachybotrys atra*. Systemic protection against *P. infestans* and reduced disease severity has been reported in studies that used *Bacillus puimilus* and *Pseudomonas fluorescens* (Zhinong *et al.*, 2002). Daayf *et al.* (2003) showed biocontrol activity by the genera *Bacillus*, *Pseudomonas*, *Rahnella* and *Serratia* against the same pathogen. For downy mildew control in grapevines, a range of alternatives including plant extracts, microorganisms and plant strengtheners that trigger the crop's natural immune system (a response referred to as systemic acquired resistance) are being tested. To date, none of them have shown economically effective control of downy mildew (Madge, 2005).

Plant, seaweed, humic and compost extract preparations applied to crop foliage or the soil have been reported to induce/activate the natural resistance of plants against fungi and also to exhibit direct antifungal activity. However, there is very limited information about their efficacy. Compost extracts or compost teas, which are filtrated solutions of mixtures of compost materials and water, have shown promising results in some cases but not in others (Kotcon and Murray, 2003). The precise mechanisms by which they work are not well known but seem to vary depending on the host/pathogen relationship and the mode of application.

Goldstein (1998) reported that composts and compost extracts activate disease resistance genes in plants that normally respond to a pathogen by generating chemical defences. Where compost extracts are applied to crop foliage, it may be too late to be effective but these disease-prevention systems may be already running in plants growing in compost (Sullivan, 2001). Brinton *et al.* (1996) tested compost teas on late blight (*P. infestans)* in potatoes and found that the key factors influencing effectiveness were the age of the compost (extracts from older composts were more effective than those from younger extracts) and the nature of the compost's feedstock.

There are many other potential alternative treatments for disease control in organic agriculture. For example, ethanol and sodium bicarbonate for post-harvest control of *Botrytis cinerea* in organically grown strawberries (Karabulut *et al.*, 2004); mustard flour and milk powder for control of common bunt (*Tilletia tritici*) in wheat and stem smut (*Rocystis occulta*) in rye (Borgen and Kristensen, 2001). Stonemeal is also used as a crop protectant by some producers and quartz meal (silica) is a required spray preparation in biodynamic agriculture. Scientific evaluation of the efficacy of such products registered for use in organic farming is essential if the goal of reducing or eliminating copper-based fungicides is to be achieved. The formulation and method of application must also be evaluated as these may affect treatment efficacy. Some microorganisms may be physically damaged by some sprayer mechanisms and directed spraying may be more beneficial than overhead application. For effective treatments it is likely that timing of the spray is more critical than the method. Speiser and Schmid (2004) give a concise overview of the evaluation of plant protection products in organic agriculture in Europe. *The Manual of Biocontrol Agents* (British Crop Protection Council, 2004) gives a comprehensive account of information on a microorganisms and natural products for disease control and most of the active ingredients are acceptable for use in organic farming systems.

18.5 Future trends

The calls for reduction in pesticide inputs irrespective of production system are likely to intensify. Various initiatives and compulsory or voluntary schemes have been developed in most countries to reduce the amounts applied and encourage integrated pest management systems in conventional production and to stimulate adoption of 'lower-input' and organic systems. Increasing costs of pesticide inputs relative to the value of crop output are another inducement. Research to fulfil these objectives will maintain a high priority and focus is likely to intensify on the identification and development of acceptable plant protection products. Increasing knowledge and understanding of non-chemical and alternative disease control strategies in organic systems will also have an impact on conventional crop production systems. This will lead to the incorporation of strategies that deliver natural systems of pest regulation with less reliance on synthetic pesticides, without incurring significant yield penalties. However, the transfer of promising results in the opposite direction, from the conventional to the organic sector, will be limited by restrictions on the use of synthetic products and genetically modified organisms.

Copper fungicides may give only partial protection against fungal and bacterial diseases or, on the other hand, give virtually complete control but with little crop yield or quality benefit. Despite the fact that they are not always beneficial, many growers are reluctant to dispense with them unless forced to do so, or until effective alternatives are available. They are considered to be a cheap, insurance option that may protect against devastating consequences. In the long term, it is clearly desirable to grow organic crops without copper-based fungicides, but many growers are concerned about the effects of their withdrawal on the economic viability of cropping. In countries or particular regions that regularly experience high levels of disease however, limited use of copper fungicides may be necessary until other effective management practices or products are available and integrated systems have been optimized. However, in both organic and conventional cropping situations when epidemics are particularly severe, nothing will control the disease and serious losses of crop yield and quality will be inevitable.

18.6 Conclusions

There are large regional differences in the impact of crop diseases on organic production and a widespread view that a copper fungicide ban will have serious consequences for organic production unless effective alternatives are available. There is no single, alternative treatment available that offers the level of control given by copper fungicides. Consequently, successful management of diseases in organic systems will rely on the integration into existing production systems of resistance management and diversification, agronomic strategies and alternative treatments. These strategies need to be specifically adapted to particular regions if economically viable crop production is to be sustained. To this end, worldwide research programmes are needed to evaluate integrated management systems that are specific to each country and region. Clearly, the challenges for managing diseases in crops grown according to organic standards are even greater than in conventional ones where fungicide programs play a key role in delaying an epidemic. Systems of production designed for particular situations and prevailing conditions are required and the level of success in delaying, slowing or preventing disease spread may vary from system to system. This demands a much more flexible approach than in conventionally grown crops, where variety choice, agronomic management, crop protection protocols and production standards are very specific in order to meet the demands of particular market outlets.

18.7 Sources of further information and advice

There is a wide range of sources about reducing or eliminating copper-based fungicides in organic agriculture. These include published materials and web-based information and advice at international, national and local level from government agriculture and food departments, state and independently financed research stations and organisations, advisory bodies and certifying authorities. The following organisations are particularly relevant:

IFOAM – International Federation of Organic Agricultural Movements (http://www.ifoam.org)

EUROPE – European Agriculture Commission (http://europa.eu.int/comm/agriculture)

ISOFAR – International Society of Organic Agriculture Research (http://www.isofar.org)

National government agriculture and food departments

National and international organic certifying authorities

In addition there are national and international research programmes dedicated to the stated objectives. Two European projects: Blight–MOP – development of a systems approach for the management of late blight in EU organic potato production (Leifert and Wilcockson, 2005) and REPCO – replacement of copper fungicides in organic production of grapevine and apple in Europe (Bengtsson *et al.*, 2006; Schweikert *et al.*, 2006), are examples of such an approach.

18.8 References

Adams, P. B. (1990). 'The potential of mycoparasites for biological control of plant diseases', *Annual Review of Phytopathology*, **28**, 59–72.

Agrios, G. N. (1997). *Plant Pathology*, Academic Press, California, USA.

Alberta Agriculture, Food and Rural Development (1999). *Copper Deficiency: Diagnosis and Correction*, Agdex 532–3.

Andrivon, D., Lucas, J.M. and Ellisseche, D. (2003). 'Development of natural blight epidemics in pure and mixed plots of potato cultivars with different levels of partial resistance', *Plant Pathology*, **52**, 586–596.

Anonymous (2002). Commission Regulation (EC) No 473/2002 of 15 March 2002. *Official Journal of the European Communities*, **L 75 EN**, 21–24.

Bailey, K.L. and Lazarovits, G. (2003). 'Suppressing soil-borne diseases with residue management and organic amendments', *Soil and Tillage Research,* **72**, 169–180.

Bengtsson, M., Jørgensen, H.J.L., Wulff, E. and Hockenhull, J. (2006). Prospecting for organic fungicides and resistance inducers to control scab (*Venturia inaequlais*) in organic apple production. Paper presented at Joint Organic Congress, Odense, Denmark, May 30–31, 2006. (http://orgprints.org/7395)

Borgen, A. and Kristensen, L. (2001). 'Use of mustard flour and milk powder to control common bunt (*Tilletia tritici*) in wheat and stem smut (*Rocystis occulta*) in rye in organic agriculture'. *Proceedings of an International Symposium on Seed Treatment Challenges and Opportunities*, February 26–27, 2001 Wishaw, North Warwickshire, UK. British Crop Protection Council Symposium Proceedings, *Seed Treatment: Challenges and Opportunities*, Biddle, A.J. (ed.), BCPC, Farnham, Surrey, UK, 141–148.

Bouma, E. (2003). 'Decision support systems used in the Netherlands for reduction in the input of active substances in agriculture', *EPPO Bulletin*, **33**(3), 461.

Brinton, W.F., Trankner, A. and Roffner, M. (1996). 'Investigations into liquid compost extracts'. *Biocycle,* **37**(11), 68–70.

British Crop Protection Council (2004). *The Manual of Biocontrol Agents*, L.G. Copping (ed.), British Crop Protection Council, Alton, Hampshire, UK, 758 pp.

Brown, S.M., Cook, H.F. and Lee, H.C. (2000). 'Topsoil characteristics from a paired farm survey of organic versus conventional farming in southern England', *Biological Agriculture and Horticulture*, **18**, 37–54.

Brun, L.A., Maillet, J., Richarte, J., Herrmann, P. and Remy, J.C. (1998). 'Relationships between extractable copper, soil properties and copper uptake by wild plants in vineyard soils', *Environmental Pollution*, **102**, 151–161.

Carnegie, S.F. and Colhoun, J. (1982). 'Susceptibility of potato leaves to *Phytophthora infestans* on potato plants in relation to plant age and leaf position', *Phytopathology*. Z, **104**, 157–167.

Chaignon, V., SanchezNeira, I., Herrmann, P., Jaillard, B. and Hinsinger, P. (2003). 'Copper bioavailability and extractability as related to chemical properties of contaminated soils from a vine-growing area', *Environmental Pollution*, **123**, 229–238.

Copper Development Association (2003). Uses of copper compounds. www.copper.org/applications/compounds/copper_sulphate02.html.

Curl, E.A. (1988). 'The role of soil microfauna in plant-disease suppression', *Critical Review of Plant Science*, **7**, 175–196.

Daayf, F., Adam, L. and Fernando, W.G.D. (2003). 'Comparative screening of bacteria for biological control of potato late blight (strain US–8), using *in-vitro*, detached-leaves, and whole-plant testing systems', *Canadian Journal of Plant Pathology*, **25** (3), 276–284.

Davies, B., Eagle, D. and Finney, B. (1997). *Soil Management*, Farming Press, Ipswich, UK.

Dumestre, A., Sauve, S., McBride, M., Baveye, P. and Berthelin, J. (1999). 'Copper speciation and microbial activity in long-term contaminated soils', *Archives of Environmental Contamination and Toxicology*, **36**, 124–131.

Feld, S.J., Menge, J. A. and Stolzy, L.H. (1990). 'Influence of drip and furrow irrigation on *Phytophthora* root rot of citrus under field and greenhouse conditions', *Plant Disease*, **74**, 21–27.

Gallagher, D.L., Johnston, K.M. and Dietrich, A.M. (2001). 'Fate and transport of copper-based crop protectants in plasticulture runoff and the impact of sedimentation as a best management practice', *Water Research*, **35**, 2984–2994.

Garrett, K.A. and Mundt, C.C. (1999). 'Epidemiology in mixed host populations', *Phytopathology*, **89**, 984–990.

Gianessi, L.P. and Puffer, C.A. (1992), *Fungicide Use in U.S. Crop Production*. Resources for the Future, Quality of the Environment Division,Washington, D.C. (variously paged).

Giller, K.E., Witter, E. and McGrath S.P (1998). 'Toxicity of heavy metals in microorganisms and microbial processes in agricultural soils: a review', *Soil Biology and Biochemistry*, **30**, 1389–1421.

Goldstein, J. (1998). 'Compost suppresses disease in the lab and on the fields', *BioCycle*, November, 62–64.

Harinath, N.P. and Subbarami, R.M. (1996). 'Effect of soil amendments with organic and inorganic manures on the incidence of dry root rot of groundnut', *Indian Journal of Plant Protection*, **24**, 44–46.

Hiltunen, L.H. and White, J.G. (2002). 'Cavity spots of carrot (*Daucus carota*)', *Annals of Applied Biology*, **141**, 201–223.

Hoitink, H.A.J., Stone, A.G. and Han, D.Y. (1997). 'Suppression of plant diseases by composts', *HortScience*, **32**, 184–187.

IFOAM (2000). *Basic Standards for Organic Production and Processing*, IFOAM, Tholey-Theley, Germany.

Jindal, K.K., Singh, H., and Meeta, M. (1988). 'Biological control of *Phytophthora infestans* on potato', *Indian Journal of Plant Pathology,* **6**, 59–62.

Karabulut, O.A., Arslan, U. and Kuruoglu, G. (2004). 'Control of postharvest diseases of organically grown strawberry with preharvest applications of some food additives and postharvest hot water dips', *Journal of Phytopathology*, **152** (4), 224–228.

Karamanos, R.E., Kruger, G.A. and Stewart, J.W.B. (1986). 'Copper deficiency in cereal and oilseed crops in Northern Canadian prairie soils', *Agronomy Journal*, **78**, 317–323.

Koike, S., Gaskell, M., Fouche, C., Smith, R. and J. Mitchell (2000). *Plant Disease Management for Organic Crops*, University of California, Division of Agriculture and Natural Resources. Publication 7252, http://anrcatalog.ucdavis.edu.

Kotcon, J. B. and Murray, W.K. (2003). 'Efficacy of compost and compost tea for suppression of early blight in organically grown tomato', *Phytopathology*, **93** (6 Supplement), June S47.

Kuepper, G. and Sullivan, P. (2004). 'Organic alternatives for late blight control in potatoes', in *Pest Management Technical Note*. http://attra.ncat.org/attra-pub/lateblight.html

Lampkin, N. (1999). *Organic Farming,* Farming Press, Ipswich, UK, 214–271.

Langer, V. (1995). 'Pests and diseases in organically grown vegetables in Denmark: A survey of problems and use of control methods', *Biological Agriculture and Horticulture*, **12** (2) 151–171.

Lapwood, D.H. (1977). 'Factors affecting the field infection of potato tubers of potato tubers of different cultivars by blight (*Phytophthora infestans*), *Annals of Applied Biology*, **85**, 23–42.

Leifert, C. and Wilcockson, S.J. (2005). Blight-MOP: Development of a systems approach for the management of late blight (caused by *Phytophthora infestans*) in EU organic potato production. Final Report, Newcastle University, UK.

Madge, D. (2005). *Organic Farming: Managing grapevinevine downy mildew*. Agricultural notes, AG1174, page 1–3 (http://www.dpi.vic.gov.au/notes).

Merrington, G., Rogers, S.L. and VanZwieten, L. (2002). 'The potential impact of long-term copper fungicide usage on soil microbial biomass and microbial activity in an avocado orchard', *Australian Journal of Soil Research*, **40**, 749–759.

Ministry of the Environment Programs and Initiatives (2001). *Copper in the Environment*. Ontario. http://www.ene.gov.on.ca/cons/4141e.htm

Mooi, J.C. (1965). 'Experiments on testing field resistance to *Phytophthora infestans* by inoculating cut leaves of potato varieties', *European Potato Journal*, **8**, 182–183.

Nelson, E.B., Burpee, L.L. and Lawton, M.B. (1994). 'Biological control of turfgrass diseases', in Leslie A.R., *Handbook of Integrated Pest Management for Turf and Ornamentals,* CRC Press/Lewis Publishers, Boca Raton, Florida, 409–427.

Pietrzak, U. and McPhail, D.C. (2004). 'Copper accumulation, distribution and fractionation in vineyard soils of Victoria, Australia', *Geoderma*, **122**(2–4), 151–166.

Quimby, P.C., King, L. R. and Grey, W.E. (2002). 'Biological control as a means of enhancing the sustainability of crop/land management systems', *Agricultural Ecosystem and Environment*, **88**, 147–152.

Rehm, G. and Schmitt, M. (2002). *Copper for Crop Production*, University of Minnesota, http://www.extension.umn.edu/distribution/cropsystems/DC6790.html

Samac, D.A., Willert, A.M., McBride, M.J. and Kinkel, L.L. (2003). 'Effects of antibiotic-producing Streptomyces on nodulation and leaf spot in alfalfa', *Applied and Soil Ecology*, **22**, 55–66.

Schneiderhan, F.J. (1933). 'The discovery of Bordeaux mixture' (three papers): I. Treatment of mildew and rot./ II. Treatment of mildew with copper sulphate and lime mixture./ III. Concerning the history of the treatment of mildew with copper sulphate. By Perre Marie Alexis Millardet 1885', A Translation from the French by Felix John Schneiderhan. *American Phytopathological Society Phytopathological Classics*, 3.

Schweikert, C., Mildner, M., Vollrath, C. and Kassemeyer, H-H. (2006). Systems for testing the efficacy of biofungicides and resistance inducers against grapevine downy mildew (REPCO project). Paper presented at Joint Organic Congress, Odense, Denmark, May 30–21, 2006. (http://orgprints.org/7492)

Sivan, A. and Chet, I. (1989). 'Degradation of fungal cell walls by Litic enzymes of *Trichoderma harzianum*', *Journal of Genetic Microbiology*, **135**, 675–682.

Speiser, B. and Schmid, O. (2004). 'Current evaluation procedures for plant protection products used in organic agriculture', Proceedings of a workshop held September 25–26 2003 at FiBL in Switzerland (http://orgprints.org/2942)

Speiser, B., Tamm, L., Amsler, T., Lambion, J., Bertrand, C., Hermansen, A., Ruissen, M.A., Haaland, P., Zarb, J., Santos, J., Shotton, P., Wilcockson, S., Juntharathep, P., Ghorbani, R. and Leifert, C. (2006). 'Improvement in late blight management in organic potato production systems in Europe: Field tests with more resistant potato varieties and copper based fungicides', *Biological Agriculture and Horticulture*, **23**(4), 393–412.

Stewart, H. E. (1990). 'Effect of plant age and inoculum concentration on expression of major gene resistance to *Phytophthora infestans* in detached potato leaflets', Mycological *Research*, **94**, 823–826.

Stockdale, E.A., Shepherd, M.A., Fortune, S. and Cuttle, S.P. (2002). 'Soil fertility in organic farming systems – fundamentally different?' *Soil Use and Management*, **18**, 301–308.

Sullivan, P. (2001). *Sustainable Management of Soil-borne Plant Diseases*. ATTRA, USDA's Rural Business Cooperative Service. www. a t t r a . o r g attra-pub/soilborne.html

Tamm, L., Smit, B., Hospers, M., Janssens, B., Buurma, J., Mølgaard, J.P., Lærke, P.E., Hansen, H.H., Bodker, L., Bertrand, C., Lambion, J., Finckh, M.R., Schüler, C., Lammerts van Bueren, E., Ruissen, T., Solberg, S., Speiser, B., Wolfe, M., Phillips, S., Wilcockson, S.J. and Leifert, C. (2004). *Assessment of the Socio-economic Impact of Late Blight and State-of-the-art Management in European Organic Potato Production Systems*, FiBL, Frick, Switzerland. (http://orgprints.org/2936).

Van-Zwieten, L., Merrington, G. and M. Van-Zwieten (2005). *Review of Impacts on Soil Biota Caused by Copper Residues from Fungicide Application.* http://www.regional.org.au

Viana, F.M.P., Kobory, R.F., Bettiol, W. and Athayde, S. C. (2000). 'Control of damping-off in bean plant caused by *Sclerotinia sclerotiorum* by the incorporation of organic matter in the substrate', *Summa Phytopathologica*, **26**, 94–97.

Visker, M.H.P.W., Keizer, L.C.P., Budding, D.J., Van Loon, L.C., Colon, L.T. and Struik, P.C. (2003). 'Leaf position prevails over plant age and leaf age in reflecting resistance to late blight in potato', *Phytopathology*, **93** (6), 666–674.

Wells, A.T., Chan, K.Y. and Cornish, P.S. (2000). 'Comparison of conventional and alternative vegetable farming systems on the properties of a yellow earth in New South Wales', *Agricultural Ecosystem and Environment*, **80**, 47–60.

Wolfe, M.S. (1985). 'The current status and prospects of multiline cultivars and variety mixtures for disease resistance', *Annual Review of Phytopathology*, **23**, 251–273.

Wotton, H.R. and Strange, R.N. (1987). 'Increased susceptibility and reduced phytoalexin accumulation in drought-stressed peanut kernels challenged with *Aspergillus flavus*', *Applied and Environmental Microbiology*, **53**, 270–273.

Yang, C.C., Wu, M.L. and Deng, J.F. (2004). 'Prolonged hemolysis and methemoglobinemia following organic copper fungicide ingestion', *Veterinary and Human Toxicology*, **46**, (6), 321–323.

Zhinong, Y., Reddy, M.S., Ryu, C.M., McInroy, J.A., Wilson, M. and Kloepper, J.W. (2002). 'Induced systemic protection against tomato late blight elicited by plant growth-promoting rhizobacteria', *Phytopathology*, **92**, 1329–1333.

19

Pre-harvest strategies to ensure the microbiological safety of fruit and vegetables from manure-based production systems

Ulrich Köpke, Johannes Krämer, University of Bonn, Germany and Carlo Leifert, Newcastle University, UK

19.1 Introduction

Nutritional quality (which includes safety aspects) was defined: 'Nutritional quality is determined by the value of the product for the consumer's physical health, growth, development, reproduction and general well-being' (Köpke, 2004). This extended definition may be divided into two terms. One term is for the effect of food determined by its substance, that is the sum of all beneficial ingredients and harmful compounds and their nutritional aspect. The other term covers the perception or psychological effect of well-being based on knowledge related to the organic label, indicating the process quality and the ethical, environmental and social values. Nutritional quality describes the inherent biological or health value of produce including the ratio of beneficial to harmful substances, taste, fragrance, freshness and shelf-life as well as the risk of pathogen contamination as important quality characteristics that govern consumer behaviour (Codex Alimentarius, 2003; Köpke, 2004).

Traditional whole food recommendations (Kollath, 1983; von Koerber *et al.*, 1999), as well as public information campaigns, such as 'Five-a-Day for Better Health', have recommended increased consumption of fresh (raw) fruit and vegetables in western diets (Hedberg *et al.*, 1994). Higher per capita consumption of fresh or minimally processed fruits and vegetables (MPF), ready-to-eat produce, as well as the increase in imports of fresh fruits and vegetables from countries where hygiene standards may be low, has resulted in heightened interest in outbreaks of human gastroenteritis which may be attributed to contaminated fresh food, particularly vegetables for salads and, to a lesser extent, fruits (Beuchat, 1996, 2002).

Pathogens capable of causing human health risks include bacteria, viruses and parasites. Health risk management can be defined as the 'scientific evaluation of the probability of occurrences of known or potential health effects resulting from human exposure to foodborne hazards' (Anonymous, 1995). Normally, individual adverse health effects related to microbial pathogens result from a single acute exposure rather than from long-term chronic exposure to a microbial hazard. The response of the human population to exposures of foodborne pathogens is highly variable depending on a broad spectrum of factors such as the virulence characteristics of the pathogens, the level of infective units ingested, the general health and immune status of the hosts and the attributes of the food that alter microbial or host status. Highest risks exist for immunocompromised individuals, the very young and the elderly (Riley *et al.*,1984; Buchanan *et al.*, 2000). Enteric bacterial pathogens (e.g. *Salmonella*, *Yersinia*, *Campylobacter* and *E. coli*) are currently considered as the greatest concern in terms of serious illness and numbers of persons at risk of gastrointestinal infection on an international scale (Fernandez-Alvarez *et al.*, 1991; Beuchat, 1996).

Fresh vegetables are naturally contaminated with microorganisms in a range of 6–7 $\log_{10}$ colony-forming units per gram (CFU/g, aerobic plate count) and 3–7 $\log_{10}$ CFU/g, (coliforms, but not *E. coli*), respectively (Jay *et al.*, 2005). Surveys of raw fruits and vegetables demonstrate that a wide range of these products can also be contaminated by human pathogens. Outbreaks of gastrointestinal illness caused by bacteria, viruses and parasites have been reported for different intact or processed vegetables and, to a less extent, fruits. However, the proportion of reported outbreaks of food poisonings attributed to fruits and vegetables is low (SCF, 2002). The products of most concern are sprouted seeds and unpasteurised juices for which the European Commission published special microbiological standards (European Commission, 2004).

Pathogen contamination can occur at every stage of the whole production chain from 'field-to-fork' (Table 19.1). Nevertheless, besides other components of infection, manures derived from livestock are considered to be an important potential preharvest contamination source (Fig. 19.1). Several disease outbreaks have been associated with raw fruits and vegetables and unpasteurised products; some have been traced back to livestock (Pell, 1997), but in the majority of the outbreaks associated with fresh produce, minimally processed products and unpasteurised products, the source of contamination remains unknown (de Roever, 1998).

Livestock wastes contain large numbers of microorganisms. Farmyard manure (faeces, excreta and bedding material) contains about 10^{10}cells /ml from which only a few may be pathogens and sufficient to cause infection depending on the immune status of the host. This makes quantification of risks difficult (Pell, 1997). Examples of pathogenic microorganisms are given in Table 19.2. However, in many cases it has not been conclusively proven that animals are the source of human infection (Mawdsley *et al.*, 1995).

Table 19.1 Sources of pathogenic microorganisms on fresh produce and conditions that influence their survival and growth (based on Beuchat, 1996)

Preharvest	*(Postharvest)*
Excreta/manures	Wild and domestic animals
Soil	Air (dust)
Irrigation water	Wash and rinse water
Green or inadequately composted manure	Sorting, packing, cutting, and further processing equipment
Air (dust)	Ice
Wild and domestic soil organisms	Transport vehicles
Human handling	Improper storage (temperature, physical environment)
Postharvest	Improper packaging (includes new packaging technologies)
Excreta/manures	Cross-contamination (other foods in storage, preparation and display areas)
Human handling (workers, consumers)	Improper display temperature
Harvesting equipment	Improper handling after wholesale or retail purchase
Transport containers (field to packing shed)	

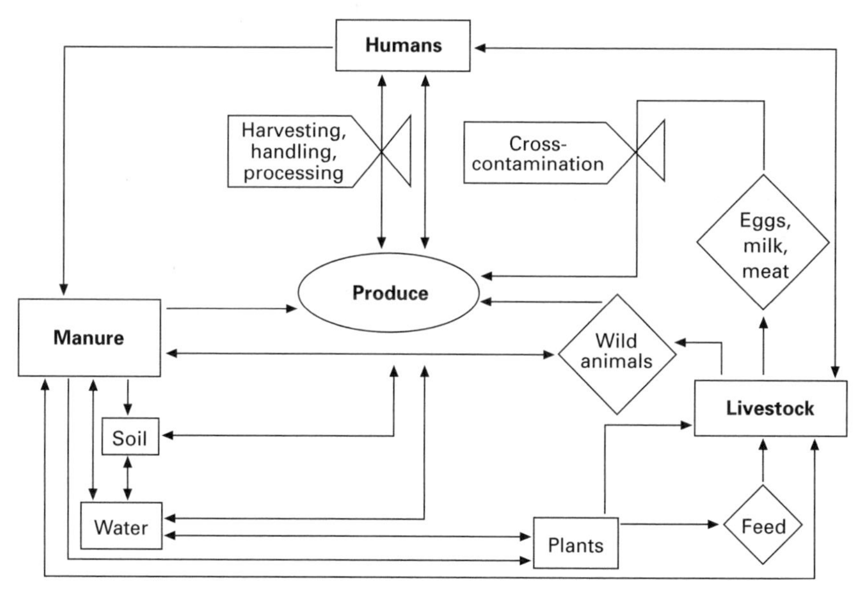

Fig. 19.1 Sources and pathways by which fresh produce can become contaminated with microbial human pathogens (based on Beuchat, 1996).

19.2 Use of manure in organic, 'low input' and conventional farming

Application of animal manures to soil serves as an important element of nutrient management. In addition to nutrients and the organic carbon recycled to maintain or increase soil organic matter content, which is a key element of soil fertility, animal manure also contains a broad spectrum of microbes, including many types of potential pathogens. Original sources of several outbreaks of human gastroenteritis have been directly traced back to animal manure used in vegetable and fruit production or to contact with that produce (Cieslak *et al.*, 1993; Nelson, 1997; Guan and Holley, 2003). Organic produce has been claimed to represent an increased risk to consumers' health, where manure is used in crops that are potentially eaten raw, and/or where no chemical treatments (e.g. fungicides to control mycotoxin-producing fungi, see also Chapter 17) are employed to reduce the microbiological loading of the raw product or to extend its shelf-life (McMahon and Wilson, 2001). Nevertheless, an extensive evaluation of 3200 samples of ready-to-eat vegetables of organic origin resulted in only 1.5% containing *E. coli* and *Listeria* spp. (no *Listeria monocytogenes*) and no other bacterial pathogens were detected (Sagoo *et al.*, 2001). It is important to note that manures are widely used in low-input and conventional farming and should therefore not be considered to be used exclusively in organic agriculture. Furthermore, in high input conventional agricultural systems (e.g. pig production systems) considerably higher livestock densities are frequently used. On such farms manure becomes a 'waste disposal problem' and is applied in extremely high levels (often in the form of slurry) when compared to organic and low-input farming.

Microbial and chemical composition of manure is highly variable and depends on manure type, storage and treatments (Hancock *et al.*, 1997). Intensified farming systems have often shifted from bedding livestock on straw and the production of farmyard manure (FYM), towards the collection of waste as semi-liquid slurry, which contains only low amounts of solid bedding material and cannot be composted but has to be aerated. On the other hand, composting of FYM or use of well-rotted FYM is still more common in organic farming. Aerobic composting makes it possible to 'disinfect' compost heaps via a self-heating-based pasteurisation process with temperatures higher than 55 °C, which is known to reduce enteric pathogen levels significantly (Hancock *et al.*, 1997; Food and Drug Administration, 1998). In contrast, slurry or fresh FYM, is known to carry higher levels of enteric pathogens, because the 'pasteurisation' associated with composting does not take place before application (Rankin and Taylor, 1969).

19.3 Risk of transfer of enteric pathogens from manure to fruit and vegetable crops

Human infections with *Salmonella* spp. and *E. coli* O157:H7 are a common worldwide phenomenon. Livestock may serve as a source of several relevant human pathogenic microorganisms (Table 19.2). The most prevalent group is the enteric pathogens which include bacteria, viruses and parasites (i.e. protozoa and helminths). Animals may shed pathogens through excreta without

Table 19.2 Human pathogens which may be found in livestock wastes or introduced by wild animals

Bacteria	Viruses	Parasites
Aeromonas hydrophila	Calcivirus	*Ascaris suum*
Bacillus cereus	Hepatitis A	*Cyclospora cayetanensis*
Campylobacter jejuni	Norwalk and	*Cryptosporidium parvum*
Clostridium botulinum	Norwalk like	*Giardia lamblia*
Clostridium perfringens	(Norovirus)	
E.coli O157:H7		
E.coli (enterotoxigenic)		
Listeria monocytogenes		
Salmonella spp.		
Shighella sonnei		
Vibrio cholerae		
Yersinia enterocoltica		

showing any clinical signs of being infected. Cow manure, for example, is a heterogeneous substrate consisting not only of faeces and urine but also of secretions from the nose, throat, vagina, blood, mammary gland, skin and placenta as well as carbon-rich bedding material (straw, sawdust, etc.) or milking centre waste or flush water (Alice, 1997).

Faeces from colonised cows may contain from 10^2 to 10^7 colony-forming units (CFU) of *Salmonella* cells per gram of faeces. Particularly calves and heifers colonised with *E. coli* O157:H7 may shed the bacteria at levels ranging from 10^2 to 10^5 CFU g^{-1} (Himathongham *et al.*, 1999). Furthermore, calves younger than four months are the main domestic animals that excrete pathogenic protozoa like *Giardia* and *Cryptosporidia.*

Application of contaminated solid manure or manure slurry poses a potential risk when applied to crops in an inappropriate fashion. There are several potential pathways for contaminating crops with pathogens from raw manure. Transmittance may take place via direct contact of plant surfaces with manure, by soil particles splashed from the soil surface via raindrops or overhead irrigation, or by small animals and soil organisms (Food and Drug Administration, 1998). Only a few outbreaks of human diseases have been traced back to inappropriate handling of manures, for example, application of raw manure applied continuously to the garden all summer (Cieslak *et al.*, 1993; Morgan *et al.*, 1988). Generally, fresh manure should not be applied to vegetables and fruits with edible parts close to the soil surface.

19.4 Agronomic strategies to minimise pathogen transfer risk

Pathogen transfer risk can be drastically reduced by incorporating manure applied to the soil with mouldboard ploughs or other mechanical tools able to invert these manures totally into the soil. Since most pathogens theoretically can survive in the soil for longer than three months (Jiang *et al.*, 2002), rotational positioning of fresh manure application can be regarded as a useful tool for growing vegetables, which is based on the positive pre-crop effect of a well-manured foregoing crop. Such a strategy could also be suitable for the heavily debated US National and Organic Standard Regulation (www.ams.usda.gov/NOP; these are NOP Regulations (Standards) and Guidelines 1990 RIN: 0581–1140) that regulate use of raw manure stating that it:

> must be composted unless it is (i) applied to land used for a crop not intended for human consumption; or (ii) incorporated into the soil not less than 120 days prior to the harvest of a product whose edible portion has direct contact with the soil surface or soil particles; or (iii) incorporated into the soil not less than 90 days prior to the harvest of a product whose edible portion does not have direct contact with the soil surface or soil particles.

Surface disinfection can be performed by solarisation before seedlings are planted although under most conditions, solarisation is restricted to a thin surface layer. Nevertheless, it is doubtful that there is a disinfecting effect, as irrigation and the associated favourable continuously wet soil conditions necessary for successful vegetable production are essential during the subsequent growing phase. Mulch layers may be regarded as an efficient physical barrier to reduce pathogen transmittance, especially when combined with the solarization effects given by plastic mulch, since direct contact with soil is then limited and splash-transmitted pathogen transfer by soil particles reduced.

One of the most important factors influencing the survival of newly introduced pathogenic bacteria into the soil environment is competition from the existing soil microflora (Lynch and Poole, 1979; Killham, 1995; Meickle *et al.*, 1995). In biologically active soils, introduced bacterial inocula were found to be reduced rapidly owing to competition. On the other hand, in soils with relatively low microbial activity, pathogen inocula may persist for much longer. Organic soil management was shown to result in increased soil biological activity (Mäder *et al.*, 2002) and would therefore be expected to result in reduced persistence of pathogens applied as manure. The soil type and matric potential (soil moisture levels) also influence the survival of introduced microorganisms (Meickle *et al.*, 1995). It was shown that a balance between water and oxygen availability (typical for well-aggregated soils and/or soils with optimised organic matter content) results in high soil microbial activity and poor persistence of introduced microorganisms (Killham, 1995). For example, for *L. monocytogenes* it was shown that environmental factors such as soil moisture content (depending also on type of soil), season, presence of plant root systems and decaying plant material may influence bacterial growth (Dowe *et al.*, 1997). Detailed further research is therefore necessary on how to minimise pathogen risks via optimisation of soil environmental conditions.

Other agronomic practices (e.g. weed control, harvesting) that may cause physical injuries to crops may also affect the risk of pathogen transfer. Once plant surface integrity is broken, pathogen growth, but also penetration into internal plant tissues may take place. Penetration of pathogens may occur from cut or destroyed cuticula or from the outer cell layers of plants (Seo and Frank, 1999; Takeuchi *et al.*, 2000). Thus, good management practice (GMP) includes careful use of machinery (cultivators, etc.), which avoids any disturbance of roots and shoots. Wounding of plant tissue may induce specific reactions that can affect the behaviour of microorganisms (Nguyen-the and Lund, 1991).

19.5 Strategies for reducing pathogen loads in manure through manure processing

Appropriate farm manure management plays a critic role in preventing pathogens from becoming infectious. Manure that is well mixed with bedding material is more likely to enhance aerobic fermentation with accompanying temperature increases when compared with slurry containing only minimal amounts of bedding material. Composting can thermally inactivate pathogens, so that the final product does not pose a threat to consumers or the environment. The extent of thermal inactivation of pathogens depends mainly on the degree and length of exposure to the given temperature. Within limits, a low temperature for a long period of time, or higher temperature for a short period of time, may be equally effective in inactivating pathogens (Haug, 1993). Several other parameters can influence the survival of pathogens. Besides temperature and time, pH, solids content, microbial content and oxidation–reduction potential are considered to be important modifying factors. Himathongham *et al.* (1999) observed exponential linear destruction of *E. coli* O157:H7 and *Salmonella typhimurium* strains in cow manure and manure slurry stored at 4, 20 or 37 °C. Time and temperature had significant effects on the survival of both tested pathogens in manure. Decimal reduction times (DRT) permitted estimations of how long manure should be held before being applied to the field.

Inoculation experiments (Kudva *et al.*, 1998) performed with *E. coli* 0157:H7, showed that inoculum levels of this pathogen can be reduced in ovine and bovine manure piles by mixing and aeration, that is by composting for 4 months or 47 days, compared to non-aerated manure piles in which the pathogen survived for more than 1 year. When manure is used, the time periods between application and vegetable harvest are therefore an important control point to decrease the risk of pathogen load. In order to make manure safe through a 10^5-fold reduction in pathogen levels, it is now recommended that manure should be stored for 105 days at 4 °C or 45 days at 37 °C, respectively.

For sewage sludge, keeping compost at 55 °C or even higher with a 100% relative humidity is considered essential by the US Environmental Protection Agency (US EPA, 1993) to destroy all pathogens, including helminth eggs. With the exposure time depending on the specific composting process, this standard was incorporated in the definition of composting listed in Part 503 under 'Processes to Further Reduce Pathogens' (US EPA, 1993; Hay, 1996). For composted plant and animal manures, USDA's National Organic Program strictly requests at least a temperature of 55 °C in its National Organic Standards, which is to be maintained for 3 days when using a vessel or static aerated pile system, or 15 days when using a Windrow composting system. Since it is possible for pathogens to survive owing to varying temperatures, especially in the surface or flat layers of windrows, repeated turnings are considered to be essential to allow all parts of the material to be heated. Clumping of solids can hinder effective disinfection; this can be avoided by shredding and sieving as well as by mixing the biosolids. These regulatory demands are similar in the different regulations given for organic agricultural practice worldwide. They confirm the optimal composting parameters defined by earlier research.

Old compost material has lower levels of pathogens, because after extended curing, a well-ripened compost is considered to have been stabilised by a different type of microbial community. Regrowth of bacterial pathogens, especially *Salmonella*, is then rarely observed, because the nutritionally more fastidious environment becomes unsuitable for enteric pathogens, which rely on readily available carbohydrate sources to persist. However, if cured stable compost is mixed with 'fresh', non-composted, organic materials, these may release, for instance, soluble sugars or other available energy sources and thereby increase the potential enteric pathogen regrowth (Hay, 1996). Generally, aerobic systems seem to be more efficient in rapidly reducing enteric pathogen numbers than anaerobic systems (Russ and Yanko, 1981).

19.6 Strategies used to reduce enteric pathogen contamination of crops via irrigation water

Pathogens that reach surface water or groundwater on farms may infect animals or humans through drinking water or ingestion of crops. Surface water taken from lakes, streams or rivers is generally considered to be of doubtful hygienic quality. Human pathogens from livestock manure can enter surface waters if the animals have access to surface waters or when manure is transported by run-off following heavy rains with flooding (Guan and Holley, 2003). Birds or other wild animals can release pathogens such as *Salmonella*, *Giardia* and *Cryptosporidia* into the waters. Pathogens may also be discharged into river systems via sewage works and run-off from fields with grazing animals. Also, freshwater sediments have been demonstrated to serve as reservoirs for faecal pathogens (Burton *et al.*, 1987; Crabill *et al.*, 1998). Thus, in several countries, agricultural use of surface water is not permitted.

Groundwater from wells located in agricultural areas can also be of doubtful quality, especially when the groundwater table is high and the soil filter function does not ensure water free from faecal bacteria (Howell *et al.*, 1995) or parasites such as *Giardia* or *Cryptosporidia*. These parasites are difficult to detect and have been responsible for several outbreaks of human diseases in agricultural regions of North America. They have also frequently been detected in surface waters where run-off from pastures, as well as wild animals, might be assumed to be the source.

Once in the groundwater, pathogens may multiply from low levels when exposed to favourable environmental conditions or available nutrients (Gagliardi and Karns, 2000). Long-term survival of pathogenic bacteria in groundwater has been demonstrated (Filip *et al.*, 1988). Irrigation wells need to be maintained properly and irrigation water sources should be monitored for human pathogenic microorganisms. Suitable indicators of faecal pathogen contamination such as *Enterobacteria* correlated directly with *Salmonella* (but not with *Giardia* and *Cryptosporidia*) have to be determined and used by producers and recorded in protocols. Irrigation water used for vegetable production should have tap water quality. When no information on water quality is available, deep wells have to be considered as the best source of irrigation water.

Although not yet confirmed by any reliable data available (Table 19.3), there remains the potential risk of transmittance of enteric pathogens from contaminated manure and soil by the splash effect of overhead irrigation, such as heavy rainfall, to the edible parts of fruits and vegetables that grow close to the soil surface. Consequently, no organic manure should be left on the soil surface; it should be incorporated into the soil with mouldboard ploughs or similar equipment able to invert the manure completely. Tap or drip irrigation instead of overhead spray irrigation might help to avoid transmittance caused by splash effects.

The nutrient management demands of vegetables with a short vegetative period often make irrigation and the resulting long periods of wet soil conditions necessary. Pathogen survival is favoured in moist environments. Available water is a critical factor for their survival in the soil as well as on plants, especially those displaying a wide surface-to-mass ratio able to form niches or cavities that may maintain a favourable microenvironment with free water or at least high humidity. The soil moisture level may be the most influential abiotic factor determining pathogen levels and survival in the soil (Islam *et al.*, 2004a, b). Soil radiation and dryness may also affect the survival of pathogens under field conditions. No information is available about whether a pre-irrigation solarisation might reduce the potential microbiological impact.

Table 19.3 Total aerobic bacterial count, coliform bacteria, Enterobacteriaceae and *Enterococcus, Salmonella* and *E. coli*, for lettuce growing in late spring and late summer 2004 (Rattler *et al.*, 2005)

Enteric bacteria		Fresh FYM (1)	Composted FYM (1)	Nettle extract (2)	CAN (3)
			$\log_{10}$ colony forming units/g		
Total aerobic bacterial count	Spring	6.76 a	6.24 a	6.33 a	6.35 a
	Summer	6.36 a	6.39 a	6.35 a	5.68 b
Coliform bacteria	Spring	5.78 a	6.07 a	5.40 a	5.89 a
	Summer	5.21 a	5.30 a	5.12 a	4.03 b
Enterobacteriaceae	Spring	6.20 a	6.08 a	5.63 a	6.01 a
	Summer	6.27 a	6.26 a	6.13 a	5.28 b
Enterococcus	Spring	2.82 a	2.67 a	2.85 a	3.08 a
	Summer	1.10 b	2.42 a	1.88 a	2.36 a
Salmonella	Spring	0*	0*	0*	0*
	Summer	0*	0*	0*	0*
E. coli	Spring	0*	0*	0*	0*
	Summer	3*	1*	2*	4*

Significance of differences (δ = 0.05) between treatments within season (lines) are denoted by different letters (Tukey-test)
* number of positive samples with >10<100 CFU/g (n = 16)
(1) FYM: Farmyard manure, (2) Extract of *Urtica dioica* shoots, (3) Calcium-ammonium-nitrate.

19.7 Strategies to reduce risk of pathogen transfer from animal grazing phases prior to planting of crops

In ley-farming systems, there is a potential for transmission of faecal pathogens from deposition onto pasture by grazing livestock and this may contaminate subsequent vegetable crops. This is especially true for *Cryptosporidium parvum*, which is a common cause of diarrhoea in neonatal calves and can result in severe health effects in immunocompromised humans via the pathways of contaminated food and drinking water. Heifers and calves younger than 24 months are also more likely to excrete *E. coli* 0157:H7 and weaned calves more likely to shed this bacteria than milk-fed calves (Pell, 1997; see also Chapter 10 by Dias-Gonzalez). Increased contamination of surface soil layers with coliforms during cattle grazing months and in deeper soil even after the field was no longer used for grazing has been reported (Faust, 1982). Usually, pathogenic bacteria and viruses decline within a few days after their introduction into soil, although they may survive in low numbers for several weeks or months depending on environmental conditions (SCF, 2002). For example, *E.coli* 0157:H7 survived in contaminated manure mixed with soil for 6 months at 15 °C or 21 °C (ICMSF, 2005). Because of this long persistence, authors recommend that a break of at least 12 months between the grazing phase and vegetable production is necessary for safe produce. However, the US National and Organic Standard Regulations regulate this case as for raw manure and state that the manure 'must be incorporated into the soil not less than 120 days prior to the harvest of a product whose edible portion has direct contact with the soil surface or soil particles'. This demand is nevertheless under strong discussion among farmers (Andrews *et al.*, 2003).

19.8 Other sources of enteric pathogen contamination

Wild and domestic animals, including mammals, birds, reptiles and insects, are likely to harbour enteric pathogens and are therefore potential transmitters to agricultural environments and produce. For instance, *Salmonella* spp. has been isolated from the intestinal tracts of most warm-blooded and many cold-blooded animals.

The spread of Asian bird flu westwards through Russia in autumn 2005 raised awareness of the fact that birds can be important vectors for diseases, because of their ability to transmit pathogens over substantial distances. Research has implicated wild birds as a source of various enteric pathogens for humans (Girdwood *et al.*, 1985, Healing *et al.*, 1992). Cytolethal distending toxin (CLDT) has earlier been described as being produced by *Shigella* spp. (Johnson and Lior, 1987) and later been reported from various *E. coli* isolates including serogroup O86 (Johnson and Lior, 1988). *E. coli* O86:K61 has long been linked to outbreaks of infantile diarrhoea (Ewing, 1986). Since post-mortem investigation of finches found dead in great numbers in the Scottish Highlands in spring 1994 and 1995 resulted in positive *E. coli* O86:K61 isolations from selected tissues of 43 out of a total of 46 finches, Foster *et al.* (1998) concluded that there is an existing potential for the role of wild birds in *E. coli* incidence in humans and animals.

Insects are considered to be another potential vector for faecal material to produce in production areas (Geldreich *et al.*, 1964). Pollinating insects might inoculate flowers with pathogens opening the pathway for fruits to internalize pathogens and eliminating their accessibility to surface decontamination (De Roever, 1999).

19.9 Strategies used to reduce enteric pathogen contamination of crops via wild animal vectors

Field vegetables are more susceptible to pathogens imported by wild animals compared to smaller-sized horticultural areas near buildings, which can be protected against small animals, but not birds, by fences. Generally, no free-range animals (e.g. fowl) that can reach horticultural plots should be kept. Nets that can keep the birds away can be used against pathogens delivered by the avifauna, as well as several noise repellents. Greenhouses do not provide complete protection, because birds, rodents and other animals cannot be kept out of these 'closed' production units.

19.10 HACCP-based systems for integrated control of pathogen transfer into organic food supply chains

To improve the ability to establish best practice with respect to manure use and prevention of pathogen transfer (especially on farms which are converting to organic farm management), a hazard analysis critical control point (HACCP)-based quality assurance system has been proposed (Haward and Leifert, 1999). This proposed four main critical control points (CCPs) (see Table 19.4).

Clearly, as in most quality assurance schemes, there is a hierarchy of CCPs, with CCPs closest to the source of contamination being the most important ('prevention at source is better than treatment'). In the case of manure management, the order of importance would be CCP1=CCP2>CCP3>CCP4. For example, CCP1 and CCP2 are clearly the most important control points, because if the animal remains healthy throughout its life and if appropriate feeding regimes minimise the risk of transfer of human pathogens that are non-pathogenic in livestock (see Chapter 10), no significant pathogen loads are excreted and pathogens do not need to be removed at a later CCP. However, since gastrointestinal diseases and shedding

Table 19.4 Critical control points for HACCP-based minimisation of pathogen transfer risk associated with the use of manure-based fertilisers (Haward and Leifert, 1999)

CCP1	Animal heath status	• Prevention of disease through appropriate livestock husbandry (including feeding regimes); (see also Chapter 10)
CCP2	Storage/processing of manure	• Manure from intensive (e.g. cage) based conventional rearing units not permitted • Composting for 6 months prescribed for manure from acceptable conventional systems • Production of slurry discouraged
CCP3	Soil biological activity and structural stability	• Long rotations, green manures • Animal manures, continuous ground cover to maintain high biological activity
CCP4	Manure addition to soil	• No application to growing crop • No manure application directly prior to crops that may be eaten raw

of human pathogens in animal faeces cannot be completely excluded, CCP2 storage/processing of manure is considered to be of equal importance to CCP1. If the manure treatment kills pathogens efficiently, soil biological activity is not essential for the removal of pathogens, and so on.

As described above, organic farming standards already provide a very structured quality assurance scheme for the use of animal manure and sludge. However, there is a need constantly to improve standards. If new information becomes available, which shows the requirement for additional regulations and/or changes to standards (e.g. concerning the processing requirements and quantities/timing of applications to horticultural and in particular glasshouse crops), these should be introduced immediately.

Critical control points not considered in the scheme described above are those relating to post-harvest stages of the food supply chain. While the use of HACCP-based quality assurance systems by vegetable packers, processors and retailers is now standard practice in most European countries (Knight and Stanley, 2000; Woteki and Kineman, 2003; see also Chapter 4 by Elzakker and Neuendorff and Chapter 23 by Brandt *et al.*), domestic food preparation and storage (what is done to food after it has been purchased by consumers) has become an increasing concern. Decreasing standards of food preparation and storage are thought to have contributed significantly to the rise in certain foodborne diseases (Woteki and Kineman, 2003; Mitakakis *et al.*, 2004;). There has been particular concern about observations that some consumers do not see the need to wash organic fruit and vegetables, because of a perception that recommendations to wash produce were made to ensure that pesticide residues are removed.

Clearly, awareness about food hygiene and proper food handling and preparation in the home are important to prevent diseases caused by foodborne pathogens. Vegetables and fruits are living non-sterile plants and further processing must begin with washing in running water. It should, however, be pointed out that, apart from eliminating the residues of soil and plant debris, washing has only a limited effect on the microflora attached to the plant surface (Nguyen-the and Carlin, 1994). Washing lettuce leaves, for example, reduced total culturable microbial density (CFU) by 0 to 0.5 log units only (Käferstein, 1976; Bomar, 1988; Jöckel and Otto, 1990). Clearly, a whole food supply chain approach is needed to minimise the risk from foodborne diseases and a particular focus in the future should be the development of strategies to improve domestic food storage, handling and preparation standards.

19.11 References

Alice N P (1997), 'Public and microbes: public and animal health problem', *J. Dairy Sci.*, **80**, 2673–2681.

Andrews N, Scheuerell S and Daeschel M (2003), *Human Pathogens from Livestock*

Manures. Proceedings from a National Summit, Oregon Tilth Inc., June 30, 2003. La Sells Stewart Center Oregon State University Corvallis, OR.

Anonymous (1995), 'Application of risk analysis to food standards issues', *Report of the Joint FAO/WHO Expert Consultation*, Geneve, Switzerland.

Beuchat L R (1996), 'Pathogenic microorganisms associated with fresh produce', *Journal of Food Protection*, **59** (2), 204–216.

Beuchat L R (2002), 'Ecological factors influencing survival and growth of human pathogens on raw fruits and vegetables', *Microbes and Infection*, **4** (4), 413–423.

Bomar T (1988), 'Mikrobiologische Bestandsaufnahme von Fertigsalaten in den Jahren 1985 und 1987', *Ernährungs-Umschau*, **35**, 392.

Buchanan R L, Smith J L and Long W (2000), 'Microbial risk assessment: dose–response relations and risk characterization', *International Journal of Food Microbiology*, **58**, 159–172.

Burton G A Jr, Gunnison D and Lanza G R (1987), 'Survival of pathogenic bacteria in various freshwater sediments', *Applied Environmental Microbiology*, **53**, 633–638.

Cieslak P R, Barrett T J, Griffin P M, Gensheimer K F, Beckett G, Buffington J and Smith M G (1993), '*Escherichia coli* O157:H7 infection from a manured garden', *Lancet*, **342**, 8867.

Codex Alimentarius Commission, WHO/FAO (2003), *Code of Hygienic Practices for Fresh Fruits and Vegetables*, CAC/RCP 53.

Crabill C, Donald R, Snelling J, Foust R and Southam G (1998), 'The impact of sediment fecal coliform reservoirs on seasonal water quality in Oak Creek Arizona', *Water Research*, **33**, 2163–2171.

De Roever C (1998), 'Microbiological safety evaluations and recommendations on fresh produce', *Food Control*, **10**, 117–143.

Dowe M J, Jackson E D, Mori J G and Bell C R (1997), '*Listeria monocytogenes* survival in soil and incidence in agricultural soils', *Journal of Food Protection*, **60** (10), 1201–1207.

European Commission (2004) Regulation (EC) 853/2004 (applicable from 1.01.2005) on microbiological criteria for foodstuffs/*Regulation on the Hygiene of Foodstuffs, Primary Production.*

Ewing W H (ed.) (1986), *Edward's and Ewing's Identification of Enterobacteriaceae*, Elsevier Science Publishing, New York.

Faust M. A (1982), 'Relationship between land-use practices and fecal bacteria in soils', *Journal of Environmental Quality*, **11**, 141–146.

Fernandez-Alvarez R M, Carballo-Cuervo S, de la Rosa-Jorge M C and Rodriguez-da Lecea J (1991), 'The influence of agricultural runoff on bacterial populations in a river', *Journal of Applied Bacteriology*, **70**, 437–442.

Filip Z, Kaddu-Mulindwa D and Milde G (1988), 'Survival of some pathogenic and facultative pathogenic bacteria in groundwater', *Water Science and Techno*logy, **20**, 227–231.

Foster G, Ross H M, Pennycott T W, Hopkins G F and McLaren I M (1998), 'Isolation of *Escherichia coli* O86:K61 producing cyto-lethal distending toxin from wild birds of the finch family', *Letters in Appl. Microbiology*, **26**, 395–398.

Food and Drug Administration (US) (1998), *Guide to Minimize Microbial Food Safety Hazards for Fresh Fruits and Vegetables.*

Gagliardi J V and Karns J S (2000), 'Leaching of Escherichia coli O157:H7 in Diverse Soils under Various Agricultural Management Practices', *Applied and Environmental Microbiology*, **66** (3), 877–883.

Geldreich E E, Kenner B A and Kabler P W (1964), 'The occurrence of coliforms, fecal coliforms and streptococci on vegetation and insects', *Applied Microbio*logy, **12**, 63–69.

Girdwood R W A, Fricker C R, Munro D, Shedden C B and Monaghan P (1985), 'The

incidence and significance of salmonella carriage by gulls (*Larus* spp.) in Scotland', *Journal of Hygiene Cambridge*, **95**, 229–241.
Guan T Y and Holley R A (2003), 'Pathogen survival in swine manure environments and transmission of human enteric illness: A review', *Journal of Environmental Quality*, **32** (2), 383–391.
Hancock D D, Rice D H, Herriott D E, Besser T E, Ebel E D and Carpenter L V (1997), 'Effect of farm manure-handling practices on *Escherichia coli* O157H7 prevalence in cattle', *Journal of Food Protection*, **60**, 363–366.
Haug R T (1993), *The Practical Handbook of Compost Engineering*, Boca, Raton, Lewis Publishers, Florida.
Haward, R and Leifert, C (1999) 'Leading the way on FYM', *Organic Farming*, **64**, 20–21.
Hay J C (1996), 'Pathogen destruction and biosolids composting', *BioCycle*, June, 67–76.
Healing T D, Greenwood M H and Pearson A D (1992), 'Campylobaters and enteritis', *Medical Microbiology*, **3**, 159–167.
Hedberg C W, MacDonald K L and Osterholm M T (1994), 'Changing epidemiology of food-borne disease: a Minnesota perspective', *Clinical Infectious Diseases*, **18**, 671–682.
Himathongham S, Bahari S, Riemann H and Cliver D (1999), 'Survival of *Escherichia coli* O157:H/ and *Salmonella typhimurium* in cow manure and cow manure slurry', *FEMS Microbiology Letters*, **178**, 251–257.
Howell J M, Coyne M S and Cornelius P (1995), 'Fecal bacteria in agricultural waters of the bluegrass region of Kentucky', *J. Environmental Quality*, **24**, 411–419.
ICMSF (2005), 'Fruits and fruits products', in *Microorganisms in Foods 6*: *Microbial Ecology of Food Commodities*, Kluwer Academic/Plenum Publishers, New York.
Islam M, Morgan J, Doyle M P, Phatak, S C, Millner P and Jiang X (2004a), 'Persistence of *Salmonella enterica* Serovar *typhimurium* on lettuce and parsley and in soils which they were grown in fields treated with contaminated manure composts on irrigation water', *Foodborne Pathogens and Disease*, **1** (1), 27–35.
Islam M, Morgan J, Doyle M P and Jiang X (2004b), 'Fate of *Escherichia coli* O157:H7 in manure compost-amended soil and on carrots and onions grown in an environmentally controlled growth chamber', *Journal of Food Protection*, **67** (3), 574–578.
Jay M J, Loessner M J and Golden D A (2005), 'Vegetable and fruit products', in *Modern Food Microbiology*, Springer Science + Business Media, New York.
Jiang X, Morgan J and Doyle M P (2002), 'Fate of *Escherichia coli* O157:H7 in manure-amended soil', *Applied Environmental Microbio*logy, **68**, 2605–2609.
Jöckel von J and Otto W (1990), 'Technologische und hygienische Aspekte bei der Herstellung und Distribution von vorgeschnittenen Salaten', *Archiv fur Lebensmittelhygiene*, **41**, 129.
Johnson W M and Lior H (1987), 'Production of shiga toxin and a cytolethal distending toxin (CLDT) by serogroups of *Shigella* spp.' *FEMS Microbiology Letters*, **48**, 235–238.
Johnson W M and Lior H (1988), 'A new heat-labile cytolethal distending toxin (CLDT) produced by *Escherichia coli* isolates from clinical material', *Microbial Pathogenesis*, **4**, 103–113.
Käferstein F K (1976), 'The microflora of parsley', *J. Milk Food Technology*, **39**, 837.
Killham, K. (1995) *Soil Ecology*. Cambridge University Press, Cambridge.
Knight, C and Stanley, R (2000). *HACCP in Agriculture and Horticulture* (second edition), Campden and Chorleywood Food Research Association Group, Chipping Campden.
Koerber von K, Männle T and Leitzmann C (1999), *Vollwert-Ernährung – Konzeption einer zeitgemäßen Ernährungsweise*, Haug, Heidelberg, 309 pp.
Kollath W (1983), *Der Vollwert der Nahrung*, Haug, Heidelberg, 135 pp.

Köpke U (2004), 'Organic foods: Do they have a role?' in Elmadfa I, *Diet Diversification and Health Promotion*, Forum Nutr. Basel, Karger 2005, **57**, 62–72.

Kudva I T, Blanch K and Hovde C J (1998), 'Analysis of *Escherichia coli* O157:H7 survival in ovine or bovine manure and manure slurry', *Applied and Environmental Microbiology,* **64** (9), 3166–3174.

Lynch, J M and Poole, N J (1979) *Microbial Ecology – A Conceptual Approach.* Blackwell, Oxford.

Mäder P, Fliessbach A, Dubois D, Gunst L, Fried P, Niggli U. (2002). 'Soil fertility and biodiversity in organic farming', *Science*; **296**, 1694–1697.

Mawdsley J L, Bardgett R D, Merry R J, Pain B F and Theodorou M K (1995), 'Pathogens in livestock waste, their potential for movement through soil and environmental pollution', *Applied Soil Ecology*, **2**, 1–15.

McMahon M A S and Wilson I G (2001), 'The occurrence of enteric pathogens and *Aeromonas* species in organic vegetables', *International Journal of Food Microbiology*, **70**, 155–162.

Meickle, A *et al.* (1995) Matric potential and the survival and activity of a *Pseudomonas fluorescens* inoculum in soil. *Soil Biology and Biochemistry*, **27**, 881–892.

Mitakakis, T Z, Wolfe, R, Sinclair, M I, Fairley, C K, Leder, K and Hellard, M E (2004) Dietary intake and domestic food preparation and handling as risk factors for gastroenteritis: a case-control study. *Epidemiology and Infection*, **132**, 601–606.

Morgan G M, Newman C, Palmer S R *et al* (1988), 'First recognized community outbreak of haemorrhagic colitis due to verotoxin-producing *Escherichia coli* O157:H7 in the UK', *Epidemiology and Infection*, **101**, 83–91.

Nelson H (1997), *The Contamination of Organic Produce by Human Pathogens in Animal Manures.* Available: [http://www.eap.mcgill.ca/SFMC_1.htm] (29 Sep 2003).

Nguyen-the C and Carlin F (1994), 'The microbiology of minimally processed fresh fruits and vegetables', *Critical Reviews in Food Science and Nutrition*, **34**, 371–401.

Nguyen-the C and Lund B M (1991), 'The lethal effect of carrot on *Listeria* species', *Journal of Applied Bacteriology*, **70**, 479–488.

Papavizas G C and Lumbsden A D (1980), 'Biological control of soil born fungal propagules', *Manual Review Phytopathology*, **18**, 389–413.

Pell A N (1997), 'Manure and microbes: public and animal health problem?', *Journal of Dairy Science*, **80**, 2673–2681.

Rankin J D and Taylor R J (1969), 'A study of some disease hazards which could be associated with the system of applying cattle slurry on pastures'. *Veterinary Record*, **85**, 587–591.

Rattler S, Briviba K, Birzele B and Köpke U (2005), 'Effect of agronomic management practices on lettuce quality', in Köpke U, Niggli U, Neuhoff D, Cornish P, Lockeretz W and Willer H, *Researching Sustainable Systems*, Proceedings of the First Scientific Conference of the International Society of Organic Agriculture Research (ISOFAR), Adelaide, Sout Australia, 21–23 September 2005, 188–191.

Riley L W, Cohen M L, Seals J E, Blaser M J, Birkness K A, Hargrett N T, Martin S M and Feldman R A (1984), 'Importance of host factors in human salmonellosis caused by multiresistant strains of Salmonella', *Journal of Infectious Diseases*, **149**, 878–883.

Russ C F and Yanko W A (1981), 'Factors affecting *Salmonellae* repopulation in composted sludges', *Applied and Environmental Microbiology*, **41** (3), 597–602.

Sagoo S K, Little C L and Mitchell R T (2001), 'The microbiological examination of ready-to-eat organic vegetables from retail establishments in the UK', *Letters in Applied Microbiology*, **33**, 434–439.

Scientific Committee on Food (SCF), European Commission (2002), *Risk Profile on the Microbiological Contamination of Fruits and Vegetables Eaten Raw*, Report.

Seo K H and Frank J F (1999), 'Attachment of *Escherichia coli* O157:H7 to lettuce leaf surface and bacterial viability in response to chlorine treatment', *Journal of Food Protection*, **62**, 3–9.

Takeuchi K, Matute C M, Hassan A N and Frank J F (2000), 'Comparison of the attachment of *Escherichia coli* O157:H7, *Listeria monocytogenes, Salmonella typhimurium* and *Pseudomonas fluorescens* to lettuce leaves', *Journal of Food Protection*, **63** (10), 1433–1437.

U S Environmental Protection Agency (1993), *Standards for the Use and Disposal of Sewage Sludge*, Federal Register, 58, 32, 40 CFR Parts 257, 403 and 503.

Woteki, C E and Kineman, B D (2003) 'Challenges and approaches to reducing foodborne illness', *Annual Review of Nutrition*, **23**, 315–344.

Part IV

The organic food chain: processing, trading and quality assurance

20

Post-harvest strategies to reduce enteric bacteria contamination of vegetable, nut and fruit products

Gro S. Johannessen, National Veterinary Institute, Norway

20.1 Introduction

All fruits and vegetables carry a microbial load upon harvest. This microbial burden may contain enteric pathogens such as *Salmonella* spp., different variants of *Escherichia coli*, including shiga-toxin producing (STEC), *Shigella* spp., *Yersinia enterocolitica*, *Campylobacter* spp. and other pathogens. These bacteria will mainly be on the surface of the product, but internalization of pathogens has been demonstrated in seedlings and in mature plants where the microorganisms have been introduced through cuts on the surface (Jablasone *et al.*, 2005; Janisiewicz *et al.*, 1999a; Solomon *et al.*, 2002; Warriner *et al.*, 2003a, 2003b).

There has been an increased awareness of foodborne diseases associated with fruits and vegetables in recent years. These are products that are considered safe and beneficial to human health. Not surprisingly, health authorities throughout the industrialized world are urging the population to increase their consumption of fruits and vegetables. Organisms that have been implicated in foodborne diseases from fruit and vegetables have typically been *Salmonella* spp., *E. coli* O157:H7, *Shigella* spp., the parasites *Cryptosporidium, Cyclospora* and *Giardia* in addition to hepatitis A and Noro virus (Table 20.1). The list of organisms involved in outbreaks of disease also occasionally includes *Bacillus cereus, Listeria monocytogenes, Campylobacter* spp., *Clostridium* spp. and *Vibrio cholerae*. Many of the potential foodborne pathogens are natural residents in the intestines of warm-blooded animals and will consequently be isolated from faecal material of livestock.

The microflora of fruits and vegetables at harvest reflects the environment that the product is cultivated in. Contamination of such products may occur in the field from soil and fertilizer, wild and domestic animals, irrigation water, equipment and human activities in the field. Contaminated seed may also be a source of pathogenic bacteria where seeded sprouts, such as mung bean sprouts, alfalfa sprouts and other types, are most at risk. Post-harvest contamination by human handling, water and ice used for washing and processing and the various equipment used, is also a possibility. Questions regarding the bacteriological quality of organic fruits and vegetables have been raised. Only a few surveys have been carried out in order to investigate the bacteriological quality of organic vegetables, but the quality seems to be generally acceptable (Loncarevic *et al.*, 2005; McMahon and Wilson, 2001; Sagoo *et al.*, 2001). However, a recent study by Mukherjee *et al.* (2004) from the USA showed that the prevalence of *E. coli* in conventional and certified organic produce was not significantly different. When comparing the prevalence of *E. coli* in certified and non-certified organic products with conventional products, significantly more *E. coli* was isolated from organic than from conventionally produced vegetables. Nevertheless, only one outbreak of foodborne disease has so far been traced back to organic vegetables. This was an outbreak of Shiga-toxin producing *Citrobacter freudii* traced to the consumption of sandwiches with green butter made from organic parsley fertilized with swine manure (Tschäpe *et al.*, 1995). To date, no documentation is currently available that establishes that there are differences in the hygienic quality of conventional and organic fruit and vegetables. Since there is a variety of sources of contamination on vegetables, there are several bacteriological, parasitological and virological pathogens that are of concern. This chapter will focus on enteric pathogenic bacteria.

20.2 Processing strategies used

The most important factor for safe fruit and vegetable production is that the raw material is free from contaminants that are potentially hazardous to

Table 20.1 Selected outbreaks of foodborne diseases associated with fruit and vegetables

Product	Organism	Reference
Unpasteurized apple juice	*E. coli* O157:H7	Cody *et al.*, 1999
Tomatoes	*Salmonella* Baildon	Cummings *et al.*, 2001
Lettuce	*E. coli* O157:H7	Ackers *et al.*, 1998
Lettuce	*E. coli* O157:H7	Hilborn *et al.*, 1999
Lettuce	*Shigella sonnei*	Kapperud *et al.*, 1995
Ruccola salad	*Salmonella* Thomson	Aavitsland and Nygård, 2004
Pre-cut salad vegetables	*Salmonella* Newport	Sagoo *et al.*, 2003
Sprouts	*E. coli* O157:H7	Michino *et al.*, 1999
Sprouts	*Salmonella* Stanley	Mahon *et al.*, 1997
Cantaloupe	*Salmonella* Poona	Anonymous, 2002
Raspberries	*Cyclospora cayetanensis*	Herwaldt and Ackers, 1997

humans. Raw fruits and vegetables may not be totally free from potential pathogens, but there are preventive measures that can be taken to reduce the risk. Such actions and programmes are described elsewhere in this book and will not be further discussed in this chapter.

Traditionally, a variety of strategies has been used to reduce the risk of becoming ill from consuming fresh produce. Conventional methods have included different additives or treatment with various chemicals in order to disinfect the product. Removing or inhibiting the growth of the background microbiological flora on fruits and vegetables by use of disinfectants may increase the shelf-life of the products. The packing of vegetables, such as lettuce, in modified atmosphere, may enhance the shelf-life of the salad, but this may lead to unwanted effects, such as growth of potential pathogenic bacteria. Furthermore, the use of disinfectants on fruits and vegetables may have undesirable side effects such as discolouration, bad flavour or other sensory qualities, and production of harmful by-products.

However, the trend today is towards foods that are minimally processed and more 'natural' and consequently there is a demand from the consumers that less chemicals and additives are used in food processing. Modified atmosphere packing is commonly used and increases the shelf-life of pre-cut ready-to-eat (RTE) products by inhibiting the development of spoilage microorganisms. Hurdle technology is used throughout the world, both in developed and developing countries for the gentle, but effective, preservation of foods. Hurdle technology exploits the deliberate combination of hurdles that improves microbiological stability of the products. The most important hurdles are temperature, water activity, acidity, redox potential, preservatives and competitive microflora (Leistner, 2000). There are also other options, but many are expensive and difficult to integrate in a processing line with the current technology. Several of these technologies are briefly discussed below.

20.3 Differences in organic and conventional processing standards

Food processing is generally a means to enhance the shelf-life of a product and, concurrently, the processing may also reduce the presence of pathogens in a food item. However, the process employed to achieve these ends may vary from producer to producer. The key difference between organic and conventional food processing is the more widespread use of additives and processing aids in the latter. However, there are strict regulations regarding the use of such agents both internationally and nationally. This section specifically focuses on the organic processing standards and the basic principles in the standards from the International Federation of Organic Movements (IFOAM, 2005).

The IFOAM standard (Basic Standards for Organic Production and Processing), which was the first set of standards that was set up in the 1980s, forms the basis for organic standards together with national regulations. In organic food processing, the general principle is that organic food provides consumers with nutritious, high quality supplies of organic products while providing organic farmers with a market, without compromising the organic integrity of their products (IFOAM 2005). The IFOAM standard also states that:

- Organic foods are processed by biological, mechanical and physical methods in a way that maintains the vital quality of each ingredient and the finished product.
- Processors should choose methods that limit the number of non-organic additives and processing aids.
- Unnecessary packaging material should be avoided and the food should be packaged in reusable, recycled, recyclable and biodegradable packaging.

This means that the use of additives should be limited and organic food processors should carefully design the lay-out of their processing facilities in order to prevent the usage of, and contamination with, non-permitted substances or non-organic ingredients. However, this does not mean that additives and processing aids are totally prohibited for use in organic food processing, but that their use should be limited. The IFOAM standard (2005) lists a number of additives that are permitted in organic food processing, where many are natural substances such as different gums that are used as emulgators, thickening or gelling agents. Organic acids are permitted as preservatives or acidity regulators and different gases are allowed as preservatives. Irradiation is not permitted for use in organic food processing, whereas in conventional food processing irradiation of dried herbs and spices is allowed for disinfecting purposes.

20.4 Disadvantages of chlorine sanitation methods

Chlorine-based washing systems for fresh fruit and vegetables are commonly used throughout the world, although they are prohibited in some countries (Betts and Everis, 2005). Chlorine washing systems are dependent on factors such as pH, levels of hypochlorite present in the solution, organic matter present, temperature and type of product. In many cases the effect of the chlorine washing is only marginally better than washing with potable water (Beuchat, 1999; Beuchat and Brackett, 1990; Carlin and Nguyen-the, 1994; Li *et al.*, 2001).

The bactericidal effect of hypochlorite is due to the presence of available chlorine. This is hypochlorous acid (HOCl) and hypochlorite ions (OCl^-), where HOCl is the most effective disinfectant (Betts and Everis, 2005). The dissociation of HOCl is dependent on pH and at 20 °C and between pH 6.0 and 8.0 the percentages of HOCl is approximately 97% and 23%, respectively (Beuchat, 1998). This suggests that the pH should be as low as possible, but as the pH is reduced, the chlorine solution may have a corrosive effect on equipment Therefore, the pH in a chlorine-based washing system is normally around 6.0– 7.5. This also shows that it is important for the industry to check the pH in the washing water regularly. The presence of organic matter is also a disadvantage for the use of chlorine-based washing systems because organic matter binds available chlorine, thereby reducing the disinfecting capacity. This was illustrated when Beuchat and Ryu (1997) inoculated alfalfa sprouts and fresh-cut cantaloupe cubes with a cocktail of *Salmonella* strains. Chlorine treatment (2000 ppm for 2 min) resulted in less than 1 $\log_{10}$ reduction of *Salmonella* on the cantaloupe cubes, whereas on alfalfa sprouts the bacteria were not detected. The authors explain that the very high level of organic matter in the cantaloupe juices neutralizes the chlorine before any biocidal effect may work.

Another important factor to be considered is the effect of temperature in the washing process. Chlorine has its maximum solubility at 4 °C, but it has been shown that the chlorinated washing water should ideally have a temperature that is at least 10 °C higher than that of the product. Results from a study by Zhuang *et al.* (1995) showed that a significantly higher number of *Salmonella* cells was taken up by the core tissue when tomatoes at a temperature of 25 °C were dipped in a chlorine solution held at 10 °C.

Potential hazards to humans may be connected with the use of chlorine-based washing systems. Apart from the direct occupational hazard, there are increasing concerns about the formation of potentially harmful by-products. Chlorine may react with organic compounds on the products and form hazardous organochlorines that are considered to be potential carcinogens (Beuchat, 1998; Betts and Everis, 2005). Betts and Everis (2005) suggest that a ban on the use of hypochlorite systems is likely in the future.

20.5 Methods used to study the efficacy of disinfection methods

20.5.1 Bacterial methods

Several approaches have been used to study the efficacy of different disinfection methods and the results from these experiments may be difficult to compare. Since fruits, vegetables and nuts are a very heterogeneous product group, different approaches are needed for each type of product. There is a large difference between how leafy vegetables and apples are contaminated and disinfected and the efficacy of the disinfection. Beuchat *et al.* (2001) have proposed a standardization of the method to determine the efficacy of disinfection methods, which are supposed to inactivate human pathogenic microorganisms on raw fruits and vegetables. According to the authors, there are several critical points that should be standardized, namely type of produce, pathogen of interest, procedure for inoculation, procedure for evaluating test conditions, retrieval of pathogens and reporting results. One single protocol will certainly not be sufficient to cover all types of vegetables and fruits, pathogens or disinfectants, so a basic protocol that may be modified to the different types of products, pathogens and disinfectants tested would be desirable.

20.6 Alternative strategies to the use of chlorine for disinfection

Although the use of chlorine in different forms has been the traditional and preferred disinfection method so far, there are several other options that might be applied, such as ozone and other gas treatments, UV-irradiation, different organic acids, essential oils and other treatments, including barrier or hurdle technology and mild heat treatment (steaming/pasteurization). The advantages and disadvantages of these methods are summarized in Table 20.2 and will be further discussed below.

Table 20.2 Alternative disinfection methods with their advantages and disadvantages

Treatment	Advantages	Disadvantages
Ozone	Effective disinfectant, kills rapidly	Must be produced on-site, harmful to humans
Hydrogen peroxide (H_2O_2)	Potential as disinfectant	Affect sensory qualities of some products, harmful to humans, not applicable to all products
Irradiation	Very effective disinfectant	May affect sensory qualities, harmful to humans, not permitted in organic food production
Organic acids	Effective alone or on combination with other sanitizers, simple products such as lemon juice or vinegar may be used	Not useful for all products, may have adverse effects on sensory qualities, may lead to loss of germination percentage when used on seeds
Essential oils (EOs)	Most effective for gram-positive bacteria	Gram-negative bacteria are more resistant, adverse sensory effects
High temperatures	Successful disinfection method	Not applicable for all products types consumed raw
Biocontrol and non-thermal processes	Only to a limited extent tested on fruit and vegetable products	Need further research, high costs

20.6.1 Ozone and other gas treatments

Ozone

Has been used in the water industry since the early 1900s, but it was not until recently that ozone was 'generally recognised as safe' (GRAS) and was approved for use in the food industry (Sharma, 2005). In the fruit and vegetable industry, ozone is often used as a disinfectant in the washing step. Ozonation includes the on-site production of low concentrations of the gas, which is immediately injected into water where it dissolves. The resulting ozonated water has a bactericidal effect on the fresh produce. The mechanism of this effect is believed to be damage to different cellular components, such as unsaturated lipids in the microbial cell envelope, the lipopolysaccharides layer of Gram-negative bacteria, intracellular enzymes and microbial genetic material (Kim *et al.*, 2003). These actions again lead to rapid destruction of the cell wall, which causes a quick collapse of the cells.

Ozone kills very rapidly, which is in contrast to other chemicals that need to be transferred across the cell membrane in order to cause damage. Ozone has to be produced on-site in the processing plant owing to its short half-life in an aqueous state and this production may be achieved by one of the following three methods: electrical discharge methods, electrochemical and UV-irradiation (Mahapatra *et al.*, 2005). Studies have shown that ozone is an effective disinfectant that gives a considerable reduction in the numbers of pathogenic bacteria on different types of fruit and vegetables (Kim *et al.*, 2003; Mahapatra *et al.*, 2005; Sharma, 2005). The results from a study by Rodgers *et al.* (2004) showed that five minutes after treatment with 3 ppm ozone, the numbers of *L. monocytogenes* and *E. coli* O157 were reduced from $\log_{10}$ 6 to < $\log_{10}$ 1 for all the products tested: whole apples, sliced apples, whole lettuce, shredded lettuce, strawberries and cantaloupe. Ozone also showed a significantly lower LRT ($\log_{10}$ reduction time) than the other disinfectants tested (peracetic acid, CTP chlorine 100 ppm, CTP chlorine 200 ppm and chlorine dioxide 3 ppm) with the exception of chlorine dioxide 5 ppm which gave LRT values similar to ozone.

However, ozone may have some undesirable effects. There have been a few reports of changes in aroma and surface colour of some fruits and vegetables (Kim *et al.*, 1999; Perez *et al.*, 1999). Ozone can also be hazardous to humans. A concentration above 0.1 ppm in air has a strong odour that causes irritation of the nose, throat and skin (Sharma, 2005). In addition, long-term exposure to the gas may lead to mutagenic effects and even death.

Other gas treatments

Other gas treatments may include chlorine dioxide (ClO_2) both as a gas and in an aqueous solution. The potential for use of this gas in disinfection of fruits and vegetables is mainly due to the fact that it is not affected by pH and organic matter contents to the same degree as chlorine (Betts and Everis, 2005; Busta *et al.*, 2001). However, ClO_2 must be generated on-site and it may be explosive when concentrated. There are only a few reports on the efficacy of chlorine dioxide used to disinfect fruits and vegetables. Sy *et al.* (2005a) concluded that the use of gaseous ClO_2 has promise as a sanitizer for small fruits such as blueberries, raspberries and strawberries, while Lee *et al.* (2004) reported that a ClO_2 gas sachet was effective at killing pathogens (e.g. *E. coli* O157 and *Salmonella typhimurium*) on lettuce leaves. The results of Lee *et al.* (2004) are in contrast to those of Sy *et al.* (2005b) who observed substantial reductions in pathogens on apples, tomatoes and onions, but not on peaches or fresh-cut lettuce and carrots.

Experiments with gaseous acetic acid have been performed for disinfection of seeds intended for the production of bean sprouts (Delaquis *et al.*, 1999). *Salmonella typhimurium* and *E. coli* O157:H7 were eradicated from the surface of mung bean seeds and it was reported that the seed germination loss was not too large. The seeds have often been the suspected source of contamination in sprout-associated outbreaks and sanitation methods that do not interfere with the germination of the seeds are appreciated. Other volatile chemical treatments have been tested for lethality to *Salmonella* spp. on alfalfa seeds and sprouts with varying results both on the efficacy of the disinfectant and its effect on sensory qualities (Weissinger *et al.*, 2001).

Modified atmosphere packaging (MAP)

Packaging in modified atmosphere (MA or MAP) is a technology designed to prolong the shelf-life of minimally processed foods and may have both bacteriostatic and bactericidal effects. The modified atmosphere has increased levels of CO_2 (3–10%) and reduced levels of O_2 (3–10%) (Day, 2005). The use of MA inhibits the growth of most spoilage microorganisms, hence increasing the shelf-life of the minimally processed product. It is important that the level of O_2 is not too low, because then undesirable side effects such as off-odours and off-flavours and destruction of plant tissue may occur (Busta *et al.*, 2001). However, by reducing the number of spoilage bacteria, pathogenic bacteria present may grow as a result of the decrease or absence of natural competitors.

Studies have shown that the hygienic quality of MAP products is generally good (Johannessen *et al.*, 2002; Szabo *et al.*, 2000), but that the presence of pathogenic bacteria does occur. In the UK a national outbreak of *Salmonella* Newport associated with pre-cut bagged lettuce was discovered during a study of the hygienic quality of bagged pre-cut salads (Sagoo *et al.*, 2003). In addition, it is well known that *L. monocytogenes* may be a risk in such products, as the organism grows under the conditions that are applied in MAP (Carlin and Nguyen-the, 1994; Farber *et al.*, 1998; Jacxsens *et al.*, 1999). Experiments with vegetables spiked with *E. coli* O157:H7 and *Salmonella* spp. have indicated that MAP does not eliminate these pathogens. It has been shown that the storage atmosphere has little effect on the growth and survival of *E. coli* O157:H7 (Abdul-Raouf *et al.*, 1993; Diaz and Hotchkiss, 1996). *Salmonella*, however, tend to show a reduction in numbers when stored in lower concentrations of O_2 (Finn and Upton, 1997). Information on the survival of other enteric pathogens such as *Campylobacter* spp. and *Y. enterocolitica* in MAP is limited; however, it is speculated that the reduced concentration of O_2 in MAP may be favourable to the survival of microaerophilic *Campylobacter* (Busta *et al.*, 2001).

20.6.2 Hydrogen peroxide (H_2O_2)

Hydrogen peroxide (H_2O_2) has been applied as an antimicrobial agent both in vapour form and as a liquid (Betts and Everis, 2005). The bactericidal effect of H_2O_2 is mainly due to its properties as an oxidant. In addition, H_2O_2 also possesses sporicidal activity which, together with the rapid breakdown of the substance, makes it a desirable sanitation agent (Busta *et al.*, 2001). H_2O_2 is classified as GRAS for use in the food industry and in the IFOAM standards. H_2O_2 is permitted for use as cleanser and sanitizer in direct contact with food items (IFOAM, 2005). Results from several experiments have indicated that hydrogen peroxide, depending on the concentration, has potential as a sanitizing agent. Taormina and Beuchat (1999) showed that treatment with 0.2% H_2O_2 significantly reduced the number of *E. coli* O157:H7 on alfalfa seeds. However, undesirable side effects such as browning and bleaching of products have been observed. Thus, it seems that H_2O_2 is not applicable as a sanitizer to all types of fruits and vegetables.

20.6.3 Ionizing irradiation

Ionizing irradiation has been shown to be an effective, but controversial, disinfectant of foods. Irradiation of foods is not a new technology and was patented for food preservation in 1905 by British scientists. The use of irradiation in food processing is split into three doses, namely low, medium and high doses, where low doses inhibit sprouting, delay ripening and eliminate insects, medium doses eliminate spoilage and pathogenic microorganisms and high irradiation doses equal sterilization (Korkmaz and Polat, 2005). Ionizing irradiation changes the structure of nucleic acids and cell membranes, which interferes with metabolic enzyme activities and cell division and eventually leads to cell death. The bactericidal effect of irradiation is influenced by environmental factors, condition of the target and type of target (Busta *et al.*, 2001).

The use of ionizing irradiation on fruits and vegetables has not been intensively studied, but a few experiments have suggested that irradiation has potential for such an application (Allende *et al.*, 2006; Zhang *et al.*, 2006). Results from Buchanan *et al.* (1998) indicated that low-dose irradiation could easily eliminate *E. coli* O157:H7 from fresh apple juice, but they also showed that acid-adapted strains needed higher doses of irradiation in order to obtain a 5 *D* inactivation in juice (*D*-value: decimal reduction time, i.e. the amount of time it takes at a certain temperature to kill 90% of the organisms being studied (Wikipedia, 2006)). In experiments where sprouts were irradiated prior to inoculation with separate cocktails of *Salmonella* and *E. coli* O157:H7, *Salmonella* was not detected. Sprouts that originated from seeds naturally contaminated with *Salmonella* were also negative after irradiation of the sprouts (Rajkowski and Thayer, 2000). An additional advantage with the irradiation was the increase in shelf-life of the sprouts. Another likely application of irradiation is to treat water intended for use in food processing. UV-irradiation of water is common worldwide and especially in the case where reuse of process water is desirable. UV-treatment together with a pre-filter that removes organic material has potential for water treatment. However, a study by Robinson and Adams (1978) resulted in no differences in numbers of bacteria on celery that had been irrigated with untreated and irradiated water.

Ionizing irradiation is an effective tool for decontamination of fruits and vegetables, but the main problem with this sanitation method is consumer acceptance. The use of irradiation may have adverse effects on the sensory quality of fruits and vegetables. The ionizing radiation can be harmful to those who conduct the process and proper protective measures need to be in place when using ionizing irradiation in food processing. Irradiation is currently not permitted in organic food processing.

20.6.4 Organic acids

Different organic acids, primarily lactic acid, have been successfully used for decontamination of whole livestock carcasses, and the application of different organic acids used for decontamination has also been tested in the fruit and vegetable industry. Organic acids other than lactic acid that are known to have bactericidal effects are acetic, benzoic, citric, malic, propanoic, sorbic, succinic and tartaric acids (Betts and Everis 2005). The antimicrobial action is due to a reduction in the pH in the bacterial environment, disruption of membrane transport, anion accumulation or a reduction in the internal pH in the cell (Busta *et al.*, 2001). Many fruits contain naturally occurring organic acids. Nevertheless, some strains, for example *E. coli* O157, are adapted to an acidic environment. Its survival, in combination with its low infective dose, makes it a health hazard for humans.

Several studies have investigated the potential of organic acids, often in the form of simple compounds such as lemon juice or vinegar and found that they may be effective, either alone or in combination with other sanitizers. The efficacy of acetic acid and vinegar for control of *Y. enterocolitica* inoculated on parsley was investigated by Karapinar and Gonul (1992). They observed that no aerobic bacteria were recovered after 30 min in 5% (v/v) acetic acid, whereas vinegar reduced the number of viable aerobic bacteria by 3–6 $\log_{10}$ units depending on concentration and treatment time. Liao *et al.* (2003) investigated the antimicrobial action of acetic acid on *Salmonella* bacteria inoculated onto the surface of pre-cut apple slices and found that washing with a mixture of acetic acid and H_2O_2 was the most effective method for removing *Salmonella* bacteria from cut-surfaces of apples. The use of gaseous acetic acid has also been tested for disinfection of mung bean seed with promising results (Delaquis *et al.*, 1999).

Weissinger and Beuchat (2000) studied the efficacy of different chemical treatments, including different concentrations of acetic, citric and lactic acids to eliminate *Salmonella* on alfalfa seeds. When testing a 5% solution of the separate organic acids, a substantial reduction in numbers of *Salmonella* was observed, but there was also a rather large loss of germination percentage. This suggests that organic acids may not be applicable for all decontamination purposes. Wu *et al.* (2000) observed that the use of vinegar containing 7.6% acetic acid reduced initial populations of around 7 $\log_{10}$ units of *Shigella sonnei* to undetectable levels and suggested that treatment with vinegar is a simple and inexpensive method for decontamination. The use of household sanitizers or other fluids that contain organic acids has been shown to be effective for sanitation of fruits and vegetables (Sengun and Karapinar, 2004; Sengun and Karapinar, 2005a; Sengun and Karapinar, 2005b; Vijayakumar and Wolf-Hall, 2002). Although the use of different organic acids has potential as sanitizers for fruits and vegetables, there may be adverse effects with the application of these compounds as well, for example, losses in sensory quality such as taste.

20.6.5 Essential oils

Essential oils (EO) are complex aromatic substances derived from plants that are mainly composed of terpenes and other compounds, namely aldehydes, fatty acids, phenols, ketones, esters, alcohols, nitrogen and sulphur compounds (reviewed by Ippolito *et al.*, 2005). Their effect against microorganisms is related to increased permeability of the cell membrane causing leakage of the cell contents (Piper *et al.*, 2001). The antimicrobial activity of EOs against different bacteria has been demonstrated and it seems that Gram-negatives (e.g. *Enterobacteriaceae*) are more resistant than Gram-positive bacteria (reviewed by Burt, 2004). The use of EOs as a disinfectant has been tested in food systems, mainly in meat, and only a few experiments have been carried out on vegetables. Some of these studies indicate only a slight to moderate reduction of bacterial numbers. In these studies the EO was either applied to the washing water or used in fumigation of alfalfa seeds (Burt, 2004; Park *et al.*, 2000).

In a study by Skandamis and Nychas (2000), where oregano EO in different concentrations was added to homemade eggplant salad at different pH and various temperatures, a reduction of at least 1 $\log_{10}$ unit of *E. coli* O157:H7 was observed in each case. The death rate of the pathogen depended on all the factors tested. Although *Aeromonas* spp. are not considered to be enteric pathogens, they may be a potential risk in vegetables, and a study by Uyttendaele *et al.* (2004) indicated that *Aeromonas* spp. on minimally processed vegetables could be controlled by applying thyme EO. However, adverse effects, such as off-odours and -flavours and softening of tissue from the use of EOs, have been shown, but this effect varies between the compounds applied. Although EOs are generally regarded as safe and approved for use as food flavourings, there are some research data that indicate irritation and toxicity. Burt (2004) recommends that more safety studies are carried out before EOs are more widely used in food processing than at present.

20.6.6 Other technologies

In addition to the actions listed above, which may be used to reduce risks of transfer of enteric pathogens post-harvest and during processing, there are several other possibilities that are currently being investigated. These technologies may include treatment with other chemicals, biocontrol, non-thermal processes, high temperature treatment and combination methods and hurdles. In order for an alternative method to really be an alternative, it must be easy to use, be cost effective, require minimal changes to the factory equipment already present, achieve at least the equivalent reduction in the undesired treatment of the target organism, be approved for use in a food environment and finally provide no concerns over safety issues (Betts and Everis, 2005). Treatments with chemicals other than chlorine compounds may not be desirable owing to the fact that ideally in organic production, chemicals and synthetic disinfects are not permitted or should be avoided as much as possible. The use of other chemicals will not be discussed further in this section, but other alternative technologies will be briefly discussed.

High temperature treatment

A technology that has been successfully implemented in other food industries, such as the dairy industry, is the use of high temperatures, that is pasteurization, to kill unwanted bacteria. Heat may be applied as hot water, steam, vapour heat, hot dry air and via more recent technologies such as far-infrared radiation or electromagnetic energy (Geysen *et al.*, 2005). This is a safe and effective way to eliminate most pathogens, but might be of limited use in the production of minimally processed fruit and vegetables where the goal is to have the products as fresh as possible with little processing. However, unpasteurized liquid products such as fruit juices may be risk products as they are not heat treated in any way after pressing. There have been outbreaks of foodborne disease associated with unpasteurized fruit juices. In 1996 an outbreak of STEC O157:H7 infection, with 70 people ill and one death, was traced to consumption of unpasteurized commercial apple juice (Cody *et al.*, 1999). The contaminated juice was traced to one producer and the strains of STEC O157:H7 isolated from both patients and apple juice were shown to be indistinguishable by a fingerprinting method (PFGE – pulsed field gel electrophoresis). It can be speculated that the whole outbreak could have been avoided if the apple juice had been pasteurized. Normally, the acidity of apple juice would reduce the number of pathogenic bacteria, but STEC O157:H7 and other STEC are particularly acid resistant and inoculation studies have shown that STEC may survive for several days in unpreserved apple cider/juice where the pH is below 4 (Zhao *et al.* 1993).

Heat treatments have been successfully used to reduce pests and decay in fruits and vegetables since the late 1920s (Geysen *et al.*, 2005), but high temperature treatments are unsuitable for highly perishable products, such as leafy vegetables, and certain fruits and berries, as their shelf-life will be reduced and other properties will be altered. Traditional pasteurization is an effective way to eliminate pathogenic bacteria in liquid fruit and vegetable products, such as juices and purées. Jams and jellies are normally heat treated and would be considered as safe with regard to enteric pathogens. However, there is always a risk of recontamination of the products.

In a study by Annous *et al.* (2004) surface pasteurization of whole fresh cantaloupes was tested. The results indicated that the reduction in numbers of test organisms, which were *Salmonella* Poona and *E. coli*, was significantly different than in the uninoculated controls and that the shelf-life of the pasteurized cantaloupes also increased compared with the untreated controls.

Biocontrol

It is well known that specific microorganisms can be used to control other specific microorganisms. This fact may be used in biocontrol where 'good' microorganisms are added to foods in order to kill the 'bad bugs', either by competition or by production of compounds (bacteriocins) that are toxic to the undesired bacteria, that is pathogen or spoilage bacteria, which might be present in a food stuff. Biocontrol may be described as 'an inhibitory effect of one organism on the growth and survival of another' (Betts and Everis, 2005) or, as described by Breidt and Fleming (1997), 'the objective in using biocontrol cultures is not to ferment foods, but to control the microbial ecology through competitive inhibition of pathogenic bacteria'. Only a few reports have been published on the use of biocontrol to prevent growth of pathogenic bacteria on fruit and vegetables. Vescovo *et al.* (1996) reported no growth of *Aeromonas hydrophila, L. monocytogenes, Salmonella* Typhimurium or *Staphylococcus aureus* on salad products after co-inoculation of the pathogens with lactic acid bacteria (LAB) isolated from fresh vegetable salad ingredients. Others have observed that *Pseudomonas syringae* prevented growth of *E. coli* O157:H7 in apple wounds (Janisiewicz *et al.*, 1999b), whereas Leverentz *et al.* (2001) studied the possible use of species-specific bacteriophages to control populations of *Salmonella* on fresh-cut fruits. Wilderdyke *et al.* (2004) isolated LAB from alfalfa sprouts and demonstrated their usefulness as competitive inhibitors *in vitro*. However, more research is needed in this field to develop safe systems that are applicable to the processing of fruit and vegetables.

Other non-thermal processes

Non-thermal processes other than ionizing irradiation are processes such as ultrasound, high hydrostatic pressure (HHP) and pulsed electric fields (PEFs) (Raso and Barbosa-Canovas, 2003; Ross *et al.*, 2003).

Ultrasound (US) disrupts biological structures and may lead to death when applied with sufficient intensity (Betts and Everis, 2005). The use of ultrasound to disinfect fruit and vegetables has not been described frequently in the literature. The results from a study by Seymour *et al.* (2002) indicated that a combination of ultrasound and chlorinated water reduced the numbers of both *Salmonella* Typhimurium and *E. coli* from iceberg lettuce. However, the authors concluded that the cost of such a method is high and that the combination does not completely remove pathogens from fresh produce. Therefore, this is probably not a well-suited alternative method for the decontamination of fruit and vegetables.

High hydrostatic pressure (HHP) and pulsed electric fields (PEFs) are rather novel technologies where relatively little research has been carried out. These methods have mostly been tested in combination with other decontamination methods and also mostly on either microorganisms *in vitro* or on foods of animal origin (e.g. milk, eggs and meat) (Raso and Barbosa-Canovas, 2003; Ross *et al.*, 2003). More research is needed in order to find out if such methods are applicable to the fresh produce industry.

Hurdle technology

The hurdle concept is an old technology and is used in all countries for the gentle, but effective, preservation of foods (Leistner, 2000). Hurdle technology is a combination of preservation methods. However, the same set of hurdles cannot be used for all types of food products and the specifications of the hurdles are dependent on which food type it is used for. For example, in dairy production, the addition of starter culture will effectively reduce the pH to levels where common milkborne pathogens cannot survive, thus serving as a preservation factor. A good example from fruit and vegetable processing is the production of pre-cut RTE salad vegetables. The production of such vegetables requires several hurdles or barriers in order to avoid contamination of the end product. As mentioned above, all raw vegetables carry a bacterial load where potential pathogens may be present.

The first step in the production of RTE salads is sorting and removing of outer leaves, peeling and so on, after which the raw materials are cut and washed several times. In some countries it is permitted to add disinfectants to this washing water. In Norway, however, this is not allowed and only potable water is used. Recycling of the washing water should be practised, but is not always feasible owing to the contamination of the water. In order to reuse washing water, it might be a good idea to purify it by simple processes such as filtration or irradiation by UV light. After washing, the pre-cut ingredients are mixed according to the recipe and packed in MAP. The MAP will increase the shelf-life of the product by reducing the spoilage bacterial flora. However, some pathogens survive and even grow in MAP, and raw materials without potential pathogens are imperative to the production of safe RTE products. Additionally, the whole production of RTE salads is often carried out in production facilities where the temperature is between 2 and 4 °C, which will reduce the potential for growth of human pathogenic bacteria. This production utilizes several different hurdles or barriers in order to make a product that is minimally processed, but still safe.

Hurdle technology may be considered to be a combination of different preservation technologies, where both traditional and newer technologies are applied and where the combinations are thought to give a synergistic effect. Non-thermal preservation methods, such as HHP, US, PEFs and irradiation, have been tested together with more traditional methods, but these experiments have mainly been tested *in vitro* and on some food products like milk and meat (reviewed by Raso and Barbosa-Cánovas, 2003 and Ross *et al.*, 2003). However, more research is needed on methods and technologies that are applicable to the varieties of products of fruits and vegetables throughout the world.

20.7 Integration of strategies to minimize pathogen transfer risk during processing into organic and 'low input' standard systems

The most important factor for minimizing pathogen transfer from raw fruits and vegetables to the finished product is to prevent their introduction into the processing line after harvest. If the products are heavily contaminated with pathogenic agents after harvest, it is very difficult to remove the pathogens through disinfection steps during processing. Furthermore, the production of safe fruit and vegetable products is a continuous process where all the steps are dependent on each other. However, during processing, actions can be taken that reduce the risk of spreading potential pathogenic microorganisms. During the production of fresh-cut, RTE salads it is important to ensure that the water used in processing is of satisfactory microbiological quality and that the processing facilities are kept at low temperatures (2–4 °C) where most pathogenic organisms do not grow. This will not necessarily remove potential pathogens, but if the cooling chain is maintained, the potential for growth of microorganisms is reduced. For organic and low input products, recycling of resources is important. Reuse of processing water is an option instead of using 'new' water all the time; a simple disinfection step could be included for reuse of water in the production process. For some product types, the introduction of a disinfection step with a sanitizing agent is needed. It would then be up to the producer to decide which sanitizer is most suitable for the product. For other products, such as fruit juices, pasteurization is a very important step for reducing the risk of transfer of foodborne pathogens. Enteric pathogens such as *Salmonella* spp., *E. coli, Yersinia* spp., *Shigella* spp.

and *Campylobacter* spp. do not survive normal pasteurization (72 °C, 15 s) and this type of heat treatment does not affect the sensory qualities of the beverage. However, it must be emphasized that one single alternative method may not be useful and applicable for all product types. The individual differences in the products must be taken into consideration when choosing the appropriate disinfection method.

20.8 Conclusions

To summarize, there are several disinfection or decontamination methods that could successfully be implemented in organic or low input production systems as an alternative to hypochlorite-dependent systems. Some of these alternatives have been researched, but several need further work. Individual differences between the product types must be considered when choosing a system for decontamination. However, the most important factor in reducing the risk of becoming ill from foods is that the raw products are of satisfactory quality. Recontamination of products may of course take place, but theoretically the processing systems should be designed to prevent this. All in all, there is no reason to believe that there is a greater risk of becoming ill from eating certified organic products than conventional products.

20.9 Sources of further information and advice

There are several useful websites for information and advice on organic agriculture and the addresses for some of these are listed below.
www.ifoam.org
www.fao.org
www.codexalimentarius.net
www.soilassociation.org
www.ams.usda.gov
europa.eu.int/pol/food/index_en.htm

The national organic certification bodies will also provide useful information on food processing standards in organic food production. In addition, there are publications that have reviewed decontamination processes and disinfection procedures used in the fresh produce industry. Some of these are also listed below.

Jongen, W. (2005), *Improving the Safety of Fresh Fruit and Vegetables*, Woodhead Publishing, Cambridge, UK.

USDA (2001) *Analysis and Evaluation of Preventive Control Measures for the Control and Reduction/Elimination of Microbial Hazards on Fresh and Fresh-cut Produce*. US Food and Drug Administration, Center for Food Safety and Applied Nutrition, September 30, 2001. www.cfsan.fda.gov

Beuchat, L.R. (1998) *Surface Decontamination of Fruits and Vegetables Eaten Raw: A review*. WHO, Food Safety Unit. WHO/FSF/FaS/98-2. www.who.int/foodsafety/publications/fs_management/surfac_decon/en/

20.10 References

Aavitsland, P. and Nygård, K. (2004) *Outbreak* of Salmonella *Thompson Infections*. MSIS - Norwegian Surveillance System for Communicable Diseases.

Abdul-Raouf, U.M., Beuchat, L.R. and Ammar, M.S. (1993) 'Survival and growth of *Escherichia coli* O157:H7 on salad vegetables'. *Applied and Environmental Microbiology*, **59**, 1999–2006.

Ackers, M.-L., Mahon, B.E., Leahy, E., Goode, B., Damrow, T., Hayes, P.S., Bibb, W.F., Rice, D.H., Barrett, T.J., Hutwagner, L., Griffin, P.M. and Slutsker, L. (1998) 'An outbreak of *Escherichia coli* O157:H7 infections associated with leaf lettuce consumption'. *The Journal of Infectious Diseases*, **177**, 1588–1593.

Allende, A., Mcevoy, J.L., Luo, Y.G., Artes, F. and Wang, C.Y. (2006) 'Effectiveness of two-sided UV-C treatments in inhibiting natural microflora and extending the shelf-life of minimally processed 'Red Oak Leaf' lettuce'. *Food Microbiology*, **23**, 241–249.

Annous, B.A., Burke, A. and Sites, J.E. (2004) 'Surface pasteurization of whole fresh cantaloupes inoculated with *Salmonella* Poona or *Escherichia coli*'. *Journal of Food Protection*, **67**, 1876–1885.

Anonymous (2002) 'Multistate outbreaks of *Salmonella* serotype Poona infections associated with eating cantaloupe from Mexico – United States and Canada, 2000-2002'. *MMWR Morbidity and Mortality Weekly Reports*, **51**, 1044–1047.

Betts, G. and Everis, L. (2005) 'Alternatives to hypochlorite washing systems for the decontamination of fresh fruit and vegetables'. in Jongen, W. *Improving the Safety of Fresh Fruit and Vegetables*, Woodhead Publishing, Cambridge, 352–372.

Beuchat, L.R. (1998) *Surface Decontamination of Fruits and Vegetables Eaten Raw: A Review*. World Health Organization. Food Safety Issues, WHO/FSF/FOS/98.2, 1–49.

Beuchat, L.R. (1999) 'Survival of enterohemorrhagic *Escherichia coli* O157:H7 in bovine feces applied to lettuce and the effectiveness of chlorinated water as a disinfectant'. *Journal of Food Protection*, **62**, 845–849.

Beuchat, L.R. and Brackett, R.E. (1990) 'Survival and growth of *Listeria monocytogenes* on lettuce as influenced by shredding, chlorine treatment, modified atmosphere packaging and temperature'. *Journal of Food Science*, **55**, 755–758.

Beuchat, L.R. and Ryu, J.-H. (1997) 'Produce handling and processing practices'. *Emerging Infectious Diseases*, **3**, 459–465.

Beuchat, L.R., Harris, L.J., Ward, T.E. and Kajs, T.M. (2001) 'Development of a proposed standard method for assessing the efficacy of fresh produce sanitizers'. *Journal of Food Protection*, **64**, 1103–1109.

Breidt, F. and Fleming, H.P. (1997) 'Using lactic acid bacteria to improve the safety of minimally processed fruits and vegetables'. *Food Technology*, **51**, 44–&.

Buchanan, R.L., Edelson, S.G., Snipes, K. and Boyd, G. (1998) 'Inactivation of *Escherichia coli* O157:H7 in apple juice by irradiation'. *Applied and Environmental Microbiology*, **64**, 4533–4535.

Burt, S. (2004) 'Essential oils: their antibacterial properties and potential applications in foods–a review'. *International Journal of Food Microbiology*, **94**, 223–253.

Busta, F.F., Beuchat, L.R., Farber, J.N., Garrett, E.H., Harris, L.J., Parish, M.E. and Suslow, T.V. (2001) *Analysis and Evaluation of Preventive Control Measures for the Control and Reduction/Elimination of Microbial Hazards on Fresh and Fresh-cut*

Produce. US Food and Drug Administration, Center for Food Safety and Applied Nutrition.

Carlin, F. and Nguyen-the, C. (1994) 'Fate of *Listeria monocytogenes* on four types of minimally processed green salads'. *Letters in Applied Microbiology* **18**, 222–226.

Cody, S.H., Glynn, M.K., Farrar, J.A., Cairns, K.L., Griffin, P.M., Kobayashi, J., Fyfe, M., Hoffman, R., King, A.S., Lewis, J.H., Swaminathan, B., Bryant, R.G. and Vugia, D.J. (1999) 'An outbreak of *Escherichia coli* O157:H7 infection from unpasteurized commercial apple juice'. *Annals of Internal Medicine*, **130**, 202–209.

Cummings, K., Barrett, E., Mohle-Boetani, J.C., Brooks, J.T., Farrar, J., Hunt, T., Fiore, A., Komatsu, K., Werner, S.B. and Slutsker, L. (2001) 'A multistate outbreak of *Salmonella enterica* serotype Baildon associated with domestic raw tomatoes'. *Emerging Infectious Diseases*, **7**, 1046–1048.

Day, B. P. F. (2005) 'Modified atmosphere packaging (MAP) and the safety and quality of fresh fruit and vegetables', in, Jongen, W., *Improving the Quality of Fresh Fruit and Vegetables*, Woodhead Publishing, Cambridge, 493–512.

Delaquis, P.J., Sholberg, P.L. and Stanich, K. (1999) 'Disinfection of mung bean seed with gaseous acetic acid'. *Journal of Food Protection*, **62**, 953–957.

Diaz, C. and Hotchkiss, J.H. (1996) 'Comparative growth of *Escherichia coli* O157:H7, spoilage organisms and shlef-life of shredded iceberg lettuce stored under modified atmospheres'. *Journal of the Science of Food and Agriculture*, **70**, 433-438.

Farber, J.M., Wang, S.L., Cai, Y. and Zhang, S. (1998) 'Changes in populations of *Listeria monocytogenes* inoculated on packaged fresh-cut vegetables'. *Journal of Food Protection*, **61**, 192–195.

Finn, M.J. and Upton, M.E. (1997) 'Survival of pathogens on modified-athmosphere-packaged shredded carrot and cabbage'. *Journal of Food Protection*, **60**, 1347–1350.

Geysen, S., Verlinden, B. E. and Nicolaï, B. M. (2005) 'Thermal treatments of fresh fruit and vegetables'. in, Jongen,W., *Improving the Safety of Fresh Fruit and Vegetables,* Woodhead Publishing, Cambridge, 429–453.

Herwaldt, B.L. and Ackers, M.L. (1997) 'An outbreak in 1996 of cyclosporiasis associated with imported raspberries'. The Cyclospora Working Group. *New England Journal of Medicine*, **336**, 1548–1556.

Hilborn, E.D., Mermin, J.H., Mshar, P.A., Hadler, J.L., Voetsch, A., Wojtkunski, C., Swartz, M., Mshar, R., Lambert-Fair, M.-A., Farrar, J.A., Glynn, K. and Slutsker, L. (1999) 'A multistate outbreak of *Escherichia coli* O157:H7 infections associated with consumption of mesclun lettuce'. *Archives of Internal Medicine*, **159**, 1758–1764.

IFOAM (2005). Norms. Final revision draft of the *IFOAM Basic Standards for Organic Production and Processing*. Version: 20th May, 2005. www.ifoam.org/about_ifoam/standards/norms/ibsrevision/FinaldraftIBS02_Version050520.pdf

Ippolito, A., Nigro, F. and de Cicco, V. (2005) 'Natural antimicrobials for preserving fresh fruit and vegetables', in, Jongen,W., *Improving the Safety of Fresh Fruit and Vegetables*, Woodhead Publishing Limited, Cambridge, 513–555.

Jablasone, J., Warriner, K. and Griffiths, M. (2005) 'Interactions of *Escherichia coli* O157:H7, *Salmonella* Typhimurium and *Listeria monocytogenes* plants cultivated in a gnotobiotic system'. *International Journal of Food microbiology*, **99**, 7–18.

Jacxsens, L., Devlieghere, F., Falcato, P. and Debevere, J. (1999) 'Behavior of *Listeria monocytogenes* and *Aeromonas* spp. on fresh-cut produce packaged under equilibrium-modified atmosphere'. *Journal of Food Protection*, **62**, 1128–1135.

Janisiewicz, W.J., Conway, W.S., Brown, M.W., Sapers, G.M., Fratamico, P. and Buchanan, R.L. (1999a) 'Fate of *Escherichia coli* O157:H7 on fresh-cut apple tissue and its potential for transmission by fruit flies'. *Applied and Environmental Microbiology*, **65**, 1–5.

Janisiewicz, W.J., Conway, W.S. and Leverentz, B. (1999b) 'Biological control of postharvest decays of apple can prevent growth of *Escherichia coli* O157:H7 in apple wounds'. *Journal of Food Protection*, **62**, 1372–1375.

Johannessen, G.S., Loncarevic, S. and Kruse, H. (2002) 'Bacteriological analysis of fresh produce in Norway'. *International Journal of Food Microbiology*, **77**, 199–204.

Kapperud, G., Roervik, L.M., Hasseltvedt, V., Hoeiby, E.A., Iversen, B.G., Staveland, K., Johnsen, G., Leitao, J., Herikstad, H., Andersson, Y., Langeland, G., Gondrosen, B. and Lassen, J. (1995) 'Outbreak of *Shigella sonnei* infection traced to imported iceberg lettuce'. *Journal of Clinical Microbiology*, **33**, 609–614.

Karapinar, M. and Gonul, S.A. (1992) 'Removal of *Yersinia enterocolitica* from fresh parsley by washing with acetic acid or vinegar'. *International Journal of Food Microbiology*, **16**, 261–264.

Kim, J.G., Yousef, A.E. and Dave, S. (1999) 'Application of ozone for enhancing the microbiological safety and quality of foods: a review'. *Journal of Food Protection*, **62**, 1071–1087.

Kim, J.G., Yousef, A.E. and Khadre, M.A. (2003) 'Ozone and its current and future application in the food industry'. *Advances in Food and Nutrition Research*, **45**, 167–218.

Korkmaz, M. and Polat, M. (2005) 'Irradiation of fresh fruit and vegetables'. in Jongen, W., *Improving the Safety of Fresh Fruit and Vegetables*, Woodhead Publishing, Cambridge: 387–428.

Lee, S.Y., Costello, M. and Kang, D.H. (2004) 'Efficacy of chlorine dioxide gas as a sanitizer of lettuce leaves'. *Journal of Food Protection*, **67**, 1371–1376.

Leistner, L. (2000) 'Basic aspects of food preservation by hurdle technology'. *International Journal of Food Microbiology*, **55**, 181–186.

Leverentz, B., Conway, W.S., Alavidze, Z., Janisiewicz, W.J., Fuchs, Y., Camp, M.J., Chighladze, E. and Sulakvelidze, A. (2001) 'Examination of bacteriophage as a biocontrol method for *Salmonella* on fresh-cut fruit: a model study'. *Journal of Food Protection*, **64**, 1116–1121.

Li, Y., Brackett, R.E., Chen, J. and Beuchat, L.R. (2001) 'Survival and growth of *Escherichia coli* O157:H7 inoculated onto cut lettuce before or after heating in chlorinated water, followed by storage at 5 or 15 °C'. *Journal of Food Protection*, **64**, 305–309.

Liao, C.H., Shollenberger, L.M. and Phillips, J.G. (2003) 'Lethal and sublethal action of acetic acid on *Salmonella* in vitro and on cut surfaces of apple slices'. *Journal of Food Science*, **68**, 2793–2798.

Loncarevic, S., Johannessen, G.S. and Rorvik, L.M. (2005) 'Bacteriological quality of organically grown leaf lettuce in Norway'. *Letters in Applied Microbiology*, **41**, 186–189.

Mahapatra, A.K., Muthukumarappan, K. and Julson, J.L. (2005) 'Applications of ozone, bacteriocins and irradiation in food processing: a review'. *Criticul Reviems Food Science and Nutrition,* **45**, 447–461.

Mahon, B.E., Pönkä, A., Hall, W.N., Komatsu, K., Dietrich, S.E., Siitonen, A., Cage, G., Hayes, P.S., Lambert-Fair, M.-A., Bean, N.H., Griffin, P.M. and Slutsker, L. (1997) 'An international outbreak of *Salmonella* infections caused by alfalfa sprouts grown from contaminated seeds'. *Journal of Infectious Diseases*, **175**, 876–882.

McMahon, M.A.S. and Wilson, I.G. (2001) 'The occurrence of enteric pathogens and *Aeromonas* species in organic vegetables'. *International Journal of Food Microbiology*, **70**, 155–162.

Michino, H., Araki, K., Minami, S., Takaya, S., Sakai, N., Miyazaki, M., Ono, A. and Yanagawa, H. (1999) 'Massive outbreak of *Escherichia coli* O157:H7 infection in schoolchildren in Sakai City, Japan, associated with consumption of white radish sprouts'. *American Journal of Epidemiology*, **150**, 787–796.

Mukherjee, A., Speh, D., Dyck, E. and Diez-Gonzales, F. (2004) 'Preharvest evaluation of coliforms, *Escherichia coli, Salmonella* and *Escherichia coli* O157:H7 in organic and conventional produce grown by Minnesota farmers'. *Journal of Food Protection*, **67**, 894–900.

Park, C.M., Taormina, P.J. and Beuchat, L.R. (2000) 'Efficacy of allyl isothiocyanate in killing enterohemorrhagic *Escherichia coli* O157:H7 on alfalfa seeds'. *International Journal of Food Microbiology*, **56**, 13–20.

Perez, A.G., Sanz, C., Rios, J.J., Olias, R. and Olias, J.M. (1999) 'Effects of ozone treatment on postharvest strawberry quality'. *Journal of Agricultural and Food Chemistry*, **47**, 1652–1656.

Piper, P., Calderon, C.O., Hatzixanthis, K. and Mollapour, M. (2001) 'Weak acid adaptation: the stress response that confers yeasts with resistance to organic acid food preservatives'. *Microbiology*, **147**, 2635–2642.

Rajkowski, K.T. and Thayer, D.W. (2000) 'Reduction of *Salmonella* spp. and strains of *Escherichia coli* O157:H7 by gamma radiation of inoculated sprouts'. *Journal of Food Protection*, **63**, 871–875.

Raso, J. and Barbosa-Canovas, G.V. (2003) 'Nonthermal preservation of foods using combined processing techniques'. *Critical Reviews in Food Science and Nutrition*, **43**, 265–285.

Robinson, I. and Adams, R.P. (1978) 'Ultra-violet treatment of contaminated irrigation water and its effect on the bacteriological quality of celery at harvest'. *Journal of Applied Bacteriology*, **45**, 83–90.

Rodgers, S.L., Cash, J.N., Siddiq, M. and Ryser, E.T. (2004) 'A comparison of different chemical sanitizers for inactivating *Escherichia coli* O157:H7 and *Listeria monocytogenes* in solution and on apples, lettuce, strawberries, and cantaloupe'. *Journal of Food Protection*, **67**, 721–731.

Ross, A.I., Griffiths, M.W., Mittal, G.S. and Deeth, H.C. (2003) 'Combining nonthermal technologies to control foodborne microorganisms'. *International Journal of Food Microbiology*, **89**, 125–138.

Sagoo, S.K., Little, C. and Mitchell, R.T. (2001) 'The microbiological examination of ready-to-eat organic vegetables from retail establishments in the United Kingdom'. *Letters in Applied Microbiology*, **33**, 434–439.

Sagoo, S.K., Little, C.L., Ward, L., Gillespie, I.A. and Mitchell, R.T. (2003) 'Microbiological study of ready-to-eat salad vegetables from retail establishments uncovers a national outbreak of salmonellosis'. *Journal of Food Protection*, **66**, 403–409.

Sengun, I.Y. and Karapinar, M. (2004) 'Effectiveness of lemon juice, vinegar and their mixture in the elimination of *Salmonella* Typhimurium on carrots (*Daucus carota* L.)'. *International Journal of Food Microbiology*, **96**, 301–305.

Sengun, M.Y. and Karapinar, M. (2005a) 'Effectiveness of household natural sanitizers in the elimination of *Salmonella* Typhimurium on rocket (*Eruca sativa* Miller) and spring onion (*Allium cepa* L.)'. *International Journal of Food Microbiology*, **98**, 319–323.

Sengun, M.Y. and Karapinar, M. (2005b) 'Elimination of *Yersinia enterocolitica* on carrots (*Daucus carota* L.) by using household sanitisers'. *Food Control*, **16**, 845–850.

Seymour, I.J., Burfoot, D., Smith, R.L., Cox, L.A. and Lockwood, A. (2002) 'Ultrasound decontamination of minimally processed fruits and vegetables'. *International Journal of Food Science and Technology*, **37**, 547–557.

Sharma, R. (2005) 'Ozone decontamination of fresh fruit and vegetables'. in Jongen, W., *Improving the Safety of Fresh Fruit and Vegetables,* Woodhead Publishing, Cambridge, 373–386.

Skandamis, P.N. and Nychas, G.J. (2000) 'Development and evaluation of a model predicting the survival of *Escherichia coli* O157:H7 NCTC 12900 in homemade eggplant salad at various temperatures, pHs, and oregano essential oil concentrations'. *Applied and Environmental Microbiology*, **66**, 1646–1653.

Solomon, E.B., Yaron, S. and Matthews, K.R. (2002) 'Transmission of *Escherichia coli* O157:H7 from contaminated manure and irrigation water to lettuce plant tissue and its subsequent internalization'. *Applied and Environmental Microbiology*, **68**, 397–400.

Sy, K.V., McWatters, K.H. and Beuchat, L.R. (2005a) 'Efficacy of gaseous chlorine dioxide as a sanitizer for killing *Salmonella*, yeasts, and molds on blueberries, strawberries, and raspberries'. *Journal of Food Protection*, **68**, 1165–1175.

Sy, K.V., Murray, M.B., Harrison, M.D. and Beuchat, L.R. (2005b) 'Evaluation of gaseous chlorine dioxide as a sanitizer for killing *Salmonella, Escherichia coli* 0157: H7, *Listeria monocytogenes*, and yeasts and molds on fresh and fresh-cut produce'. *Journal of Food Protection*, **68**, 1176–1187.

Szabo, E.A., Scurrah, K.J. and Burrows, J.M. (2000) 'Survey for psychrotrophic pathogens in minimally processed lettuce'. *Letters in Applied Microbiology*, **30**, 456–460.

Taormina, P.T. and Beuchat, L.R. (1999) 'Comparison of chemical treatments to eliminate enterohemorrhagic *Escherichia coli* O157:H7 on alfalfa seeds'. *Journal of Food Protection*, **62**, 318–324.

Tschäpe, H., Prager, R., Streckel, W., Fruth, A., Tietze, E. and Böhme, G. (1995) 'Verotoxinogenic *Citrobacter freundii* associated with severe gastroenteritis and cases of haemolytic ureamic syndrome in a nursery school: green butter as the infection source'. *Epidemiology and Infection*, **114**, 441–450.

Uyttendaele, M., Neyts, K., Vanderswalmen, H., Notebaert, E. and Debevere, J. (2004) 'Control of *Aeromonas* on minimally processed vegetables by decontamination with lactic acid, chlorinated water, or thyme essential oil solution'. *International Journal of Food Microbiology*, **90**, 263–271.

Vescovo, M., Torriani, S., Orsi, C., Macchiarolo, F. and Scolari, G. (1996) 'Application of antimicrobial-producing lactic acid bacteria to control pathogens in ready-to-use vegetables'. *Journal of Applied Bacteriology*, **81**, 113–119.

Vijayakumar, C. and Wolf-Hall, C.E. (2002) 'Evaluation of household sanitizers for reducing levels of *Escherichia coli* on iceberg lettuce'. *Journal of Food Protection*, **65**, 1646–1650.

Warriner, K., Ibrahim, F., Dickinson, M., Wright, C. and Waites, W.M. (2003a) 'Interaction of *Escherichia coli* with growing salad spinach plants'. *Journal of Food Protection*, 1790–1797.

Warriner, K., Spaniolas, S., Dickinson, M., Wright, C. and Waites, W.M. (2003b) 'Internalization of bioluminescent *Escherichia coli* and *Salmonella* Montevideo in growing bean sprouts'. *Journal of Applied Microbiology*, **95**, 719–727.

Weissinger, W.R. and Beuchat, L.R. (2000) 'Comparison of aqueous chemical treatments to eliminate *Salmonella* on alfalfa seeds'. *Journal of Food Protection*, **63**, 1475–1482.

Weissinger, W.R., McWatters, K.H. and Beuchat, L.R. (2001) 'Evaluation of volatile chemical treatments for lethality to *Salmonella* on alfalfa seeds and sprouts'. *Journal of Food Protection*, **64**, 442–450.

Wikipedia. (2006) D-value. http://en.wikipedia.org/wiki/D-value. Accessed on 08.11.2006.

Wilderdyke, M.R., Smith, D.A. and Brashears, M.M. (2004) Isolation, identification, and selection of lactic acid bacteria from alfalfa sprouts for competitive inhibition of foodborne pathogens. *Journal of Food Protection,* **67**, 947–951.

Wu, F.M., Doyle, M.P., Beuchat, L.R., Wells, J.G., Mintz, E.D. and Swaminathan, B. (2000) 'Fate of *Shigella sonnei* on parsley and methods of disinfection'. *Journal of Food Protection*, **63**, 568–572.

Zhang, L., Lu, Z. and Wang, H. (2006) 'Effect of gamma irradiation on microbial growth and sensory quality of fresh-cut lettuce'. *International Journal of Food Microbiology*, **106**, 348–351.

Zhao, T., Doyle, M.P. and Besser, R.E. (1993) 'Fate of enterohemorrhagic *Escherichia coli* O157:H7 in apple cider with and without preservatives'. *Applied and Environmental Microbiology*, **59**, 2526–2530.

Zhuang, R.Y., Beuchat, L.R. and Angulo, F.J. (1995) 'Fate of *Salmonella* Montevideo on and in raw tomatoes as affected by temperature and treatment with chlorine'. *Applied and Environmental Microbiology*, **61**, 2127–2131.

21

Fair trade: a basis for adequate producers' incomes, farm reinvestment and quality and safety focused production

Michael Bourlakis, Brunel University, UK and
Catherine Vizard, Newcastle University, UK

21.1 Introduction

This chapter aims to illustrate the role and importance of ethical (fair) trade in the agricultural production chain. Towards that, we will initially analyse the organic market and will illustrate the role of the accreditation schemes and the local certification bodies. This is followed by a section on ethical (fair) trade discussing the evolution of the concept(s) and their association with organic production. The last sections examine the view of the stakeholders and selected supply chain members in relation to ethical (fair) trade, before drawing relevant conclusions.

21.2 Organic market

Globally, it is estimated that the organic market is equivalent to 2% of the total food market, being worth US$11 billion (IIED, 1997; Blowfield, 1999; Robins *et al.,* 2000). The world's largest organic market is within Europe, which in 1997 had an estimated value of US$5 billion (Willer and Yussefi, 2000), the most important organic product groups being vegetables, fruit, potatoes, milk products and cereals (Michelson *et al.*, 1999). In 1999, the Soil Association reported that the UK ranked fifth within Europe in terms of turnover for organic products, which stood at over half a billion US dollars. Together with Switzerland, Denmark and Sweden, the UK market has recorded the highest annual growth rates in organic sales within Europe, estimated to be in excess of 30% (Willer and Yussefi, 2000).

Of the organic food sold in the UK, 70% is imported (Soil Association, 1999). The majority of organic food imports into the UK come from other European countries. This particularly applies to cereals (with the exception of rice and milk products). However, many certified organic products originate from countries outside Europe. Developing countries supply much of the demand for vegetables and herbs, rice, fresh fruits and raw materials for beverages (tea and coffee as well as fruit juices such as orange, pineapple and mango).

Mainstream retailers have become increasingly interested in organic markets in recent years. In the UK, supermarkets are currently expanding the number of organic products, and increasing investment campaigns advertising organic food. The expansion of the organic market in Europe therefore provides a significant opportunity for producers to benefit in developing countries, especially from those countries, such as the UK, where the market for certified organic food continues to expand (Dolan *et al.,* 1999).

21.2.1 Accreditation schemes

In contrast to other food commodity networks, the need for organic certification is integral in organic food trade. Voluntary accreditation schemes such as the International Federation of Organic Agricultural Movements (IFOAM) make most policy recommendations to governments. IFOAM has a membership of about 700 organisations including research bodies, certification organisations, education bodies and growers. The aim of this international organisation is to promote the organic movement, and as such it provides a forum and publishes basic standards as well as awarding its own accreditation. Organisations and their standards of production can achieve accreditation through the IFOAM Accreditation Programme.

The development of national laws regulating organic farming has been influenced significantly by the standards set by IFOAM. This includes Regulation 2092/91 and the Codex Alimentaris guidelines that were set up by the Food and Agriculture Organisation of the United Nations (FAO) and the World Health Organisation (WHO). The aim of IFOAM is to establish international standards that will act as a minimum throughout the world.

21.2.2 Local certification bodies

Local certification has a number of advantages, not least that it is one way to reduce costs to producers in developing countries via locally determined fees reflecting local incomes (Barrett *et al.*, 2001). To be accepted by the European Union (EU), local certification bodies are required to demonstrate that their standards of organic production and inspection are equivalent to EU regulations. The standards need not necessarily be identical, however, and as such this means more locally appropriate standards can be set in place. For example, local certification bodies may well allow the use of such natural pesticides that would not normally be allowed under EU Standards (Myers, 2000).

There are a number of successful local certification bodies in developing countries, such as Bio Latina based in Peru. Compared to the certification fees charged by European agencies, those charged by Bio Latina are significantly lower. It has been estimated from the results of a study in Bolivia that, when local certification replaces overseas inspection bodies, certification costs can be reduced by 50% (UNCTAD, 1996). There are, however, few or no local certifiers in the poorest countries or in sub-Saharan Africa. In the poorest developing countries, there has been little progress made towards developing such certification protocols. The next section builds on this analysis and focuses on the fair/ethical trade movement, including the provision of the relevant associations with organic trade.

21.3 Ethical (fair) trade

21.3.1 Ethical trade

Blowfield (1999) describes ethical trade as the adoption of societally and environmentally responsible strategies within the value chain, the monitoring and verification of these strategies and the reporting of societal and environmental performance to key stakeholders. The term 'ethical trade' is, however, acknowledged as being ambiguous. In a study by Browne *et al.* (2000) where a number of stakeholders were interviewed about their perception of ethical trade, the results indicated that no strict definition of ethical trading exists. The concept of ethical was reported to mean different things depending on the respondent. The study did, however, establish that there are a number of defining principles. These can be grouped into three broad areas:

1 People centred:
   - child labour: minimum working ages taking account of local legislation and cultural tradition
   - wages: non-exploitation
   - conditions: reasonable working conditions
   - equality: equal pay for equal work
   - worker organisation: freedom of association (trade unions)
   - management systems: product quality and effective control through monitoring and auditing.
2 Environmental sustainability:
   - sustainable environmental practices
   - non-degrading environmental practices, e.g. reducing pollution, food miles, etc.
3 Animal centred – animal rights and welfare
   - no animal testing
   - non-exploitative practices.

These defining principles have been used by Browne *et al.* (2000) to develop a composite definition of ethical trading as follows:

Ethical trading is defined as trading in which the relationship between the interested parties is influenced by concern for some or all of:

- workers' pay, rights and conditions including health and safety, non-exploitative and non-discriminatory labour practices for men, women and children, and effective monitoring and auditing procedures
- producer livelihoods including fair prices and a commitment to social development
- sustainable production methods, which engender sustainable environmental and developmental practices
- animal welfare including non-exploitative practices and humane treatment.

Ethical trade is a generic term that can be applied to schemes which employ sets of social and/or environmental values within the production and marketing process. This can be in terms of *inter alia* human rights, worker welfare, producer livelihoods, sustainable production methods, animal welfare and biodiversity (Blowfield, 1999). In the UK especially, ethical trade is used to describe a specific approach in the monitoring of issues such as worker welfare and human rights. As a result, a debate has evolved concerning the similarities and differences between ethical trading and fair trade as well as other initiatives involved with the social impacts of trade. Some or all of these issues are addressed in fair trade schemes, organic schemes and in the in-house codes of major western retailers and brand owners, which will be discussed in the following sections.

21.3.2 Fair trade

The European Fair Trade Association (EFTA) describes the objectives of fair trade as aiming to contribute to the alleviation of poverty in southern countries through establishing a system of trade allowing marginalised producers in the south to gain access to northern markets (EFTA, 1998). As a result, fair trade is mainly concerned with the treatment of producers and workers within agricultural production systems. Although both social and environmental standards are encompassed which are not normally associated with conventional trade, animal welfare or production methods that are environmentally benign are not guaranteed by fair trade initiatives (Browne *et al.*, 2000). A wider concept of fair trade is, however, evolving to include not only fair trade agreements and safe working conditions for disadvantaged producers and employees, but also sustainable and environmentally safe management of natural resources (Browne *et al.*, 2000).

The fair trade movement began in the late 1950s when Oxfam started retailing products that were manufactured by Chinese refugees. In 1964, Oxfam created the first alternative trading organisation (ATO). In The Netherlands similar initiatives were established and the importing organisation, Fair Trade Organisatie, was founded in 1967. In addition, Dutch third-world groups began to retail sugar cane and in 1969 the first fair trade shop was opened (EFTA, 1998).

In 1990, EFTA was established representing 12 organisations importing fair trade products. As a proportion of fair trade imports, EFTA represents around 60% of European market. Fair trade shops expanded in Europe after the 1970s and now there are more than 3000 in 18 European countries. The International Federation of Alternative Trade (IFAT) was established in 1989, as an umbrella organisation for ATOs from Africa, Asia, Australia, Europe, Japan and North and South America (EFTA, 1998).

Fair trade exploited the mainstream channels by focusing on selling to institutional outlets. In 1988, the first fair trade label, Max Havelaar, was established in The Netherlands. This seal of approval was awarded to conventional businesses that respected fair trade standards that were open to external monitoring. Other labels such as Transfair International (co-founded by EFTA) and the Fairtrade Foundation, have also evolved. The International Fair Trade Labelling Organisation (FLO) has coordinated all fair trade labelling since April 1997 (EFTA, 1998).

The outcome of the above is that, nowadays, the fair trade organisations represent an important alternative trade market (Raynolds *et al.*, 2004). It is reported that international fair trade sales are worth over US$ 500 million and are expanding at approximately 30% per annum (Fair Trade Federation, 2003 in Raynolds *et al.*, 2004). There has been a shift in recent years as fair trade product lines have been expanded by trade networks, moving from mainly handicraft items into a range of food products. Today, sales of fair trade products are mainly in the food sector, including items such as coffee, bananas, cocoa and tea, rice, honey, sugar, fruit juices and fresh fruit (FLO, 2004 in Raynolds *et al.*, 2004). However, the most important fair trade product is reported to be coffee. Almost 16 000 tonnes of fair trade certified coffee was purchased by consumers in 17 countries in 2002. In some countries, sales of fair trade coffee are rising at approximately 50% per annum (Raynolds *et al.*, 2004). The European movements have been generally more successful than initiatives based in the United States. In 2000, 3% of the total coffee market was accounted for by fair trade certified coffee in The Netherlands, Luxembourg and Switzerland, approximately 1% of the total market in most other European countries, but merely 2% in the United States (Levi and Linton, 2003). Raynolds (2002) reports that, although the fair trade market in the United States is not as well developed as in the European market, it is expanding more rapidly.

21.3.3 Associations between organic and ethical (fair) trade

The idea of organic agriculture has different origins from the concepts of ethical and fair trade. In the UK, organic agriculture has been developing since the 1930s and, since the early 1970s, certified organic produce has been available. There are a number of principles involved in organic agriculture including concerns for safe food production, for the environment, for animal welfare and for issues of social justice (Browne *et al.*, 2000). Both crop and animal production, as well as a wide range of processed foods, are covered by organic standards. UK and European legislation incorporates these principles and, to some extent, they are also regulated by international accreditation, with the current exception of social justice (Browne *et al.*, 2000).

There is, however, an increasing overlap between ethical and organic trading. An expanding number of fairly traded goods are also organic and, in addition, there is a move by the organic sector towards the inclusion in its standards of social rights and fair trade (Quested, 1988 in Browne *et al.*, 2000). There is evidence to suggest that, in the UK, the fair trade and organic movements are moving closer together (Quested, 1988 in Browne *et al.*, 2000). However, these links are unlikely to result in a merging of organic and ethical production considering their major differences (see Table 21.1).

21.4 View of stakeholders and key supply chain members

There are a number of stakeholders involved in promoting ethical (fair) trade initiatives. Liu *et al.* (2004) suggest that the stakeholders include *inter alia* farmer organisations, private companies including leading food multinationals, government agencies, donors, aid agencies, consumer associations, trade unions, research institutes, certification bodies and other non-government

Table 21.1 Contrast between organic production and ethical trade

	Organic	Ethical
Origins	As a method of agricultural production, originating in the 1930s	As a description of trade between the developed and developing worlds, becoming widely used in the 1990s
Focus	Focus on agricultural production systems that utilise biological rather than chemical inputs	Focus on people's working conditions, especially in the developing world
Development	Not a development issue but concerned with sustainability of farming systems	A development issue and may contribute to livelihood enhancement
Standards	Universal production standards, assured through accreditation and inspection	No universal standards. Voluntary codes of conduct and self-regulation becoming more common
Certification	Yes, based on regulation by the state. Assured by legally registered labelling symbols on marketed produce	Yes for fair trade; no for ethical. No legal status for ethical claims on marketed produce

Source: Browne *et al.* (2000).

organisations (NGOs), all active in sustainable agriculture. These organisations and a number of other non-governmental organisations are largely responsible for the development of most voluntary social and environmental standards. Some of the key stakeholders and their impact on the supply chain are discussed by Blowfield (1999) as follows:

- Buyers: their purchasing policies have a direct impact on social and environmental performance of exporters (market price, quality standards, late payments);
- Government: policies affecting export industries vary from country to country;
- Banks and Financial Institutions: high interest rates and difficulty getting loans for investment in social infrastructure are real constraints faced by exporters;
- Packaging companies, input suppliers: strict standards on health and safety and the environment need to be supported by high standard chemicals and equipment, properly packaged and available in quantities that encourage safe and proper application;
- Freight companies: freight can be the largest single cost borne by many exporters, especially air freight. Exporters claim that they are not in a strong position to negotiate freight rates with airlines.

The above resulted in the formation of the Ethical Trading Initiative (ETI) in the UK in 1997. This multi-stakeholder organisation operates as a civil initiative funded by the Department for International Development and is endorsed by the Department for Trade and Industry. The aim of the ETI is to 'develop and encourage the use of a widely endorsed set of standards, embodied in codes of conduct, and monitoring and auditing methods which will enable companies to work together with other organisations outside the corporate sector to improve labour conditions around the world' (ETI, 1997 in Browne *et al.,* 2000). Although the term 'ethical' is not specifically defined by the ETI, it is related to labour and employment conditions and advocates for the implementation of commonly agreed standards on an international level (ETI, 1999 in Browne *et al.,* 2000).

21.4.1 Supermarkets

Supermarkets play a significant role in influencing the demand for ethically traded and organic goods, owing to their dominant market position in the UK food chain (Dawson, 2004). It is predicted that the majority of consumers in the UK will soon be buying goods that are described as ethically traded. This is due to the current move towards a greater level of ethics within supermarket trading steered by the ETI (Browne *et al.*, 2000). Although auditing of the regulatory and inspection framework in the short term may not be independent, consumers will be assured that they are buying ethically in their supermarket (Browne *et al.,* 2000).

Over recent years within the supermarket corporations, there has been an increased focus on the issue of ethical trading. Most supermarkets would report that they trade ethically, although some use other terminology including 'socially responsible sourcing' or 'sound sourcing'. Systems throughout the supply chain are being put into place by all supermarkets to check labour conditions and, where relevant, environmental conditions (Browne *et al.*, 2000). Here, the system of self-regulation is seen as preferable to a standard or state regulated labelling system for 'ethical' or its equivalent by supermarket corporations. Self-regulation encompasses systems assured through contracts, codes of conduct and rigorous product specification, which are backed up by complete traceability of produce (Browne *et al.*, 2000). However, this system does not guarantee transparency nor the implementation of consistent ethical standards, or its labelling (see House of Commons, 1999 in Browne *et al.*, 2000).

The main way in which the UK food retailers have responded to the pressure to change their sourcing practices in line with more ethical principles has been through their membership of the ETI, where the base code applies to suppliers of the retailers' own-brand products (Hughes, 2005). The practice of checking and monitoring producers' performance against the nine clauses of the ETI Base code is termed social auditing. This involves monitoring labour issues by checking producers' records and documentation. In-depth interviews are also undertaken with workers exploring their workplace experiences. As this requires specialist skills, independent auditors have entered the food supply chain network as a new set of actors (Barrientos, 2002). There is, however, significant variation between companies concerning the way in which the independent auditors are coordinated within the management system of retailer firms. Hughes (2005) suggests that there are three modes of organisation which appear to be used for the social auditing of retailers' global supply chain: the arm's-length approach, the coordinated approach and the developmental approach.

1 The arm's length approach: as a result of lack of funding associated with ethical trade, there is only limited use of third-party auditors. However, the partial use of independent auditors does occur mainly within the context of 'arm's length' relationships between retailers, auditors and suppliers. This can be defined as the contractual relations between companies involving competitive bidding and playing-off of suppliers. Weak social ties and detached social relations are characteristic of such relationships (Doel, 1996, 1999).
2 The coordinated approach: contracting relationships built on collaboration, trust and close interpersonal ties define the coordinated approach. This approach means that the producer inherits the responsibility and costs of social auditing as they are passed down the supply chain. This is achieved, however, within the framework of close retailer–supplier relationships. Although subcontracting exists, known networks of traceable suppliers are involved. As such, it is possible to exert considerably greater control over the ethical trading process.

3 The developmental approach: local NGOs as well as large independent auditing corporations, undertake social audits for a minority of firms. This has come about because of the limitations related to conducting overseas social audits. As a consequence, close retailer–supplier relationships are further broadened by strong links with stakeholders. As such, workers receive some representation and the respective benefits through NGOs who are involved with social auditing.

We must stress that social auditing requires that *all* factories supplying finished goods are formally assessed against the standards outlined in the ETI base code. This is achievable as a result of tight control over its supply chains. Nevertheless, the disadvantage is that suppliers themselves conduct the audits. The contracting of audits to large audit companies occurs only when site assessments of new producers entering the retailer's supply chain need to be produced and when a sample of suppliers' self-audit reports require external verification (Hughes, 2005).

21.4.2 The ethical consumer

In a bid to define the ethical consumer, Browne *et al.* (2000) proposed three categories of shoppers:

- True ethical consumers: those who will go out of their way to buy on a cause-related basis (suggested to be about 2% of the population).
- Semi-ethical consumers: those who shop at supermarkets and will buy fairtrade coffee or organic produce sometimes. This is because they accept its claims and are prepared to pay a modest premium (suggested to be between 20–30% of the population).
- Would-be-ethical consumers: shoppers who would be ethical if there was no price premium and no special effort was required (suggested to be 80%).

Browne *et al.* (20000 have also classified the concerns of ethical consumers into four groups as follows:

- What is in the food and how it could affect their own and their families' health
- How the food is produced and what the effects are on the environment
- Issues relating to animal welfare
- The exploitation of people involved in food production in the developing world.

Those who are willing to pay extra for goods traded in an ethical manner do so based on the understanding that significantly improved livelihoods result from the premium paid for fair trade and/or organic certified produce. Statements on packaging of ethically traded products relating to the impact on the livelihoods of producer communities serve to link consumption and production for the consumer.

21.4.3 Growers/producers

There is significant variation in fair trade producer groups, not only between regions, but also between different fair trade products. EFTA (1998) lists groups as federations of producers, cooperatives, family units, workshops for handicapped people, state organisations, private companies, and, increasingly, northern groups producing goods in the context of the social economy. Consequently, the size of the producer group is highly variable. Large networks can represent thousands of producers, whereas some outfits can be small workgroups of about 20 people. In principle, producers joining in the fair trade initiative receive a number of benefits including the following:

1. Direct access to an otherwise inaccessible market.
2. A fair price is guaranteed. The producers gain a larger share of the final price and the fair trade brand itself commands a price premium. Producers can decide how this should be utilised within the community. EFTA (1998) quotes benefits such as product improvement, improving the financial, technical or managerial capacity of their organisations, or improving farms or buildings that can also be used for education, health care, housing, etc.
3. An increased likelihood of a fair price for non-fair trade mainstream production. Even if a fair price is paid for only a small part of production, there is often a snowball effect with higher prices paid for the rest of production.
4. The provision of credit: as price advances from fair trade organisations are common, it means that those producing goods on a small scale can buy inputs for production. This invariably increases stability of production and income.
5. The provision of technical assistance. many ATOs offer training and development on issues such as production methods and financial and managerial affairs.
6. Opportunities for producers to network through meetings organised by fair trade organisations.

The most recognised of these benefits is the guaranteed fair price, and Bacon (2005) mentions that farmers who are members of cooperatives received higher average prices and enjoyed extra security. However, there are a few key factors dictating the success of producers and producer groups in the fair trade market. For the producers, their willingness and ability to engage successfully in fair trade is shaped by their social and economic characteristics (Raynolds *et al.*, 2004). The majority of producers have a basic schooling often comprising less than four years in education. This is coupled with limited understanding of international markets and often no ability to communicate via international market languages. As a consequence, producers often have difficulty in meeting trade expectations and maintaining viable, democratically based, organisations. In addition, Raynolds *et al.* (2004) highlight the importance of the quality of the producers' land, labour and capital resources in establishing and maintaining a fair trade initiative, as these factors influence the quality and quantity of export products. For the producer groups, Raynolds *et al.* (2004) suggest that strong external ties with organisations such as developmental NGOs and corporate buyers should be created and maintained.

21.5 Conclusions

This chapter has shed light on the increasing role of ethical (fair) trade and its relationship with organic production and, subsequently, their key associations and differences have been illustrated. The authors expect the above discussion to prove beneficial and to provide further knowledge in this field of study. We also expect further research to be undertaken that could validate and/or contradict some of the key arguments cited in this work.

21.6 References

Bacon, C. (2005). 'Confronting the coffee crisis: can fair trade, organic trade, and speciality coffees reduce small-scale farmer vulnerability in Northern Nicaragua?' *World Development*, **33**, 497–511.

Barrett, H.R., Browne, A.W., Harris, P.J.C. and Cadoret, K. (2001). 'Smallholder farmers and organic certification: accessing the EU market from the developing world'. *Biological Agriculture and Horticulture*, **19** (2), 183–199.

Barrientos, S. (2002). 'Mapping codes through the value chain: from researcher to detective', in *Corporate Responsibility and Labour Rights: Codes of Conduct in the Global Economy* Jenkins, R., Pearson, R. and Seyfang, G. Earthscan, London and Stirling, VA, 61–76.

Blowfield, M. (1999). *Ethical Trade: A Review of Developments and Issues.* Unpublished report for NRI, University of Greenwich, UK.

Browne, A.W., Harris, P.J.C., Hofny-Collins, A.H., Pasiecznik, N. and Wallace, R.R. (2000). 'Organic production and ethical trade: definition, practice and links', *Food Policy*, **25**, 69–89.

Dawson, J.A. (2004). 'Food retailing, wholesaling and catering', in Bourlakis M. and Weightman P., *Food Supply Chain Management*, Blackwell, Oxford, 116–135.

Doel C. (1996). 'Market development and organisational change: the case of the food industry', in Wrigley, N. and Lowe, M. *Retailing, Consumption and Capital: Towards the New Retail Geography* Longman, Harlow.

Doel C. (1999). 'Towards a supply chain community? Insights from governance processes in the food industry', *Environment and Planning A*, **31**, 69–85.

Dolan, C., Humphry, J. and Harris-Pascal, C. (1999). *Horticulture Commodity Chains: The Impact of the UK Market on African Fresh Vegetable Industry*, IDS Working Paper 96, University of Sussex, UK.

EFTA (1998). *Fair Trade Yearbook. Towards 2000*, Dunk in de Weer, Ghent, Belgium.

ETI (1997). *Ethical Trading Initiative – Information Pack: Everything You Need to Know about the Ethical Trading Initiative*, Ethical Trading Initiative, London.

ETI (1999). *'Learning from Doing' Review*, Ethical Trading Initiative, London.

Fair Trade Federation (2003). *2003 Report on Fair Trade Trends in US, Canada and the Pacific Rim.* Fair Trade Federation, Washington DC.

FLO (Fairtrade Labelling Organizations International) (2004). *About FLO*. http://www.fairtrade.net/sites/aboutflo/aboutflo.html.

House of Commons (1999). *Ethical Trading, Sixth Report of the Trade and Industry Committee*, The Stationery Office, London.

Hughes, A. (2005). 'Corporate strategy and the management of ethical trade: the case of the UK food and clothing retailers', *Environment and Planning A*, **37** (7), 1145–63.

IIED (International Institute for Environment and Development) (1997). *Changing Consumption and Production Patterns: Unlocking Trade Opportunities*, Robbins, N. and Roberts, S. (eds), United Nations Department of Policy Co-ordination and Sustainable Development, United Nations, New York, USA.

Levi, M. and Linton, A. (2003). 'Fair trade: a cup at a time?' *Politics and Society*, **31** (3), 407–432.

Liu, P., Andersen, M. and Pazderka, C. (2004). *Voluntary Standards and Certification for Environmentally and Socially Responsible Agricultural Production and Trade*, FAO Commodities and Trade Technical Paper No 5, Commodities and Trade Division, Food and Agricultural Organization of the United Nations.

Michelson, J., Hamm, U., Wynen, E. and Roth, E. (1999). *The European Market for Organic Products: Growth and Development. Organic Farming in Europe: economics and policy*, Vol 7, Universitat Hohenheim, Germany.

Myers, J. (2000). Independent consultant (inspection experience), personal communication.

Quested, H. (1998). 'No easy choices', *Living Earth*, **197**, 14–15.

Raynolds, L.T. (2002). *Poverty Alleviation through Participation in Fair Trade Coffee Networks: Existing Research and Critical Issues*, Background paper prepared for project funded by the Community Resource Development Program, The Ford Foundation, New York. http://www.colostate.edu/Depts/Sociology/FairTradeResearchGroup.

Raynolds, L.T., Murray, D. and Leigh Taylor, P. (2004). 'Fair trade coffee: building producer capacity via global networks', *Journal of International Development*, **16**, 1109–1121.

Robins, N., Roberts, S. and Abbot, J. (2000). *Who Benefits? A Social Assessment of Environmentally Driven Trade*, IIED, London.

Soil Association (1999). *The Organic Food and Farming Report 1999*, Soil Association, Bristol, UK.

UNCTAD (1996). *Organic Production in Developing Countries: Potential for Trade, Environmental Improvement and Social Development*, UNCTAD/COM/88, Geneva, Switzerland.

Willer, H. and Yussefi, M. (2000). *Organic Agriculture Worldwide*, Stifung Ökologie & Landbau, Bad Dürkheim, Germany.

22

Development of quality assurance protocols to prevent GM-contamination of organic crops

R. C. Van Acker, University of Guelph, Canada; N. McLean, and R. C. Martin, Nova Scotia Agricultural College, Canada

22.1 Introduction

Genetic engineering (GE) is a truly novel technology, which allows for the inclusion of almost any trait imaginable into crop plants to serve all manner of desired functions and end uses (Tolstrup *et al.*, 2003). Since the commercial introduction of GE crops (commonly referred to as genetically modified (GM) crops), the global area seeded to GM crops has risen rapidly, reaching 102 million hectares in 2006 (ISAAA, 2006). In countries such as Canada and the United States (USA), farmer adoption levels of GM crops have been high. In Canada, more than 75% of the canola grown in 2004 was GM, while GM soybean and corn crop acreages represent over 60% of total acreage. In 2004 in the USA, over 80% of the soybeans grown were GM and almost 80% of the cotton grown was GM. Although GM crops are registered for unconfined release in countries like Canada and the USA, they continue to be a concern in countries where GM crops are not yet registered for unconfined release. In addition, because GE allows for the realization of truly extraordinary traits in crop plants, it can also produce novel and unexpected risks. As GM crop development proceeds, more unique traits are introduced into crop plants, including transgenes which encode for pharmaceutical proteins (USDA, 2003). The release of these types of traits into the environment is truly novel. Most risks related to the release of GM crops are related to transgene movement, which remains relatively poorly understood and has been studied to only a very limited extent (Marvier and Van Acker, 2005). This is especially true for the intraspecific (within species) movement of transgenes within and among farming systems (NRC, 2004; Tolstrup *et al.*, 2003).

For organic farmers and low input farmers serving certain markets, there is a requirement to maintain their produce free from transgenes (GM-free) in order to meet customer expectations. For these farmers, there is a need to understand the movement of transgenes in order that they may prevent transgene movement into their systems and maintain the product qualities they are expected to deliver.

Genetic engineering holds much promise for farmers, consumers and the biotechnology industry, but the exploitation of GM crops will require responsible introduction which, in turn, requires the creation of effective and acceptable transgene confinement protocols. These protocols must be based on knowledge of the nature and interaction of those factors which contribute to transgene movement and a realistic consideration of the cooperation required to make confinement effective (Tolstrup *et al.*, 2003). The protocols must also be based on the understanding that the movement of transgenes beyond their intended destinations under current agri-food production and handling systems is a certainty and, that once transgenes have escaped into the environment, it is unlikely that they can be absolutely retracted. In order to be administered effectively, the protocols must include the assignment of responsibilities for transgene confinement, which are enforced through law.

22.2 Terminology

Preventing transgenes from GM crops appearing where they are not wanted or intended is a novel problem in agriculture. The terminology associated with this challenge can be novel in its meaning and the meaning is important because the terms are used to convey specific ideas or to support certain positions. In the context of the problem of preventing transgene movement, the following terms are important to consider.

22.2.1 Transgenes

In strict terms, transgenes include the genetic material which has been moved into the GM crop via recombinant DNA techniques (GE technology). The transgene(s) encodes for a protein (or proteins), which deliver the intended trait (GM trait). More broadly, transgenes include all of the genetic material included within the DNA construct (the genetic vehicle for transformation) and, in addition to the DNA which specifically encodes for intended GM trait-related proteins, it includes the promoter sequence(s) which regulate the transcription of the trait-related DNA and antibiotic resistance genes (Kingsman and Kingsman, 1988).

22.2.2 Containment

Containment means to hold within a volume or area. In the context of transgenes movement, this term is used when there is a high level of confidence that there will be no transgene escape from the containment space or area.

Containment also brings connotations of employing active and perhaps physical means of preventing escape as in the terms 'container' or 'containment facility'. This term might be most suitably used in the context of bioreactor facilities or self-contained, indoor research and development facilities.

22.2.3 Confinement

Confinement means to keep within bounds. In the context of the prevention of transgene movement, the word confinement has been used by the National Research Council in the USA (2004) because it is less absolute than containment. Confinement suggests the maintenance of a border around a confinement area, with the implication that borders are porous to various extents. Confinement reflects the biological reality that, when GM organisms are grown outside, there will always be some level of transgene escape from given confinement areas.

22.2.4 Contamination

In the context of GM crops, if a given crop contains transgenes which render it unusable in a manner which requires it to be destroyed, then the given crop is said to have been contaminated by the transgenes. For example, for transgenes that encode for proteins that pose a known direct threat to human or environmental health, presence in any crop outside of a confined production and processing stream would render those crops contaminated. However, for transgenic events (traits) (e.g. GM crops) that are registered for unconfined release within a given jurisdiction, transgene escape to non-GM crops within that jurisdiction does not generally make the non-GM crop completely unusable. In this scenario, a reference to contamination is not technically correct. Nonetheless, the term contamination is being used more generally to refer to the presence of unwanted transgenes in a given crop whether or not the 'contaminated' crop is, or is not, rendered completely useless by the contamination. For example, organic crops which cannot be sold as organic because of the presence of transgenes are commonly said to be 'contaminated' with transgenes (Friesen *et al.*, 2003) even though they can still be sold and used as conventional food or feed.

22.2.5 Adventitious presence

According to the government of Canada, adventitious presence in the context of GM crops is defined as 'the unintended, technically unavoidable presence of genetically engineered material in an agri-food commodity' (CFIA, 2005). The phrase 'technically unavoidable' might more accurately be replaced by the phrase 'technically difficult-to-avoid'. Adventitious presence is the technically correct term to use when transgenes which have been registered for unconfined release in a given jurisdiction have found their way into a non-GM crop within that jurisdiction. It is, however, not the correct term to use if that same non-GM crop is exported to a jurisdiction where the transgene is not registered for unconfined release. In this case, contamination is probably a more accurate term.

22.2.6 Exclusion zone

An area over which there is political jurisdiction by a ruling body and which, in the context of GM crops, has been legally ruled to be kept free from the growing of GM crops. The establishment of exclusion zones may limit transgene presence but it does not guarantee freedom from transgenes. An exclusion zone may range in scale from national (e.g. Austria) or multinational (e.g. the European Union (EU)) to sub-national (e.g. GM crop moratoria in states in Australia) and sub-regional within sub-national jurisdictions (e.g. municipalities within provinces in Canada) (Anonymous, 2005).

Globally, legally permanent GM crop or transgene exclusion zones are rare because there remain political and legal efforts challenging the validity of the arguments being used to establish such zones. In Canada, for example, GM crops are not regulated *per se* because Canada subscribes to the notion of substantial equivalence (see definition below) between GM and non-GM crops. In addition, where there are concerns about transgene (trait) movement, the regulatory body in Canada is only allowed to regulate on the basis of human health or environmental risk and not economic risk. Similar situations exist in other countries such as the USA.

22.2.7 Threshold

In the context of GM and non-GM crop coexistence or segregation, a threshold is the limit of transgene presence above which crops (or final products) are considered to no longer be 'GM-free'. Thresholds are only useful to business entities within the agri-food industry (including farms and farm organizations) if the thresholds are set within law. Thresholds for transgene presence may be set by organizations, such as organic certification agencies, but they must be recognized in law within the political jurisdiction where that agency is functioning, if there is to be any enforcement of the threshold or recourse in the event that the threshold has been exceeded. A good example of this is the fact that the EU has established a transgene threshold in law while, in Canada, the Saskatchewan Organic Directorate's right to no threshold ('zero threshold') for transgene contamination of organic crops has not yet been recognized in Canadian law (Cullet, 2005).

22.2.8 Seed purity

Traditionally, seed purity referred to the level of foreign material within a given seed lot that has been regulated by some sort of legally recognized certification body. Foreign material includes anything that was not the seed itself, including, for example, weed seeds, ergot bodies, straw and soil (CSGA, 2005). Seed purity standards (and varietal purity standards) also include lower limits for genetic purity with respect to assurances that the seed (and given variety) is sufficiently genetically pure to ensure that the seed (and variety) is 'fit for purpose'.

These definitions and regulations traditionally served the farm and seed industries well because the service requirements for seed or variety purity were related to meeting agronomic standards or processing quality end-uses. In this context, absolute (or even narrow limits of) genetic purity were not necessarily required to meet the end uses and deem the seed or variety 'fit for purpose'. With the development of GM crops, the notions of seed and varietal purity are being reconsidered because 'fit for purpose' must now include a consideration of transgene contamination. The international seed industry argues in support of an update of seed and varietal purity notions within the context of GM crops, but they also express strong concern over the ability of the seed industry to meet absolute genetic purity standards and the tremendous cost of meeting such standards (ISF, 2004). In the context of this issue it is worth noting that the International Federation of Organic Agriculture Movements (IFOAM) adopted the position in 2002 that organic certification is a certification of a process of production and, as such, does not imply an end product guarantee. In this sense, IFOAM does not necessarily support *de minimis* threshold levels ('zero thresholds' or minimal testing level thresholds). This creates a challenge for organic farmers who are trying to keep their products 'GM-free' because it is not certain what GM-free means.

22.2.9 Substantial equivalence

In countries such as Canada and the USA, the basis for not having regulation specific to GM crops stems from a decision to regulate on the basis of the product not the process. This decision is based on a belief in substantial equivalence where GM crops are deemed to be substantially equivalent to non-GM crops with respect to regulation. This belief in substantial equivalence is based on a trust in the central dogma of genetics that one gene encodes for one gene product and that GE transfers genetic material that functionally only results in the production of known gene product(s) (Skaftnesmo, 2004). However, many geneticists are moving on from this belief after recognizing some indirect effects of genetic insertions (DeVries and Wackernagel, 2002). There is also a realization that, because of an adherence to substantial equivalence in countries such as the USA and Canada where there has been the greatest adoption of GM crops, the true effects on human health of consuming GM plant material has not been sufficiently tested (Pryme and Lembcke, 2003).

22.3 Examples of transgene escape

Genetically modified crops have been extensively grown in a number of regions around the world. One of the most extensive areas of cultivation of GM crops has been North America and it is in North America where there have been a high number of documented cases of transgene escape (Marvier and Van Acker, 2005). Among all documented cases, intraspecific transgene movement in canola (*Brassica napus* L.) has been the most common. In western Canada, where canola is grown on a large number of acres, GM canola varieties are grown on the majority of those acres. There has been so much intraspecific transgene escape in canola that farmers in this region have come to expect the unintended appearance of transgenes in their canola (Van Acker *et al.*, 2004). Even after only three years of commercial production of GM canola, Hall *et al.* (2000) found that the specific transgenes encoding for different herbicide tolerance traits were stacking within individual volunteer canola plants, giving rise to multiple herbicide-resistant volunteer canola plants. By the year 2000, only 5 years after the initiation of commercial production of GM canola in western Canada, farmers began to complain about the appearance of volunteer glyphosate herbicide-tolerant canola in their fields, even when they had not intentionally sown glyphosate tolerant canola in these fields the previous year. Many of these farmers suspected that the non-glyphosate-tolerant certified canola seed they were using had adventitious presence of glyphosate-tolerant canola seed.

Independent testing of certified canola seed lots from western Canada revealed that the majority of tested seed lots contained at least trace amounts of genetically engineered herbicide tolerance traits. In fact, 97% of the seed lots (32 of 33) tested by Friesen *et al.* (2003) and 59% of the seed lots tested by Downey and Beckie (2002) (41 of 70) had foreign transgenes present at detectable levels (above 0.01%). The source of the adventitious presence of these seed lots was never determined but could have resulted from inadvertent mechanical mixing of certified seed lots during harvest or handling, or pollen-mediated gene flow occurring in earlier generations of pedigreed seed production (i.e. breeder or foundation seed). The high level of adventitious presence of unintended transgenes in pedigreed certified seed lots was disturbing because it showed that stringent seed production segregation systems were not sufficient to prevent significant transgene movement (Friesen *et al.*, 2003).

Hybridization is another means of transgene escape. Most crops have some ability to hybridize with non-crop species in at least some part of their global ranges (Ellstrand *et al.*, 1999). For example, Wilkinson *et al.* (2003) estimated that tens of thousands of canola-weed hybrids are produced each year in the United Kingdom (UK). In theory, unassisted cross-species hybridization should be relatively rare for many species, including many in the Brassicaceae family, because it requires that species mate and produce viable offspring and, if a transgene is exchanged in this mating, its introgression into the wild type is only possible if the hybrid offspring are able to backcross successfully with the original wild type species. Introgression is the introduction of genes from one species into the gene pool of another species, occurring when matings between the two produce fertile hybrids. Stable introgression requires a series of successful backcrosses. The initial hybridization most often results in ecologically unfit offspring and the chances of repeated successful backcrossing are low (Légère, 2005). The rarity of successful introgression depends on the species and there are some crop-weed examples (such as jointed goat grass (*Aegilops cylindrical* L.) and wheat (*Triticum aestivum* L.)) where successful introgression is not that rare (Morrison *et al.*, 2002).

The most troubling examples of transgene escape are those that involve human error because they are so unpredictable. In the USA there have been a number of documented cases of transgene escape involving human error. Perhaps the most famous of these is the 'Starlink' case where corn, engineered to express an insecticidal protein, was approved for animal feed but not human consumption. There was insufficient segregation oversight between food and feed streams in the US bulk commodity handling systems and the insecticidal protein was found in a number of processed foods (Marvier and Van Acker, 2005).

Three years after this discovery and after the execution of a massive recall effort, traces of the Starlink protein could still be commonly found in both food and feed handling streams in the USA (USDA, 2003). The Starlink case showed not only that human error can result in problematic transgene escape, but also that full retraction of transgenes (and their products) from complex and massive commercial food and feed systems is difficult, and perhaps impossible. Another disturbing example of human error is the Prodigene case, also in the USA, where corn that was genetically modified to produce a vaccine that prevents diarrhea in pigs was discovered in a commercial grain elevator in Iowa (Gillis, 2002). Upon investigation, the US Department of Agriculture (USDA) found that the company who owned this GM corn (Prodigene) had failed to comply with federal regulations requiring that the company destroy volunteer GM corn growing in subsequent crops. This error required that 500,000 bushels of contaminated soybean be destroyed and the company was fined US $250,000.

22.4 Implications of transgene escape

Escape of transgenes may or may not constitute risk, but it does allow for the possibility of risk. Generally, regulation relating to trait movement (or the appearance of whole organisms in places where they are not wanted or expected) is based on considerations of environmental and human health risks. There may also be economic (market) risk associated with transgene escape, but this is not generally a basis for regulation. The assessment of environmental risk relating to transgene movement is often focused on the movement and introgression of transgenes into wild species and the impact such altered wild species might have in natural settings. As mentioned, introgression of transgenes into wild species is considered unlikely in most cases, but where it is possible there can be some potentially dramatic impacts. For relatively compatible species pairs such as for example, wheat and jointed goatgrass or wild and domesticated sunflower, the potential for transgene introgression into the wild type population is significant.

The level and type of risk depends then upon the nature of the trait (transgenes). Herbicide tolerance traits conferred through GE may not create environmental risk because they do not necessarily have an ecological impact in areas where the herbicide is not used (Snow *et al.*, 1999), but physiological traits, such as drought or salt tolerance, would perhaps expand the range and invasiveness of some wild type species (Snow *et al.*, 2003). Environmental risks can also result indirectly from transgene escape, including intraspecific transgene escape. For example, Van Acker *et al.*, (2004) showed that the introduction of glyphosate-tolerant wheat in western Canada would have threatened the economics of reduced-tillage farming in this region and thus threatened the documented environmental benefits that come from farmers using these systems.

Human health risks related to transgene escape have not, generally, been considered. For GM crop regulation, the process of GE is not regulated, only the product, and GE products have not been robustly tested for effects on human health as the result of consumption (Pryme and Lembcke, 2003). In the future, there may be a more obvious direct relation of transgene escape to human health as more crop plants are transformed to produce compounds which directly affect human health, including pharmaceutical proteins. In 2002, the US Department of Agriculture (USDA) approved 20 permits for field trials (130 acres on 34 sites) involving plants engineered to produce pharmaceutical proteins (Marvier and Van Acker, 2005).

In a global marketplace where consumers demand choice, transgene escape can also create market risk. This is especially true for farmers and food processors who operate within jurisdictions where there is significant cultivation and or trade in both GM and non-GM crops and commodities. For example, while the GM crops released in Canada are deemed to be an acceptable risk to the environment, their biosafety in other countries has not been determined (Van Acker, 2003). These concerns are specifically addressed in the Cartagena Protocol on Biosafety to the Convention on Biological Diversity (Pratt, 2005). When the introduction of glyphosate-tolerant GM wheat was pending in the USA and Canada, it was reported that all of the domestic millers in both the USA and Canada would be refusing to buy North American wheat because of the market risk associated with potential adventitious presence of GM wheat in non-GM wheat shipments (Patty Rosher, Canadian Wheat Board, personal communication). In addition, in the early 2000s it was commonly reported that the European Union regularly refused commodity shipments from outside the EU if the shipments were not below GM content threshold levels. By these actions, the EU (and other jurisdictions such as Japan) supported a segregated market for GM and non-GM crops and made market risk, especially for exporting nations, a reality.

22.5 Mechanisms of transgene escape

While small-scale experiments prior to GM crop release have measured transgene flow (usually associated with pollen movement or short distance seed/volunteer movement), few studies have addressed post-release monitoring, with notable exceptions (Hall *et al.*, 2000; Downey and Beckie, 2002; Rieger *et al.*, 2002; Friesen *et al.*, 2003; Mauro and McLachlan, 2003; Ma, 2005). At the time of unconfined commercial release of Roundup Ready® (glyphosate-tolerant) canola in Canada, it was known that there was significant potential for out-crossing within the canola (*Brassica napus* L.) genome and that transgene movement from canola crop to canola crop would occur (CFIA, 1995). The eventual appearance of the transgenes conferring glyphosate tolerance in non-Roundup Ready (RR) canola seed lots (Friesen *et al.*, 2003) showed that there was an effective means of transgene movement in operation but the mechanisms of movement were not known. Some of the non-RR canola seed lots had RR transgene adventitious presence levels that were very high (approaching 5%). Given current knowledge of pollen-mediated gene flow in *B. napus*, it is unlikely that pollen flow would cause greater than 0.1% presence in a single generation of pedigreed seed production, given strict seed production protocols. Adventitious presence levels above 0.25% were likely to be the result of inadvertent mechanical mixing of certified seed lots during harvest or handling, or contamination occurring in earlier generations of pedigreed seed production (i.e. breeder or foundation seed) (Friesen *et al.*, 2003).

In the context of practical transgene confinement, the two vectors of transgene movement are pollen and seed. Gene flow via pollen tends to occur on a small scale, generally, but pollen can be carried long distances by wind or pollinators (for some species) (e.g. Rieger *et al.*, 2002). The distance for effective pollen-mediated gene flow (PMGF) depends on a great many factors including to what extent the species will outcross, the size and weight of its pollen, the size of the pollen source and the weather (in relation to movement of the pollen as well as effects on receptivity of the female). There have been some attempts to measure and model PMGF (Gustafson *et al.*, 2005) but, for almost all crop species, good models of PMGF do not yet exist (Tolstrup *et al.*, 2003). This is because there are few (and sometimes no) measurements of PMGF taken from full commercial scale scenarios either because the scenario does not exist (GM crop has not yet been registered for commercial release) or because the effort has not yet been made to collect the data. In the absence of accurate and precise PMGF models, those wanting to establish protocols to prevent transgene escape have relied on traditional isolation distances for given crops. These isolation distances were established to prevent PMGF to an extent sufficient to protect varietal purity in seed production scenarios (Downey and Beckie, 2002; Tolstrup *et al.*, 2003). Within this context there was a requirement to keep PMGF levels low, but perhaps only below 0.1%, and to protect the integrity of complex polygenic traits it might have been important to only keep PMGF levels below 5%. However, to maintain stringent confinement of transgenes, traditional isolation distances may not be sufficient, because even low levels of PMGF (levels below 0.01%) can be problematic owing to the risk of bioaccumulation within metapopulation scenarios (Claessen *et al.*, 2005).

Seed movement is another means of transgene movement. Admixture of seed can occur at a great many points within a given grain handling system (Le Bail, 2003; Pedersen *et al.*, 2003). Genes may travel great distances when crop seeds are transported by humans either knowingly or unknowingly (Marvier and Van Acker, 2005) and with the assistance of the seed and grain movement infrastructure, transgene movement can potentially occur on a global scale (Colbach *et al.*, 2001). In addition, because some seeds are persistent, seed movement can facilitate gene flow over time (Hall *et al.*, 2000).

The persistence of seeds of GM crops is an important consideration for transgene escape and movement. After a crop has been harvested, volunteer and feral GM crop populations can appear in subsequent years and act as a place for the transgenes to come from or escape to. In this sense, for crop species which have large and robust volunteer and feral populations, and especially for crops that produce very persistent seed (or propagule) banks, a metapopulation for a given transgene may arise within a given region. Depending on the extent and robustness of the subpopulations within such a metapopulation (especially the volunteer and feral populations) and the frequency of the given crop species in rotations in time and space within a region, the containment of a transgene, once it has entered a metapopulation, may become impossible unless some drastic action is taken, such as exclusion of all cultivation of a given crop in a region for a given time (length of its seed bank duration) and the eradication of all volunteer and feral individuals of that species within the same region. This would be impractical and perhaps would only be required if the escaped transgene was particularly dangerous. One must also consider that, if farmers do not save their own seed, and if the transgene is selectively neutral in both agricultural and natural settings, then although a transgene may persist within a metapopulation, the frequency of that transgene within that metapopulation is likely to remain very low (i.e. below 0.1%). An exception to this may occur if a given GM crop is very commonly grown within a region and if it produces a very robust and common feral population. For example, in western Canada there is some evidence that a very high proportion of feral (non-field) populations of canola are GM and that some of these populations are accumulating multiple GM traits (Stephane McLachlan, personal communication).

The mechanism of transgene movement among crops within a region is a complex function of crop biology and ecology and the environmental and agronomic conditions under which the crops are grown. The species characteristics and agronomic conditions interact to create opportunities for genes to move from crop to crop. For example, empirically the characteristics and conditions which combined to create effective transgene movement for the Roundup Ready® trait in canola in western Canada included:

- a very large number of acres of Roundup Ready® (e.g. 2.25 million ha in 2003) and non-Roundup Ready® canola (e.g. 2.45 million ha in 2003) (Lawton® 2003) grown in fields across western Canada in a temporal and spatial randomly stratified fashion (Van Acker *et al.*, 2004). This may be the most important condition because it allowed for many opportunities for transgene escape;
- the relatively high frequency of canola in crop rotations in western Canada (e.g. on average 1 in 4 years on any given field in the province of Manitoba) (Thomas *et al.*, 1999);
- the large volunteer canola population in fields in western Canada (Leeson *et al.*, 2005);
- volunteer canola commonly survives to flowering at significant occurrence densities in a significant proportion of fields in western Canada (Leeson *et al.*, 2005);
- volunteer canola can persist, emerge in and flower in subsequent canola crops (Simard *et al.*, 2002; Légère *et al.*, 2001; Leeson *et al.*, 2005);
- plant to plant out-crossing rates in canola are relatively high (Cuthbert and McVetty 2001);
- the current canola pedigreed seed production system was designed to maintain varietal purity standards related to performance and end-use function. The system was not designed to prevent gene flow at the level required to prevent the problematic appearance of the Roundup Ready® traits in non-Roundup Ready® canola varieties.

Transgene movement (escape) may also be facilitated by human error. Experiences with GM crops in North America suggest that breakdown of containment may be inevitable and that these breakdowns frequently result from human error. Examples of human error include accidental commingling of GM with non-GM seeds or food products, accidental release of unapproved transgenes into commercial seed and the failure of industry and growers to follow protocols for field trials (Marvier and Van Acker, 2005). Given the complexity of the mechanisms of transgene escape and movement and the potential global nature of this movement, there are many opportunities for human error to facilitate transgene escape and movement. A hazard analysis critical control point (HACCP) system (CFIA, 2005) may well be warranted to reduce the potential for human error.

Protocols to measure and mechanistically model and predict transgene escape and movement are in the preliminary stages of development (Tolstrup *et al.*, 2003; Marvier and Van Acker, 2005; Mauro and McLachlan, 2003, Colbach *et al.*, 2001, Gustafson *et al.*, 2005). However, GM technology developers lack the incentive to measure adventitious transgene presence and movement post-release, because there is currently no regulatory requirement for them to do so. Additionally, while the means are available to test for the presence of given transgenes, their presence does not provide information on the points of transgene escape or the pathways of transgene movement. Currently, industry is attempting to mitigate concerns regarding transgene escape by providing stewardship plans for farmers. However, the effectiveness of these plans, especially if they are only voluntary, may be very limited (Van Acker *et al.*, 2004).

22.6 Managing coexistence

Any considerations of transgene confinement must be realistic with respect to the many complex routes of transgene escape. The challenges in managing the coexistence of GM and non-GM crops (transgene confinement) are many and they include the fact that the transgenes are invisible, requiring specific and possibly expensive detection methods (Tolstrup *et al.*, 2003; Marvier and Van Acker, 2005). In addition, if there is a functional metapopulation for given transgene(s) within a given crop in a region, confinement becomes even more difficult because the points of transgene escape are stratified in time and space. In this type of scenario it is very hard to predict when and from where the transgene will arrive. Finally, effective transgene confinement is highly dependent upon detection and eradication of the transgene at reception points. This is a critical consideration for those devising coexistence plans because the reception points for transgenes, in the context of preventing transgene escape, are the fields, farms and business operations of those people not wanting the transgene (the 'receptor'). In this sense, it is the receptor (often non-adopters of GM technology) who is most critically responsible for ensuring the confinement of transgenes (Van Acker, 2003). Effective coexistence management will also rely on *a priori* effective risk assessment-based regulation. For example, it may be prudent to restrict the production of plants producing high-risk proteins to fully contained facilities or to disallow, altogether, their production in food and feed plants (NRC, 2004).

Segregation of GM crops from non-GM crops has not been attempted within generalized bulk commodity production and handling systems. Small-scale segregation is managed for seed (although not always successfully; Friesen *et al.*, 2003) and for speciality products of high value. Currently, without regulations or incentives, handlers have little or no motivation to segregate GM crops effectively. The required stringency of a given transgene confinement system (plan) will depend upon the threshold level and the facility for transgene escape and movement. The latter depends fundamentally upon the nature of the crop species (e.g. it is most difficult to confine transgenes from species which are highly outcrossing and form persistent seed banks). If threshold levels are high and the crop in question is an obligate selfing species (the same individual is both the mother and the father; obligate selfing individuals carry two sets of genes and offspring are a combination of those genes), which does not produce a persistent seed bank, then confinement will be less difficult. However, GM technology developers often suggest that voluntary stewardship plans are suitable for managing the coexistence of GM and non-GM crops (Lawton, 2003). These plans are functionally problematic because the technology developers have little ability to demand, monitor, or enforce adherence to such plans. In the case of non-adopters ('receptors'), technology developers have no ability to demand adherence to these plans because the non-adopter is not bound by contract to do so.

To be effective, coexistence plans must encompass a broad range of characteristics (Van Acker, 2003). These plans must be based on realistic, science-based, robust and tested models of transgene movement. The plans also need to extend beyond individual fields or farms. Experience with movement of the Roundup Ready trait in canola in western Canada showed that volunteer canola existed as a metapopulation with respect to the Roundup Ready transgene, and that transgene confinement would have required a coexistence plan which encompassed the entire cropping system (not just the canola crop) and was operational across the entire region of western Canada. Management for confinement within a given field and for a given crop alone will be insufficient to achieve coexistence. In this regard, there needs to be a specific recognition in these plans of the fact that, in the absence of genetic technology preventing PMGF, transgene flow has to be controlled at the receptor crop. This poses a particular challenge for transgene confinement when receptor crops are grown by non-adopters of GM technology. In this respect as well, the plans must reflect a realistic expectation of commitment from farmers to implement the plan, given economic constraints and capacity limits. A comprehensive plan would also incorporate formal mechanisms for dealing with non-compliance and recognition of the jurisdiction and responsibilities of the various stakeholders. Issues of liability assignment must also be dealt with and formalized compensation mechanisms will be required. These mechanisms will provide recourse for those affected by transgene escape and movement as well as incentives for compliance. Enforcement of compensation and liability actions are only possible if coexistence plans are supported by regulations arising from legislation.

Genetic engineers have argued that they can devise technological solutions to the problem of transgene escape and movement. The most common technological solutions are (i) transformation of chloroplasts rather than nuclear DNA, (ii) technology which terminates the germinability of seeds possessing the escaped transgene(s), and (iii) cytoplasmic male sterility (mitochondrial genes that prevent production of functional pollen). At current levels of development, not one of these technological approaches will provide absolute transgene confinement (NRC, 2004). In the USA, the National Research Council suggested that confinement might be facilitated more by a careful selection of the species which are transformed rather than relying on genetic technology to prevent PMGF because all crop species do not present the same level of risk of transgene escape. Ultimately, approaches that employ a multitude of tactics in an integrated fashion would be most effective.

For farmers wishing to avoid transgene presence in their crops, or on their farms, there are a number of management practices they need to implement (Riddle, 2004). Given the complexity of mechanisms leading to transgene escape and its stochasticity, farmers and business operators who wish to guarantee and deliver GM-free product will need to employ all methods available to them in order to prevent contamination by transgenes and assure confinement (NRC, 2004).

Perhaps the most important management practice to prevent transgene presence on given fields or farms is to use clean seed. This is fundamental, because even low levels of adventitious presence of transgenes within a seed lot can result in significant levels of adventitious presence in the harvested product (Friesen *et al.*, 2003). And, for species which produce a persistent seed bank, a single seeding with an impure seed lot can lead to years of problematic transgene presence via self-replicating and self-disseminating volunteers. The use of clean seed is a problem both for farmers who save seed and for those who purchase certified seed because there are currently no regulations requiring seed certification agencies to ensure the genetic purity of seed (ISF, 2004). The EU and New Zealand are the only jurisdictions globally in which discussions in this regard are ongoing within governments. The International Seed Federation has demonstrated that the costs of ensuring genetic purity rise exponentially as threshold levels decline below 1%, to the point where costs are prohibitive at thresholds below 0.5–0.3% (ISF, 2001). If there is no regulation requiring seed certification agencies to guarantee genetic purity of seed, or if the genetic purity standards are set relatively high (e.g. above 0.5%), then farm-saved seed, or speciality suppliers may be the only option for farmers wanting to start with genetically pure, GM-free seed.

The involvement of farmers and their attitudes towards transgene confinement protocols and the risk associated with the production of GM crops has not commonly been considered by regulators, yet cooperation between neighbouring farmers may be a fundamental requirement for transgene confinement (Mauro and McLachlan 2003; Riddle, 2004; Tolstrup *et al.*, 2003). The 'human' or cultural element of coexistence management is difficult to characterize and control (Mauro and McLachlan, 2003) and this makes coexistence success difficult to predict. Open communication between neighbours, either formal or informal, is an essential element of a successful coexistence plan (Riddle, 2004).

Physical isolation is a traditional means of limiting PMGF in crop breeding programmes and it can be exploited as an aid to help limit transgene escape or arrival. Physical isolation, however, is not an absolute protection from transgene invasion and farmers must realize that traditional isolation distances were established to guarantee seed purity not genetic purity. Isolation distance must be suited to the nature of the species and the threshold (Tolstrup *et al.*, 2003). PMGF at low levels has been recorded at very long distances for species such as canola (3 km) (Rieger *et al.*, 2002) and creeping bentgrass (21 km) (Watrud *et al.*, 2004). Pollen traps (or barriers) are sometimes promoted as an additional means for achieving physical isolation to limit PMGF, but in order to be effective these pollen traps must be significantly taller than the targeted crop at the time of flowering (and/or pollen release) of that crop. In western Canada, maize (*Zea mays* L.) has sometimes been used ineffectively as a pollen barrier to prevent pollen escape from confined GM-wheat trails because, in the cooler climate of western Canada, maize is only as tall as normal wheat plants at time of flowering (early July) (personal observation).

The effective transfer of transgenes between volunteers and cropped plants can be minimized by creating isolation in time between these cohorts. A long and diverse crop rotation allows farmers to reduce seed banks for volunteer species before that same species is cropped again within a given field. Isolation in time is traditionally used by plant breeders and seed growers to help them to facilitate the maintenance of seed purity (CSGA, 2005). A diverse crop rotation facilitates the detection of volunteers in subsequent crops and allows for effective roguing or control with herbicides (CSGA, 2004). Although many common annual crop species do not produce persistent seed banks, there are some exceptions. For example, in the United Kingdon (UK) canola seed (*Brassica napus*) has been shown to persist for up to 12 years (Lutman, 2003).

Farming equipment is a common vector for the dispersal of weed seeds and propagules (e.g. rhizomes) (Shirtliffe *et al.*, 2000) and it can be a common vector for the movement of crop seed from field to field. Cleaning farm equipment, as well as storage, transport and grain handling facilities, including farm-based seed cleaning facilities, is a critical part of coexistence management (Riddle, 2004). It can be difficult to completely clean some pieces of equipment, such as commercial combine harvesters, and farmers who wish to maintain their operations free from even trace transgene sources might consider never using difficult to clean equipment which they cannot guarantee to be GM-free. This may preclude non-GM farmers from renting equipment or from sharing equipment with farmers who do not guarantee their operations to be GM-free.

Tolerance of GM seeds in non-GM varieties ranges from 0 to 5% in different countries (ISF, 2004). It is widely agreed that it is impossible to maintain or guarantee that seed from a crop such as maize, soybeans, canola or cotton is absolutely free of GM seeds. It should be reiterated that IFOAM opposes mandatory testing for GM contamination and proposes that organic farmers do not have to prove that their crops are GM-free (IFOAM, 2002). The International Seed Federation (ISF) proposed that its 70 member countries agree on a 1% threshold for GM content in non-GM varieties, until data was collected and analysed from different crop species to allow specific, realistic thresholds for particular crops. Adoption has not yet taken place. ISF maintains that a threshold of less than 1% 'would be extremely difficult to achieve at a reasonable cost' (ISF, 2001).

The International Seed Testing Association (ISTA) has conducted a series of proficiency tests involving 90 seed testing laboratories in 30 countries. ISTA is evaluating the laboratories' abilities to determine whether control or 'spiked' seed lots contained GM seeds and to determine the percentage of GM seeds in individual seed lots (Kahlert, 2005a). Results from qualitative determinations of the presence or absence of GM seeds were highly successful, ranging from 70–85% of laboratories reporting no false results (Kahlert, 2005a). The laboratories had less success in determining whether seed lots contained less than or equal to 1% GM seed. Ten out of 20 laboratories, which reported quantitative evaluations of GM content in maize in the Third ISTA Proficiency test, reported results of less than or equal to 1% for samples that had been spiked with 2% GM seeds (Kahlert, 2005b).

ISTA does not dictate testing methods; however, quantitative tests are based on PCR (polymerase chain reaction) amplification of specific sequences of transgenic DNA and fall into two groups: the subsampling method which divides the samples into subsamples and evaluates frequencies of positive results from the total number of subsamples tested and the numbers of seeds per subsample; and the real-time PCR method which estimates GM content based on the rate of amplification of GM-specific DNA sequences (Kahlert, 2005b). Qualitative tests for absence or presence can be based on either protein tests or PCR tests. The proficiency tests are proposed to accelerate the accreditation process for laboratories when GM testing becomes included in the ISTA accreditation programme (Kahlert, 2005c).

Liability is an issue for any operation trying to assure specifications on a delivered product and liability is an issue for farmers wishing to deliver GM-free product. Whether or not there is law and formal recourse for the presence of transgenes in products meant to be free from transgenes, farmers who wish to assure delivery of GM-free product will need to establish clear lines of liability and support these with legal and binding contracts. The maintenance of a liability position also requires excellent record keeping and complete and full documentation of all transactions and guarantees from suppliers of equipment, inputs and seed, and from those handling transport, storage and processing. Testing to assure GM-free status will also be required and maintaining an area or operation free from transgenes will require that area or operation to be constantly tested for the presence of transgenes. It is critical for farmers to understand that, when GM crops are grown in quantity across a given landscape, transgene escape will happen and therefore, the prevention of transgene escape relies upon action at the receptor point (the GM-free operation). Of course, even with a stringent testing and assurance program, it is not certain what happens when non-GM farmers find transgenes in their crop or on their farm. This is because it is not necessarily easy to determine the source of transgenes, and because, for example, PMGF can be very unpredictable, the source may never be known. In this respect, when GM crops are registered for unconfined release within a given jurisdiction there is no formal or currently effective recourse for those affected by transgene escape resulting in adventitious presence in non-GM crops or on non-GM farms.

22.7 Coexistence legislation

Denmark, by being the first state in the world to introduce legislation for the coexistence of GM and non-GM crops (Danish Ministry of Food, Agriculture and Fisheries, 2004), is the world leader in building a context in which rights are protected to some extent for those who choose not to grow or consume GM products. The coexistence legislation passed by the Danish parliament in 2004 has many good functional characteristics including: a confinement training requirement for GM crop farmers, a clear statement that the GM crop farmer is responsible for confinement, a mandatory requirement for open information access on GM crop sites, a formalized compensation mechanism for adventitious presence of transgenes which is based on thresholds, the assignment of criminal liability to those found responsible for transgene escape resulting from negligence and search and seizure rights to facilitate the investigation of transgene escape incidents. However, the existence of such legislation is not enough to fully protect non-GM farmers from problems associated with transgene escape. Because there is now legislation in place in Denmark, Danish farmers can now allow GM crops to be grown in Denmark and these farmers must consider how they will use the legislation. Will they wait until someone downstream (a customer) discovers transgenes in Danish organic products or will Danish organic farmers establish a comprehensive testing programme which will allow them to protect themselves and use the legislation should a problem arise? In this context it is important to remember that the critical control point for transgene confinement is the receptor crop (the crop grown by the non-GM farmer), so even with legislation in place, the real responsibility for protecting a non-GM crop from transgene escape belongs to the non-GM farmer and the cost for doing so also belongs to the non-GM farmer. Therefore, coexistence legislation in Denmark will give non-GM farmers additional responsibilities and added costs.

The Danish legislation makes no mention of technology developers. It absolves the ones who own the patent on the GM crop from any responsibility for transgene escape. The next wave of GM crops is coming quickly and one must consider how legislation is positioned to deal with the risks, liabilities and costs associated with managing the coexistence of these crops. In Canada, the government is reluctant to allow production of pharmaceutical proteins in food or feed crops, but there is no formal mechanism or agreement to stop this from happening. Regulators must consider what the threshold need would be for these traits and whether they would be different (lower) than the 'normal' threshold for GM in non-GM crops or food. Will there be a trait-specific threshold? Would it be simpler just to keep these traits out of the food and feed stream altogether?

Percy Schmeiser is a farmer from Saskatchewan, Canada who was sued by Monsanto under Canadian patent law for infringement relating to the presence of Roundup Ready® (GM) canola on his land. His case has many interesting legal and liability implications for coexistence management. Mr. Schmeiser lost his case in final appeal to the Supreme Court of Canada, which ruled in May 2004 that Monsanto could retain the full rights and privileges of patent ownership, as well as the right to sue farmers for the possession of this transgene, regardless of how it came to be in their possession and regardless of whether or not they profited from possessing it (Supreme Court of Canada, 2004). The ruling is problematic because it does not explicitly consider the case of innocent infringement and because the Roundup Ready® (GM) transgene is now present in the majority of certified non-Roundup Ready® canola seed sold in western Canada. Any farmer in this 40 million ha region who grows canola has a better than 50% chance that Roundup Ready canola will be on their land even if they purposely choose not to grow it. Monsanto can choose to sue any one of these farmers, yet Monsanto is not bound by any responsibility for the uncontrolled movement of their Roundup Ready® transgene.

In an academic legal assessment of this case, Cullet (2005) makes special note that the outcome of the case points to a real need to assign liability and responsibility in regard to transgene ownership and the effects of transgene escape, and that in the current context all burdens resulting from transgene escape are shifted to the users of GM crops and those potentially affected by their unconfined cultivation. In the absence of formal coexistence legislation which clearly assigns liability and responsibility, recourse for damage suffered by transgene escape will be difficult to achieve. For example, recently in Canada, the Saskatchewan Organic Directorate (SOD, an association of organic farmers in the province of Saskatchewan) attempted to set precedence in civil law in Canada by suing in class action, Monsanto Canada Ltd and Bayer CropScience for the ubiquity of GM transgenes in canola in western Canada and the resultant inability of organic farmers in this region to produce GM-free organic canola. The case was denied class action status by a Canadian federal court judge because the plaintiffs had failed to prove adequately that the entire class (all farmers in SOD) was suffering damage (Smith, 2005).

22.8 GM-free regions

Several north American jurisdictions are addressing the issue of GM and non-GM coexistence by developing legislation to become GM-free regions, that is to isolate, or perhaps insulate, non-GM crops. Notable examples are Mendocino (Geniella, 2004) and Marin (Heath, 2004) counties in California. In Canada's smallest province, the Prince Edward Island Certified Organic Producers Cooperative is assessing a market for agriculture products produced in an Island GMO-free grow zone (PEI COPC, 2005). Whether such GM-free regions can prevent adventitious presence or contamination has yet to be determined. However, such regions are changing the concept of coexistence from spatial differentiation at the farm level to the county level and to larger more isolated regions, including islands.

22.9 Future research needs

The National Research Council in the USA (2004) cites lack of relevant data as the single most significant factor limiting effective bioconfinement. In no jurisdiction to date has the regulation of GM crops included a comprehensive assessment of the possibility and probability of transgene escape and movement. This is due in part to the fact that, when a GM crop has been registered for unconfined release, the regulatory body certifies that the GM crop poses no environmental or human health risk, even if there is transgene escape. However, the lack of consideration of transgene escape mechanisms and probability in GM crop regulation is also due to the lack of understanding of how transgene escape works and the lack of real field-based data which could be used to obtain such an understanding. To consider quality assurance management fully with respect to maintaining GM-free crops and products, there is a need for realistic, reliable deterministic and probabilistic models of intraspecifc gene flow (which include a consideration of both physical and pollen-mediated mechanism of movement), and these models must be based on real field data. These models must account for the contribution of volunteer and feral populations to transgene movement and they must include good estimations of long distance PMGF (Rieger *et al.*, 2002) since it is often low-level long distance PMGF tails that produce the most unexpected transgene escapes (Watrud *et al.*, 2004). With these types of models it would become possible to more accurately identify critical control points for gene movement and to more accurately vet the risk of gene movement in different scenarios and with different crops and traits (Colbach *et al.*, 2001).

22.10 Conclusion

Genetically engineered traits are being introduced into new crops (e.g. alfalfa and strawberries) and new GM events (traits), which may have a greater impact on human health (e.g. pharmaceutical traits) and the environment (e.g. drought tolerance and plant made industrial products) are being developed.

Before these new GM crops reach broad scale release, procedures should be in place to monitor, model and mitigate transgene escape and movement and its consequences, including the economic impact on farmers wishing to assure and deliver non-GM products. The absolute containment of transgenes is not possible, especially when they are engineered into plants grown in quantity outdoors. In addition, completely recalling escaped transgenes from agri-food systems and the environment is not possible. For non-GM crop farmers, the implications of these realities are grave and action on the part of these farmers worldwide is required now in order to protect the integrity of the non-GM industry. Transgene escape and movement occur by a myriad of means, including human error. Programmes for assuring that given crops and products are GM-free must be comprehensive and include as many strategies and tactics as possible. The means to predict transgene movement accurately are still limited, even for GM crops that have been commercially grown for a decade in some jurisdictions (e.g. GM canola in Canada). Farmers wishing to deliver an assured non-GM product must protect their farms from transgene escape. Even in jurisdictions where there is GM–non-GM coexistence legislation (e.g. Denmark), it is the responsibility of those who have something to lose from transgene escape to be the most vigilant in protecting themselves from transgene escape and its effects.

22.11 Sources of further information and advice

DIAS (Danish Institute of Agricultural Sciences) (www.agrsci.dk).
Genewatch UK (www.gmcontaminationregister.org)
ISAAA (International Service for the Acquisition of Agri-Biotech Applications). (www.isaaa.org)
IFOAM (International Federation of Organic Agriculture Movements) (www.ifoam.org)
ISF (International Seed Federation), (www.worldseed.org)
NRC (US National Research Council) (www.nationalacademies.org/nrc/)
SOD (Sakatchewan Organic Directorate), (www.saskorganic.com)
US Department of Agriculture National Organic Program (www.ams.usda.gov/nop/indexIE.htm)
US National Organic Standards Board (www.ams.usda.gov/nosb/)

22.12 References

Anonymous (2005). 'Activists push for no-GM zones', *The Western Producer*, June 9, 75.

CFIA (Canadian Food Inspection Agency) (1995). *Decision Document DD95-02: Determination of Environmental Safety of Monsanto Canada Inc.'s Roundup® Herbicide-tolerant* Brassica napus *Canola Line GT73*, Canadian Food Inspection Agency, Ottawa,

Ontario, Canada. http://www.inspection.gc.ca/english/plaveg/bio/dd/dd9502e.shtml. Accessed February 20, 2007.
CFIA (Canadian Food Inspection Agency) (2005). *Food Safety Enhancement Program HACCP Curriculum Guidelines*. www.inspection.gc.ca/english/fssa/polstrat/haccp/manue/tablee.shtml. Accessed February 20, 2007.
Claessen, D.C., Gilligan, A. and van den Bosch, F. (2005). 'Which traits promote persistence of feral GM crops? Part 2: implications of metapopulation structure', *Oikos*, **110**, 30–42.
Colbach, N., Clermont-Dauphin, C. and Meynard, J.M. (2001). 'GeneSys: a model of the influence of cropping system on gene escape from herbicide tolerant rapeseed crops to rape volunteers II. Genetic exchanges among volunteers and cropped populations in a small region', *Agric. Ecosystems and Environ*, **83**, 255–270.
CSGA (Canadian Seed Growers Association) (2004). *CSGA Bulletins and Regulatory Changes re:Volunteer Wheat. Technical Update*: Circular 6 changes and updates for 2004. www.seedgrowers.ca/regulationsandforms/circular.asp Accessed June 5, 2005.
CSGA (Canadian Seed Growers Association) (2005). *Regulations and Forms*. www.seedgrowers.ca/regulationsandforms/index.asp?lang=e Accessed February 20, 2007.
Cullet, P. (2005). 'Case law analysis. Schmeiser v Monsanto: A landmark decision concerning farmer liability and transgenic contamination', *J. Environ. Law*, **17**, 83–108.
Cuthbert, J.L. and McVetty, P.B.E. (2001). 'Plot-to-plot, row-to-row and plant-to-plant out-crossing studies in oilseed rape', *Can. J. Plant Sci.*, **81**, 657–664.
Danish Ministry of Food, Agriculture and Fisheries (2004). *Act on the Growing etc. of Genetically Modified Crops*, Act No. 436 of 9 June 2004. Copenhagen, Denmark. pp. 5.
DeVries, J. and Wackernagel, W. (2002). 'Integration of foreign DNA during natural transformation of *Acinetobacter* sp. by homology-facilitated illegitimate recombination', *Proc. Natl. Acad. Sci. USA*, **99**, 2094–2099.
Downey, R.K., and Beckie, H.J. (2002). *Isolation Effectiveness in Canola Pedigreed Seed Production*. Internal Research Report, Agriculture and Agri-Food Canada, Saskatoon Research Centre, Saskatoon, SK, S7N 0X2, Canada.
Ellstrand, N.C, Prentice, H.C. and Hancock, J.F. (1999). 'Gene flow and introgression from the domesticated plants into their wild relatives', *Annu Rev Ecol Syst*, **30**, 539–563.
Friesen, L.F., Nelson, A.G. and Van Acker, R.C. (2003). 'Evidence of contamination of pedigreed canola (*Brassica napus*) seedlots in western Canada with genetically engineered herbicide resistance traits'. *Agron J*, **95**, 1342–1347.
Geniella, M. (2004). *Mendocino County Voters Ban Biotech Crops*. THE PRESS DEMOCRAT. www.oacc.info/Issues/mendocino_ban_mar04.html Accessed July 12,
Gillis, J. (2002). 'Soybeans mixed with altered corn; Suspect crop stopped from getting into food', *The Washington Post*, November 13, p. E01.
Gustafson, D.I., Horak, M.J., Rempel, C.B., Metz, S.G., Gigax, D.R. and Hucl, P. (2005). An empirical model for pollen-mediated gene flow in wheat. *Crop Sci.*, **45**, 1286–1294.
Hall, L., Topinka, K., Huffman, J., Davis, L. and Good, A. (2000). Pollen flow between herbicide-resistant *Brassica napus* is the cause of multiple-resistant *B. napus* volunteers. *Weed Sci*, **48**, 688–694.
Heath, C. (2004). *GMO-Free Marin's Impact Goes Beyond County*. Organic Consumers Association. www.organicconsumers.org/biod/marinimpact31105.cfm Accessed July 12, 2005.
IFOAM (International Federation of Organic Agriculture Movements) (2002). *Position on Genetic Engineering and Genetically Modified Organisms* (adopted May 2002). www.ifoam.org/press/positions/pdfs/IFOAM-GE-Position.pdf. Accessed July 19, 2005.
ISAAA (International Service for the Acquisition of Agri-Biotech Applications) (2006).

ISAAA Briefs 35-2006. Global status of commercialized biotech/GM crops – 2006. www.isaaa.org, accessed February 20, 2007.

ISF (International Seed Federation) (2001). *Motion on Adventitious Presence of GM Material in non-GM Seeds.* ISF position paper. www.worldseed.org. Accessed June 3, 2005.

ISF (International Seed Federation) (2004). *Coexistence of genetically modified, conventional and organic crop production.* ISF position paper. www.worldseed.org. Accessed June 3, 2005.

Kahlert, B. (2005a). 'ISTA proficiency test on GMO testing: a brief overview', *Seed Testing International*, **129**, 17–18. www.seedtest.org/upload/cms/user/STI1291.pdf. Accessed July 18, 2005.

Kahlert, B. (2005b). 'Third ISTA proficiency test on GMO testing on *Zea mays* (MON819+T25): summary of results', *Seed Testing International*, **129**, 10–12. www.seedtest.org/upload/cms/user/STI1291.pdf, Accessed July 18, 2005.

Kahlert, B. (2005c). 'Fourth ISTA proficiency test on GMO testing on *Glycine max* (L.) Merr.: summary of results', *Seed Testing International*, **129**, 12–16. www.seedtest.org/upload/cms/user/STI1291.pdf. Accessed July 18, 2005.

Kingsman, S.M. and Kingsman, A.J. (1988). *Genetic Engineering: An Introduction to Gene Analysis and Exploitation in Eukaryotes*, Blackwell Scientific Publications, Oxford, UK, pp 522.

Lawton, M. (2003). *Management of Herbicide Tolerant Crops and Future Research*, Canadian Food Inspection Agency, Plant Products Directorate, Plant Biosafety Office Technical Workshop on the Management of Herbicide Tolerant (HT) Crops Report. www.inspection.gc.ca/english/plaveg/bio/consult/herbtolrepe.shtml#7. Accessed June 17, 2005.

Le Bail, M. (2003). 'GMO/non GMO segregation in supply zone of country elevators', in Boelte, B. *Proceedings of the 1st European Conference on the Coexistence of Gentically Modified and Conventional and Organic Crops*, Denmark, Nov 13–14 2003. Danish Institute of Agricultural Sciences, Research Centre, Flakkebjerg, Denmark, 125–127.

Leeson, J.Y., Thomas, A.G., Hall, L.M., Brenzil, C.A., Andrews, T., Brown, K.R. and Van Acker, R.C. (2005). *Prairie Weed Surveys of Cereal, Oilseed and Pulse Crops from the 1970s to the 2000s*, Weed Survey Series Publication 05-1 (and CD), Agriculture and Agri-Food Canada, Saskatoon Research Centre, Saskatoon, SK, 395 pp.

Légère, A. (2005). 'Risks and consequences of gene flow from herbicide-resistant crops: canola (*Brassica napus* L) as a case study', *Pest Manag Sci.*, **61**, 292–300.

Légère, A., Simard, M.J., Thomas, A.G., Pageau, D., Lajeunesse, J., Warwick, S.I. and Derksen, D.A. (2001). 'Presence and persistence of volunteer canola in Canadian cropping systems', *Proceedings of the British Crop Protection Council Conference–Weeds*, held Nov 12–15, 2001, Brighton, UK. BCPC, Surrey UK, 143–148.

Lutman, P.J.W. (2003). 'Coexistence of conventional, organic and GM crops – role of temporal and spatial behaviour of seeds', in Boelte, B. *Proceedings of the 1st European Conference on the Coexistence of Genetically Modified and Conventional and Organic Crops*. Denmark, Nov 13–14, 2003. Danish Institute of Agricultural Sciences, Research Centre, Flakkebjerg, Denmark, 33–43.

Ma, B.L. (2005). 'Frequency of pollen drift in genetically engineered corn', *ISB News Report*, February, 2005. http://www.isb.vt.edu/news/2005/news05.feb.html#feb0502 Accessed July 12, 2005.

Marvier, M. and Van Acker, R.C. (2005). Can crop transgenes be kept on a leash? *Front. Ecol. Environ*, **3**, 93–100.

Mauro, I.M. and McLachlan, S.M. (2003). *Risk Analysis of Genetically Modified Crops on the Canadian Prairies.* Canadian Food Inspection Agency, Plant Products Directorate, Plant Biosafety Office Technical Workshop on the Management of Herbicide Tolerant (HT) Crops Report. www.inspection.gc.ca/english/plaveg/bio/consult/herbtolrepe.shtml#7. Accessed June 17, 2005.

Morrison, L.A., Crémieux, L.C. and Mallory-Smith, C.A. (2002). 'Infestations of jointed goatgrass (*Aegilops cylindrical*) and its hybrids with wheat in Oregon'. *Weed Sci.*, **50**, 737–747.

NRC (National Research Council) (2004). *Biological Confinement of Genetically Engineered Organisms*, The National Academies Press, Washington, DC.

Pedersen, S.M., Gylling, M. and Sondergaard, J. (2003). 'A method to assess the costs of avoiding admixture in a combined GM and conventional processing system', in Boelte, B. *Proceedings of the 1st European Conference on the Coexistence of Gentically Modified and Conventional and Organic Crops*, Denmark, Nov 13–14 2003. Danish Institute of Agricultural Sciences, Flakkebjerg, Denmark, 131–133.

PEI COPC (Prince Edward Island Certified Organic Producers Coop) (2005). *COPC Undertakes Non GMO Markets Study*, www.organicpei.com Accessed July 12, 2005.

Pratt, S. (2005). GMO treaty talks end with no deal. *The Western Producer*, June 9, 75.

Pryme, I.F. and Lembcke, R. (2003). *In vivo* studies on possible health consequences of genetically modified food and feed – with particular regard to ingredients consisting of genetically modified plant material, *Nutrition and Health*, **17**, 1–8.

Riddle, J.A. (2004). *A Plan for Co-existence: Best Management Practices for Producers of GMO and non-GMO Crops*, University of Minnesota, St. Paul, MN, USA, 4 pp.

Rieger, M.A., Lamond, M., Preston, C., Powles, S.B. and Roush, R.T. (2002). Pollen-mediated movement of herbicide resistance between commercial canola fields, *Science*, **296**, 2386–2388.

Shirtliffe, S.J., Entz, M.H. and Van Acker, R.C. (2000). *Avena fatua* development and seed shatter as related to thermal time.' *Weed Sci.*, **48**, 555–560.

Simard, M.J., Légère, A., Pageau, D., Lajeunesse, J. and Warwick, S. (2002). The frequency and persistence of volunteer canola (*Brassica napus*) in Quebec cropping systems, *Weed Technol.*, **16**, 433–439.

Skaftnesmo, T. (2004). 'Generne er ikke laengere, hvad de har vaeret', *Proceedings of the 2004 Okologi-Kongres*, Odense, Denmark, Nov 16–17, 185–187.

Smith, G.A. (2005). *Larry Hoffman, L.B. Hoffman Farms Inc and Dale Beaudoin v. Monsanto Canada Inc. and Bayer CropScience Inc. Queen's Bench for Saskatchewan.* Citation: 2005 SKQB 225 Date: 20050511 Docket: Q.B.G. No. 67/2002, Judicial Centre: Saskatoon, SK, Canada.

Snow, A.A., Andersen, B. and Jorgensen, R.B. (1999). Costs of transgenic herbicide resistance introgressed from *Brassica napus* into weedy *B. rapa*. *Mol. Ecol.*, **8**, 605–615.

Snow, A.A., Pilson, D., Rieseberg, L.H., Paulsen, M.J. Pleksac, N., Reagon, M.R., Wolf, D.E. and Selbo, S.M. (2003). Bt transgene reduces herbivory and enhances fecundity in wild sunflowers, *Ecol. Appl.*, **13**, 279–286.

Supreme Court of Canada (2004). *Percy Schmeiser and Schmeiser Enterprises Ltd v. Monsanto Canada Inc and Monsanto Company*. File no. 29437. January 20, 2004: May 21. Ottawa, Canada.

Thomas, A.G., Leeson, J.Y. and Van Acker, R.C. (1999). *Farm Management Practices in Manitoba, 1997 Manitoba Weed Survey Questionnaire Results*, Weed Survey Series Publication 99–3. Agriculture and Agri-Food Canada, Saskatoon, SK, 296 pp.

Tolstrup, K., Andersen, S.B., Boelt, B., Buus, M., Gylling, M. Holm, P.B., Kjellsson, G., Pedersen, S., Ostergard, H. and Mikkelsen, S. (2003). *Report From the Danish working Group on the Coexistence of Genetically Modified Crops with Conventional and Organic Agriculture*, Danish Institute of Agricultural Sciences Report, Plant Production no. 94. DIAS, Tjele, Denmark, 275 pp.

USDA (United States Department of Agriculture) (2003). *Starlink Test Results*, November 19, 2003. United States Department of Agriculture, Grain Inspection, Packers and Stockyards Administration, Washington DC, USA.

Van Acker, R.C. (2003). *Potential Approaches for the Management of Herbicide Tolerant Crops*, Canadian Food Inspection Agency, Plant Products Directorate, Plant Biosafety

Office Technical Workshop on the Management of HerbicideTolerant (HT) Crops Report. www.inspection.gc.ca/english/plaveg/bio/consult/herbtolrepe.shtml#7. Accessed June 17, 2005.

Van Acker, R.C., Brule-Babel, A.L. and Friesen, L.F. (2004). 'Intraspecifc gene movement can create environmental risk: the example of Roundup Ready® wheat in western Canada', in Breckling, B and Verhoeven, R. *Risk, Hazard, Damage – Specification of Criteria to Assess Environmental Impact of Genetically Modified Organisms*, Naturschutz und Biolische Viefalt, Bundesamt fur Naturshutz, Bonn, Vol **1**, 37–47.

Watrud, L.S., Lee, E.H., Fairbrother, A., Burdick, C., Reichman, J.R., Bollman, M., Storm, M., King, G. and Van de Water, P.K. (2004). 'Evidence for landscape-level, pollen-mediated gene flow from genetically modified creeping bentgrass with *CP4 EPSPS* as a marker', *Proc. Natl. Acad. Sci.*, **101**, 14533–14538.

Wilkinson, M.J., Elliott, L.J., Allainguillaume, J., Shaw, M.W., Norris, C. Welters, R., Alexander, M., Sweet, J. and Mason, D.C. (2003). 'Hybridization between *Brassica napus* and *B. rapa* on a national scale in the United Kingdom', *Science*, **302**, 457–459.

23

Integration of quality parameters into food safety focused HACCP systems

Kirsten Brandt and Lorna Lück, Newcastle University, UK; Unni Kjærnes, National Institute for Consumer Research (SIFO), Norway; Gabriela S. Wyss, Swiss Research Institute of Organic Agriculture (FiBL), Switzerland; and Annette Hartvig Larsen, Aarstiderne, Denmark

23.1 Introduction

This chapter describes how quality control of organic food throughout the food chain can be improved by adopting some of the concepts from the hazard analysis by critical control points (HACCP) procedure, which is commonly used in processing enterprises to control food safety, or to ensure consistently high quality in the form of QACCP (quality analysis by critical control points). The concept here differs from the standard HACCP by covering the entire supply chain and by using the concepts for a wide range of qualities that are valued by consumers, including product quality (taste, nutrients, etc.) and ethical values (authenticity, fairness, etc.).

The text summarises the activities and outcome of the Organic HACCP project (www.organichaccp.org) that was completed in 2005, how a database of critical control points (CCPs) was developed for some representative supply chains and how this was used to define a set of recommendations that were then developed into leaflets with advice to producers, processors, retailers or consumers, respectively. The chapter will thus explain how companies at every step of the production chain can utilise the concepts to improve customer satisfaction in a cost-effective manner. Finally, it will describe an example of implementation in a group of collaborating companies and suggest where additional activities are needed in order to develop the concept further.

23.2 Need to integrate and focus control systems for quality and safety

The rules and regulations presently governing organic production are intended to assure consumers that producers adhere to certain organic principles and standards. The organic principles are summarised as health, ecology, fairness and care (IFOAM, 2005). These principles comprise statements that 'organic agriculture is intended to produce high quality, nutritious food that contributes to preventive health care and well-being' and 'that precaution and responsibility are the key concerns in management, development and technology choices in organic agriculture'. These principles form the basis for the more detailed basic standards that the producers are required to follow and for the generally even more detailed regulations developed by regional, national or local organisations or governments. However, the specific rules generally focus on what should be avoided, in terms of practices that are considered unnatural and/or potentially harmful to the environment or to the welfare of humans or animals, such as the use of synthetic pesticides or unnecessary food additives and keeping animals under conditions that are more stressful than necessary.

In contrast, there is little positive guidance for a careful producer or processor who wants to promote the consumers' satisfaction, health and well-being, beyond what can be achieved by eliminating recognised risks. In particular, it is a challenge when it requires coordination of actions taken at every stage along the chain, from the primary producer to the retailer. For factors important for quality and safety, such as hygiene standards and control of temperatures during storage and transport, organic producers, processors and retailers are generally required to follow the same legal regulations and guidelines that are used by conventional producers. In other words, there is a clear expectation in the organic movement that organic producers demonstrate special care and responsibility for the food quality, in addition to the minimum standards required by law for all producers. However, there is a need for additional advice and recommendations for tools to use in order to fulfil this expectation, in particular, when the product passes through several enterprises before it reaches the consumer. This is, in turn, crucial for consumers' trust in foods labelled as organic.

23.3 Hazard analysis by critical control points

HACCP is one of the minimum standards that is often required in a food processing enterprise to ensure that products do not contain harmful levels of biological, physical or chemical hazards such as pathogenic microorganisms or toxins. The overall idea of HACCP is to identify specific CCPs, which are those steps in the production process where the safety of the final product can be controlled most efficiently, and then define systematic procedures for monitoring and corrective action, to ensure that the risk is controlled at each

CCP. This is much more efficient than just relying on general good practice for workmanship and hygiene, since it ensures that, if something goes wrong, the problem is detected and solved before any contaminated food reaches the consumer and that, for every critical point, it is clear who is responsible and what the responsible person has to do. The implications of this type of control system for organic certification are described in more detail in Chapter 4 by van Elzakker. HACCP consists of seven principles, which can be briefly described as follows:

Principle 1: *Conduct a hazard analysis.*

The production steps covered by the HACCP plan are described systematically in the form of a flow diagram. All hazards that may jeopardise the safety of the food are listed and it is decided which ones should be included in the analysis.

This procedure is formalised as six steps:

- Step 1: Assemble a HACCP team
- Step 2: Describe the product
- Step 3: Identify the product's intended use
- Step 4: Construct a flow diagram
- Step 5: Confirm the flow diagram on-site
- Step 6: List all potential hazards associated with each step, conduct a hazard analysis to estimate the severity of each of these hazards and consider any measures to control identified hazards.

Principle 2: *Determine the critical control points (CCPs).*

Each step in the flow diagram is analysed to determine if it is suitable as a CCP for one or more of the hazards found using Principle 1. A production step is a CCP if it fulfils all of the following three requirements: the step contains a control measure for the hazard; lack of control at this step could lead to this hazard reaching an unacceptable level; and subsequent steps are not going to eliminate the hazard.

Principle 3: *Establish critical limit(s).*

The level of a measure necessary to ensure that a hazard is fully controlled is determined and called the critical limit. For example, how high the temperature must be to ensure that a pathogen has been eliminated, or how low the humidity must be to prevent the growth of mould fungi.

Principle 4: *Establish a system to monitor control of the CCP(s).*

A plan is draw up detailing how checks will be made to ensure that the critical limits are not exceeded. It shows how often the checks are made, how and by whom, so that it is clear who is responsible that it is done correctly. It also specifies regular inspections and calibrations of measuring equipment such as thermometers.

Principle 5: *Establish the corrective action to be taken when monitoring indicates that a particular CCP is not under control.*

The plan must provide information for the operators on what to do if the critical limits are exceeded, to ensure that the possibly defective food is identified and retained and that further production is stopped until the problem has been rectified.

Principle 6: *Establish procedures for verification to confirm that the HACCP system is working effectively*

This part of the HACCP plan describes how to verify that the checks are done correctly and that assumptions are met, for example, if an incident has occurred where the critical limits have been exceeded, as well as procedures for handling changes to the product or the process. Another type of verification involves unannounced inspections by independent auditors, which is a requirement for certified schemes.

Principle 7: *Establish record keeping and documentation procedures.*

All decisions must be recorded, as well as an explanation of the decision, to provide background information if changes are needed in the future. Also, all monitoring activities must be documented, so that it is clear who has done what and when. The documentation is used to demonstrate compliance in relation to certification, and in case of problems it is used to try to find out if a problem was due to human error or a technical failure.

There is an abundance of literature and courses about HACCP. On the internet, an overview can be found at FAO's website (FAO, 1997), while the National Advisory Committee on Microbiological Criteria for Foods (NACMCF) (1997) has made a more thorough set of guidelines available. From the start, the concept of HACCP was foreseen as being applied throughout the food chain from primary production to final consumption (FAO, 1997). However, standard HACCP systems normally cover only one enterprise, for the practical and legal reasons explained in more detail in Section 23.5. QACCP is still mainly used in proprietary systems developed specifically for members of an organisation or other fee-paying users. However, a basic QACCP guideline can be constructed from HACCP guidelines simply by replacing the word 'hazard' with 'quality defect' or equivalent expressions.

23.4 Introducing the Organic HACCP project

The objective of the Organic HACCP project (Organic HACCP, 2005) was to utilise some of the concepts described above in order to improve the control of quality and safety of organic food in a broader context than standard HACCP. One aspect of broadening was to extend the control beyond the individual enterprise and try to identify the CCPs at the level of the entire supply chain, covering all enterprises involved, from the primary producer to the retailer. Another aspect was to extend the types of risks that could be controlled in this way, encompassing all those qualities where a loss can pose a substantial threat to the supply chain, even when there is no direct threat to the health of the consumer. For example, all organic supply chains depend on unbroken authenticity throughout the chain. If the consumer has the impression that the integrity of the product has been jeopardised at any one of the enterprises that have produced, transported, processed, stored or sold the product, the authenticity is lost for the entire chain.

In this context, assurance of authenticity is similar to safety control for a product that will support pathogen proliferation if exposed to high temperature. Here an unbroken cool chain throughout the supply chain is necessary to protect the consumer against a biological hazard, and loss of temperature control at any step will jeopardise safety. The project demonstrated that the principles already in operation for ensuring food safety could be used to improve the control of authenticity. In both cases it is important that the cost of control does not jeopardise the product's affordability. So, increasing the level of control at a critical point is only an improvement if other points can be identified where the workload for control can be reduced and the revised system or combination of systems is more dependable than before, or cheaper, or both.

The Organic HACCP project reviewed studies of consumer concerns and preferences in relation to organic production systems. It collected information about typical production chains for seven commodities: cabbage, apple, tomato, milk, eggs, wheat bread and wine, each in three or more regions across Europe. For each of the criteria listed below, the information was analysed to identify CCPs, defined as the steps in supply chains where the qualities of the final product can be controlled most efficiently. These qualities were microbial toxins and abiotic contaminants, potential pathogens, natural plant toxicants, freshness and taste, nutrient content and food additives, fraud, and social and ethical aspects. The new concept was thus to improve how consumer concerns are addressed, through the use of the CCP concept for a wide range of criteria, not only safety.

23.5 Benefits and drawbacks of using CCP-based systems at the level of a supply chain

From the consumer's perspective, cost is an important characteristic of the food and the best buy is the product that is considered to have the best ratio of quality and price. So consumers should welcome any measure that will improve quality at no extra cost or reduce the cost at no reduction in quality. In principle, using a CCP-based system at the level of a supply chain is supposed to do just that, since it will ensure that the controls are taking place at the steps where they can be implemented most effectively, thus eliminating superfluous double or triple testing of the same thing, or unnecessary waste when food is discarded because its safety is uncertain.

For example, protection of the milk supply from contamination with antibiotics can be done at a relatively low cost at the farm, where the main costs are recurrent costs related to ensuring that the milk from treated cows is separated and discarded. This would include a few cases of costs of 'corrective action' (HACCP, Principle 5), which in this case would be to discard all of the milk in the tank if it is discovered that an error has occurred and that the tank milk may have been contaminated by milk from a treated cow. At the dairy, in principle it is possible to ensure the same level of quality control by analysing samples after the milk arrives there. However, here the costs will be a relatively high recurrent cost of the analyses, as well as a very high intermittent cost of discarding all the milk in a production batch if an error occurs, even if the error is just that the sample for analysis was lost in the mail.

If both the farms and the dairy are covered by a reliable control system, the everyday monitoring will take place at the optimal CCP at the farm and analysis of milk samples at the dairy will only be needed at low frequency for verification of the system (Principle 6). This will ensure that there are no antibiotic residues in the milk sold by the dairy, with minimal expenses for control. In fact, the most expensive item will be the cost of the unannounced inspections at the farms to monitor the integrity of the system. However, if not all the farms are covered by a sufficiently good control system, the dairy will have to add the extra costs of its own independent system, in order to be able to take responsibility for this quality aspect. So the main benefit of a supply chain-based system is that it provides full control at the lowest cost. The main drawback is that the more entities that are involved, the greater is the risk that one of them will experience a system failure and this can have disastrous consequences for all those other entities that rely on the defaulting entity for their product control.

HACCP is most often used in relation to a legal requirement or a certification scheme, where the primary objective is to be able to document that the company has done everything that is required to prevent problems, to minimise the risk of being blamed for negligence in case something goes wrong. In this type of context it can be seen as a disadvantage for an enterprise to participate in a control system that includes additional parts of the chain, since its reputation becomes dependent on the actions of other enterprises, which it does not control directly.

In the milk antibiotics example, the dairy's management may consider the risk of having to take responsibility for an undetected error at its supplier as being so large that it prefers to pay for an extra level of control, even though this is not cost effective from the supply chain perspective. Even though the risk is very low, a scandal that reaches the media often jeopardises the survival of a company. Because of this, many customers would prefer to choose a company that uses particularly stringent quality controls, even if it makes the products slightly more expensive and even if there is no clear need for this stringency. This will also be seen as an insurance against possible problems in the next step in the chain, for example, for a wholesaler or retailer. In other words, the statement that 'we do everything possible to protect the quality' will very often be perceived as more trustworthy than an argument that a company has discontinued certain analyses and other control procedures, because an analysis showed that they were not cost effective, even when there are no objective indications that 'everything' is any better than 'enough'.

Merging the systems for several enterprises to cover a large proportion of a supply chain in a single system can easily lead to increasing complexity, as the number of processes and people increases. If no specific precautions are taken, this complexity leads to decreased transparency, which increases the risk that a problem will be overlooked. This means that, in order to benefit from the advantages of a supply chain-orientated control of quality and safety, a system must be organised in a way that puts special emphasis on preservation of transparency and traceability and convincingly prevents any of the participating entities from evading their responsibilities.

23.6 Concerns about social and ethical values among consumers of organic food

As detailed in Torjusen *et al.* (2004), central social and ethical concerns among consumers of organic food are related to:

- the need for an environmentally sustainable system of food production and distribution
- the social impacts of food production in the functioning of local communities
- the need for 'fair trading' and equity of relations between different enterprises in the food system
- the need for safe technologies of production and processing
- additional moral aspects of the relationship between man and nature, such as animal welfare
- transparency, as assurance that all actors in the supply chain are accountable for taking care of these aspects.

These concerns match very well with the generally agreed principles of organic agriculture (IFOAM, 2005) and most consumers of organic food expect that organic food is the outcome of a supply chain where all these concerns are addressed, at least to a greater extent than for other foods. Some of these concerns are relatively easy to reformulate as quality criteria that can be integrated into systems based on safety-orientated concepts, for example, transparency. Others can be handled in this way with manageable effort if relevant procedures are agreed on, for example, environmental sustainability, while some aspects of what consumers find important about organic food quality are not suited to systems focusing on CCPs in the supply chain, for example, the impact on local communities.

23.7 Providing assurance that consumer concerns are met

There are, in principle, two strategies for handling the social and ethical concerns of consumers of organic food. One is to pledge to allow the consumer access to exhaustive information about the status of each of these concerns and describe how much has been achieved and, where relevant, explain why some goals have not yet been met, or how choices have been made where demands based on such concerns are contradictory. The other strategy is to try to convince the consumer that the organic certification scheme that is used provides full insurance that all reasonable concerns have been met, as far as it is practically feasible, so the consumer does not have to worry about these matters. The second strategy is much simpler for the consumer, since in complex distribution systems consumers are not able to check for themselves how the food has been produced and handled. However, it requires he/she to be convinced of the trustworthiness of the scheme and the simplified information, so that consumers will often show a strong negative reaction to information about any weaknesses or contradictions in the control schemes. This is one of the reasons why the involvement of a third party, an auditor or an inspection body, is important to ensure the accountability of such systems and is a requirement for organic certification. In contrast, consumers will be more likely to accept that a particular expectation will not be fulfilled if they are kept fully informed about how and why this is the case. In practice, most real systems use a combination of these two strategies, with varying degrees of transparency and information for consumers. One of the factors determining this balance is the character of the supply chain. Small-scale provisioning and personal relations will more easily allow exhaustive exchange of non-formalised information, while larger enterprises and complex chains will more often apply more anonymous certified assurance and labelling schemes.

When relating this to a control system that aims to minimise the risk of disappointing the consumer, we need to distinguish the following kinds of problems that can arise from the need to address the social and ethical concerns of consumers:

1. problems relating to situations where some suppliers fail to live up to existing agreed standards
2. problems relating to situations when national standards differ between countries, where certifying bodies within any given country have different standards or where standards, norms and expectations differ between the stakeholders within any given market
3. problems relating to situations where the standards which consumers (or other stakeholders) would like to see upheld or which they believe are upheld, have not yet been developed, formulated or commonly agreed upon between suppliers of organic food or regulators of the organic market
4. problems relating to situations when some concerns of consumers (or other stakeholders) do not lend themselves to standardisation.

A condition for formulating CCPs is that requirements are standardised and codified. Agreed definitions must be either already available, or sufficiently well defined, so that standards can be described by the team that is analysing the supply chain, to allow it to define the CCPs. So, for the above-mentioned four kinds of problem, only the first one, failure to live up to existing agreed standards, can therefore be readily controlled at one or more CCP. Failure to prevent problems of this kind may manifest themselves as food scandals, unfortunate experiences or bad reputations, which in turn can undermine consumer trust in organic suppliers and products, with effects extending far beyond the specific supply chain that was actually involved. The identification of CCPs and the implementation of controls can serve to reduce risks of this kind considerably, on condition that there is consensus about standards and expectations and that the control system is designed to ensure that products are traceable and responsibilities are clear. The consequences of this topic for the design and implementation of organic certification schemes is described in Chapter 4.

The second and third types of problem define situations where additional effort is needed in order to implement the first and second HACCP principles, since everyone involved (in the controlled parts of a supply chain, as well as the customers relying on the control system) must agree on what the control system is controlling. It is meaningless to try to define CCPs for a quality defect if the expectations differ along the supply chain. Depending on the specific situation, standards can be reached using one of the following strategies (or a combination):

- Try to obtain general agreement on as broad a base as possible about what will be required for a particular concern, for example, animal welfare or a processing technology. This is useful where there are relatively uniform ideals, which only need to be crystallised into workable standards. An example of this is the Delphi enquiry (Kretzschmar and Schmid, 2005), which investigated views about which processing methods should be considered to be properly organic, as a step in the development of agreed standards on this topic.
- Allow different producers to use different standards by defining several categories of products, where the customer obtains clear information about the category each product belongs to, and explaining which standards have been used. This is useful where different consumers have different priorities, which can be qualitative (vegetarian, halal, etc.) or some aspect of price versus quality. For example, a shop can be selling several versions of the same organic food, corresponding to different varieties or breeds, or of different origins. Usually the cheapest type is one that has the least information about origin, variety and so on.
- An enterprise can define its own standards, which it enforces by requesting suppliers and, in some cases, retailers to follow specific guidelines in order to be considered for contracts, and by merging the brand image with the image of the quality standards in marketing to consumers. This

approach is most useful for companies with a strong market position, for example it is common among supermarkets with integrated supply chains.

Animal welfare is an issue that is, at the moment, subject to these kinds of differentiation regarding standards and assurance schemes (Roux and Miele, 2005). A number of initiatives have been taken both within and in parallel to organic production systems, all with the aim of improving farm animal welfare, but with rather diverse standards and expectations.

Still, some consumer concerns are difficult to address as standardised quality norms. Issues such as impact on the environment, heritage and biodiversity as well as fair economic relations, each encompass a wide range of different issues, where even experts may disagree on which method is best and where the best method for one enterprise may not be useful for another enterprise. In these cases the key issue is to ensure efficient flow of reliable information along the supply chain. Each enterprise can define and describe how they address the topics which consumers are interested in and then ask its customers to convey this information along the supply chain until it reaches the consumer. Many consumers are willing to pay more for food produced under certain conditions, when they experience this as an indication that their interests are being taken into account (European Commission, 2005; Torjusen *et al.*, 2004), in contrast to the use of methods merited only by the lowest cost per kilogram. This is one of the main drivers for farm shops, where the consumers can see for themselves how the plants and animals are grown and talk to people who are seen as taking personal responsibility for the food. Some of the trust that is inherent in the situation when food is bought directly from the producer can also be retained in more complex supply chains, if the person you buy the food from can still answer all the questions you would have asked the producer, or if the producer is clearly identified and accessible. So, rather than standardising the procedures and making all the food identical, the food is designed to display individuality, familiarity and personal responsibility. Typical examples are individualised labels displaying the name and telephone number of the producer. Few consumers will actually contact the producer, but the fact that they are able to do so may be perceived as a very strong indicator of quality and authenticity. Another option is the use of provenance as an encompassing quality concept, while not relating directly to a person, it still provides more choice and familiarity than a completely anonymous product. Provenance is a strategy that dominates in southern Europe, but much less so in the north.

Many of these strategies can be used together, combining a certified origin label with the name of the producer or with the supermarket's own brand. The important issue is that the quality of the food is perceived to be something that is taken seriously and that it is clear to the consumer that substantial emphasis is placed on ensuring and monitoring this quality. If this is done well, many consumers will find it acceptable to pay a substantial premium compared with anonymous 'bulk' products.

It must be kept in mind that any of these strategies for increased standards or transparency, or both, also have their drawbacks, in addition to the direct costs of quality control, individualised labelling, multiple product lines and so on. From the retailer's point of view, advertising the use of a transparent supply chain restricts the freedom to change supplier at short notice in order to take advantage of a good bargain. From the producer's point of view, this is, of course, an asset, since more stable relationships or even contracts between producers and retailers allow the producer to optimise the production planning and thus minimise waste. However, downstream integration will also increase the producers' dependence on one retailer, potentially enhancing retailer-led marketing as well as increasing pressures on increased efficiency.

23.8 How identification of quality-focused CCPs in organic food production chains was carried out in the Organic HACCP project

In the Organic HACCP project, the project members and associated experts interviewed representatives of enterprises at every step of a supply chain, from producer to retailer. In some chains this was the same person, other chains included four or five enterprises. The chains were selected considering that it had to be possible to interview all enterprises in the chain, and the commodities chosen were relatively simple, with wheat bread and wine being the most processed ones, so that the project did not attempt to analyse very complex supply chains such as those for highly processed products made from multiple ingredients sourced across the world. The seven commodities chosen: cabbage, apple, tomato, milk, eggs, wheat bread and wine, provide representatives of product groups covering most of the organic foods on the market. For each supply chain, a flow diagram was constructed showing all the major steps that were relevant to this supply chain (Fig. 23.1).

Major steps included, for example, plant production, transport, processing and retail. Each major step was subdivided into activities called 'substeps', which were defined in common categories. For example, the major step 'transport from processor to retailer' could contain substeps in the following categories: certification/inspection; management; labour; packaging/labelling; equipment; pest control; contact to supplier/customer.

The project group designed a questionnaire with 286 questions, structured to cover each substep of all the types of major steps encountered in the supply chains. If an enterprise covered more than one major step, this meant that more than one interview was carried out, with more than one person or more interviews with the same person. For each enterprise, the interviewees typically answered 150–200 questions about 1–3 major steps. The questions were designed by experts to describe the principles and intentions of the

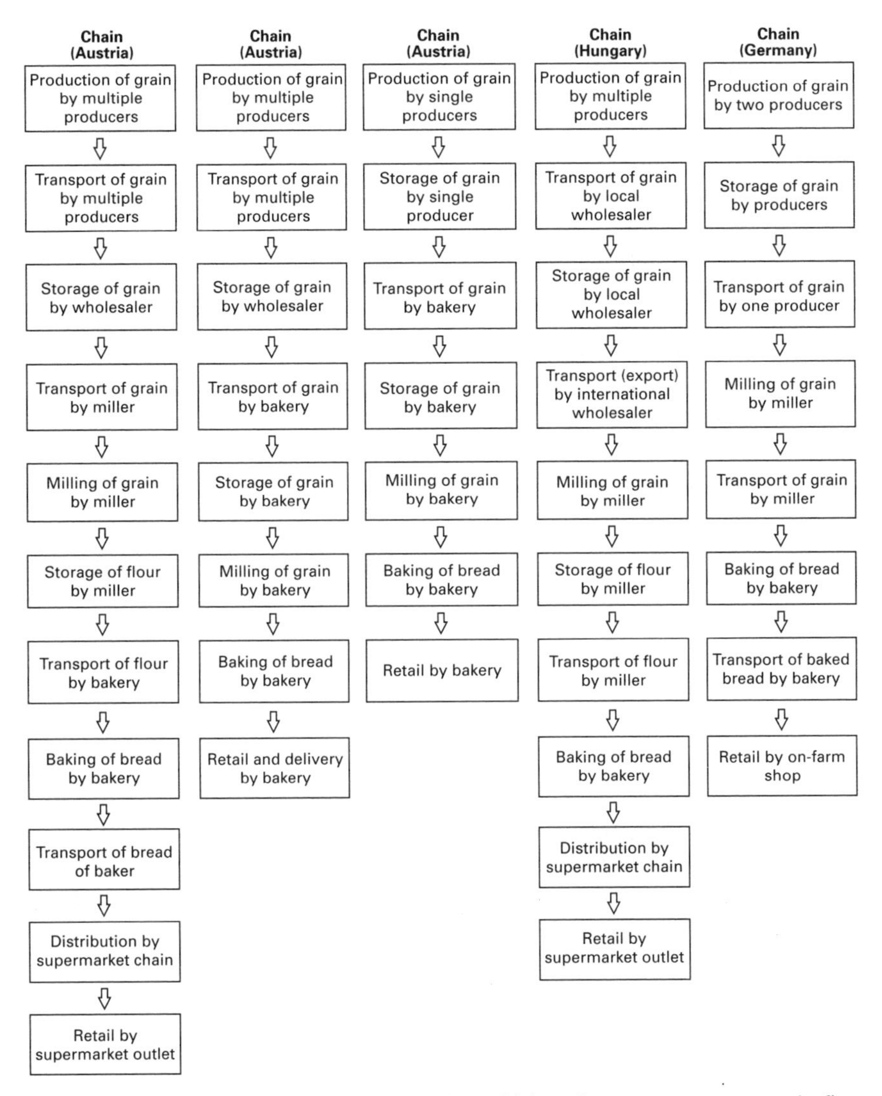

Fig. 23.1 Schematic flow diagrams showing which major steps were present in five different supply chains for organic wheat bread; from leaflets no. 9 and 10, Organic HACCP (2005).

actions by each participant in the supply chain, in other words, what this person considered it important that he or she did in order to secure quality and safety of the final product.

Several types of software were developed. One programme contains the questionnaire and allows each interviewer to upload all the answers from each interview to an internal common password-protected database accessible via the internet. Another programme displays an overview of each substep of each major step in each supply chain on a webpage, and by clicking on a substep a text relating to this substep can be entered into an analysis database. This text then becomes visible in the output section of this database. The output from the analysis database can be accessed without a password.

After all the answers from the interviews had been uploaded, an expert analysed each supply chain for each of the seven defined criteria for quality and safety: microbial toxins and abiotic contaminants; potential pathogens; natural plant toxicants; freshness and taste; nutrient content and food additives; fraud; social and ethical aspects. For example, an expert on freshness and taste would check each major step in a supply chain for tomatoes to determine if it fulfilled the definition of a CCP (HACCP, Principle 2) in relation to freshness and taste for this commodity. If the step was considered to be a CCP, the answers in the questionnaire that related to relevant substeps at this step would be reviewed, to assess the control procedures that were in use for this CCP. The expert would then fill in the text field, structuring the input to consist of the following points:

- explanation why there is a CCP (in relation to the other steps)
- description of which measures can be used to control the risk at this step in the chain
- estimation of the quantitative risk relative to other chains in the analysis (either as a percentage of incidence or 'medium', 'high' or 'low'), explanation of the assessment done and if relevant, comparison with data from other studies
- description of possibility of alleviating at a later stage, if relevant.

The project's homepage (Organic HACCP, 2005) contains instructions on how simple quality control procedures can be implemented in enterprises where CCPs occur, either because these enterprises perform the most relevant step for such control measures or because the potential problems are not adequately controlled by other enterprises.

23.9 Examples of identified CCPs

All the analysed supply chains were well controlled with regard to serious food safety issues (pathogens, toxins), since there are compulsory regulations which must be followed by all producers, whether organic or not. Still, the project members were surprised at the scale of differences between national food safety standards of countries in Europe. For example, temperatures considered to be adequate for safe storage of eggs differ between 5 and 25 °C, with no apparent explanation for these large differences. However, for other aspects of food quality there were many cases where more systematic controls could substantially improve the assurance that consumer expectations would be met. In most cases, the responsible persons were aware of what they should do to fulfil these expectations and were generally undertaking the appropriate actions, but did not have any structured plans to ensure that they would be done properly every time. In other words, there was a substantial risk that a disturbance to the routine, such as illness or unusually bad weather, could lead to a situation becoming out of control with the ensuing impact on quality standards. For example, to ensure the highest nutritional value of eggs, hens should be fed with fresh green plant material and many consumers also consider this an advantage for animal welfare. However, a year-round supply of fresh green plant feed requires substantial effort and planning. In these cases it was also relatively often noticed that some of the relevant information about the quality of the food was not passed on to the consumer. Sometimes this meant that producers did not receive credit for the full extent of the quality of the produced food. In other cases consumers probably believed that the standards were better than what was actually the case. In a few cases, a producer or processor appeared to be unaware of the relationship between their actions and the quality of the food.

For example, egg whites become more fluid with time and this process happens substantially faster at room temperature than when chilled. However, in one of the analysed chains each producer used a different temperature for on-farm storage, ranging from controlled temperature storage starting within 4 hours of collection, to keeping the eggs at ambient temperature for up to one week. This certainly cannot have been the most cost-effective quality control when viewed from a supply chain perspective. In this case, clearly the producers could either save money (by all abolishing any attempts to chill the eggs) or improve the quality of the product (if all the eggs are kept chilled) by agreeing to use the same temperature. A uniform raw material will also help the egg packer and the retailer to optimise their procedures in order to achieve the best product on the shelves, including defining the optimal shelf-life.

With regard to transparency and authenticity, in many of the chains the retailer appeared to be a weak link in conveying information to the consumer, in particular for products sold through supermarkets. Several supermarkets, but not all, do train their employees how to inform consumers about organic certification in general, for example by supplying leaflets explaining the definitions of organic food. However, there seems to be less interest in focusing awareness on the specific qualities that each producer or processor has provided in a product, such as the use of traditional varieties/breeds and ensuring optimal conditions for the products, the people involved and the environment. It could be speculated that, in some cases, emphasising other qualities of the organic products than simply that it is organic might be seen as competition with other high-quality products in the supermarket, including own brands. This may be one of the reasons why anonymous organic 'bulk' products appear to be more common in supermarkets than in speciality shops. Still, examples were found where supermarkets did promote such special qualities, both by allowing individualised labelling and by emphasising these aspects in advertising. These retailers stated that this was appreciated by the consumers and supported the overall strategy of the chain. Interestingly, one of these examples was a 'no-frills' supermarket chain that focuses on very price-attentive customers. This chain uses locally produced and/or organic products as a slightly more expensive product option than the standard products, advertising the premium qualities of these products as an expression of its desire to ensure a good product with very good value for money for its customers.

23.10 Organisational and educational requirements for utilising CCPs across real supply chains

The successful establishment of supply chain-orientated control of quality and safety requires that each involved entity experiences advantages from engaging in this concept and taking responsibility for the part of the supply chain that it covers. This means that structures must be in place to ensure that the benefits of compliance are greater than the costs for every enterprise and that the safeguards against cheating are so good that the complying enterprises are confident that their good reputation is well protected.

In practice, this means that the terms of the agreement must be understood and agreed by everyone involved and that compliance with the agreed terms must be monitored in a way that ensures credibility. It must be kept in mind that a supply-chain-wide CCP-based quality control system can be more or less complex, and that it is imperative for success that the effort spent on control is proportional to the purpose of the system. The credibility in short, often personalised, chains will have a different foundation from that required in complex chains producing labelled, prepackaged foods (Kjærnes *et al.*, 2006).

If the purpose of a system is to ensure that eggs are kept within the same agreed temperature range from collection until sale, in order to provide a uniform, dependable quality for the consumer, then the control system can be very simple.

In this case the necessary components would be:

- installation of cooling equipment and temperature recorders in all rooms used for storage of eggs;
- agreed limits for which temperature range is acceptable and how often the temperature recorder must be checked;
- an agreed procedure about what to do if a temperature recorder or the cooling equipment breaks down (including reporting of every such incident to the group);
- an arrangement with a trusted independent person or organisation (an auditor) to make a specified number of unannounced visits each year to randomly selected enterprises, to check that the temperature limits were adhered to, that the recorder was working and that eggs were not stored in rooms without temperature control;
- and last, but not least, agreed sanctions against enterprises that do not follow the rules.

This minimal model can be used in many situations where one clearly defined factor is crucial for the quality of the product. It can be organised as a voluntary agreement between independent enterprises, or it can be imposed by a dominant enterprise such as a wholesaler or retailer, as a condition for obtaining a contract to sell products to this enterprise.

It is generally recognised that HACCP systems work best if the people who actually work with the food are directly involved in the decision-making process, so that a sense of ownership can be achieved (NACMCF, 1997). This is, of course, also the case for control systems that aim to ensure high standards for other aspects of food quality than health hazards. In particular, the process of analysing the relevant part of the supply chain, or all of it, almost inevitably leads to innovations or adaptations that improve quality and/or productivity. This happens simply because counterproductive habits tend to accumulate, as long as no one has any occasion to question why something is done in a particular way, and until they are exposed by the very process of describing every step in the process. Additional benefits are obtained in the form of easier integration of new staff, who can use the description of the steps in the supply chain as manuals for what they are supposed to do, in particular, in busy situations where experienced staff have less time to supervise new or temporary staff.

Because of these benefits, there is a tendency for a successful system to increase in complexity, as additional topics and enterprises become covered. This can be a positive development. However, it is very important to keep a focus on the basic purpose of the control system and to ensure that the effort is proportional to this purpose. If substantial numbers of the people using the system perceive it to be more of a burden than a help, then it is necessary to review the system critically to identify ways of simpling it and making it more transparent. Fortunately, it is feasible to use the HACCP principles to control this problem, simply by including loss of motivation among staff as one of the hazards in the analysis, which must be monitored and controlled, and where corrective action, such as a renewed analysis, must be taken if the level falls below a pre-set limit. The key feature of this concept, which is often missing from standard employee satisfaction surveys, is that the management in advance commits itself to specific actions if this element of satisfaction among staff turns out to be lower than the preset level. So, once the people involved in a supply chain become familiar with the concepts in the HACCP principles, ample opportunities arise for continued improvement.

23.11 Example of successful integration of the HACCP concept in a vegetable supply chain to control product quality as well as safety

In Denmark the company Aarstiderne ('The seasons', www.aarstiderne.com) has established a highly successful box scheme, based on communication with consumers via the internet. Several farms are part of the company and other farms and processors supply foods, mainly vegetables, fruit and bread, through more or less integrated operations and contracts with the main company. The company has experienced a period of rapid growth, starting out in 1999 with three people supplying approximately 100 local customers, which in 5 years increased to 110 employees distributing to 35 000 households across the country. The customers order prepacked boxes via the internet, each box contains a selection of organic foods and recipes for their use. Vegetables and fruit are the main products, but cheese, wine, bread, fish or meat as well as mixed boxes with entire meals are increasingly popular. Each box is delivered to the doorstep, each postcode is served once a week and orders can be entered or changed up to two days before delivery.

From the start, the company's management was committed to ensuring high standards of quality and transparency. Problems arose as soon as the number of employees rose to a level where the key people in management were no longer able personally to take part in all operations. It is one thing to be committed to ensure that ideals are followed and that all expertise gained from previous experience (including errors) is constructively implemented in the form of improved procedures; it is quite a different thing actually to make it happen.

So the company hired a consultant, who organised a full analysis of all the operations in the company, from production planning on the farms to collection and cleaning of the empty boxes after delivery to the consumers. For each aspect of quality, the CCPs were identified and critical limits were defined, which ensured that quality was maintained. For example, the temperature during storage and transport must be neither too high nor too low, with different optima for different products and there is also a critical limit for the length of time from when a product is received until it is delivered. For each CCP, procedures were established to record and monitor these data, including actions when the figures are approaching the limits, together with what actions to take if, despite all efforts, limits are exceeded. Complete documentation is kept of the results of the measurements and of all actions taken. The status and developments are communicated with the employees through posters and at internal meetings and on courses. Since the company is in control of every step of the supply chain, it is relatively easy to persuade employees in one division to carry out a monitoring task that provides information for another division. This would have been much more of a barrier if the divisions had been independent enterprises. It was an important advantage of this process that the company was so new, so there had not been time to establish traditions and privileges and the need for structure and guidance was obvious to everyone, so the new system was overwhelmingly perceived as an asset rather than as a burden.

The CCP analysis and other HACCP principles were originally meant to be a once-off process intended to provide the management of 'Aarstiderne' with a set of tools to control the most critical processes during chaotic periods of high activity. However, the success meant that additional areas and ideals were included and the value of the process as a continued activity became recognised. The employees' timesheets show that the time used for quality control corresponds to more than one full-time post and this is considered a worthwhile continued investment. The system is used exclusively within the company, it is not mentioned in marketing or documentation of the company's products, so there has not been any need for external monitoring and certification. The employees understand the relevance and need for each of the prescribed procedures and are continuously encouraged to suggest further improvements. Additionally, the system is not used to compare the achievements of individual employees, which helps to ensure that all employees are committed to make it work well.

23.12 Future research and development needs and trends

Consumer concerns about food quality are continually changing. There is an urgent need to understand better the psychological and social processes that shape the demands and preferences for food and to incorporate present and future knowledge of this topic into targeted development of foods and food production systems that fulfil the expectations of different groups of consumers.

For organic supply chains there is a need for more knowledge about the role and organisation of monitoring, certification and audit systems for quality assurance schemes in small- and large-scale organic supply chains. Specifically, we need to develop new tools to handle issues and expectations that are not easily standardised and communicated, in particular, in cases where these may have a substantial influence on consumer preferences. Improved understanding of social and ethical concerns, as reflected and communicated through different steps and types of supply chains, are a key part of this process.

Definitions of food safety and food quality also need to be more precisely understood. Present safety regulations and guidelines are based on the use of extrapolations and safety limits to control uncertainties about very small risks. This often leads to otherwise unnecessary conflicts with other quality aspects. For example, the use of concrete-covered outdoor runs in organic egg production to minimise contamination with dioxins may be counter-productive for consumer health, since we do not know if the nutritional advantages of eggs from hens eating grass and worms might outweigh the harmful effects of the additional dioxins. Clearly, many consumers act as if they believe this and no scientific reports appear to have addressed this and similar hypotheses, despite their central position in consumer (and producer) perception of health effects. The recent advances in nutrition science and toxicology should provide a good selection of the basic tools for this type of interdisciplinary research.

A third path for research is to study associations between system change and improvement, on the one hand, and public and consumer responses, on the other. In particular, what are the conditions for the development and maintenance of trustworthiness and satisfaction through organisational and communicative procedures? With better knowledge of consumer demands and improved quantification of risks and benefits to health, the future trend is likely to point towards increased differentiation, with different products targeting consumer groups with different priorities. Some consumers will demand more convenience, while others will invest in participatory community farm schemes that allow direct involvement in the production and preparation of their food.

23.13 Sources of further information and advice

The Organic HACCP project produced a wide range of materials to assist companies and consumer groups to get started in this area. Overviews of the analysed supply chains, as well as the identified CCPs, can be accessed without password from the website (Organic HACCP, 2005), together with instructions on how to use the databases and how to establish simple quality control procedures within enterprises. The website also contains links to each of 14 leaflets, with concrete recommendations for all involved in organic supply chains, from producers to consumers. The set of leaflets has presently been translated into five languages in addition to English and can be copied and printed free of copyright charges. A literature review of European consumers' conceptions about organic food (Torjusen *et al.*, 2004) can be downloaded from the website and a more detailed account of the concept is being prepared in the form of a book specifically on this topic (Brandt *et al.*, in press).

There are a large number of books and courses available in most major languages regarding HACCP procedures in the traditional sense that relate specifically to hazards to human health. The newer concept of QACCP, which corresponds closely to the general procedures described in the present chapter, is also included in some of the courses offered by educational and consultancy organisations in many countries. Owing to the large number of providers, the most relevant option to use in order to find the relevant ones will normally be to search on the internet using a setting that shows only links from your own country.

23.14 References

Brandt, K., Lück, L. and Langton, S. (eds) (2007) *Organic Food Production: ensuring quality and safety*, Blackwell Publishing, Oxford, in Press.

European Commission (2005). *Attitudes of Consumers towards the Welfare of Farmed Animals*, Special Eurobarometer 229, wave 63.2. Brussels: DG SANCO.

FAO (1997) *Hazard Analysis and Critical Control Point (HACCP) System and Guidelines for its Application.* http://www.fao.org/DOCREP/005/Y1579E/y1579e03.htm#bm3
IFOAM (2005) *Principles of Organic Agriculture.* http://www.ifoam.org/organic_facts/principles/pdfs/Principles_Organic_Agriculture.pdf
Kjærnes, U., Harvey, M. and Warde, A. (2006). *Trust in Food in Europe,* Palgrave Macmillan, London.
Kretzschmar, U. and Schmid, O. (2005). *Approaches Used in Organic and Low Input Food Processing – Impact on Food Quality and Safety. Results of a Delphi survey from an expert consultation in 13 European Countries.* http://orgprints.org/7032/
NACMCF (National Advisory Committee on Microbiological Criteria for Foods) (1997) *Hazard Analysis and Critical Control Point Principles and Application Guidelines* http://vm.cfsan.fda.gov/~comm/nacmcfp.html
Organic HACCP (2005) www.organichaccp.org
Roux, J. and Miele, M. (2005). *Farm Animal Welfare Concerns. Consumers, Retailers and Producers*, Welfare Quality Reports No.1. Cardiff University, School of City and Regional Planning, Cardiff.
Torjusen, H., Sangstad, L., O'Doherty Jensen, K. and Kjærnes, U. (2004). *European Consumers' Conceptions of Organic Food: a review of available research*, Professional Report no. 4, SIFO (National Institute for Consumer Research), Oslo, Norway (available for download at http://www.organichaccp.org).

Index